FREE RADICALS IN DIAGNOSTIC MEDICINE

A Systems Approach to Laboratory Technology, Clinical Correlations, and Antioxidant Therapy

ADVANCES IN EXPERIMENTAL MEDICINE AND BIOLOGY

Recent Volumes in this Series

Volume 360
ARTERIAL CHEMORECEPTORS: Cell to System
Edited by Ronan G. O'Regan, Philip Nolan, Daniel S. McQueen, and David J. Paterson

Volume 361
OXYGEN TRANSPORT TO TISSUE XVI
Edited by Michael C. Hogan, Odile Mathieu-Costello, David C. Poole, and Peter D. Wagner

Volume 362
ASPARTIC PROTEINASES: Structure, Function, Biology, and Biomedical Implications
Edited by Kenji Takahashi

Volume 363
NEUROCHEMISTRY IN CLINICAL APPLICATION
Edited by Lily C. Tang and Steven J. Tang

Volume 364
DIET AND BREAST CANCER
Edited under the auspices of the American Institute for Cancer Research;
Scientific Editor: Elizabeth K. Weisburger

Volume 365
MECHANISMS OF LYMPHOCYTE ACTIVATION AND IMMUNE REGULATION V: Molecular Basis of Signal Transduction
Edited by Sudhir Gupta, William E. Paul, Anthony DeFranco, and Roger Perlmutter

Volume 366
FREE RADICALS IN DIAGNOSTIC MEDICINE: A Systems Approach to Laboratory Technology, Clinical Correlations, and Antioxidant Therapy
Edited by Donald Armstrong

Volume 367
CHEMISTRY OF STRUCTURE–FUNCTION RELATIONSHIPS IN CHEESE
Edited by Edyth L. Malin and Michael H. Tunick

Volume 368
HEPATIC ENCEPHALOPATHY, HYPERAMMONEMIA, AND AMMONIA TOXICITY
Edited by Vicente Felipo and Santiago Grisolia

A Continuation Order Plan is available for this series. A continuation order will bring delivery of each new volume immediately upon publication. Volumes are billed only upon actual shipment. For further information please contact the publisher.

FREE RADICALS IN DIAGNOSTIC MEDICINE

A Systems Approach to Laboratory Technology, Clinical Correlations, and Antioxidant Therapy

Edited by

Donald Armstrong
State University of New York at Buffalo
Buffalo, New York

PLENUM PRESS • NEW YORK AND LONDON

Library of Congress Cataloging in Publication Data

On file

Proceedings of an International Symposium on Free Radicals in Diagnostic Medicine: A Systems Approach to Laboratory Technology, Clinical Correlations, and Antioxidant Therapy,
held October 7–9, 1993, in Buffalo, New York

ISBN 0-306-44981-1

A Division of Plenum Publishing Corporation
233 Spring Street, New York, N. Y. 10013

Printed in the United States of America

ACKNOWLEDGEMENTS

Support from the following sponsors are gratefully acknowledged:

MAJOR CORPORATE SPONSORS

Marion Merrell Dow
Institute de Recherches Internationales Servier and Servier International, France
Eastman Kodak Company, Clinical Products Division
La Haye Laboratories, Inc.
DDI Pharmaceuticals, Inc.
Upjohn Laboratories, Clinical Research Division
Research Institute on Addictions
Sanofi Winthrop Pharmaceuticals
Roche Diagnostic Systems, Inc.
Pfizer Pharmaceutical Company
Otsuka America Pharmaceutical Company
Ono Pharmaceutical Company, Japan
Merck & Company, Inc., Human Health Division
Hoffman LaRoche, Inc.
The Henkel Corporation, Fine Chemicals Division
Hamamatsu Photonics Systems, Japan
Eisai Company, Ltd., Japan
Cayman Chemical Company
Boehringer Mannheim Corporation
3M Pharmaceuticals
Clinico PharmacoKinetics Laboratory, The Millard Fillmore Hospital, Buffalo, NY

OTHER CORPORATE SPONSORS

Empire Imaging Systems
FRESA BioMedical Laboratories, Inc.
JM Science, Inc.
Miles, Inc., Diagnostic Division
Shimadzu Scientific Instruments, Inc.
SLM - American Instruments and Milton Roy Company
Roswell Park Cancer Institute
VERIS Vitamin E Research and Information Science

ORGANIZATIONS

University at Buffalo
Roswell Park Cancer Institute
American Association of Clinical Chemistry, Upstate NY Section

The Organizing Committee wishes to thank G. Alan Stull, Dean of the School of Health Related Professions, University at Buffalo and James Karr, Chief of the Office of Scientific Administration, at Roswell Park Cancer Institute for their introductory remarks on the relevance and timeliness of this topic and to the following individuals who graciously gave of their time to serve as Moderators of the various sessions: Harold Box, Ph.D., Chairman, Department of Biophysics, Roswell Park Cancer Institute; David Hohnadel, Ph.D., President, National Academy of Clinical Biochemistry; Joseph Izzo, M.D., Chairman, Department of Medicine, The Millard Fillmore Hospitals and University at Buffalo;

Gerald Louge, M.D., Chief of Staff, The Veterans Administration Medical Center of Buffalo; John Naughton, M.D., Vice President for Clinical Affairs and Dean of the School of Medicine and Biomedical Sciences, University at Buffalo; Kyu Shin, M.D., Chief of Radiation Medicine, Roswell Park Cancer Institute; John Wright, M.D., Chairman, Department of Pathology, University at Buffalo and Lloyd Horrocks, Ph.D., Department of Biochemistry, The Ohio State University and Editor of the Journal of Molecular and Cellular Neurochemical Pathology. We also extend our special appreciation to Richard Stockton, Ph.D., Gail Bersani and Arthur Michalek, Ph.D. and Debbie Holden for assistance with other organizational matters, to LuAnn Kaite and Debbie Murello for typing and their help with numerous details before, during and after the Symposium and to Marion Merrell Dow, Inc. for a generous educational grant to produce the conference program and abstract booklet. The University at Buffalo Health Science Library, serving as a resource for the Middle Atlantic Region of the National Network Libraries and Medicine, provided a MEDLINE search for participants.

PREFACE

An **International Symposium on Free Radicals in Diagnostic Medicine** was co-sponsored by the State University of New York at Buffalo, Roswell Park Cancer Institute, and the Upstate NY Section of the American Association of Clinical Chemistry. The theme was "A Systems Approach To Laboratory Technology, Clinical Correlations And Antioxidant Therapy." The symposium was held on October 7-8, 1993 at the Hyatt Hotel and on October 9 at Roswell Park Cancer Institute, Buffalo, New York. This proceedings volume contains chapters from platform presentations, poster sessions and from invited special lectures in the areas of basic science, clinical applications and efficacy of treatment.

A *Special Lecture on* the **relevance of free radical analysis to clinical medicine** was presented by Professor Kunio Yagi of Japan. The Yagi procedure to measure thiobarbituric acid (TBA) reaction reflects the amount of reactive substances, lipid peroxides and aldehydes, in the sample. For example, normal subjects will have less than 4 nmol/ml of serum lipid peroxides, while a person with diabetes generally has equal or greater than 5.0 and a diabetic person with vascular complications often exceeds 7.5 nmol/ml. Serum TBA is a clinically important measure that relates to aging, gender and estrogen as an antioxidant, in the prognosis for vascular disorders, and in pathological conditions relative to the amount of lipid peroxidation.

The BASIC SCIENCES portion of the program examined: "Mechanisms of Action, Pathophysiology and Laboratory Tests" in six presentations.

Free radicals in normal physiology, phagocytosis and necrosis was presented by Professor Mary Treinen Moslen who emphasized that free radicals are continually formed in biology and most abundantly by the mitochondrial-electron transport chain reactions. Free radical generation is enhanced by radiation, inflammation, and by the ischemic-reperfusion process. Three stages identified are: initiation, detoxification and membrane propagation reactions, and decomposition of membrane into smaller aldehydes.

Chromosomal damage in Bloom's Syndrome was examined by Professor Thomas Nicotera as a cellular model to study chronic superoxide anion stress. Hypotheses were raised of cells deficient in enzyme activities to detoxify superoxide and/or overprotection of superoxide radical. Clinical correlates are seen in: neoplasia, neurologic disease, short stature, sensitivity to sunlight and predisposition to diabetes.

Professor Donald Armstrong presented an **analysis of free radicals and related compounds in the clinical laboratory.** He emphasized that it is nearly impossible to read the literature today without finding involvement of oxidative stress and oxygen free radical (OFR) generation. Dr. Armstrong gave a detailed presentation of assays and methodologies within the province of the modern hospital clinical laboratory to approach the quantitation of free radical biology status and the numerical documentation of oxidative stress. Thus, clinical laboratories have an important role in the application of free radical measurement methods.

State of the art of free radical testing was discussed by Dr. Charles Pippenger. Enzymes relating to OFR generation and antioxidant defense systems are present in a wide variety of tissues and reflect genetic characteristics; their levels are modified by environmental and disease factors. Laboratory quality control and standards were emphasized with recognition of a certain biologic variability. This type of clinical laboratory data applied to population studies can be utilized as information to improve health care. Examples include: (1) Ashkenazi Jews carry low SOD levels, (2) OFR mediated disease precursor that takes years to develop may be present in "normal" populations, and (3) after 60 years of age, OFR scavenging enzymes decline even with vitamin supplementation.

In comment, Professor Joseph L. Izzo, Jr. Chairman of this session, observed that there will be resistance to reimbursement of OFR clinical laboratory studies until we demonstrate that the test result alters therapy and physician decision making behavior to change what we do for the patient.

Two CLINICAL SCIENCE sections addressed **organ specific disorders** and **systemic involvement.** Dr. Domenico Pellegrini-Giampietro began with a talk on **Neurological Disease.** The brain is vulnerable to OFR damage because it: (1) has 1/5 of cardiac output, (2) is lipid rich, (3) poor in catalase activity, and (4) several areas of brain are rich in iron but cerebral spinal fluid transferrin is low. Radical production is associated with many neurologic diseases including: Parkinson's, tardive dyskinesia, Schizophrenia, Down's and Alzheimer's. OFR production is linked to glutamate receptor excitation. If a molecule can be designed that is a glutamate receptor antagonist and also has scavenger properties this would be useful to prevent neuronal cell death in ischemia.

Ocular Disease was presented by Professor Robert Anderson. Cataracts, retinopathy, light damage, age-related macular degeneration (ARMD) are all OFR associated. Lens clarification depends on maintaining a reducing environment. Cigarette smoking increases cataract risk (there are 10^{17} radicals in one puff of a cigarette). High oxygen levels may damage the retina and this is of special concern in pre-term infants who are deficient in vitamin E (with 10% of full term levels). With aging there is a greater susceptibility to OFR damage. ARMD is the leading cause for blindness in persons over 50 years of age.

Endocrine Disease was discussed by Professor Paresh Dandona who in discussing thyroid hormone experiments raised the precautionary note that in vitro chemistry data may show striking differences from in vivo biological data. For example, an in vitro antioxidant may cause increased oxygenation in vivo through metabolic and other factors. In commenting on other diseases he noted that diabetics have much higher 8-hydroxy quanosine (4 to 5 x) than normal controls and also greater DNA damage in mononuclear cells. Oophorectomy increases OFR levels and estrogen hormone replacement therapy (HRT) decreases OFR levels. When small amounts of endotoxin (4ng/kg) are given to volunteers there is an increased plasma TNF alpha and IL-6, and increased OFR generation by chemiluminescence not withstanding a concomitant increased glucocorticoid level.

Cutaneous Disease was discussed by Professor Alice Pentland to complete the first of the three-day symposia. This presentation dealt with oxygen free radicals (OFR), the electromagnetic spectrum, clinical considerations, photochemistry and molecular aspects of tissue injury and repair.

Cardiac Disease was presented by Professor Roberto Ferrari. This presentation included discussion of defenses against OFR toxicity, sources of OFR, detection and oxidative damage. Problems relating to measurements (indices) of oxidative stress in the clinical setting were discussed together with the need for more specific data and knowledge to most effectively conduct clinical therapy trials.

Vascular Disease was presented by Professor Peter Reaven who examined the role of oxidized low density lipoprotein (LDL) together with the dynamic/kinetic dimensions they activate which leads to cytokine release, macrophage accumulation, passage of LDL through the endothelial barrier and finally to endothelial cell damage and plaque formation. This presentation also examined factors that affect LDL oxidation and discusses methodologies to determine in vivo oxidation.

Hemolytic Disease was the subject of the presentation by Professor Edward J. Lesnefsky, Jr. This discourse looked at clinical syndromes, target organs, and examined in substantial detail the two categories of iron pools: (1) transferrin bound and (2) non-transferrin pool bound to ammonium citrate. The former is non-toxic and the latter produces disease since iron is insoluble without ligand. Organ dysfunction, increased susceptibility to ischemia-reperfusion injury and methods of therapy by phlebotomy, chelation and antioxidants were also discussed.

Infertility was discussed by Professor Claude Gagnon who presented the role of reactive oxygen species (ROS) in male infertility. The mode of action of ROS was outlined and a series of experiments examining both beneficial and detrimental aspects of ROS on spermatozoa was covered. Toxic levels of ROS reduce sperm function and fertility. Yet ROS has beneficial effects to produce capacitated activity (CA) or hyperactivity (HA) giving zigzag movement to sperm and greater torque, thus greater ova membrane penetration. Leukocytes in the vagina do harm and are one cause for infertility.

Renal Disease and reactive oxygen species contribution was presented by Professor Leonard Feld with both common features and differences emphasized among clinical syndromes. Resident cell populations (mesangial, endothelial, epithelial) and infiltrating cells (neutrophils, macrophages, platelets) may be involved at different renal sites (vascular, glomerular, tubular). Three models to study renal disease are: (1) anti-glomerular basement membrane (anti-GBM) which features proteinuria and beneficial effects of catalase and disferoxamine but not DMSO, and (2) membranous glomerulonephritis with decreased proteinuria from DMTU and DMSO but not catalase or SOD which suggest that hydroxyl radical is involved, and (3) puromycin nephrosis in which hypoxanthine is an intermediate step in ROS injury. Allopurinol and SOD reduce the injury but not catalase or DMSO. Free radicals in one form or another are important in each renal disease state.

Liver, Pancreas and Gastric Disease were reviewed by Professor Kunio Yagi to wrap-up the second day's platform presentations. One can identify gastric mucosal injury associated with a rise in plasma thiobarbituric acid (TBA). An experimental preparation with infusion of hypoxanthine and oxygen caused detectable mucosal injury and increased TBA. Better protection is obtained by SOD plus catalase than cimetidine and H-2 receptor blocker.

THE CLINICAL SCIENCES II SESSIONS on Systemic Involvement addressed:

Autoimmune Disease which was presented by Professor Jonathan Leff who commented on markers of oxidative damage in conditions such as sepsis, ARDS, AIDS, rheumatoid arthritis, hepatitis, congestive heart failure and hemolytic anemia. Polarographic catalase assay, superoxide dismutase (MnSOD) assay which is elevated prior to development of ARDS, breath hydrogen peroxide, lipid peroxide products present in blood or synovial fluid of particular relevance to rheumatoid arthritis, and reduced levels of serum antioxidants for example in AIDS were discussed.

Cancer and Chemotherapy was presented by Professor Peter O'Brien who discussed mechanisms by which dietary and natural antioxidants have anti-carcinogenetic actions. All antioxidants inhibit P-450 enzyme systems. Categories of effects include: precursor, blocking and suppressors. Dietary constituents mentioned were: flavenoids in fruits and vegetables, medicinal plants, green tea, garlic, onions, leaks, strawberries, cabbage, etc.

Free radical induced drug reactions was addressed by Dr. Charles Pippinger who focused on idiosyncratic drug reactions (IDR) which appear to be related to and controlled by genetic a profile. Specific examples include: adriamycin toxicity, valproic acid and acute hepatic toxicity in children under 10 years, and halothane malignant hyperthermia. He brought out the concept of free radical deficiency and risk for disease. Markers for clinical laboratories are needed to identify and prevent potential IDR.

Shock and Multiple Organ Dysfunction Syndrome (MODS) was presented by Professor Patricia Abello who brought together a coordinated presentation which connected clinical events, trigger mechanisms, acute phase responses, with the interactions and consequences of OFR. MODS is the leading cause of death in Critical Care Units. Gene expression results in acute phase and heat shock responses. In sepsis hypermetabolism precedes MODS. It is important to understand extracellular (eg.,SOD, catalase) from intracellular (e.g., DMSO, allopurinol) protective mechanisms and agents.

Alcoholism and fetal alcohol syndrome was presented by Professor Consuelo Guerri who commented on mechanisms of alcohol-induced free radical (FR) formation and discussed pathways involving the cytosol, cellular fractions and the cytochrome P-450 system. The role of acetaldehyde in ethanol-induced FR production, the effects of liver anti-oxidants and oxidative stress were commented on. The fetal alcohol syndrome includes: metabolic effects, astrocyte development impairment and mental retardation.

Oxidative stress and aging was presented by Dr. Richard Cutler who discussed this topic in terms of: 1) the problem that humans may be living longer than genetically programmed with advances in life span compared to previous periods over past 50,000 years, 2) organ function peaks at the onset of sexual maturity and underpoes a steady rate of decline thereafter, 3) elimination of major causes of death effects aging. and 4) comparative life spans in nature and models that correlate oxidative stress and antioxidant defense mechanisms.

THERAPEUTIC INTERVENTIONS: Inhibitors of Free Radical Reactions.

Vitamin E - Professor Jeffrey Blumberg. There are age-related changes to decrease immune mechanisms such as T-cell responses which translates to a greater incidence of infectious disease and risk of morbidity and mortality. Vitamin E improves the immune response index. There is evidence of vitamin E deficiency in certain elderly populations.

Gliclazide - Professor Paul Jennings. Diabetes Mellitus is associated with cardiovascular complications in which prevalence correlations with duration of disease. Therapeutic strategies to prevent premature death include: diet, exercise, no smoking, and glucose control. Gliclazide (glyburide), a sulfonyl urea, increases insulin secretion by β cell and decreases platelet adhesiveness; the actions decrease platelet aggregation, increase tissue plasminogen activator (tPA) and fibrinolysis, decrease oxidative stress and lipid peroxidation.

Beraprost and aminoguanide - Professor Paresh Dandona. The culprit cell in the atherosclerotic plaque is the monocyte that undergoes adhesion and penetration into the vascular wall, imbibes oxidized low-density lipoprotein (LDL), is transformed into the foam cell with capacity for excreting reactive materials; this forms the fatty streak. A further stage of endothelial damage will result in the beginning of a mural thrombus. Oxidation of LDL is a calcium - dependent process. Beraprost and Aminoguanidine are inhibitory to free radical generation. LDL oxidation is inhibited by Aminoguanidine. Clinical trials are in progress to test the potential beneficial effects of these drugs.

Carotenoids and other Vitamins - Dr. Wolfgang Schalch. There is epidemiological evidence for cardiovascular disease risk reduction associated with increased dietary carotene; 30% decrease in coronary heart disease risk (NEJM 328;1450,1993), and 50% decrease in major cardiovascular events (Circ. Supplement 82: 111 -201, 1990). There is also potential mechanism of cancer risk reduction and decreased cataracts. In China a Nutrition International Trial with B carotene, vitamin E and selenium was initiated (J. National Cancer Institute 85: 1483 - 92, 1993).

Flavenoids - Professor Elliott Middleton. At low concentrations flavenoids affect mitogenesis, secretory function, and inflammatory cells. Secretagogues stimulating histamine can be inhibited by flavenoids; they also have scavenger activity against the hydroxyl and the lipid peroxide radicals. Protection against free radical damage shown with vitamin C and E, selenium, and flaverroids. Flavenoids have vitamin C sparing activity; they are a relatively unexplored group of compounds.

PEG-SOD - Dr. Mark Saifer. Polyethylene glycol - superoxide dismutase is more effective with local administration that systemic; indications include rheumatology and urology disorders. SOD catalyzes superoxide radical. Veterinary indications include soft tissue inflammation in horses and vertebral column disorders in the canine. After intravenous administration clearance is greater the 95% at 2 hours; there is little tissue uptake except by kidney. There is a small but definite immunicity.

A Special Presentation was made by Dr. James Mitchell of the National Cancer Institute who gave an informative exposition entitled, "Fight a radical with a radical."

Closing Remarks - Professor Michael Wilson. This first **International Symposium on Free Radicals in Diagnostic Medicine** has been an outstanding success in: 1)Quality of presentations, 2) intensity of discussions, 3) information presented, and 4) controversies generated. It sets the standard for future congresses. We are indebted to the organizing team, to the presenters and all participants in attendance.
The program covered: basic concepts, organ-specific disorders, systemic involvement and therapeutic interventions. The sessions have been tremendously informative and enjoyable with special applause to Professor Yagi for all of his participation including a poignant after dinner song entertaining us in Japanese with English translation about memories of youth.

It is appropriate to hold this international conference in Buffalo with the rich architectural history of this city including many buildings listed in the National Heritage Preservation Archives and the four Frank Lloyd Wright Homes.

Finally, it is germane to remember that the modern clinical laboratory is equipped to give clinical data on Free radical biology and to document oxidative stress. Future clinical trials can benefit from the elegant information presented in this symposium and from close clinical-basic science interaction

Michael F. Wilson, M.D.,F.A.C.C.
Professor of Medicine
Director / Cardiovascular Medicine
The Millard Fillmore Hospitals and University at Buffalo

CONTENTS

Pathophysiology and Analysis

Organ Specific Disorders

Systemic Involvement

Therapeutic Intervention

Poster Presentations

LIPID PEROXIDES AND RELATED RADICALS IN CLINICAL MEDICINE

Kunio Yagi

Institute of Applied Biochemistry
Yagi Memorial Park
Mitake, Gifu 505-01
Japan

INTRODUCTION

In 1980, when the author organized an international symposium on "Lipid Peroxides in Biology and Medicine"[1], he was convinced that the research on the significance of lipid peroxides and their related free radicals in medicine had gotten off to a good start. Thereafter, many valuable results on this topic have been accumulated, and the problem has become of practical importance in clinical medicine. Thus, the present symposium seems to be timely organized.

Nowadays, the term "free radical", which means an independent chemical species having an unpaired electron(s), is a common word in medicine. The medical importance of free radicals is obviously due to their high reactivity, through which they can denature biomolecules such as proteins, lipids, and nucleic acids, resulting in injury to cells, organs, and tissues. It might not be an overstatement to say that all diseases involve free radicals in their pathogenesis.

When any free radical species is generated in our body, lipid component polyunsaturated fatty acids, which are reactive with free radical species, react with the former to become lipid radicals. For example, when X-ray irradiation produces hydroxyl radicals (•OH) from water, they react with polyunsaturated fatty acids to produce lipid radicals. Radical species such as those contained in oxidants, when absorbed in the body, give the same results. Radical-generating substances such as carbon tetrachloride also give the same. When active oxygen species formed by the leakage of electrons from the mitochondrial electron-transfer system under special conditions such as ischemia-reperfusion generate •OH, these radicals would then attack polyunsaturated fatty acids contained in the lipids in the mitochondrial membranes. However, the primary formation of lipid radicals from polyunsaturated fatty acids by metal-catalyzed hydrogen abstraction seems to be more important. Polyunsaturated fatty acid-containing lipids are the main component of biomembranes that exist ubiquitously in our body, and the formation of lipid radicals from polyunsaturated fatty acids is provoked by catalysis with transition metal, if the latter comes into contact with the former. The most realistic metal to be involved is iron. In our body, usually iron is shielded by proteins and

Free Radicals in Diagnostic Medicine, Edited by
D. Armstrong, Plenum Press, New York, 1994

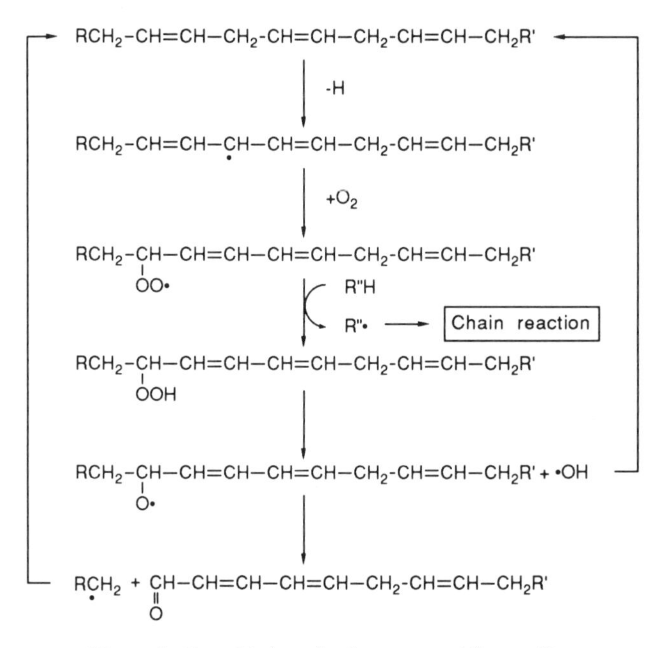

Figure 1. Peroxidation of polyunsaturated fatty acids.

membranes so that it cannot react with lipids. However, if iron becomes liberated from iron-containing proteins upon denaturation and forms its chelate with suitable substances in our body, it can then react with polyunsaturated fatty acids in biomembranes. Incidentally, electron-transfer systems resident in the membrane can reduce the metal so that metal-catalyzed reduction is realized. Thus, catalysis by iron coupled with electron transfer is the most important event in initiation and propagation of lipid peroxidation, as demonstrated by the pioneering work of Hochstein and Ernster[2]. Accordingly, the most realistic case for this scenario involves initiation of lipid peroxidation by injury to biomembranes by either physical or chemical causes such as mechanical shock, burn injury or metabolic disorders.

Although many comprehensive reviews on these problems have appeared, it might be appropriate to summarize here the general feature of the initiation and propagation of lipid peroxidation.

Figure 1 presents the series of reactions involved in lipid peroxidation. As shown in the upper part of the figure, lipid peroxidation is initiated by hydrogen abstraction. As is well known in oil chemistry, this can be done by light irradiation of oils. However, light irradiation is a rather rare occurrence in our body. The iron-catalyzed hydrogen abstraction mentioned above is a more realistic case. When hydrogen is abstracted, an unpaired electron remains on a carbon atom, and thus a carbon-centered radical (lipid radical) is formed. Since oxygen is usually dissolved in lipids, this radical reacts readily with oxygen to result in the formation of lipid peroxyl radicals, which then react with neighboring substances. If these substances are antioxidants or biological substances such as proteins, the lipid peroxyl radicals react with them to become lipid hydroperoxides, and the reaction process is tentatively terminated, since lipid hydroperoxides are relatively stable. We regard the lipid hydroperoxide as the primary product of lipid peroxidation. However, if such scavengers of the peroxyl radical do not exist, this radical may attack intact polyunsaturated fatty acids to abstract their hydrogen, and thereby a chain reaction occurs.

The primary product, lipid hydroperoxide, when it occurs and accumulates to some extent in cells, leaks from the cells into the bloodstream and migrates to other cells, organs, and tissues because of its relative stability. This means the transference of lipid peroxidation from one site to others in our body.

When lipid hydroperoxides are subjected to a catalytic reaction with iron-chelate, alkoxyl radicals and •OH are formed. This reaction has been suspected for a long time, but direct demonstration of •OH was never made until recently, when we used a spin trapping method to confirm the occurrence of •OH[3]. This result will be described later in more detail. It is well recognized that •OH are highly reactive with any substance, and so •OH, if they occur in our body, would be highly toxic[4].

As also shown in Figure 1, the alkoxyl radical changes to an aldehyde with a long carbon chain and an alkyl radical that abstracts hydrogen from other substances to become an alkane. If the hydrogen donor is a polyunsaturated fatty acid, a chain reaction again occurs. Aldehydes with a long carbon chain and alkanes thus formed are rather stable. So, we regard these aldehydes and alkanes as the secondary product in the lipid peroxidation reaction.

As can be seen from the reaction process shown in Figure 1, lipid peroxidation in the body can be estimated by measuring any substance involved in the reaction process. Since radical species are the actual agents to injure cells, their direct measurement would be preferable for clinical diagnosis. However, radical species disappear very rapidly to yield the primary and secondary lipid peroxidation products. Although radical species can be measured by use of trapping agents under special conditions such as in vitro experiments, this cannot be well applied to the clinical specimens, since the radical reaction occurring in the specimens has already been terminated by the time of analysis. Therefore, we are forced to measure relatively stable substances, viz., the primary peroxidation product, lipid hydroperoxides, and/or the secondary peroxidation product, aldehydes with a long carbon chain and alkanes.

For the estimation of the level of both lipid hydroperoxides and aldehydes with a long carbon chain, the thiobarbituric acid (TBA) reaction carried out under the conditions specified by us[5,6] is recommendable. For the estimation of lipid hydroperoxides, our methylene blue method can be recommended[7,8]. A detailed description of these methods will be made later.

Other methods are also available for the estimation of lipid peroxidation and radical formation in the body such as analysis of pentane and ethane in exhalation[9] or determination of urinary 8-hydroxy-2'-deoxyguanosine[10] formed from the reaction of guanine residues in nucleic acids with •OH[11]. Comparing with these methods, ours might be simpler.

ASSAY FOR SERUM LIPID PEROXIDE LEVEL BY THE TBA REACTION

When the author sought to measure the lipid peroxide level in the blood in the 1960's, no reliable assay method was available. Thus, he decided to devise one himself. At that time, it was supposed that the amount of lipid peroxides in the blood might be small and that these substances might be decomposed by prolonged pretreatments such as organic solvent extraction of these substances. Among assay methods available for the estimation of the lipid peroxide level in oils at that time, *viz.*, determination of peroxide value, diene conjugate, or TBA-reactive substances, the latter one seemed to be more preferable than the former two for the application to serum or plasma samples without extraction of lipids with organic solvents. In addition, the TBA reaction was known to be sensitive to lipid peroxides[12,13], though its specificity was rather low. On the basis of these considerations, the author decided to utilize the TBA reaction for the assay of lipid peroxide levels in serum or plasma. As a matter of fact, lipid peroxides in these samples needed to be isolated by a simple and rapid procedure from other TBA-reactive substances prior to reaction with TBA. After a systematic investigation, we devised a reliable method to determine lipid peroxide levels in blood serum[14]. Our success was mainly due to the isolation of lipids by precipitating them along with serum

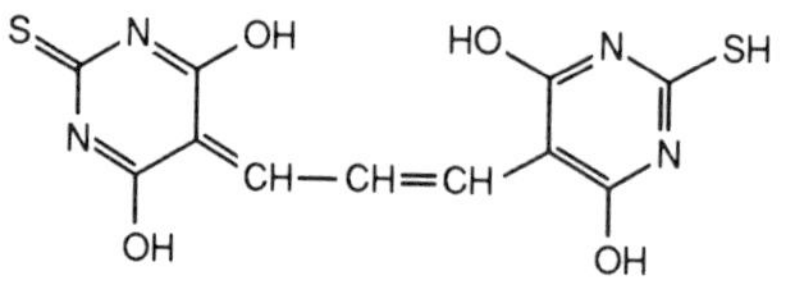

Figure 2. Structure of the product obtained from the reaction of lipid peroxides with TBA.

proteins by use of a phosphotungstic acid-sulfuric acid system from other TBA-reactive substances such as glucose and water-soluble aldehydes in the serum and to the reaction of TBA with these lipids in acetic acid solution to avoid its reaction with substances such as sialic acid. After the TBA reaction, the product was determined by the absorption at 532 nm. Although this method proved useful, the author was able to improve it further, since he found that the TBA reaction product is a fluorescent substance[5]. By use of fluorometry, he could increase the specificity of the assay method, and could apply this method to a small volume of the blood, and, therefore, to infants and small animals. In this chapter, basic data for the establishment of this assay method, and its recommended standard procedure will be described.

Basic Background Data for Assay Procedure

Properties of the TBA Reaction Product. As was reported by Bernheim et al.[12], oxidation products of lipids react with TBA upon being heated in acidic solution. Sinnhuber et al.[13] reported that the products have the structure shown in Figure 2. Malondialdehyde is known to react with TBA and the reaction product is the same as that of lipid peroxide with TBA. The absorption spectrum of the product is shown in the inset of Figure 4. As mentioned earlier, the author found that the product, a red pigment, is a fluorescent substance[5]. Figure 3 shows the excitation and emission spectra of the pigment.

Conditions for the TBA Reaction. Since it was already known that the reaction of lipid hydroperoxides with TBA in an aqueous solution requires heating at nearly 100°C, we decided to adopt a temperature of 95°C for the reaction, and examined the effect of pH and duration of heating on the reaction. We found that the maximum formation of the product could be attained at around pH 3.5 for linoleic, linolenic, or arachidonic acid hydroperoxides[15]. The time course of the reaction of these lipid hydroperoxides with TBA at pH 3.5 indicated that 60 min was necessary for the product level to reach plateau[15].

Separation of Lipid Peroxides from Water-soluble TBA-Reactive Substances by a Simple Procedure. Water-soluble substances, which react with TBA to yield the same product as lipid peroxides, must be removed by a simple procedure. For this purpose, we made extensive preliminary experiments, and reached a conclusion that such substances can be separated from lipid peroxides by precipitating the latter along with serum proteins in a phosphotungstic acid-sulfuric acid system. Under the optimum reaction conditions mentioned in the previous section, the precipitates were reacted with TBA. The amount of the product was compared with that obtained from the reaction of TBA with chloroform-methanol extract of the precipitates. By repeated extractions, most of the TBA-reactive substances were recovered in the organic phase. This means that the precipitates contained organic solvent-soluble TBA-reactive substances, *viz.*, lipid hydroperoxides and related aldehydes with a long carbon chain.

Another substance that reacts with TBA is sialic acid, which cannot be separated by the above-mentioned precipitation procedure. We found that in trichloroacetic acid solution sialic

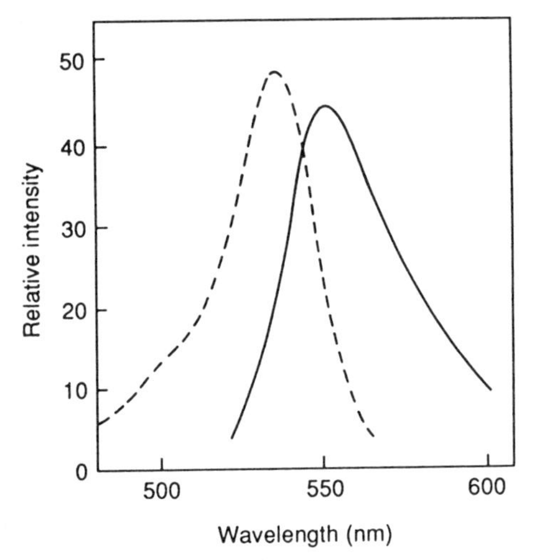

Figure 3. Excitation and emission spectra of the product obtained from the reaction of malondialdehyde with TBA. Dotted line: excitation spectrum monitored at 575 nm; solid line: emssion spectrum obtained with excitation at 515 nm.

acid reacts with TBA to give a substance(s) having the absorption spectrum shown in Figure 4, but that sialic acid cannot react with TBA in acetic acid solution. Therefore, the use of trichloroacetic acid adopted by some researchers[16,17] is not applicable to samples that contain sialic acid. Even when samples do not contain sialic acid, trichloroacetic acid solution is not preferable, since it decreases the reaction of TBA with lipid hydroperoxides[15]. When measured by fluorometry, the product formed from sialic acid and TBA in trichloroacetic acid does not interfere with the measurement of the product from TBA and serum lipid peroxides, but the author cannot recommend the use of trichloroacetic acid for the above reason.

Related to this problem, it should be mentioned that the situation is the same for the

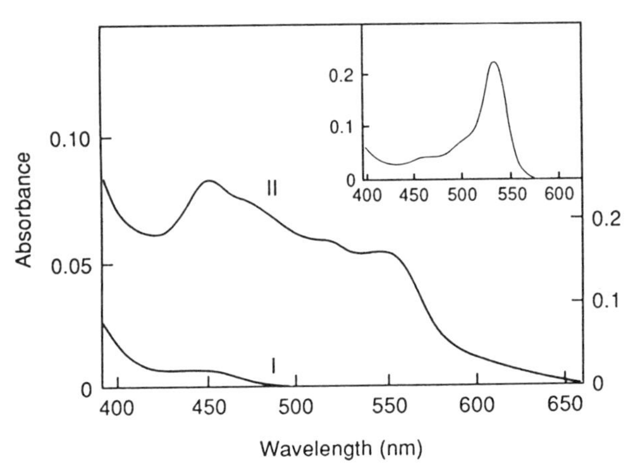

Figure 4. Absorption spectrum of the product obtained from the reaction of sialic acid with TBA. I: reacted in acetic acid solution (left scale); II: reacted in trichloroacetic acid solution (right scale). The concentration of sialic acid: 0.3 μg/reaction mixture. Inset: product obtained from the reaction of malondialdehyde with TBA.

measurement of lipid peroxide levels in organs and tissues by the TBA method. Therefore, the author recommends the procedure described by Ohkawa et al.[18], but not that using trichloroacetic acid (see also the chapter on "Lipid peroxides in hepatic, gastrointestinal, and pancreatic diseases" in this volume).

Bilirubin, if it occurs in the blood, reacts with TBA under the conditions mentioned above. Fortunately, however, the fluorescence of this product is different from that of the reaction product of lipid peroxides with TBA[19]. Therefore, bilirubin does not disturb the assay of lipid peroxides, if the fluorescent intensity is measured at 553 nm.

Relationship between the Amount of Lipid Peroxides and Fluorescence Intensity. The amount of product obtained from the reaction of lipid peroxides contained in the phosphotungstic acid-sulfuric acid precipitates with TBA was checked by the fluorescence intensity of this product. Linearity was found up to 0.04 ml of serum[19].

Since lipid peroxides are plural substances and their reactivity with TBA is similar but not identical, we should use a proper standard to evaluate the level of lipid peroxides in serum or plasma. As such standard substance, we decided to use malondialdehyde. However, as malondialdehyde is somewhat unstable, we practically adopted tetramethoxypropane or tetraethoxypropane, either of which is stable and converts quantitatively to malondialdehyde upon heating in acidic solution. When the relationship between the amount of malondialdehyde and the fluorescence intensity of the product was examined by use of tetramethoxypropane, a linear relationship was found up to 1.0 nmol of malondialdehyde under the conditions mentioned above.

Lipid peroxides were also found to give the fluorescent product in a concentration-dependent manner. However, thier reactivity with TBA is not stoichiometric; e.g., one mole of linoleic acid hydroperoxide gives only one-twentieth of the malondialdehyde equivalent. Accordingly, it should be noted that lipid peroxide levels measured by the above-mentioned method are relative values.

Standard Procedure

In most clinical investigations, blood serum or plasma would be used as the sample. For separation of plasma, both heparin and citrate can be used without any effect on the assay. The standard procedure is presented in Scheme 1. If only a small volume of the blood is available, as in the case of infants, plasma lipid peroxide level/ml blood can be measured. For this purpose, standard procedure presented in Scheme 2 is recommendable.

The value obtained by this method gives relative amounts of lipid hydroperoxides and/or aldehydes with a long carbon chain, but not their absolute quantities, as mentioned before. Therefore, the author recommends to describe the value as "lipid peroxide level expressed in terms of malondialdehyde". The description "TBA reactive substances (TBARS)" is often used by some researchers. In the case of the measurement by the present method, TBARS that are not related to lipid peroxides are eliminated; and, therefore, the obtained value should better be described as "lipid peroxide level", and not as "TBARS".

SERUM OR PLASMA LIPID PEROXIDE LEVELS IN DISEASES

Serum lipid peroxide levels of normal subjects were determined by the above-mentioned method with a relatively large number of people[20]. As summarized in Table 1, the mean value increases with advancing age, but does not exceed 4 nmol/ml. It was found that the level for women of middle age is lower than that for their male counterpart. We thought that this difference might explain the lower incidence of coronary heart disease in women than in men[21,22], and ascribed this difference to the antioxidant effect of female hormone[23~25].

Scheme 1. Standard procedure for the assay of serum or plasma lipid peroxide levels.

1. Twenty microliters of serum or plasma are placed in a glass centrifuge tube.
2. To this tube, 4.0 ml of N/12 H_2SO_4 is added and mixed gently.
3. Then, 0.5 ml of 10% phosphotungstic acid is added and mixed. After allowing to stand at room temperature for 5 min, the mixture is centrifuged at 3,000 rpm for 10 min.
4. The supernatant is discarded, and the sediment is mixed with 2.0 ml of N/12 H_2SO_4 and 0.3 ml of 10% phophotungstic acid. The mixture is centrifuged at 3,000 rpm for 10 min.
5. The sediment is suspended in 4.0 ml of distilled water, and 1.0 ml of TBA reagent is added. TBA reagent is a mixture of equal volumes of 0.67% TBA aqueous solution and glacial acetic acid. The reaction mixture is heated at 95°C for 60 min in an oil bath.
6. After cooling with tap water, 5.0 ml of *n*-butanol are added and the mixture is shaken vigorously.
7. After centrifugation at 3,000 rpm for 15 min, the *n*-butanol layer is taken for fluorometric measurement at 553 nm with excitation at 515 nm.
8. Taking the fluorescence intensity of the standard solution, which is obtained by reacting 0.5 nmol of tetramethoxypropane with TBA by steps 5-7, as F and that of the sample as f, the lipid peroxide level can be calculated and expressed in terms of malondialdehyde:

Serum or plasma lipid peroxide level

$$= 0.5 \times \frac{f}{F} \times \frac{1.0}{0.02} = \frac{f}{F} \times 25 \text{ (nmol/ml of serum or plasma)}$$

Scheme 2. Standard procedure for the assay of lipid peroxide levels in a small volume of blood.

1. Using a pipet for determination of blood cells, 0.05 ml of the blood is taken (e.g., from the ear lobe).
2. The blood is put into 1.0 ml of physiological saline in a centrifuge tube, and shaken gently.
3. After centifugation at 3,000 rpm for 10 min, 0.5 ml of the supernatant is transferred to another centrifuge tube.

Then, steps 2-8 of the procedure in Scheme 1 are followed.

Plasma lipid peroxide level

$$= 0.5 \times \frac{f}{F} \times \frac{1.05}{0.5} \times \frac{1.0}{0.05} = \frac{f}{F} \times 21 \text{ (nmol/ml of blood)}$$

In this section, the description is limited to the changes in serum or plasma lipid peroxide levels in vascular diseases, implication of lipid hydroperoxides in these diseases, and the changes in the lipid peroxide level in thermal injury to emphasize the pathogenicity of lipid peroxides in diseases, since the author expects that the role of lipid peroxides in each disease would be amply discussed by each specialist in other chapters in this volume.

Table 1. Serum lipid peroxide levels of normal subjects.

Age (years)	Lipid peroxide level			
	Male		Female	
≤ 10	1.86 ± 0.60	(10)	2.08 ± 0.48	(8)
11 - 20	2.64 ± 0.60	(10)*	2.64 ± 0.54	(9)
21 - 30	3.14 ± 0.56	(10)	2.98 ± 0.50	(9)
31 - 40	3.76 ± 0.52	(11)**	3.06 ± 0.50	(9)***
41 - 50	3.94 ± 0.60	(11)	3.16 ± 0.54	(10)***
51 - 60	3.92 ± 0.92	(8)	3.30 ± 0.74	(10)
61 - 70	3.94 ± 0.70	(10)	3.46 ± 0.72	(10)
≥ 71	3.76 ± 0.76	(12)	3.30 ± 0.78	(10)
Mean	3.42 ± 0.94	(82)	3.10 ± 0.62	(75)***

Lipid peroxide levels were measured according to Yagi[5, 6] and expressed in terms of malondialdehyde (nmol/ml serum). Mean ± SD is given. The number of subjects is given in parentheses. Significant difference: *$p<0.05$ vs. the age group ≤ 10; **$p<0.05$ vs. the age group 21-30; ***$p<0.05$ vs. the corresponding group of males.

Serum or Plasma Lipid Peroxide Levels of Patients with Vascular Diseases

Plasma lipid peroxide levels of patients with diabetes mellitus were significantly higher than those of normal subjects[26]. The patients with angiopathy showed higher lipid peroxide levels as compared with those of the patients without angiopathy, indicating the intimate relationship between the increase in lipid peroxide levels in the blood and angiopathy.

Plachta et al.[27] found that plasma lipid peroxide levels of patients suffering from atherosclerosis were significantly higher than those of normal subjects.

Kawamoto et al.[28] measured serum lipid peroxide levels of patients with apoplexy and found that the levels were high in both cases of cerebral bleeding and cerebral infarction at the time of hospitalization. The levels remained high in those patients who died, whereas the levels decreased to normal within 1 month in the patients who survived.

All these clinical investigations mentioned above indicate that an intimate relationship exists between an increase in lipid peroxides in the blood and development of vascular diseases. However, this result alone does not tell us whether the former is the cause or the result of the latter, and, therefore, we performed experimental studies.

Implication of Lipid Hydroperoxides in Vascular Disorders

Since the primary target of lipid peroxides increased in the blood would be the intima of blood vessels, we examined whether lipid peroxides increased in the blood would indeed provoke an injury to the endothelial cells of the vessels. To clarify this point, we injected linoleic acid hydroperoxide into a rabbit through an ear vein[29]. After the injection, the lipid peroxide level in the serum increased immediately and then decreased gradually for about 10 h. Twenty-four hours after the injection of the hydroperoxide, the lipid peroxide level in the aorta increased two-fold. This suggests that the hydroperoxide was transferred to the aorta through the bloodstream. Then morphological examination of the endothelium of the aorta was carried out by scanning electron microscopy. In the rabbit injected with linoleic acid as the control, the endothelial surface of the thoracic aorta was normal and smooth.

Upon the injection of linoleic acid hydroperoxide, a marked damage to the endothelial cells was seen. That is, the arrangement of the endothelial cells became irregular; many holes were seen in the cells; some cells were enucleated; in some areas subendothelial fibrous tissue was exposed; and to the injured site, platelets appeared to be adhering. Blood platelets are known to aggregate and adhere to sites of endothelial injury. Such aggregation of platelets and their adherence are prevented by prostacyclin[30]. However, the biosynthesis of prostacyclin is inhibited by lipid peroxides[31,32], and thereby lipid peroxides promote atherogenesis.

It is well known that platelet aggregates produce migration and proliferation factors that affect smooth muscle cells in the aortic media[33]. These muscle cells are regarded as one of the origins of lipid-laden cells, viz., foam cells, which are characteristic cells appearing in atherogenesis. Our experimental results showed that uptake and internalization of low density lipoproteins (LDL) by cultured smooth muscle cells were enhanced by linoleic acid hydroperoxide[34]. This finding was further verified by morphological observation[35]. The macrophage is another source of the foam cells[36]. We found by morphological observation on cultured macrophages derived from monocytes that the pretreatment of LDL with the hydroperoxide is essential for the formation of typical foam cells[35].

We also recognized that lipid peroxides increased the net synthesis of collagen by arterial smooth muscle cells[37], which is obviously essential for the formation of the atheromatous plaque.

These results indicate that lipid peroxides intervene in atherogenesis not only in the initiation but also in the propagation and completion of atherosclerosis.

Increase in Lipid Peroxides in Burn Injury

Considering that the increased lipid peroxides in the blood are the cause of vascular disorders, any factor that increases the level of lipid peroxides in the blood would be regarded as an indirect cause of vascular disorders. Relating to this problem, the author paid special attention to the increase in serum lipid peroxide levels in the case of burn injury.

We reported the changes in serum lipid peroxide levels of five human subjects after thermal injury[38]. The levels of three cases were higher at the time of admission as compared with the levels of normal subjects; whereas in the other two, they were normal at the time of admission and then increased. Thus, it was obvious that the level increased without exception in the early post-burn period.

When thermal injury was provoked in rats by pouring boiling water on the skin of the back, the lipid peroxide level in the serum increased in accordance with the increase in that in the skin[39]. Thereafter, the level in the spleen was found to be significantly increased. The levels in the liver and kidney also tended to increase. The elevated lipid peroxide levels in these organs are thought to be a consequence of the increase in serum lipid peroxides. It is considered, therefore, that the increased lipid peroxides in the serum would bring about secondary disorder in other intact organs and tissues. In fact, we observed that glutamate oxalacetate transaminase, alkaline phosphatase, and lactate dehydrogenase increased significantly following the burn injury, indicating liver injury[39]. Accordingly, we should consider that when lipid peroxides increased in any site in our body leak into the bloodstream, and they can injure intact cells, organs or tissues; in other words, lipid peroxide-mediated injury may be transferred from one site to others in our body.

SIGNIFICANCE OF LIPID HYDROPEROXIDES IN SERUM OR PLASMA

As described in the introductory part of this paper, lipid peroxidation yields relatively stable primary and secondary products. As seen in thermal injury, lipid peroxides formed are transferred via the bloodstream. Therefore, we must pursue the effect of these substances.

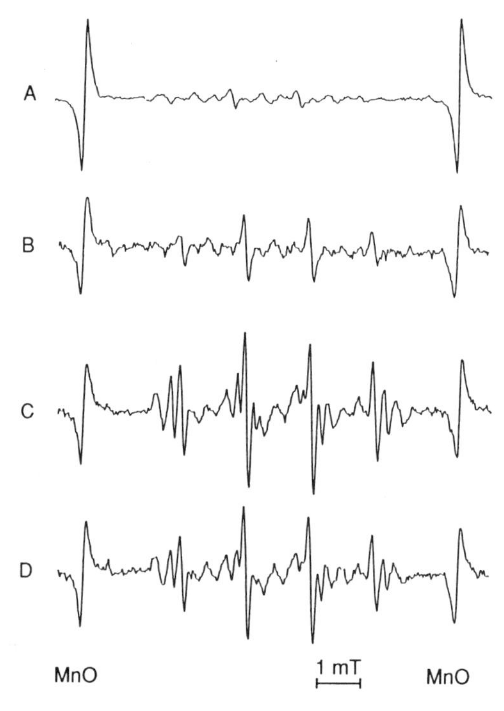

Figure 5. ESR spectra of DMPO-radical adducts obtained from linoleic acid hydroperoxide. The reaction mixture contained the following agents in a final volume of 0.3 ml: A, 50 mM potassium phosphate buffer (pH 7.4), 2.5 mM linoleic acid hydroperoxide, and 613 mM DMPO; B, A + 100 mM $FeCl_3$; C, A + 100 mM $FeSO_4$; D, A + 5 mM epinephrine + 100 mM $FeCl_3$.

When we examine the reaction process involved in lipid peroxidation, we should notice that lipid hydroperoxides can generate radicals, whereas aldehydes with a long carbon chain do not. Accordingly, lipid hydroperoxides seem to have special meaning, because, if they react with ferrous iron, they generate •OH.

Generation of •OH from Lipid Hydroperoxides

The generation of •OH from lipid hydroperoxides has been suspected for a long time. However, nobody demonstrated their generation. Recently we studied this problem with purified linoleic acid hydroperoxide[3]. When ferrous iron was added to the reaction mixture containing linoleic acid hydroperoxide and spin trap 5,5-dimethyl-1-pyrroline-N-oxide (DMPO), we observed the typical electron spin resonance (ESR) signals characteristic of the spin adduct of DMPO and •OH, as shown in Figure 5. When ferric iron was used instead of ferrous iron, the signals were not detected. When epinephrine was added together with ferric iron, however, the generation of •OH was actually found. This indicates that ferrous iron formed from ferric iron by reduction with epinephrine decomposes linoleic acid hydroperoxide to generate •OH. Further, we demonstrated that •OH can be generated from lipid hydroperoxides contained in oxidatively modified LDL in the presence of an epinephrine-ferric iron mixture, as shown in Figure 6[40]. We assumed that •OH generated in the LDL further denature its apoprotein and that the cytotoxicity of oxidatively modified LDL is attributable, at least in part, to the hazardous effect of •OH generated from the oxidatively modified LDL.

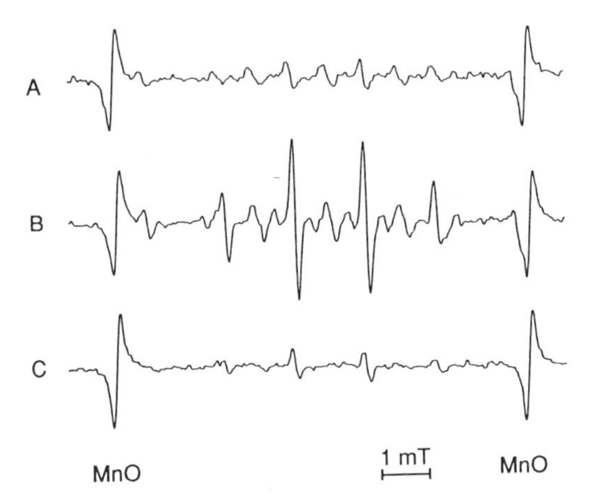

Figure 6. ESR spectra of DMPO-radical adducts obtained from oxidatively modified LDL. The reaction mixture contained the following agents in a final volume of 0.2 ml: A, 2 mM Tris-HCl buffer (final pH, 7.4) containing 30 mM KCl, oxidatively modified LDL (1.49 mg protein/ml), and 690 mM DMPO; B, A + 0.1 mM epinephrine-$FeCl_3$ (molar ratio, 1;1); C, B - oxidatively modified LDL.

Specific Measurement of Lipid Hydroperoxides

In the light of the toxicity of lipid hydroperoxides, it is desirable to measure specifically lipid hydroperoxides in serum or plasma. Earlier we had already developed a simple colorimetric method for the specific assay of lipid hydroperoxides[7,8]. This method, the so-called Hb-MB method, is based on the reaction of lipid hydroperoxides with a leuco-methylene blue derivative (10-N-methylcarbamoyl 3,7-dimethylamino-10H-phenothiazine) in the presence of a catalyzer, hemoglobin, as shown in Figure 7. The amount of methylene blue formed was measured spectrophotometrically at 675 nm to calculate the amount of lipid hydroperoxides with cumene hydroperoxide used as the external standard. When the amount of purified linoleic acid hydroperoxide was determined by this method, the result obtained coincided well with the peroxide value as well as with the value determined by the glutathione peroxidase reaction.

For application of this method directly to serum or plasma, however, pretreatment of the sample is necessary. Namely, lipids associated with proteins should become accessible to the reagent, and the disturbing substances must be eliminated. For the former purpose, addition

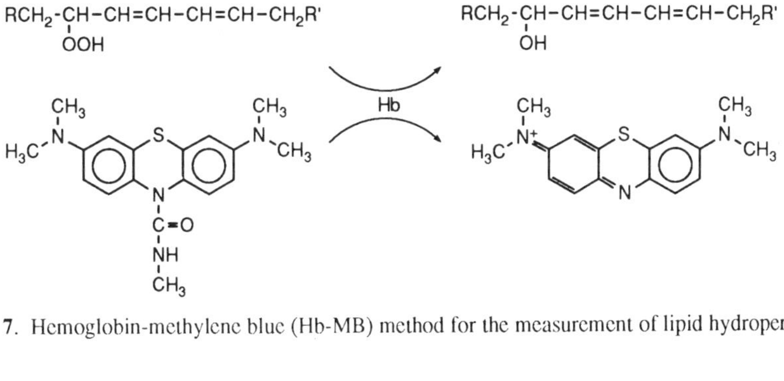

RCH$_2$-CH(OOH)-CH=CH-CH=CH-CH$_2$R' → RCH$_2$-CH(OH)-CH=CH-CH=CH-CH$_2$R'

Hb

Figure 7. Hemoglobin-methylene blue (Hb-MB) method for the measurement of lipid hydroperoxides.

Scheme 3. Standard procedure for the assay of serum or plasma lipid hydroperoxide levels by Hb-MB method.

1. One-hundred microliters of serum, plasma, or standard solution is placed in a glass tube.
2. To this tube, 1.0 ml of pretreatment reagent containing ascorbate oxidase (10 U/ml), lipoprotein lipase (1.3 U/ml), Triton-X100 (1 mg/ml) and triethylenetetramine hexaacetic acid (4.5 mg/ml) in Good buffer (pH 5.8) is added and the reaction mixture is incubated at 30°C for 2~5 min.
3. Then, 2.0 ml of reaction reagent containing leucomethylene blue (0.04 mM) and hemoglobin (67.5 μg/ml) in Good buffer (pH 5.8) is added and incubation is continued at 30°C for 10 min.
4. Absorbance at 675 nm is measured.
5. Taking the absorbance of the standard solution (0.1 ml of cumene hydroperoxide, 50 nmol/ml) as A and that of the sample as a, the lipid hydroperoxide level can be calculated.

Serum or plasma lipid hydroperoxide level

$$= \frac{a}{A} \times 50 \text{ (nmol/ml of serum or plasma)}$$

Reagents for Hb-MB method are available from Kyowa Medex Co., Ltd., Tokyo, Japan.

of lipoprotein lipase and a detergent, Triton-X100 followed by incubation at 30°C for 2 min is necessary. Then, we found that both holotransferrin and ascorbic acid disturb the reaction. To eliminate their effects, we examined the pretreatment with triethylenetetramine hexaacetic acid and ascorbate oxidase at 30°C for 2 min and showed this step to be satisfactory.

Scheme 3 presents the procedure of the Hb-MB method for serum or plasma. By use of this method, serum lipid hydroperoxide levels of normal human subjects were found to be less than 1 nmol/ml (Table 2). In the serum or plasma of patients having a lipid peroxide level of over 4 nmol/ml in terms of malondialdehyde obtained by the TBA reaction, the value obtained by the Hb-MB method gives a higher value than 1 nmol/ml. It seems that the increment of lipid peroxide level is largely due to lipid hydroperoxides.

For the determination of serum or plasma lipid hydroperoxide level, several other methods were reported. Table 2 summarizes serum or plasma lipid hydroperoxide levels of normal human subjects determined by various methods. Warso and Lands[41] adopted a prostaglandin H synthase-oxygen electrode system and reported the hydroperoxide concentration of 0.5 nmol/ml for normal plasma. Yamamoto et al.[42] described that the concentrations of free fatty acid hydroperoxides and cholesteryl ester hydroperoxides were 0.056 and 0.317 nmol/ml, respectively. Miyazawa et al.[43] reported that the level of phosphatidylcholine hydroperoxide in normal human plasma ranged from 0.05 to 0.43 nmol/ml plasma. Since the Hb-MB method we developed is much simpler than other methods, it would be more useful for clinical diagnosis.

CONCLUDING REMARKS

In this chapter, a method devised by the author for the assay of lipid peroxide level in serum or plasma by use of the TBA reaction was described in detail. Some data on

Table 2. Serum or plasma lipid hydroperoxide levels of normal subjects.

Assay method	Lipid hydroperoxide level (nmol/ml)		Ref.
	Serum	Plasma	
Hb-MB	≤ 1		
Prostaglandine H synthase		0.5	(41)
HPLC-chemiluminescence			
Fatty acid-OOH		0.056	(42)
Cholesteryl ester-OOH		0.317	(42)
Phosphatidylcholine-OOH		0.05~0.43	(43)

implication of lipid peroxides in diseases were also presented. Among lipid peroxides, lipid hydroperoxides seem to have much significance in lipid peroxide-mediated pathogenesis. Measurement of lipid hydroperoxides in serum or plasma is, therefore, significant especially for the prediction of secondary disorder provoked by them, in other words, for prognosis of diseases. Accordingly, our simple method, the so-called Hb-Mb method, for the assay of lipid hydroperoxides in serum or plasma was also described.

Although the assay method using TBA measures the level of both lipid hydroperoxides and aldehydes with a long carbon chain, samples having the values above 4 nmol/ml, which is pathological, give high values of hydroperoxides measured by the hydroperoxide-specific Mb-HB method. Thus, a value higher than 4 nmol/ml indicates an increase in the level of lipid hydroperoxides. It should also be noted that even in the case of less than 4 nmol/ml, which is normal, the value is still useful for evaluation of our physical conditions, since a lower level is preferable as evidenced by the fact that a lower "normal" level is found for women than for men in their 30's~50's and women are less prone to coronary heart disease.

Clinical application of these two assay methods is useful to reveal the abnormal situation of lipid peroxides and related radicals in patients, which data may aid in the diagnosis and prognosis of diseases.

REFERENCES

1. K. Yagi (ed.), "Lipid Peroxides in Biology and Medicine", Academic Press, New York, London (1982).
2. P. Hochstein and L. Ernster, ADP-activated lipid peroxidation coupled to the TPNH oxidase system of microsomes, *Biochem. Biophys. Res. Commun.* 12: 388 (1963).
3. K. Yagi, N. Ishida, S. Komura, N. Ohishi, M. Kusai, and M. Kohno, Generation of hydroxyl radical from linoleic acid hydroperoxide in the presence of epinephrine and iron, *Biochem. Biophys. Res. Commun.* 183: 945 (1992).
4. G. Czapski, Reaction of •OH, *Methods Enzymol.* 105: 209 (1984).
5. K. Yagi, A simple fluorometric assay for lipoperoxide in blood plasma, *Biochem. Med.* 15: 212 (1976).
6. K. Yagi, Assay for blood plasma or serum, *Methods Enzymol.* 105: 328 (1984).
7. N. Ohishi, H. Ohkawa, A. Miike, T. Tatano, and K. Yagi, A new assay method for lipid peroxides using a methylene blue derivative, *Biochem. Int.* 10: 205 (1985).
8. K. Yagi, K. Kiuchi, Y. Saito, A. Miike, N. Kayahara, T. Tatano, and N. Ohishi, Use of a new methylene blue derivative for determination of lipid peroxides in foods. *Biochem. Int.* 12: 367 (1986).
9. C. J. Dillard and Al. L. Tappel, Volatile hydrocarbon and carbonyl products of lipid peroxidation: A comparison of pentane, ethane, hexanal and acetone as in vivo indices, *Lipids* 14: 989 (1979).

10. M. K. Shigenaga, C. J., Gimeno, and B. N. Ames, Urinary 8-hydroxy-2'-deoxyguanosine as a biological marker of in vivo oxidative DNA damage, *Proc. Natl. Acad. Sci. USA* 86: 9697 (1989).
11. H. Kasai and S. Nishimura, Hydroxylation of deoxyguanosine at the C-8 position by ascorbic acid and other reducing agents, *Nucleic Acids Res.* 12: 2137 (1984).
12. F. Bernheim, M. L. C. Bernheim, and K. M. Wilbur, The reaction between thiobarbituric acid and the oxidation products of certain lipides, *J. Biol. Chem.* 174: 257 (1948).
13. R. O. Sinnhuber, T. C. Yu, and Te. C. Yu, Characterization of the red pigment formed in the 2-thiobarbituric acid determination of oxidative rancidity, *Food Res.* 23: 626 (1958).
14. K. Yagi, I. Nishigaki, and H. Ohama, Measurement of serum TBA-values, *Vitamins* (Japan) 37: 105 (1968).
15. H. Ohkawa, N. Ohishi, and K. Yagi, Reaction of linoleic acid hydroperoxide with thiobarbituric acid, *J. Lipid Res.* 19: 1053 (1978).
16. T. F. Slater, Overview of methods used for detecting lipid peroxidation, *Methods Enzymol.* 105: 283 (1984).
17. J. A. Buege and S. D. Aust, Microsomal lipid peroxidation, *Methods Enzymol.* 52: 302 (1978).
18. H. Ohkawa, N. Ohishi, and K. Yagi, Assay for lipid peroxides in animal tissues by thiobarbituric acid reaction, *Anal. Biochem.* 95: 351 (1979).
19. K. Yagi, Assay for serum lipid peroxide level and its clinical significance, *in* : "Lipid Peroxides in Biology and Medicine," K. Yagi, ed., Academic Press, New York, London (1982).
20. T. Suematsu, T. Kamada, H. Abe, S. Kikuchi, and K. Yagi, Serum lipoperoxide level in patients suffering from liver diseases, *Clin. Chim. Acta* 79: 267 (1977).
21. W. P. Castelli, Epidemiology of coronary heart disease: The Framingham study, *Am. J. Med.* 76, 2A: 4 (1984).
22 R. H. Knopp, Arteriosclerosis risk. The roles of oral contraceptives and postmenopausal estrogens, *J. Reprod. Med.* 31: 913 (1986).
23. K. Yagi and S. Komura, Inhibitory effect of female hormones on lipid peroxidation, *Biochem. Int.*, 13: 1051 (1986).
24. K. Yoshino, S. Komura, I. Watanabe, Y. Nakagawa, and K. Yagi, Effect of estrogens on serum and liver lipid peroxide levels in mice, *J. Clin. Biochem. Nutr.* 3: 233 (1987).
25. K. Sugioka, Y. Shimosegawa, and M. Nakano, Estrogens as natural antioxidants of membrane phospholipid peroxidation, *FEBS Letters* 210: 37 (1987).
26 Y. Sato, N. Hotta, N. Sakamoto, S. Matsuoka, N. Ohishi, and K. Yagi, Lipid peroxide level in plasma of diabetic patients, *Biochem. Med.* 21: 104 (1979).
27. H. Plachta, E. Bartnikowska, and A. Obara, Lipid peroxides in blood from patients with atherosclerosis of coronary and peripheral arteries, *Clin. Chim. Acta* 211: 101 (1992).
28. M. Kawamoto, M. Kagami, and A. Terashi, Serum lipid peroxide level in apoplexia, *J. Clin. Biochem. Nutr.* 1: 1 (1986).
29. K. Yagi, H. Ohkawa, N. Ohishi, M. Yamashita, and T. Nakashima, Lesion of aortic intima caused by intravenous administration of linoleic acid hydroperoxide, *J. Appl. Biochem.* 3: 58 (1981).
30. R. J. Gryglewsky, S. Bunting, S. Moncada, R. J. Flower, and J. R. Vane, Arterial walls are protected against deposition of platelet thrombi by a substance (prostaglandin X) which they make from prostaglandin endoperoxides, *Prostaglandins* 12: 685 (1976).
31. A. Szczeklik and R. J. Gryglewski, Low density lipoproteins (LDL) are carriers for lipid peroxides and inhibit prostacyclin (PGI_2) biosynthesis in arteries, *Artery* 7: 488 (1980).
32. Y. Sasaguri, M. Morimatsu, T. Nakashima, O. Tokunaga, and K. Yagi, Difference in the inhibitory effect of linoleic acid hydroperoxide on prostacyclin biosynthesis between cultured endothelial cells from human umbilical cord vein and cultured smooth muscle cells from rabbit aorta, *Biochem. Int.* 11: 517 (1985).
33. R. Ross, E. W. Raines, and D. F. Bowen-Pope, The biology of platelet-derived growth factor, *Cell*, 46: 155 (1986).
34. I. Nishigaki, M. Hagihara, M. Maseki, Y. Tomoda, K. Nagayama, T. Nakashima, and K. Yagi, Effect of linoleic acid hydroperoxide on uptake of low density lipoprotein by cultured smooth muscle cells from rabbit aorta, *Biochem. Int.* 8: 501 (1984).

35. K. Yagi, T. Inagaki, Y. Sasaguri, R. Nakano, and T. Nakashima, Formation of lipid-laden cells from cultured aortic smooth muscle cells and macrophages by linoleic acid hydroperoxide and low density lipoprotein, *J. Clin. Biochem. Nutr.* 3: 87 (1987).
36. R. G. Gerrity, The role of the monocyte in atherogenesis, *Am. J. Pathol.* 103: 181 (1981).
37. I. Nishigaki, H. Yanagi, Y. Sasaguri, M. Morimatsu, T. Hayakawa, and K. Yagi, Synthesis of collagen by cultured smooth muscle cells from rabbit aorta is increased in the presence of linoleic acid hydroperoxide, *J. Clin. Biochem. Nutr.* 11: 183 (1991).
38. M. Hiramatsu, Y. Izawa, M. Hagihara, I. Nishigaki, and K. Yagi, Serum lipid peroxide levels of patients suffering from thermal injury, *Burns* 11: 111 (1984).
39. I. Nishigaki, M. Hagihara, M. Hiramatsu, Y. Izawa, and K. Yagi, Effect of thermal injury on lipid peroxide levels of rat, *Biochem. Med.* 24: 185 (1980).
40. K. Yagi, S. Komura, N. Ishida, N. Nagata, M. Kohno, and N. Ohishi, Generation of hydroxyl radical from lipid hydroperoxides contained in oxidatively modified low-density lipoprotein, *Biochem. Biophys. Res. Commun.* 190: 386 (1993).
41. M. A. Warso and W. E. M. Lands, Presence of lipid hydroperoxide in human plasma, *J. Clin. Invest.* 75: 667 (1985).
42. Y. Yamamoto, M. H. Brodsky, J. C. Baker, and B. N. Ames, Detection and characterization of lipid hydroperoxides at picomol levels by high-performance liquid chromatography, *Anal. Biochem.* 160: 7 (1987).
43. T. Miyazawa, K. Yasuda, K. Fujimoto, and T. Kaneda, Presence of phosphatidylcholine hydroperoxide in human plasma, *J. Biochem.* 103: 744 (1988).

REACTIVE OXYGEN SPECIES IN NORMAL PHYSIOLOGY, CELL INJURY AND PHAGOCYTOSIS

Mary Treinen Moslen

Department of Pathology
University of Texas Medical Branch
Galveston, TX 77555-0605

INTRODUCTION

This presentation will examine major processes by which biological systems cope with reactive oxygen species (ROS). Peroxidation of membrane lipids will be considered as one example of the destructive processes that occur when ROS are not controlled by the protection systems that detoxify reactive species. In contrast, enzymatic peroxidation of arachidonic acid initiates the biosynthesis of chemotactic leukotrienes. Phagocytic cell formation of ROS by the "oxidative burst" will be examined because of its important role in the defense against invading organisms. The focus will be on the processes that stringently control activation of the NADPH oxidase of phagocytic cells.

FORMATION AND DESTRUCTIVE CAPACITY OF ROS

Oxygen metabolism by biologic systems is inevitably associated with formation of small amounts of $\bullet O_2^-$, H_2O_2 and $\bullet OH$ as well as various unstable oxidized lipids which are the entities commonly known as ROS. The electron transport chain of mitochondria is one of the major processes for production of ROS under physiologic conditions. Usually, mitochondria reduce 95% of the O_2 consumed by cells to H_2O by sequential transport of four electrons. Under normoxic conditions an estimated 1-2% of the mitochondrial electron flow "leaks" off to form $\bullet O_2^-$ (Grisham, 1992). Ubiquinone and NADH oxidase are two sites of $\bullet O_2^-$ production along the mitochondrial electron transport chain. Once formed, $\bullet O_2^-$ anions dismutate to H_2O_2 either spontaneously or 4 orders of magnitude faster when catalyzed by the enzyme superoxide dismutase (SOD).

H_2O_2 reduction to $\bullet OH$ is readily facilitated by the transition metals Fe^{2+} or Cu^{1+} which serve as electron donors. Cells and tissues are equipped to cope with H_2O_2 by proteins that bind Fe and Cu (e.g. transferrin, ferritin, ceruloplasmin and metallothionein) and by the enzymes catalase and glutathione (GSH) peroxidase that safely reduce H_2O_2 to H_2O. However as shown by the following pathway, $\bullet O_2^-$ can lead to the production of other ROS:

$$\bullet O_2^- \xrightarrow{SOD} H_2O_2 \xrightarrow{Fe} \bullet OH$$

Additional processes which produce ROS include: lipid peroxidation, radiation, metabolism of quinones by redox-recycling, and reactions of the enzymes xanthine oxidase, amine oxidase, cytochrome P450 and prostaglandin synthase. Phagocytic cells produce $\bullet O_2^-$ by NADPH oxidase and HOCl by myeloperoxidase.

Potential capacity of the ROS for destruction of cells and tissues cannot be overstated. The cartoon in Figure 1 depicts some of the ways that a crew of ROS could wreck components of a cell. For example, the very reactive •OH would act near the site of generation and could break up plasma membrane, nuclear membrane, chromosomes or other structures. Lipid peroxyl radicals (LOO•) which have half lives of seconds (Grisham, 1992) could travel to disrupt plasma membranes. The stable, lipophilic oxidant H_2O_2 is depicted in Figure 1 as the leader of a group of other ROS in a "blasting zone" since H_2O_2 will readily cross membranes and then give rise to more reactive species. Hypochlorous acid (HOCl) and its unprotonated anion (OCl^-) are very powerful oxidizing agents which will react at multiple sites on cellular molecules.

Figure 1. Cartoon depicting multiple destructive actions of a wrecking crew of reactive oxygen species on cellular membranes, organelles and chromatin. Reproduced with permission with slight modifications from Hooper, C. "Free radicals: Research on biochemical bad boys comes of age" (1989), *J. NIH Research* 1:101.

LIPID PEROXIDATION AND DETOXIFICATION REACTIONS

Polyunsaturated fatty acids of membrane phospholipids are readily attacked by oxidizing free radicals (Cheeseman and Slater, 1993). As illustrated in Figure 2, the peroxidation initiated by the attack of a radical species on a polyunsaturated fatty acid leads to oxidation of the carbon chain, cleavage and shortening of the carbon chain, and release of small aliphatic products. Such alterations of a membrane phospholipid would disrupt the hydrophobic core of the membrane. Some of the released aliphatic products, such as 4-hydroxynonenal and malondialdehyde, are documented cytotoxins that can destructively interact with other molecules, for example, by forming adducts with nucleosides or by cross linking two lipids or a lipid to a protein (Esterbauer et al, 1991).

A complimentary series of detoxification reactions can halt or limit the process of lipid peroxidation, particularly the destructive propagation and decomposition reactions. As shown in Figure 2, radical scavenging by Vitamin E could limit propagation reactions instituted when a lipid peroxide radical (LOO•) abstracts a hydrogen from the fatty acid of another phospholipid. The interaction of LOO• with Vitamin E is considered a radical scavenging reaction since the Vitamin E radical is stable, relatively harmless, and can be recycled back to a nonradical form by enzyme systems (Buettner, 1993). Cleavage of the fatty acid peroxide from the phospholipid by phopholipase is considered the first step in a mechanism to repair peroxidized phospholipids (van den Berg et al, 1993). Proposed aspects of this rapid repair mechanism include: activation of phospholipase A_2 by lipid peroxidation; preferential cleavage of oxidized fatty acids by phospholipase A_2; and alterations in the molecular conformation of oxidized phospholipids that facilitate access of the phospholipase to the cleavage site (van den Berg et al, 1993). The cytotoxic alkenals released by decomposition reactions can be detoxified by either glutathione transferase or aldehyde dehydrogenase (Alin et al, 1985; Mitchell and Petersen, 1987).

HETEROGENEOUS DISTRIBUTION OF PROTECTION SYSTEMS

Protection systems are not uniformly distributed in extracellular fluids, tissues or cellular organelles. Glutathione concentrations are about 100-fold higher in human alveolar lining fluid than in plasma, which may be important in the protection of the lung against ROS (Cantin et al, 1987). Vitamin E levels are 30 to 50 times higher in human red blood cells than mononuclear or polymorphonuclear cells (Ogihara el al, 1989). Immunohistochemical studies in hamster tissues have documented wide variations in the amounts of catalase and SOD among tissues, including higher amounts of SOD in heart than liver (Oberley et al, 1990). Heterogeneities within tissues were also immunohistochemically evident including higher activities of catalase and SOD in kidney proximal tubular cells versus glomerular cells and in stomach surface mucous cells versus gland neck cells.

Cellular organelles exhibit a nonuniform distribution of protection systems (Moslen, 1992). Mitochondria are defended against ROS by Vitamin E, GSH, SOD, GSH peroxidase, GSH transferase and aldehyde dehydrogenase. Vital material of the nucleus is protected by Vitamin E, GSH and metallothionein. Differences in the distribution of protection systems likely reflect physiological conditions,
such as a greater need for protection against ROS in the mitochondrion versus the nucleus and in the heart with its high tissue O_2 versus the liver and other sites with lower tissue O_2.

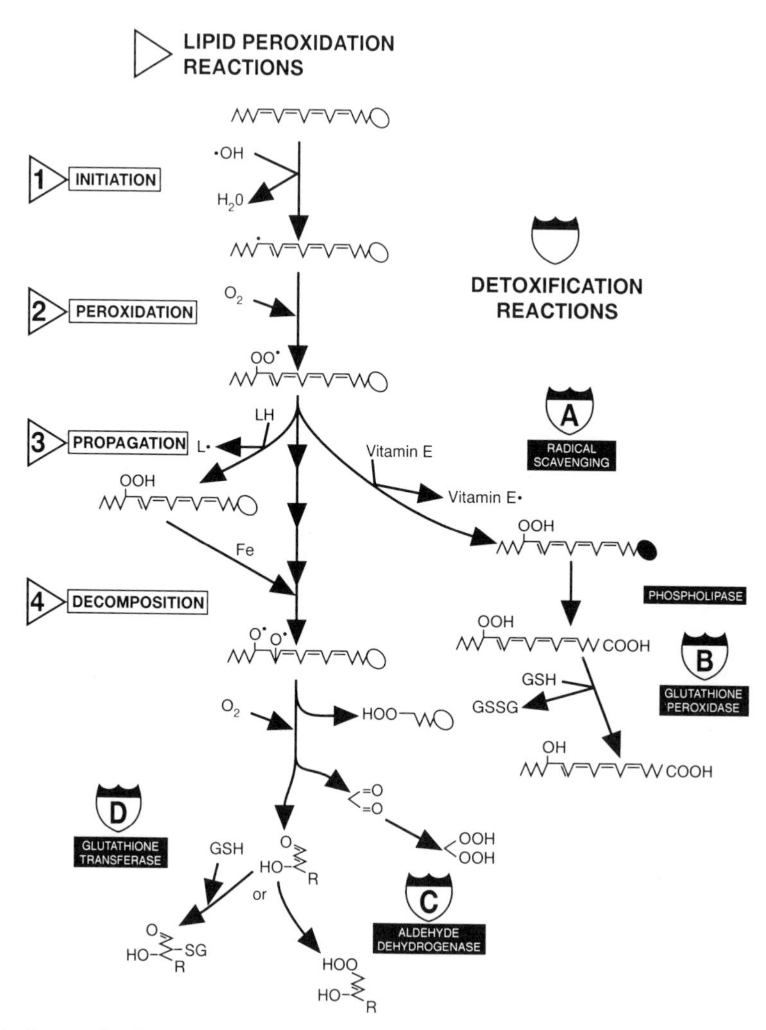

Figure 2. Schematic of four stages in the peroxidation of a membrane polyunsaturated phospholipid and detoxification reactions A to D which serve to prevent or control the propagation and decomposition stages. Note that the detoxification reaction B by glutathione peroxidase is preceded by a phopholipase-mediated cleavage of the phospholipid to a free fatty acid. Cytotoxic hydroxyalkenals can be detoxified either to acids by aldehyde dehydrogenase (reaction C) or to glutathione conjugates (reaction D).

OXIDATIVE BURST OF PHAGOCYTIC CELLS

Phagocytic inflammatory cells kill invading organisms by the following series of reactions which is known as the "oxidative burst" because of the associated 4 to 100-fold increase in consumption of O_2.

$$\text{1) } NADPH + O_2 \xrightarrow{\text{NADPH OXIDASE}} \bullet O_2^- + NADP^+ + H^+$$

$$\text{2) } \bullet O_2^- + \bullet O_2^- + 2H^+ \xrightarrow{\text{SOD}} H_2O_2 + O_2$$

$$\text{3) } H_2O_2 \xrightarrow[Cl^-]{\text{MYELOPEROXIDASE}} H_2O + HOCl$$

Types of phagocytic cells capable of the oxidative burst are leukocytes (chiefly neutrophils) as well as monocytes and eosinophils. Our present understanding of both the importance and the control of the oxidative burst has evolved largely from studies of a usually fatal condition in children known as chronic granulomatous disease (CGD) (Smith and Curnutte, 1991).

CGD is a rare genetic disease characterized by recurrent catalase-positive bacterial infections and granulomatous lesions in many organs. Phagocytic cells in persons with CGD have minimal or no capacity for the oxidative burst due to an impaired ability to produce $\bullet O_2^-$ by the NADPH oxidase. Some capacity for killing catalase-deficient organisms is retained since CGD phagocytes can "borrow" microbe-generated H_2O_2 for biosynthesis by myeloperoxidase of the potent killing entity HOCl. The particular vulnerability of CGD patients to catalase-positive organisms stems from the ability of these organisms to maintain such low H_2O_2 levels that the CGD phagocyte cannot borrow microbe-generated H_2O_2 for biosynthesis of HOCl (Curnutte, 1993). The invading organisms that cannot be killed are "walled up" within clusters of macrophages and lymphocytes that form the characteristic granulomas of CGD.

Studies of CGD neutrophils by numerous investigators in the last decade have identified components of the NADPH oxidase as well as several processes involved in its activation (see reviews by Smith and Curnutte, 1991; Curnutte, 1993; and Thelen et al, 1993). Multiple defective components were suspected as causes of CGD because clinical observations indicated variable inheritance patterns with approximately two-thirds of the cases X-linked and one-third autosomal and because in vitro complementation assays indicated different defects in cytosolic components and phagolysosomal membrane components of NADPH oxidase. The schematic of Figure 3 indicates known components of the NADPH oxidase and changes that occur when the latent state is converted to the activated state after stimulation of a phagocytic cell. Unequivocally identified components of NADPH oxidase are given the designation "*phox*" which indicates that the protein is part of the *ph*agocytic *ox*idase.

Multiple molecular defects have been characterized in CGD cases (see reviews by Smith and Curnutte, 1991; Curnutte, 1993; and Thelen et al, 1993). Many X-linked CGD cases were found to lack the phagolysosomal membrane cytochrome b_5 that transfers electrons from NADPH to O_2 and specifically to lack the 91-kDa-phox subunit of NADPH oxidase. Autosomal CGD cases were found to have deficiencies in other components of the membranous cytochrome b_5 or other subunits of the NADPH oxidase which are found in the cytosol of phagocytic cells (e.g., 22-phox, 47-phox and 67-phox). Activation of NADPH oxidase from its latent state was found to involve:

* Assembly of cytosolic components and their translocation to the phagolysosomal membrane.
* Phosphorylation of the 47-phox subunit.
* Translocation of a GTP-binding protein called "rac 2" that is found in the cytosol with GDP-bound in a latent state.

The nature of these changes likely serves to restrict the inadvertent activation of the NADPH oxidase.

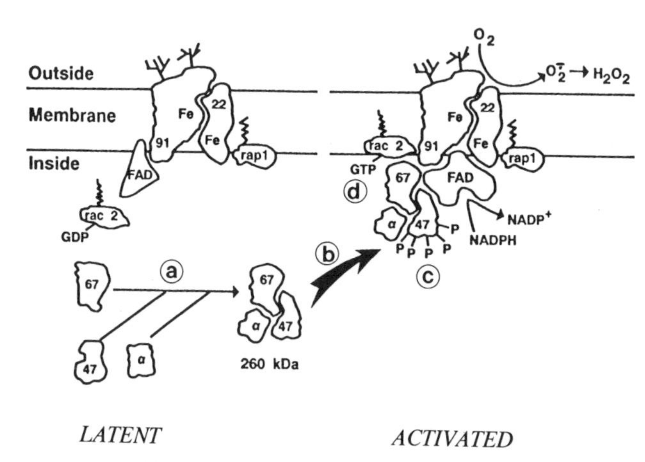

Figure 3. Hypothetical model of NADPH oxidase activation. Activation is associated with multiple changes including: a) assembly of cytosolic subunits, b) translocation of the assembled cytosolic subunits to the phagolysosomal membrane to form a complex with membrane subunits, c) polyphosphorylation of the 47-kDa subunit, and d) movement of the cytosolic rac 2 subunit to the membrane and an alteration of its bound-GDP to bound-GTP. Reproduced with slight modifications from Curnutte, J.T. "Chronic granulomatous disease; the solving of a clinical riddle at the molecular level" (1993), *Clin. Immunol. Immunopathol.* 67:S2.

BIOSYNTHESIS OF ARACHIDONIC ACID TO CHEMOATTRACTANTS

Killing of invading organisms by ROS from oxidative burst reactions generated by phagocytic cells is just a late stage in the mammalian response to such an insult. Phagocytic cells must first, as depicted in Figure 4, emigrate from the vasculature and then move to a site of injury in response to chemoattractants.

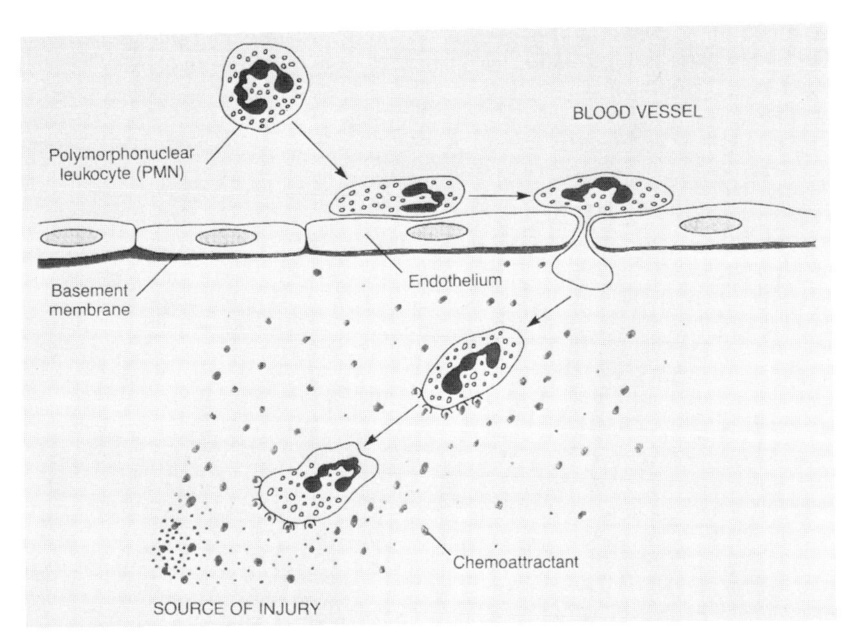

Figure 4. Cartoon of sequential events in the adherence and emigration of leukocytes in response to chemoattractants. Leukocytes adhere to the vascular endothelium, move towards a junction in the endothelium, and emigrate through the junction. Leukocytes then emigrate in response to chemoattractants that emanate from the source of the injury. Reproduced from Cotran et al (1989), Robbins Pathologic Basis of Disease, W.B. Saunders Co., Philadelphia.

Powerful chemoattractants including 5-HETE and LTB_4 are produced by the enzymatic conversion of arachidonic acid (5,8,11,14-eicosatetraenoic acid) into hydroperoxy derivatives by the lipoxygenase pathway (Figure 5). This biosynthetic pathway involves free radical intermediates which are produced by hydrogen abstraction reactions and addition of molecular oxygen to lipid radicals (Yamamoto et al, 1991). 5-HPETE is peroxidized to 5-HETE, a potent chemotactic stimulus for neutrophils. 5-HPETE also yields leukotriene products which were given the name "leukotrienes" because of their conjugated triene structure and their initial isolation from leukocytes. Leukotriene B_4 (LTB_4) is a powerful chemotactic agent and causes the aggregation and adhesion of leukocytes to venular endothelium which must precede the emigration of leukocytes into the extravascular space (Figure 4).

CONTROL OF OXIDATIVE BURST

As depicted in Figure 6, the phagocytic cells attach particles from invading organisms and other offensive materials, and engulf these particles into phagocytic vacuoles. The phagocytic vacuole is fused with granules containing myeloperoxidase, proteases and other destructive entities while the membrane of the phagocytic vacuole is the site of NADPH oxidase. Although engulfment and phagocytosis serve to contain the factors released, there is some escape of factors to the extracellular environment as depicted in Figure 6. A potential consequence of this escape is localized tissue damage. Babior (1984) described the offensive strategy of neutrophils and other professional phagocytes for killing as "like Attila the Hun, deploying a battery

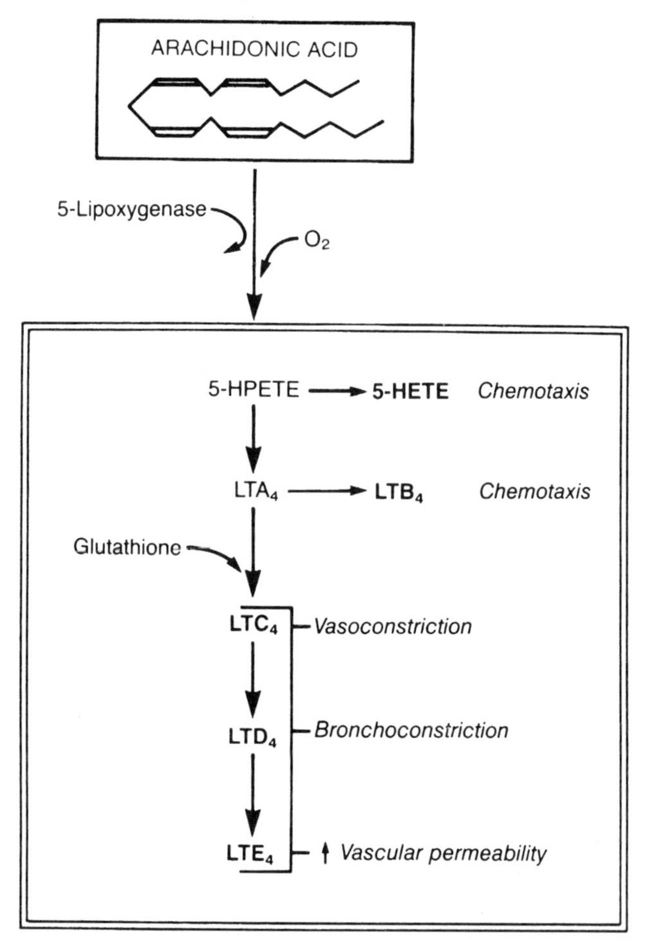

Figure 5. Lipoxygenase pathway for the biosynthesis of compounds with effects on chemotaxis, vasoconstriction, and vascular permeability. The starting substrate, arachidonic acid, is derived from membrane phospholipids by phospholipase cleavage. Multiple free radical intermediates are involved in the biosynthesis pathways as a result of hydrogen abstraction reactions and additions of molecular oxygen. Reproduced with slight modifications from Cotran et al (1989), Robbins Pathologic Basis of Disease, W.B. Saunders Co., Philadelphia.

of weapons that lay waste to both the targets and nearby landscape with the subtlety of an artillery barrage". Phagocytic-derived oxidants have, indeed, been linked to many chronic inflammatory problems including arthritis (Winrow et al, 1993).

However, at least five ways of providing some control over the ROS released by the oxidative burst of phagocytes have been identified.

First: As indicated by the schematic of Figure 3, conversion of NADPH oxidase from its latent state to its activated state requires assembly and translocation of separated components, and phosphorylation of at least one subunit. Also the activation requires complexing with a GTP binding protein which suggests that an endogenous GTPase could "serve to diminish the extent of the activation" (Curnutte, 1993).

Second: Contents of granules released by the phagocytes include lactoferrin which would bind up free Fe to limit Fe-mediated reduction of H_2O_2 to •OH (Grisham, 1992).
Third: Phagocytes are able to defend themselves from uncontrolled self destruction by a variety of protection systems including antioxidants, SOD, catalase, glutathione peroxidase, and a high concentration of taurine which has been proposed as a trap for HOCl (Babior, 1984).
Fourth: Some kinds of phagocytes, neutrophils for example, lead kamikaze assaults where the phagocytic attacker dies within 24-48 hrs (Babior, 1984).
Fifth: Fluids and cells surrounding the targets have a variety of protection systems. Heterogeneous distributions of the protection systems could account for the particular vulnerability of some regions to destruction by phagocytic cells.

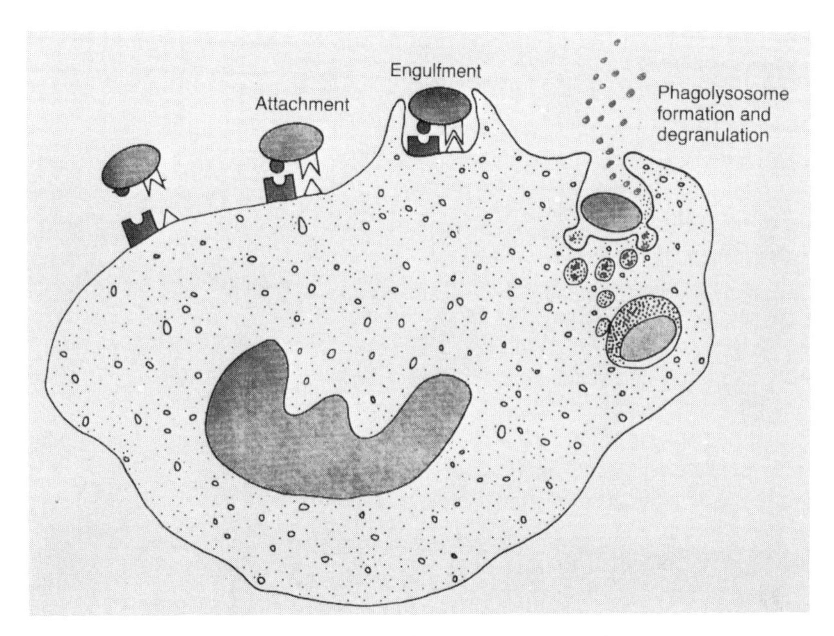

Figure 6. Cartoon of sequential events in the attachment, engulfment and phagocytosis of a particle by phagocytes. Granules fuse with the phagocytic vacuole and release their contents into the phagolysosome. Contents of these granules include lactoferrin, lysozyme, collagenase, elastases, phopholipase A_2, and myeloperoxidase. Bacterial particle attachment would also stimulate the activation of the NADPH oxidase that is located to the phagolysosomal membrane. Note that during phagocytosis, granule contents escape to the extracellular environment. Reproduced from Cotran et al (1989), Robbins Pathologic Basis of Disease, W.B. Saunders Co., Philadelphia.

SUMMARY

Formation of free radicals and other ROS is a continuous aspect of life. Examples include the free radical intermediates which are formed by the nonenzymatic peroxidation of polyunsaturated fatty acids of membrane lipids in a destructive process and which are also formed by the enzymatic peroxidation of arachidonic acid in the biosynthesis of potent chemoattractants. Organisms cope with these reactive species by a variety of strategies that limit formation of ROS or remove cytotoxic products. Oxidative burst reactions that yield ROS provide an effective, vital process for killing invading organisms. Research on why ROS formation is impaired in phagocytic cells of people with chronic granulomatous disease has provided new insights into the complexity of the factors that prevent inadvertent activation of this destructive force.

ACKNOWLEDGMENTS

I thank the graduate students in my Pathobiology of Disease class for stimulating my interest in chronic granulomatous disease and Dr. Armond S. Goldman for generously sharing his expertise in the clinical and molecular aspects of this serious disease which affects his patients.

REFERENCES

Alin, P., Danielson, U.H., and Mannervik, B. 1985, 4-Hydroxyalk-2-enals are substrates for glutathione transferase. *Proc. Meet. Fed. Eur. Biochem. Soc.* 179:267.

Babior, B.B. 1984, Oxidant from phagocytes: Agents of defense and destruction. *Blood* 64:959.

Buettner, G.R. 1993, The pecking order of free radicals and antioxidants: lipid peroxidation, α-tocopherol, and ascorbate. *Arch. Biochem. Biophys.* 300:535.

Cantin, A.M., North, S.L., Hubbard, R.C., and Crystal, R.C. 1987, Normal alveolar epithelial lining fluid contains high levels of glutathione. *J. Appl. Physiol.* 63:152.

Cheeseman, K.H., and Slater, T.F. 1993, An introduction to free radical biochemistry. *Brit. Med. Bulletin* 49:481.

Curnutte, J.T. 1993, Chronic granulomatous disease: the solving of a clinical riddle at the molecular level. *Clin. Immunol. Immunopathol.* 67:S2.

Esterbauer, H., Schaur, R.J., and Zollner, H. 1991, Chemistry and biochemistry of 4-hydroxynonenal, malonaldehyde and related aldehydes. *Free Radic. Biol. Med.* 11:81.

Grisham, M.B. 1992, "Reactive Metabolites of Oxygen and Nitrogen in Biology and Medicine", R.G. Landes Company, Austin.

Mitchell, D.Y., and Petersen, D.R. 1987, The oxidation of α-β unsaturated aldehydic products of lipid peroxidation by rat liver aldehyde dehydrogenases. *Tox. Appl. Pharmacol.* 87:403.

Moslen, M.T. 1992, Protection against free radical-mediated tissue injury. *in:* "Free Radical Mechanisms of Tissue Injury", M.T. Moslen and C.V. Smith, eds., CRC Press, Boca Raton.

Oberley, T.D., Oberley, L.W., Slattery, A.F., Launchner, L.J. and Elwell, J.H. 1990, Immunohistochemical localization of antioxidant enzymes in adult syrian hamster tissues and during kidney development. *Am. J. Pathol.* 137:199.

Ogihara, T., Miyake, M., Kawamura, N., Tamai, H., Kitagawa, M., and Mino, M. 1989, Tocopherol concentrations of leukocytes in neonates, *Ann. N.Y. Acad. Sci.* 570:487.

Smith, R.M., and Curnutte, J.T., 1991, Molecular basis of chronic granulomatous disease. *Blood* 77:673.

Thelen, M., Dewald, B., and Baggiolini, M. 1993, Neutrophil signal transduction and activation of the respiratory burst. *Physiol. Rev.* 73:797.

van den Berg, J.J.M., Op den Kamp, J.A.F., Lubin, B.H. and Kuypers, F.A. 1993, Conformational changes in oxidized phospholipids and their preferential hydrolysis by phospholipase A_2: a monolayer study. *Biochemistry* 32:4962.

Winrow, V.R., Winyard, P.G., Morris, C.J., and Blake, D.R. 1993, Free radicals in inflammation: Second messengers and mediators of tissue destruction. *Brit. Med. Bulletin* 49:506.

Yamamoto, S. 1991. "Enzymatic" lipid peroxidation: reactions of mammalian lipoxygenases. *Free Radic. Biol. Med.* 10:149.

FREE RADICAL MECHANISMS FOR CHROMOSOMAL INSTABILITY IN BLOOM'S SYNDROME

Thomas M. Nicotera

Biophysics Department
Roswell Park Cancer Institute
Elm and Carlton Streets
Buffalo, New York 14263-0001

INTRODUCTION

Bloom's syndrome (BS) is an extremely rare autosomal recessive disorder initially identified in Ashkenazi Jews and later found to be widespread in the general population (1). The distinguishing clinical features of this disorder are numerous and include severe growth deficiency, sun-sensitive facial skin lesions, areas of hypo- and hyper-pigmentation of the skin and immunodeficiency, leading to infections of the respiratory tract and chronic lung disease (1). A newly recognized complication is the development of diabetes mellitus (1). The most prominent feature of BS is a marked predisposition towards the development of cancer (1,2). The age of onset is considerably earlier than that of the general population and approximately 40% of the clinically verified cases develop neoplasms with a mean age of less than 25 years at the time of diagnosis (2). Nearly one third of the surviving cancer patients develop multiple primary tumors. There is no consistent pattern in the occurrence of the cancer type or its location.

Cells derived from BS patients exhibit a considerable degree of spontaneous chromosomal instability. It is presumed that chromosomal instability is the cause for the high incidence of neoplasia. The type of chromosomal damage observed has been extensively catalogued and include endoreduplication, quadriradial formations, chromatid and chromosome gaps (3) and breaks as well as micronuclei (4). Homologous exchange of sister chromatids (SCE) is the most prominent cellular feature and is used for clinical verification of this disorder (5). In addition to these chromosomal aberrations, BS cells

Free Radicals in Diagnostic Medicine, Edited by
D. Armstrong, Plenum Press, New York, 1994

exhibit spontaneous mutagenesis both in cell lines (6) and in patient-derived blood cells (7). Thus, BS provides one of the best correlations in a genetic disorder exhibiting chromosomal instability and an increased risk to malignant tumor formation. Identification of the mechanism(s) responsible for the chromosomal instability is therefore of considerable interest and could impact in the basic understanding of neoplastic processes.

At present there is little agreement as to the potential cause(s) of the disturbances reported in BS. In spite of early evidence indicating relatively normal rates of DNA repair in response to UV light and DNA damaging agents, defective repair processing has been presumed to explain the chromosomal instability in BS (8). This concept originated with the discovery by Cleaver (9) that the chromosomal instability in xeroderma pigmentosum is derived from a deficiency in the repair of UV-induced DNA damage. While this paradigm has been useful in delineating stepwise repair of genomic damage in cells, DNA repair defects alone cannot explain the wide phenotypic diversity of Bloom's syndrome. A more interesting observation that is consistent with the observed chromosomal changes is the increasing evidence for an elevated rate of homologous recombination and its extension to the molecular level (2).

Recent evidence that DNA from Bloom patients is homozygous for a polymorphic DNA marker in an intron at the FES locus, which maps to the chromosome band 15q26.1 (10), provides a potential breakthrough in understanding the molecular basis for BS. However, hybridization of Bloom cells with a normal chromosome 15 complemented only 50% of the SCE (11). This data combined with the existence of BS-derived cells with an intermediate number of SCE (12) indicates that additional gene(s), possibly located on a different chromosome, may contribute to the chromosome instability. Cell hybrids between BS patients of diverse ethnic origin and controls indicate the existence of only one complementation group (13).

Emerit and Cerutti reported that a small molecular weight clastogenic factor present in the media of BS cells or from plasma of BS patients caused chromosomal damage in normal cells (14). Addition of exogenous superoxide dismutase (SOD) reduced the induced level of chromosomal damage (14). It was suggested thereby, that a deficiency in the removal of oxygen radicals plays a role in the chromosomal damage observed in BS cells. Our laboratory has investigated the biochemical mechanisms potentially responsible for the chromosomal instability and have used SCE as a diagnostic tool to identify these processes. We (15) and others have (16) reported that SOD activity is elevated in all cell lines examined. These data and several other lines of indirect evidence have led to the proposal that oxidative stress takes place as a consequence of overproduction of oxygen radicals rather than from a deficiency in their removal (15).

Our approach began by testing whether endogenously generated superoxide radicals can induce the same type of chromosomal damage observed in BS cells (17). Paraquat was selected for these simulation studies since it perhaps the best characterized superoxide-generating compound available. Its capacity to redox cycle is dependent on its ability to enter the cell and interact with endogenous NADPH (18). The purpose of this chapter is to review the evidence for oxidative stress and its potential aberrant regulation in BS cells. The potential relationship(s) between oxidative stress and chromosomal instability will be discussed.

METHODS

Cell Culture The BS lymphoblastoid cell lines GM3403, GM3299 and GM1953 were obtained from the Human Genetic Mutant Cell Repository (Camden, N.J.) and HG1525 was a gift from Dr. James German (New York Blood Center, New York, NY). All cells were cultured in RPMI-1640 medium supplemented with 10% fetal bovine serum.

Chromosome Aberrations Don cell lung fibroblasts were incubated with paraquat for 4 hr and arrested in metaphase with colcemid (0.4 μg/ml) for 2 hr. Combinations of paraquat and the superoxide dismutase inhibitor diethyldithiocarbamate (DDC) were treated as indicated. Cells were harvested and treated for 10 min in KCl (0.075 M) and fixed in acetic acid/methanol (1:3) and washed 2x in the same solution. Chromosomes were prepared on slides and then stained in Giemsa (3% in 0.05 phosphate buffer, pH 6.8 for 15 min). One hundred cells were analyzed per point.

Sister Chromatid Exchange Log phase cells were incubated with 5-bromo-2-deoxyuridine (10 μg/ml) and the compound to be tested for approximately 2 cell cycles (46 hr), followed by mitotic arrest with colcemid (0.15 μg/ml) for 2 hr. The cells were treated in hypotonic KCl (0.075 M) for 20 min at 37°C, then fixed and washed 3 times in methanol-glacial acetic acid (3:1). Slides were prepared at least 24 hr prior to staining. Chromosomes for SCE analysis were stained using a modified fluorescence plus Giemsa technique (19). Twenty-five metaphases were analyzed for SCE analysis per treatment. Statistical analysis was performed by the Student's t-test.

Superoxide Dismutase Assay SOD activity in cellular extracts were determined by measuring the decrease in the rate of superoxide-dependent reduction of cytochrome C (20). Superoxide was generated by the xanthine/xanthine oxidase reaction and the enzymatic activity determined in 0.005 M phosphate at pH 7.8 and 0.1 mM EDTA, 50 μM xanthine and 6 μM ferricytochrome C in a 1.0 ml total reaction volume. The rate of cytochrome C reduction will be monitored at 25°C using a thermostatted Guilford 2600 spectrophotometer at 417 nm. One unit of SOD activity is defined as the amount of enzyme required to inhibit the rate of cytochrome C reduction by 50%. Activity is expressed as units/mg protein of the prepared cell extract with one unit defined as the amount of enzyme required to inhibit the reduction of cytochrome C by 50% under standard conditions. Protein determination was performed by the Lowry method (21). Inhibition of CuZnSOD by cyanide allows for the differential quantitation of both CuZnSOD and MnSOD.

Chemiluminescence Cells were grown in RPMI-1640 media (GIBCO, Grand Island, NY) containing penicillin, streptomycin and 10% fetal calf serum (Hyclone, Logan, UT). Log phase cells were adjusted to a concentration of 4 x 10^5 cells/ml in fresh media. Aliquots (0.5 ml) of this cell suspension were placed into a cuvette, to which 50 ul of fresh centrifuged plasma and 15 ul of luminol (10 mM in 0.1 M borate buffer) were added. The cell preparation was incubated for 2 min and N-formyl-methionine-leucine-phenylalanine (fMLP) dissolved in 1.0 to 5.0 μl of DMSO was added. Incubation proceeded for the length of time required for the agonist to effect the maximal level of chemiluminescence, which occurred at 3 min. The assays were conducted at 37°C with continuous stirring. For studies using the NADPH oxidase inhibitor diphenylene iodonium (DPI). Non-denatured and heat-denatured Proteinase K was preincubated with plasma in the assay. SOD and catalase were also tested for inhibition of chemiluminescence. Peak chemiluminescence was obtained by monitoring continuously with a Chrono-Log Model 800 Lumi-Aggregometer (Havertown, PA).

Phospholipase A_2 Assay A PLA_2 assay originally designed for intact cells was adapted to avoid problems inherent in a two phase system and used DMSO in order to generate a true solution phase (22). The assay makes use of the fluorescent phospholipid substrate, 1-acyl-2-(N-4-nitrobenzo-2-oxa-1,3-diazole) aminocaproyl phosphatidyl choline (C_6-NBD-PC), which yields the C6-NBD-labelled fatty acid that is not further metabolized nor is it incorporated into cells (23). Cells were centrifuged at 1000 rpm for 5 min at room temperature, washed twice with PBS and the pelleted material was resuspended in 10 mM Tris buffer at pH 7.4. Approximately 10^6 cells were sonicated using a Branson sonifier for 30 sec at 50 Watts. The sonicate (620 μl) was aliquoted into tubes containing different amounts of substrate (2 μM to 14 μM) in 160 μl of DMSO and 20 μl of 0.24 M $CaCl_2$. The assay mixture was incubated for 1 hr. at 37°C with mixing. The reaction was stopped and unreacted substrates extracted by the addition of 3.0 ml aliquot of isopropyl alcohol/ isopropyl ether/heptane (1.2/1.6/1.0, v/v/v) with mixing. The reaction mixture was centrifuged for 5 min. at 1000 rpm and the top organic layer removed. This extraction procedure was repeated twice. The remaining products (C6-NBD) were solubilized by the addition of chloroform/ methanol/saline in 0.2N HCl, (1:1:0.05, v/v/v) with mixing and the mixtures centrifuged for 5 min. at 1000 rpm.

One ml aliquots were pipetted from the bottom organic layer and the fluorescence immediately measured at 540 nm with an excitation wavelength of 450 nm. A caproic acid standard was used as a calibration standard. Fluorescence was converted to picomoles of substrate released per microgram of protein, protein being determined by the Lowry method (21). Kinetic data was calculate using a non-linear least squares computer program, NFIT (Island Products, Galveston, TX). Standard errors were determined by Student's t-test.

RESULTS

Chromosomal Aberrations, SCE and SOD Induction Chinese hamster lung fibroblasts (Don) demonstrated a near linear dose-dependent increase in total chromosome aberrations when treated with increasing concentrations of paraquat (Table 1). The aberrations were primarily of the chromatid-type (gaps and breaks) although some chromosome breaks and dicentrics were also observed. At 10 mM paraquat, a quadriradial formation is observed at approximately the same a 1% frequency observed in Bloom's cells. Furthermore, treatment of Don cells with a combination of paraquat and the CuZnSOD inhibitor DDC resulted in a synergistic generation of total chromosome aberrations (Fig. 1). Due to the high rate of aberrations generated in combination, a shorter incubation time was used.

Table 1. Frequency of chromosomal aberrations in Don cell line fibroblasts treated for 4 hr with paraquat during log phase growth. Reprinted from ref. (17).

Paraquat [mM]	Chromatid Gaps	Chromatid Breaks	Chromosome Breaks	Dicentrics	Quadriradials	Total Aberrations
0	4	0	0	0	0	4
3	15	3	2	2	0	22
5	34	1	4	3	0	42
10	64	10	2	7	1	84

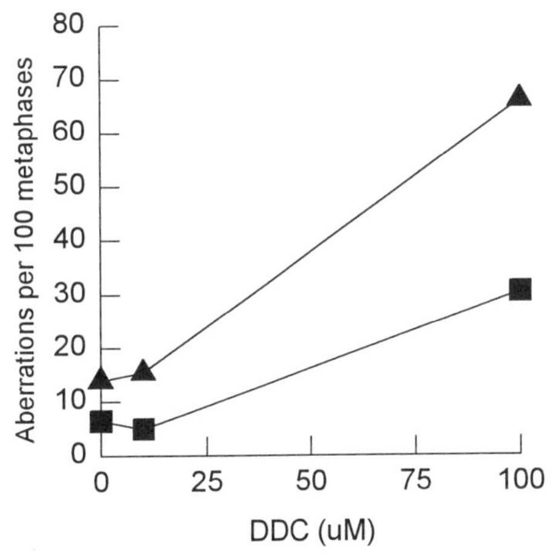

Fig. 1. Induction of chromosomal aberrations in Don cell line fibroblasts treated with (■) 2.5 and (▲) 5.0 mM paraquat and two concentrations of diethyldithiocarbamate (DDC) for 2 hr and followed by a 2 hr colcemid arrest.

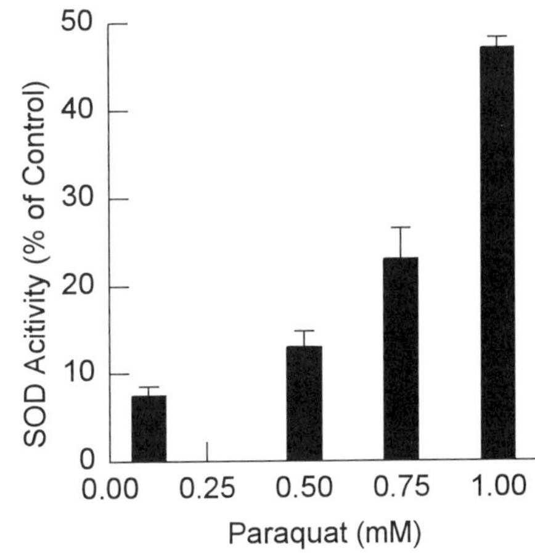

Fig. 2. CuZnSOD activity in Don cell line fibroblasts treated with paraquat for 24 hr. MnSOD activity was negligible. Control value for SOD activity is 115.6 ± 1.9 (mean ± SD) units/mg sonicate. Reprinted from ref. (17).

Treatment of Don cells with paraquat resulted in a dose-dependent increase in SCE (Table 2). The maximum level attained was 3-fold above the untreated level and was similar to that observed in human cells (15). In contrast, SCE in BS cells are found to be 10 to 15-fold higher than controls. Induction of SCE occurred at paraquat concentrations that also induced SOD activity (Fig. 2). These data provide the first indications that paraquat-generated oxygen radicals correlate with increases in SCE, chromatid-type aberrations and chromosomal aberrations similar to those observed to take place spontaneously in BS cells.

Table 2. SCE frequency in Don cell line fibroblasts treated for two cell cycles at various doses of paraquat. Reprinted from ref. (17).

Paraquat (mM)	Number	SCE/Metaphase ± SE	Range
0	25	6.48 ± 0.60	2-14
0.1	25	12.12 ± 0.73	7-19
0.5	25	14.08 ± 0.80	8-21
0.75	14	17.07 ± 0.80	9-24

Modulation of SCE in Bloom's cells Based on our previous demonstration that paraquat-generated superoxide can induce SCE in Don cells and in human lymphocytes (15) we asked whether SCE in BS cells could be modulated by redox regulation. The compounds used were the SOD mimetic copper(II)diisopropylsalicylate (CuDIPS) and α-tocopherol a potent inhibitor of lipid peroxidation. Treatment of the BS cell line GM3403 with nanomolar concentrations of CuDIPS resulted in a dose-dependent increase in SCE, while treatment with α-tocopherol resulted in a dose dependent inhibition, whether incubated for 2 cell cycles or preincubated for an additional 3 days (Fig. 3). The maximum level of inhibition with α-tocopherol was approximately 35%.

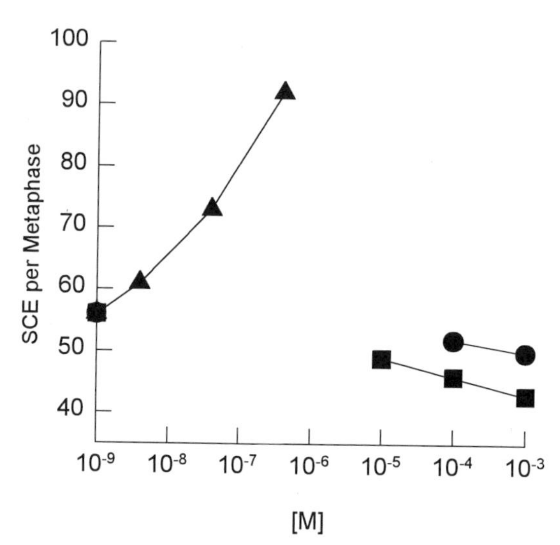

Fig 3. SCE frequency per metaphase chromosome in the BS lymphoblastoid cell line GM3403 treated for 2 cell cycles with CuDIPs (▲), α-tocopherol (●) α-tocopherol including a 3 day pretreatment prior to SCE analysis (■). Reprinted from ref. (15).

Chemiluminescence Chemiluminescence was used as a measure of oxygen radical production following the activation of cells. Cells were activated following treatment of B-lymphoblastoid cells with the chemotactic tripeptide fMLP and a small amount of fresh of plasma (8%). In all cases BS cells produced a greater level of chemiluminescence as compared to control cells and at maximal stimulation a 3-fold difference was observed (Fig. 4). The calcium ionophore A23187 and the tumor promoter TPA also produced a greater level of chemiluminescence in BS cells, however, at maximal stimulation produced only a 50% increase (24). Though plasma was required for full activation of cells, it did not contribute to the differential response between BS and control cells (24). The plasma factor is likely to be a protein since it is inactivated either by heat or proteolysis (24). Evidence for participation of plasma the membrane-bound NADPH oxidase in the oxidative burst that occurred after stimulation of cells was provided by the inhibition of the enzyme by DPI (25). Control cells were inhibited nearly 3-fold more effectively than were BS cells, with control cells exhibiting an IC_{50} of 40.1 ηM and BS cells, 111.3 ηM (Fig. 5).

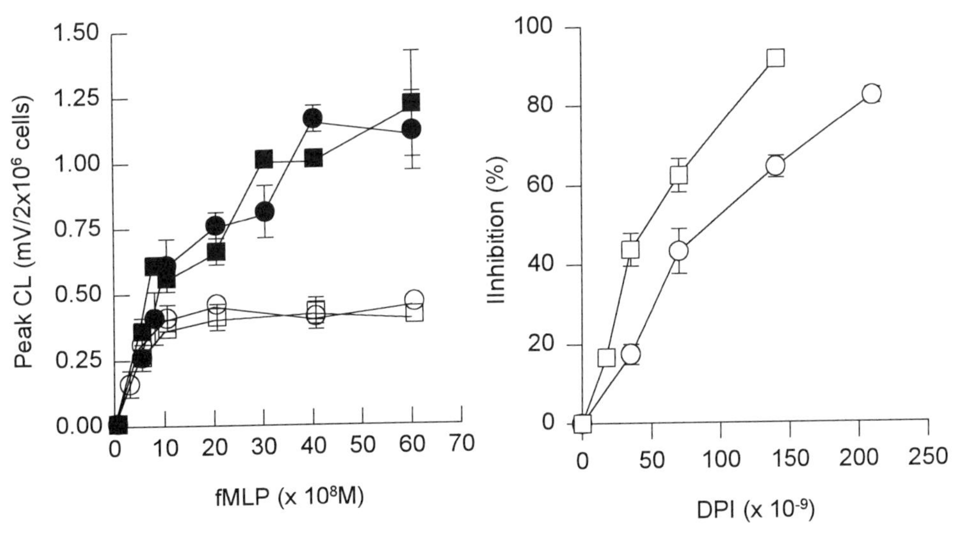

Fig. 4. Extent of chemiluminescence (CL) in BS cell lines GM3403 (●) and HG 1525 (■) and in controls GM3299 (O) and GM1953 (□) in response to treatment with fMLP and plasma. Each point is the average of triplicate determinations ± SE. Reprinted from ref. (17).

Fig. 5. Inhibition of chemiluminescence by DPI in the BS cell line GM3403 (O) and control cell line GM3299 (□) activated with fMLP and plasma. Each point is the average of duplicate determinations ± SE. Reprinted from ref. (17).

Phospholipase A_2 Activity PLA_2 activity was measured in sonicates of two BS lymphoblastoid cell lines, GM3403, and HG1525 and two control lymphoblastoid cell lines GM3299 and GM1953. Comparison of these parameters demonstrated a 3-fold higher Km in normal cells as compared to Bloom's cell lines (Table 3). Kinetic data obtained from the two cell types did not reveal a cause for the increased activity in the BS cells, which may be due to the presence of an altered enzyme, or the synthesis of a new enzyme or cofactor.

PLA_2 has been proposed to play a role in the repair of peroxidized lipids through the preferential hydrolysis rom membrane lipids (26). The relationship between lipid peroxidation and PLA_2 activity was investigated with the use of α-tocopherol, a well known inhibitor of lipid peroxidation (27). Incubation of BS cells with 10^{-4} M α-tocopherol for 24 hr inhibited PLA_2 activity nearly to the level observed in control cells (Table 3), demonstrating concomitant increases in both the Km and Vmax. The changes in PLA_2 kinetics occurred at a concentration that have been previously demonstrated to reduce SCE in a BS cell line (15). Thus, we have established a modest correlation between the inhibition of PLA_2 activity and the inhibition of SCE formation.

Table 3. Apparent kinetic parameters for phospholipase A_2 in BS (GM3403 and HG1525), controls (GM3299 and GM1953) and α-tocopherol treated cell lines. At least 3 experiments were performed containing 5 data points per cell line. Km and Vmax were calculated using a computer generated non-linear least squares fit. All data sets demonstrated a regression coefficient (R-square) of >0.98 and statistical significance was calculated using the Student's t-test.

Treatment	Cell Line	Vmax*	Km (μM)
--	GM3403	5.88	2.88
--	HG1525	5.56	3.16
--	GM3299	7.45[a]	9.47[a]
--	GM1953	7.98	10.19
10^{-4} α-tocopherol	GM3403	7.26[b]	5.31[a]
10^{-4} α-tocopherol	HG1525	6.56	5.04

*picomoles product/μg protein/hr.
[a]Statistically significant difference compared to BS controls at $p \leq 0.02$.
[b]Statistically significant difference compared to BS controls at $p \leq 0.05$.

It should be noted that inhibition of SCE in BS cells either by inhibition of lipid peroxidation with α-tocopherol (Fig. 3) or by inhibition of PLA_2 with the manoalide analog BMY30204 (Table 4), did not exceed 35-40%. Comparison of PLA_2 inhibition characteristics in BS and control cells with BMY30204 shed considerable light towards the elucidation of the relationship PLA_2 and SCE. Inhibition of PLA_2 derived from control cells reached 95% at 50μM BMY30204, while that of BS-derived PLA_2 reached only 55% maximally at the same concentration of BMY30204 (Fig 5). At 10μM BMY30204, a 30% inhibition of PLA_2 activity was observed and corresponds to a similar

Table 4. Inhibition of SCE in the BS cell line GM3403 for 2 cell cycles with the PLA_2 inhibitor BMY30204.

BMY30204 (μM)	Cell No.	SCE/Metaphase ± SE	Mitotic Index*
0	>25	61.9 ± 2.7	0.80
5	>25	--	0.82
10	>25	47.8 ± 1.3	1.4
50	25	56.9 ± 2.5	1.1

$$*\text{Mitotic Index} = \frac{\text{Metaphase Cells}}{\text{Total Cells}} \times 100$$

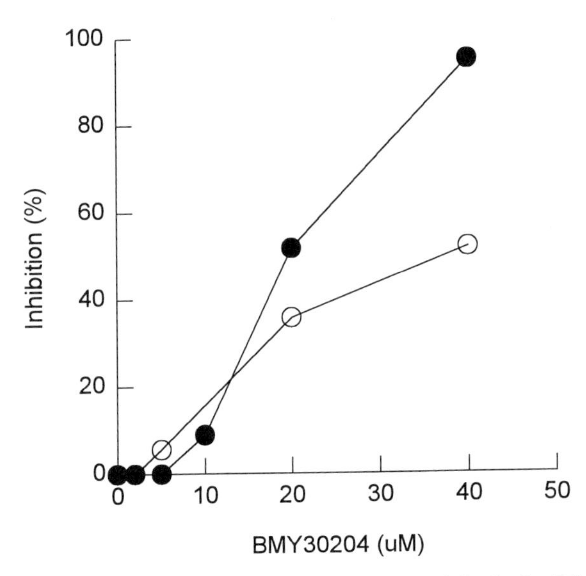

Fig. 6. Percent inhibition of phospholipase A_2 activity in the BS cell line GM3403 (○) and in the control cell line GM3299 (●) following treatment with BMY30204 for 24 hr.

level of inhibition in SCE. Thus, a positive albeit weak correlation between SCE and PLA_2 activity has been established. It is possible that more effective inhibition of PLA_2 activity in BS cells could lead to a more significant level of inhibition of SCE. However, the decreased effectiveness of BMY30204 in its capacity to inhibit the BS-derived PLA_2 as compared to that of controls further implicates the presence of a novel or modified enzyme. Perhaps the identification of more effective inhibitors may prove useful in characterizing this novel enzymatic activity and in understanding its role in SCE.

DISCUSSION

The data presented in this chapter attempts to link the chromosomal instability associated with Bloom's syndrome to the observed superoxide stress. It has been demonstrated in the Chinese hamster Don fibroblast cell line that paraquat-generated superoxide radicals produce nearly all the types of chromosomal damage observed in Bloom's cells. The aberrations observed are primarily chromatid gaps and breaks although some chromosome gaps and breaks are also observed. Total aberrations are produced in a linear dose-response curve and are further potentiated by the addition of the CuZnSOD inhibitor DDC. At 10 mM paraquat a quadriradial is observed at approximately the same 1% frequency as is observed in BS cells.

SCE formation has been correlated with oxidative stress in two ways. [i] They are induced by paraquat-generated superoxide radicals in a dose-dependent fashion in both Don fibroblasts and in human B-lymphoblastoid cell lines at concentrations that also induce SOD activity and [ii] they can be modulated directly in BS cells by prooxidant and antioxidant compounds. The partial inhibition of SCE by α-tocopherol provides a model by which the role of lipid peroxidation can be investigated in SCE formation. α-Tocopherol is a powerful inhibitor of lipid peroxidation (27) and the enzyme PLA_2 has been proposed to play a protective role in membranes through the selective removal of peroxidized lipids (26). We therefore investigated whether α-tocopherol can modulate PLA_2 activity and whether PLA_2 also plays a role in SCE formation.

We have established that BS cells exhibit a higher level of PLA_2 activity as compared to control cells. Furthermore, α-tocopherol inhibits PLA_2 activity in BS cells, reducing it to a level approaching that of controls, while control cells do not exhibit this α-tocopherol-inhibitable PLA_2 activity[1]. The mechanism for inhibition of PLA_2 is presently unclear (28). Whether inhibition PLA_2 takes place via inhibition of lipid peroxidation or via a direct mechanism is not known. The differential inhibition of PLA_2 in BS and control cells by the manoalide analog BMY30204 further underscores the presence of an altered PLA_2 activity in BS cells. These data provide the first evidence that SCE formation correlates with an altered PLA_2 activity. This approach may therefore provide a means by which the mechanisms leading to distinct types of genomic damage can be segregated. That is, chromatid gaps and breaks may derive from oxygen radicals whereas SCE may be a consequence of an altered PLA_2 activity.

Superoxide overproduction in BS cells is apparently mediated by the membrane-associated enzyme NADPH oxidase as demonstrated by its near-complete inhibition by DPI. The concentration of DPI required for 50% inhibition is 3-fold higher for BS cells as compared to controls. Whether NADPH oxidase is altered, aberrantly expressed, or perhaps aberrantly modulated has not been established. While similarities in the oxidative burst of B-lymphocytes and neutrophils are apparent, lymphocytes produce only 1/20th the superoxide level as neutrophils and its function is not known (29).

The mechanism for SCE is speculative; however, its visualization by microscopy is readily accomplished by the labelling of individual DNA strands with bromodeoxyuridine

---

[1]Unpublished data

followed by staining of chromosomes on slides. Sister chromatid exchange is recognized as a reciprocal switch of a newly replicated chromatid with its sister chromatid at apparently homologous loci and thought to reflect damage which has occurred during the period of DNA replication (30). Consequently, the cell's repair function for different type of damage may be of importance in determining the SCE response of the cell (30). An SCE is likely to be of little consequence if the exchange between chromatids is precisely homologous. However, it has been proposed that the high rate of mutagenesis combined with the high rate of recombination (SCE) that takes place in BS can lead to loss of heterozygosity (31). It is further suggested that chromatid gaps and breaks found to occur in close proximity to the site of exchange (32) may lead to unequal SCE, and hence, loss of heterozygosity (2). The proposed relationship between chromosome aberrations and SCE, though controversial (33), is provided in support of the premise for unequal SCE formation (2). Under these circumstances, these otherwise innocuous cytogenetic phenomena, may prove lethal in combination.

In summary, it appears that an altered PLA_2 activity present in BS cells participates in, and possibly, is responsible for the constitutive genomic instability. A major product of PLA_2 activity is the release of arachidonic acid from membrane phospholipids, which serves as a precursor for eicosanoid biosynthesis as well as having signal transducing functions (34). One such function is its ability to act as a positive modulator of NADPH oxidase in the release of oxygen radicals (35). The implications for the data presented are that eicosanoid products may participate in SCE, while the arachidonate-enhanced oxygen radicals release may participate in the induction of chromatid-type aberrations. Hence, we propose that a higher PLA_2 activity can explain most of the chromosomal damage observed in BS cells. This proposal can be readily tested and can be further expanded upon through the comparison of spontaneous levels of oxidative damage in DNA of BS and control cells.

REFERENCES

1. J. German, J. and E. Passarge, Bloom's syndrome. XII. Report of the Registry for 1987. *Clin. Genet.* 35:57-69 (1989).
2. J. German, Bloom Syndrome: A Mendelian Prototype of Somatic Mutational Disease. *Medicine* 72:393-406 (1993).
3. E.M. Kuhn, and E. Therman, Cytogenetics of Bloom's syndrome. *Cancer Genet. Cytogenet.*22:1-18 (1986).
4. M.P. Rosin, and J. German, Evidence for chromosome instability in Bloom' syndrome: increased numbers of micronuclei in exfoliated cells. *Human Genet.* 77:187-191 (1985).
5. R.S.K. Chaganti, S. Schonberg, and J. German, A manyfold increase in sister chromatid exchanges in Bloom's syndrome lymphocytes. *Proc. Natl. Acad. Sci. USA* 71:4508-4512 (1974).
6. S.T. Warren, R.A. Schultz, C-C. Chang, M. Wade, and J.E. Trosko, Elevated spontaneous mutation rate in Bloom's syndrome fibroblasts. *Proc. Natl. Acad. Sci. USA* 78:3133-3137 (1981).
7. Vijayalaxmi, H.J. Evans, J.H. Ray, and J. German, Bloom's syndrome: Evidence for an increased mutation frequency in vivo. *Science* 221:851-53 (1983).

8. T.M. Nicotera, Molecular and Biochemical Aspects of Bloom's syndrome. *Cancer Genet. Cytogenet.* 53:1-13 (1991).
9. J.E.Cleaver, Defective repair replication of DNA in Xeroderma Pigmentosum. *Nature* 28:652-656 968).
10. S. Mathew, V.V.V.S. Murty, J. German and R.S.K. Chaganti, Confirmation of 15q26.1 as the site of FES protooncogene by fluorescence in situ hybridization. Cytogenet. Cell Genet. 63:33-34 (1993).
11. L.D. McDaniel and R.A. Schultz, Elevated sister chromatid exchange phenotype in Bloom's syndrome cells is complemented by human chromosome 15. *Proc. Natl. Aced. Sc. USA* 89:7968-7972 (1992).
12. J. German, S. Schonberg, E. Louie. and R.S.K. Chaganti, Bloom's syndrome IV. Sister-chromatid exchanges in lymphocytes. *Am. J. Hum.* Genet. 29:248-255 (1977).
13. R. Weksberg, C. Smith, L. Anson-Cartwright and K. Maloney Bloom Syndrome: A single complementation group defines patients of diverse ethnic origin. Am. J. Hum. Genet. 42: 816-824 (1988).
14. I. Emerit and P. Cerutti, Clastogenic activity from Bloom's syndrome cultures. *Proc. Natl. Aced. Sc. (USA)* 78:1868-1872 (1981).
15. T.M. Nicotera, J. Notaro, S. Notaro, J. Schumer and A.A. Sandberg, Elevated superoxide dismutase activity in Bloom's syndrome: A genetic condition of oxidative stress. *Cancer Res.* 49: 5239-5243 (1989).
16. K-H. Lee, S. Abe, Y. Yanabe, I. Matsuda, and M.C. Yoshida, Superoxide dismutase activity in cultured chromosome instability syndrome cells. *Mutat. Res.* 244: 251-256 (1990).
17. T.M. Nicotera, A.W. Block. Z. Gibas, and A.A.Sandberg, Induction of superoxide dismutase, chromosomal aberrations, and sister chromatid exchanges by paraquat in Chinese hamster fibroblasts. *Mutation Res.* 151:263-268 (1985).
18. J.C. Gage, The action of paraquat and diquat on the respiration of rat livers. *Biochem. J.* 109:757-761 (1968).
19. P. Perry and S. Wolff, New Giemsa method for differentiation staining of sister chromatids. *Nature (London)* 251:156-158 (1974).
20. M.L. Salin, and J. McCord, Superoxide dismutases in polymorphonuclear leukocytes. *J. Clin. Chem.* 54:1005-1009 (1974).
21. O.H. Lowry, N.J. Rosebrough, A.L. Farr and R.J. Randall, Protein measurement with the folin phenol reagent. *J. Biol. Chem.* 176:265-275 (1951).
22. L.R. Ballou and W.Y. Chung, Marked increase in human platelet phospholipase A_2 activity in vitro and demonstration of an endogenous inhibitor. *Proc. Natl. Aced. Sc. (USA)* 80:5203-5207 (1983).
23. A. Dagan and S.A. Yegdar, A facile method for determination of phospholipase A_2 activity in intact cells. *Biochem. Int.* 15:801-808 (1987).
24. T. Nicotera, K. Thusu and P. Dandona, Elevated production of active oxygen in Bloom's syndrome cell lines. Cancer Res. 53:5104-5107 (1993).
25. A.R. Cross and O.T.G. Jones, The effect of the inhibitor diphenylene iodonium on the superoxide producing oxidase of neutrophils. *Biochem. J.* 237:111-116 (1986).
26. F.J.G.M. van Kuijk, A. Sevanian, G.J. Handleman and E.A. Dratz, A new role for phospholipase A_2: Protection of membranes from lipid peroxidation damage. *Trends in Biochem. Sc.* 12:31-34 (1987).
27. G.W. Burton and K.U. Ingold, Vitamin E as an *in vitro* and *in vivo* anti-oxidant. *Anal. N. Y. Aced. Sc.*, 570: 7-22 (1989).

28. A.P. Pentland, A.R. Morrison, S.C. Jacobs, L.L.Hruza, J.S. Hebert and L. Packer, Tocopherol analogs suppress arachidonic acid metabolism via phospholipase inhibition. *J. Biol. Chem.* 267:15578-15584 (1992).

29. O.T.G. Jones, J.T. Hancock and L.M. Henderson, Oxygen radical production by transformed B lymphocytes. *Molec. Aspects Med.* 12, 87-92 (1991).

30. Sasaki, M.S. Sister chromatid exchange as a cellular reflection of DNA repair. in: Sister Chromatid Exchange, A.A. Sandberg, ed., A.R. Liss NY (1982).

31. Knudson, A.G. Genetic predisposition to cancer. in: Genetic susceptibility to environmental mutagens and carcinogens. A.D. Bloom, L.Spatz and N.W. Paul, eds. March of Dimes, White Plains, (1990).

32. N.R.Schneider, R.S.K. Chaganti and J.German, Analysis of a BrdU sensitive site in the cactus mouse (Peromyscus eremicus) Chromosomal breakage and sister-chromatid exchange. *Chromosoma* 77:379-389 (1980).

33. N.C. Popescu and J.A. Dipaolo, Relevance of sister chromatid exchange to the induction of neoplastic transformation in: Sister chromatid Exchange, A.A. Sandberg, ed., A.R. Liss NY (1982).

34. W.L. Smith, Prostanoid biosynthesis and mechanisms of action. *Am. J. Physiol.* 263: F181-F191 (1992).

35. T. Rubinek and R. Levy, Arachidonic acid increases the activity of the assembled NADPH oxidase in cytoplasmic membrane and endosomes. *Biochim. Biophys. Acta* 1176:51-58 (1993).

THE ANALYSIS OF FREE RADICALS, LIPID PEROXIDES, ANTIOXIDANT ENZYMES AND COMPOUNDS RELATED TO OXIDATIVE STRESS AS APPLIED TO THE CLINICAL CHEMISTRY LABORATORY

Donald Armstrong and Richard Browne

Department of Clinical Laboratory Science
University at Buffalo
462 Grider Street
Buffalo, NY 14215

INTRODUCTION

It is almost impossible to read through a medical journal, or even the newspaper and not encounter an article that deals with oxidative stress, or with antioxidant involvement in a disease process. Indeed, free radicals, their reactive intermediates, low molecular weight aldehyde by-products derived from lipid peroxidation and antioxidant status are important measurements we can utilize to provide a more comprehensive understanding of pathologic mechanisms (1-8). All subcellular organelles normally generate superoxide ($O_2 \cdot -$), hydrogen peroxide and a variety of free radicals ie; hydroyl ($OH^\cdot$), perhydroxy($HO_2 \cdot$), carbon and nitrogen centered. It has been estimated that 10 billion of these radicals are produced daily via autoxidation and metabolic reactions. In cellular injury, increased amounts of $O_2 \cdot -$ radicals and peroxides can arise from the mitochondrial electron-transport system during hypoxia and following reperfusion, they can arise primarily through the activation of NADPH oxidase in phagocyte plasma membranes or from platelet derived endoperoxides of arachidonic acid, from the conversion of xanthine dehydrogenase to xanthine oxidase in tissue and from the generation of $OH^\cdot$ radicals in iron-catalyzed reactions involving hemoproteins (9). The most current review by Chaudiere covers theoretical and factual site-specific formation and damage (10). Liochev and Fridovich (11) have recently demonstrated that $OH^\cdot$ is produced in vivo through the Fenton reaction and Haber-Weiss pathways (FIG.1 reaction 1-2) and Yagi et al have shown that $OH^\cdot$ can also be formed when lipid hydroperoxide (LHP) is substituted for hydrogen peroxide (12). Superoxide reduces divalent metals to the reactive valency (reaction 3). Antioxidants are important physiological regulators and recently, long-term vitamin E intake has been correlated with a lowered incidence of coronary disease (13, 14). Because of the high level of albumin in plasma and interstitial fluid and its high binding capacity for lipids (including LHP) and iron it can regulate their extracellular concentration and thereby act as an antioxidant. Albumin also appears to be an effective scavenger of free radicals(15) by virtue of its sulfhydryl groups and peroxidase-like activity (16). The importance of free radicals to structure and function in pathology and medicine is well documented in the various chapters of this book.

Free Radicals in Diagnostic Medicine, Edited by
D. Armstrong, Plenum Press, New York, 1994

Since 1977, there has been a progressive increase in the number of papers in clinical chemistry journals that describe the application of certain tests to human patient samples of whole blood, leukocytes, erythrocytes, serum/plasma, urine, cerebrospinal/synovial fluid and breath (17-28). The modern hospital laboratory is currently well equipped with fluorometers, gas and high performance liquid chromatography, nephelometers, liquid scintillation counters and luminometers and are therefore capable of conducting much of the analytical work we do in the research laboratory. For more sophisticated analysis, excellent procedures describing gas and high performance liquid chromatography-mass spectroscopy are published (29, 30). In addition to $O_2{\cdot}^-$ and $OH^{\cdot}$, other major free radical species are those associated with LHP. Alkoxy radicals participate in propagation reactions with unsaturated polyunsaturated lipids in membranes or in lipoprotein particles. Polyunsaturated lipids with greater than 3 number double bonds give rise to toxic aldehydes of malondialdehyde (MDA) and 4-hydroxynonenal (4-HNE). They are produced during the iron catalyzed hydrolysis of LHP that are esterified to phospholipids and cholesterol. The various pathways leading to oxidative modification of lipids, illustrating initiation (reaction 4), propagation (reaction 5), degradation (reaction 6) and protection by antioxidants (reaction 7) are summarized below:

$O_2^{\cdot -}$ + Peroxides $\longrightarrow$ $OH^{\cdot}$ + OH^- + O_2	Haber-Weiss	(1)
Fe^{2+}/Cu^+ + Peroxides $\longrightarrow$ $OH^{\cdot}/LO^{\cdot}$ + OH^- + Fe^{3+}/Cu^{2+}	Fenton	(2)
Fe^{3+}/Cu^{2+} + $O_2^{\cdot -}$ $\longrightarrow$ Fe^{2+}/Cu^+	Recycling	(3)
$LH + In^{\cdot} \longrightarrow L^{\cdot} + InH$	Initiation	(4)
$L^{\cdot} + O_2 \longrightarrow LOO^{\cdot}$		
$LOO^{\cdot} + LH \longrightarrow L^{\cdot} + LOOH$	Propagation	(5)
$LOOH \xrightarrow{M^{n+}} LO^{\cdot}$ + 4HNE + MDA + $CH_3CH_2^{\cdot}$	Degradation	(6)
$FR^{\cdot}$ +antioxidants $\longrightarrow$ termination reaction	Protection	(7)

The objective of this chapter is to present analytical procedures which are adaptable to a hospital or reference laboratory. These tests allow the laboratory a full range of options for detecting free radical biochemical abnormalities in clinical specimens.

METHODS and RESULTS

Chemiluminescence. The quantum yield of photons for intrinsic (non-stimulated) or native reactions are low and consequently, an amplifier such as luminol, which reacts with all species of oxygen derived radicals or oxidants, is used to enhance the signal (31). Other compounds such as lucigenin and a luciferin analog 2-methyl-6-(p-methoxy phenyl)-3, 7-dihydromidazol [1,2-a] pyracin-3-one (MCLA) are specific for detecting $O_2{\cdot}^-$ and OH· respectively (32-34). LHP, phospholipid and cholesterol hydroperoxides can also be detected by chemiluminescence assay if cytochrome c-heme is added prior to luminol (35).

The easiest and most informative method of detecting free radical activity is to pre-incubate (0.5 min at 37°C) a 0.5 mL sample of whole blood with 1 mg of opsonized zymosan (OZ) particles, or 1 μg of phorbol myristate acetate and 50 μM of the appropriate luminophore (36). The intensity of emitted light from the sample is recorded at 15 min. and compared to the unstimulated control. Results can be expressed in mV if a luminometer is used, or photons/second if a liquid scintillation counter in the out-of-coincidence mode is used.

Isolated buffy coats preparation resuspended in physiologic saline to a concentration of 10^7 cells/2 mL are pre-incubated for 3 min. and the reaction initiated by adding 1.6 mg of OZ and 0.5 μM of MCLA. The light emission at 37°C is recorded from a luminescence reader and plotted against a standard curve using hypoxanthine and xanthine oxidase as a source of $O_2{\cdot}^-$ (33). The MCLA method can also be used to detect phospholipid peroxides in chloroform-methanol (2:1 v/v) extracts of heparized whole blood containing 0.002% butylhydroxytoluene (34). In this assay, MCLA at 300 nM is mixed with cytochrome c (10 μg/mL) in the port-column eluant. Phosphatidylcholine hydroperoxide standards prepared by photo-oxidation elute at 8-10 min. and sensitivity is reported at the 10 pmol level (35). If LDL particles are isolated by sequential ultracentrifugation, the same procedure may be used (37).

Aromatic Hydroxylation. In the presence of OH·, the benzene ring of salicylic acid can be attacked to produce stable hydroxylated products (38). However, only the 2,3-dihydroxybenzoic acid isomer appropriately reflects the OH· concentration in vivo (39).

A baseline sample of heparinized blood is taken and then the subject under study is given 1 gram of aspirin and 3 hours later, a second sample is collected (40). Plasma (500 μL) is acidified by the addition of 25 μL of 1 M HCl, mixed, extracted into 5 mL of diethyl ether, centrifuged at 2000 x g for 1 min., the organic phase collected and evaporated to dryness under nitrogen, reconstituted in 500 μL of citrate/acetate/methanol and 50 μL injected onto a Lichrosorb-RP-18 column for reversed phase HPLC. The eluate is monitored by an amperometric (electrochemical) detector to identify the 2,3 isomeric dihydroxbenzoate (RT= 5.8 to 6 min) and the results expressed in nanomoles.

Diene Conjugation. In the presence of free radicals, an electron is removed from polyunsaturated fatty acids, followed by an electronic rearrangement forming a conjugated system of double bonds (40). This change can be analyzed directly on chloroform-methanol extracts of serum/plasma by scanning between 200-300 nm in a spectrophotometer and then quantified from a standard curve of lipid peroxide. Second derivative spectra enhance resolution and discrimination. LHP standards can be prepared according to the method of Armstrong et al (42).

Volatile Alkanes. Ethane and pentane provide a non-invasive approach to determining lipid peroxidation. After one complete exhaled breath into a gas-tight bag, a 50 mL volume is removed with a syringe and injected onto a Chromosorb 102 stainless steel column and analyzed by gas chromatography (27). Alternatively, alveolar breath samples can be obtained using a Haldane-Priestly tube (43). Specimens are stable for 10 hours. The elution times for ethane and pentane are approximately 3 and 9 min. respectively.

Thiobarbituric Acid Reacting Substances. MDA forms a 1:2 adduct with thiobarbituric acid (TBA) that can be measured by spectrophotometry or by fluorometry (44). Although much controversy has appeared in the literature regarding the specificity of TBA towards compounds other than MDA, it remains the most widely employed assay used to determine lipid peroxidation. The biochemistry, metabolism, quantitative analysis and relationship of MDA to lipid peroxidation is provided in an excellent review by Janero (45). If lipoprotein fractions are first acid precipitated from the sample, interfering soluble TBA-reacting substances (TBARS) are minimized and the test becomes quite specific for lipid peroxidation. Nevertheless, there is always the possibility that some amount of non-lipid MDA would be present and so, TBARS seems to be a more descriptive term for this assay. It should perhaps be considered as an index of oxidative stress, but one that represents primarily lipid peroxidation (46).

We routinely use a modification of the Yagi method (44) in our laboratory for plasma/serum (47), or tissue homogenates (48) and have found it to be reproducible as described here in detail.

A. Preparation

Plasma. Centrifuge 5 mL of heparinized whole blood at 3,000 x g for 10 min. at 5-10°C, carefully remove plasma and place on ice for immediate analysis, or freeze several aliquots at -80°C for later analysis. Samples can be safely stored for 1-2 months.

Serum. Collect 7 mL of whole blood in a vacutainer containing no anticoagulant, let stand for 10 min. at room temperature for clot formation, centrifuge and process as described for plasma.

Tissue. Dilute 100 mg of trimmed tissue with 10 volumes of distilled, deionized water (DDW), disrupt in Potter-Elvejhem glass homoginzer, sonicate for 15 sec. at 40 V setting and use uncentrifuged whole homogenate for analysis, or extract with Folch reagent.

B. Assay Conditions

MDA standards using 1,1,3,3-tetramethoxypropane (TMP) from Sigma Chem. Co. are prepared fresh before each assay as follows: into a 50 mL volmetric flask, dilute 82 μL of TMP to 40 mL with distilled, deionized water, add 40 drops of conc. HCl and fill to the mark. A standard curve is prepared by withdrawing 1 mL from the stock solution and diluting it to 100 mL. The final concentration of this standard is equivalent to 100 nmol/mL of MDA. Fourteen serial, two-fold dilutions from this standard are performed which gives a concentration of 0.3 pmol/mL in the lowest standard.

Lipids with greater unsaturation, yield higher TBARS values. For example, at 20 μM concentrations, TBARS for the hydroperoxides of linoleic acid (18:2), linolenic acid (18:3), arachidonic acid (20:4) and docosahexaneoic acid (22:6) were 5,10, 20, and 40-45 mmol/mL respectively. Therefore, it must be remembered that biological samples are mixtures of LHP and consequently, the recorded value of a sample is a composite. The TBARS assay is thus a good screening procedure but should be followed up by defining a LHP profile by HPLC.

Standards or sample (200 μL) are mixed sequentially with 200 μL of 8.1% (w/v) lauryl sulfate, 1.5 mL of 20% (v/v) glacial acetic acid pH 3.5, 1.5 mL of 0.53% (w/v) TBA (Sigma Chem. Co.) and 600 μL of iron free DDW. If the unknown sample volume needs to be increased or decreased, adjustments can be made in the water aliquot as long as the final volume is maintained at 4 mL. A marble is placed on each tube to restrict evaporation, the samples heated at 95°C for 1 hour in a rotating water bath, removed and cooled in a tray of crushed ice, 1 mL of DDW is added, the reaction extracted by vortex mixing for 1 min. with 5 mL of butanol-pyridine (15:1 v/v), followed by centrifugation at 10,000 x g for 5 min. The upper solvent phase containing the fluorophore is decanted, returned to the ice tray and quantified in a Shimadzu RF 5000 U spectrofluorometer and FDU-3 CRT monitor/controller unit at 552 nm with the excitation wavelength set at 535 nm. Regulation of tube temperature at 4°C increases sensitivity and reproducibility.

To accommodate the widest range of detection, we utilize two standard curves. The high curve is used for plasma/serum and is linear from 0.4 to 12.5 nmol/mL whereas the low curve for tissue is linear from 0.3 pmol/mL to 0.2 nmol/mL. The mean concentration of plasma/serum TBARS in normal individuals is 2.2 ± 0.3 nmol/L and patients with uncontrolled diabetes may reach levels of 10 nmol/mL. Tissue levels are expressed in nmol/mg protein.

Recent articles using synchronous fluorescence spectroscopy (49) or solvent extraction-flow injection analysis (50) provide instrumentation adaptations that improve sensitivity and offer semiautomated methodology. To eliminate the high temperature step in color development, a leucomethylene blue derivative has been proposed in serum (51) which is described in detail in chapter 1. A horseradish peroxidase, dimethyl-p-phenylenediamine, hydrogen peroxide coupled reaction in erythrocyte lipid extracts is amenable to automation (52).

Several methods for the HPLC analysis of free MDA, or the MDA-TBA adduct have been reported in biological fluids (22-26, 53). These techniques are obviously more time consuming, but are quite specific since interfering chromophores are removed during chromatography. A C18 column is preferred and methanol or acetonitrile are suitable mobile phase solvents. A comparison study based on 3 published procedures for serum and urine is recommended to the reader (54). The MDA can be detected by spectrophotometry directly at 237 nm and the TBA-MBA adduct at 530 nm, or by fluorometry at 535 excitation/553 nm emission.

Aldehydes. These by-products of lipid peroxidation arise in the presence of iron from beta-cleavage and exhibit powerful biological activity. Besides MDA, acetaldehyde, butanol, propanal hexanal, heptanal and 4-HNE are all present in human plasma (54). Aldehydes of 2-10 carbons react wi1 1,3-cyclohexanedione (0.25% w/v) to form fluorescent derivatives. It is essential to add butylhydroxytoluene and an iron chelator to retard autoxidation. After 1 hour of incubation at 60°C, samples are cooled, proteins precipitated with 0.5 mL of methanol, the sample centrifuged at 3000 x g for 10 min, one mL of the supernate applied to a Sep-Pak C18 cartridge, eluted with 2 mL of methanol and 20 μL separated on a LiChrosphere RP 18 column and analyzed by HPLC fluorescence at an excitation wavelength of 380 nm and emission at 445 nm. Retention times range from 8 to 40 min., with 4-HNE eluting at around 26 min. and the detection limit is about 2 pmol. Modifications include reaction of the aldehyde with dinitrophenol hydrazine and the yellow chromogen measured at 360 to 380 nm, or the native aldehyde can be measured at 220 nm using a photodiode array UV-VIS spectrophotometer.

Lipid Hydroperoxides. In addition to using conjugated dienes and TBARS as indirect measures of lipid peroxidation, HPLC techniques can provide a direct analysis of lipid (free fatty acids, triglyceride and phospholipid hydroperoxides), cholesterol and cholesterol ester peroxides and organic peroxides (55). Plasma (0.5 mL) is extracted vigorously with 2 mL of methanol followed by 10 mL of hexane, centrifuged at 1000 x g for 10 min., the hexane layer taken to dryness under vacuum, the residue dissolved in 0.45 mL of methanol: t-butanol (50:50 v/v), chromatographed on a HPLC LC-18 column and a solution of isoluminol (177 mg) and microperoxidase (25 mg) used for a post-column chemiluminescence reaction. This fraction contains triglyceride and cholesterol ester hydroperoxides. The methanol phase is chromatographed on a $LCNH_2$ column and contains fatty acids, phospholipid hydroperoxides and organic peroxides.

Oxidized Lipoprotein. Low and high density lipoproteins transport phospholipid and cholesterol ester hydroperoxides, as well as antioxidants (56). They are also modified in the presence of MDA (57) and 4-HNE (58) which alter the apoprotein portion and/or glycoprotein receptor thereby increasing uptake of oxidized lipoprotein by the macrophage, and by Cu^{2+}, lipoxygenase and LHP radicals (59). Assays to detect increased peroxidation include diene conjugation and TBARS on isolated fractions (60) antibody titre (61) and HPLC (62).

A simplified method used in our laboratory adds 100 μL of sodium heparinate to 3 mL of plasma serum and precipitates LDL/VLDL with 150 μL of manganese chloride, leaving HDL in solution. After 10 min. at room temperature the sample is centrifuged at 12,000 x g for 15 min., the plasma/serum decanted and the pellet resuspended in 3 mL of normal saline. TBARS is measured on each fraction.

The susceptibility of lipoproteins to oxidation or their modification in vivo can also be measured by recording oxidizability indices (62). These are defined as a lag time, the rate of oxidation and the maximum amount of diene or TBARS formed in 3 hours. The lag time appears to be the more important since it reflects the amount and type of unsaturated lipids in the sample and the amount of antioxidant protection. Rapid preparation and addition of butyl hydroxytoluene are essential to minimize autoxidation.

Thiols. Glutathione (GSH) is an important co-enzyme for glutathione peroxidase activity and also provides protection for sulphydryl groups of proteins from oxidation. During these reactions, GSH is oxidized to GSSG and then recycled back by glutathione reductase. In this way, LHP and hydrogen peroxide levels are regulated in the cell and the ratio of GSH to

GSSG is indicative of oxidative stress. A semiautomated method has been reported that uses enzymatic recycling of GSSG and masking of GSH with 2-vinyl pyridine to measure both GSH and GSSG (63). For the analysis of GSH, the Cobas Fara II centrifugal analyzer is programmed to pipet 30 μL of sample, 30 μL of DTNB (5,5'-dithiobis-(2-nitrobenzoic acid) which protects against GSH autoxidation and 210 μL of the cofactor NADPH into the rotor cuvette where they are mixed, and then 30 μL of glutathione reductase is added to initiate the reaction. The mixture is incubated at 37°C, readings taken at 412 nm over the linear portion of the curve (30 sec. to 5 min.) and concentration determined from a standard curve of GSH. For the analysis of GSSG, 30 μL of 2-VP is added to the initial mixture and left at room temperature for 1.5 hours for derivatization with GSH to occur then incubated at 37°C, where readings are taken from 1 to 7 min and concentration determined from a standard curve of GSSG.

A fluorometric method is also available to measure GSH and GSSG (65). In this procedure, one sample is treated with 100 μg of the fluorophore o-phthalaldehyde at pH 8, incubated at room temperature for 15 min. and fluorescence intensity of the GSH-OPT adduct recorded at 420 nm after excitation at 350 nm. To determine the GSSG content, 200 μL of N-ethylmaleimide is added to inhibit glutathione reductase, allowed to stand for 30 min., than incubated and recorded as described for GSH. These are extremely sensitive assays with detection limits down to 1 ng.

Knowledge of the thiols such as cystine, cysteine, homocysteine and disulfides is equally important in defining the total oxidative protection afforded to tissues (66). Whole blood or plasma (0.2 mL) are diluted 1:2 with 5% metaphosphoric acid, incubated at room temperature for 15 min., centrifuged at 14,000 x g for 10 min. and the supernatant injected onto a C18 (ODS) column for HPLC analysis. Detection is carried out with a dual electrochemical detector positioned in series.

Trace elements. Certain transition metal ions are important biological mediators of lipid peroxidation, or are essential co-factors for metallo-antioxidant enzyme activity. Those that catalyze $OH^{\cdot}$ formation via the Fenton reaction are ferryl and cuprous salts: Fe^{2+}/Cu^{+} + hydrogen or LHP $\rightarrow Fe^{3+}/Cu^{2+}$ + ROH + $OH^{\cdot}$, then ascorbate and $O_2{\cdot}-$ reduce the ferric/cupric state and recycle the active prooxidant species. Selenium is a cofactor for glutathione peroxidase (GSH-P), while Cu, Zn and Mn are necessary for the isoforms of superoxide dismutase (SOD).

Because of the potential for Fenton-type reactions in the formation of free radicals and subsequent peroxidation of lipids, the measurement of iron is important (67). Both ferrous and ferric iron can be analyzed. Non-transferrin bound and low-molecular weight iron-citrate complexes can be detected in plasma/serum using a bleomycin-TBA assay (67-69), or by HPLC (70). The former assay consists of mixing 0.1 mL of sample with 0.5 mL DNA (1 mg/mL), 50 μL of bleomycin sulfate (1mg/mL), 0.1 mL Mg Cl_2 (50 mM), 50 μL HCl (10 mM), 0.1 mL ascorbic acid and 0.1 mL DDW. After incubation at 37°C for 2 hours, 1 mL of 0.1 M EDTA is added to stop the reaction, 1 mL of 1% (w/v) TBA reagent mixed with the sample, heated at 100°C for 15 min., cooled and recorded at 532 nm. This method is specific for ferrous iron not bound to transport proteins or enzymes, but is only sensitive to levels above 40 μmol/L. Therefore, bleomycin detectble iron in normal patients is not measurable. HPLC analysis requires preparation of a protein-free ultrafiltrate where 1 mL of plasma is applied to a Amicron Centrifree device, centrifuged at 1800 g for 1 hour and ferric iron in the sample eluted (RT=8.25 nm) with sodium acetate: methanol on a Anachem S5 OD52 column. Identification is made at 340 nm.

Copper, zinc and selenium are usually measured by atomic absorption spectrometry (71), with the latter ion requiring a graphite furnace. A simplified colorimetric method for measuring selenium is described by Alfthan (72). One mL of nitric acid is added to 0.5 mL serum, whole blood or urine, left overnight until all organic selenium species are digested, then 0.4 mL of sulfuric: perchloric acid (1:20 v/v) mixed and the sample heated at 120°C in a heating block for 20 min., raised to 150°C for 60 min. and finally to 180°C for 1.5 hour. Each tube contains anti-bumping granules and is wrapped with foil to avoid condensation. Samples are cooled, 30% hydrogen peroxide added, heated for 10 min. at 150°C, cooled, 1 mL of 6M HCl added and re-heated at 110°C for 10 min., buffered with 1 mL of 6M formic acid plus 1.5 mL EDTA, the pH adjusted to 1.5-2 with 4M ammonia, tubes wrapped to protect from light, 1mL of 2,3-diaminonaphthalene solution added (0.1 g/100 mL of 0.1 M HCl), the tubes placed in a 50°C waterbath for 30 min., cooled, extracted with 2.5 mL of cyctohexane, mixed vigorously for 30 sec and the organic phase fluorescence recorded at 518 nm with excitation at 369 nm. This method agrees well with electrothermal atomic absorption (r=0.997) and is sensitive to 0.45 ng of selenium.

Vitamins. The importance recently afforded to tocophenols, carotenoids and retinoids as measures of oxidative protection suggests that these assays should also be offered by the clinical laboratory. Since all are lipid soluble, extraction with n-hexane and analysis by HPLC provides a straight forward technique for identification and quantitation (73). Plasma/serum (0.5 mL) are deproteinized with an equal volume of ethanol, extracted with 2 mL of n-hexane, centrifuged at 2000 x g for 2 min., the organic lower phase recovered, evaporated under nitrogen, the residue re-dissolved in 50 μL of tetrahydrofuran made up to 20 μL with ethanol and 50 μL injected onto a Spherisorb 0051 column and eluted with two mobile phase solvents ie; 100 mM ammonium acetate in methanol: acetonitrile (80:2 v/v) followed by 100 mM ammonium acetate in DDN. Selective monitoring is performed by using two detectors coupled to a dual recording integrator and recording the major metabolites at 325 nm for retinol , 292 nm for α-tocophenol and 450 nm for ß-carotene.

Ascorbate, an important adjunct in the recycling of vitamin E, can be analyzed by spectrophotmetry, fluorescence, gas chromatography and HPLC. A new automated, colorimetric method for use with plasma/serum or urine is inexpensive and rapid (74). Proteins in 200 μL of sample are precipatied with 800 μL of 10% trichloroacetic acid, centrifuged within 30 min. of sample collection to minimize decomposition and 20 μL of the supenate mixed with 200μL of ferrozine (1.44 mmol/L). Readings are taken at 550 nm after 4.5 sec. of incubation, the concentration calculated from a standard curve of reduced ascorbic acid and expressed in μmol/L.

Antioxidant Enzymes. Assays for SOD, GSH-P, catalase and glutathione-S-transferase have been automated.* However, it should be mentioned here that a procedure for measuring both SOD and O_2·- in a single sample of whole blood has also been reported (75). Preparation of the sample requires hemolysis, hemoglobin extraction with chloroform-methanol (3:5 v/v) and centrifugation at 1400 g x 30 min. at 4°C to obtain a clear supernate. The reaction mixture for SOD contains 100 μL of sample, 100 μmol/L xanthine, 150 μg of fatty acid-free serum albumin, 25 μmol/L nitroblue tetrazolium (NBT) and sodium carbonate buffer (50 nmol/L, pH 10.2). The mixture is preincubated for 10 min at 25C, then 30μg of xanthine oxidase is added to initiation the reaction. After 20 min., the reaction is stopped with 100 μL of cupric chloride (6 mmol/L), recorded at 560 nm and compared to a blank. SOD activity is

* FRESA Biomedical Laboratories, Inc., 2239 152nd Ave., NE, Redmond, WA 98052-5519.

express as the amount of enzyme required to inhibit NBT in the blank by 50%. For $O_2\cdot-$, 100 μL of sample is preincubated for 5 min. at 37°C with 1 mL of normal saline, 2g of albumin and 30 μg of SOD and then 100 μL of 1.2 mmol/L ferricytochrome c added to start the reaction. Readings are taken at 550 nm and compared to a blank. The amount of $O_2\cdot-$ is expressed in nmol of ferricytochrome c reduced/min. Methodology involving gel electrophoretic and immunoblotting techniques for SOD are valuable to determine the CuZn containing isoform found in the cytosol and the Mn containing which is a mitochondrial enzyme (76). Samples are applied to non-dissociating, 12.5% polyacrylamide gels without SDS, subjected to 16 hr. of electrophoresis at 55V, incubated for 40 min. with 2.45 mM NBT, rinsed in DDW, reacted for 40 min. with a solution of 28 mM tetra-methylenediamine and 28 μM riboflavin at pH 7.8, exposed for 30 min. to fluorescent illumination or until the desired contrast between the blue background and white reaction sites is achieved and then scanned in a REP densitometer (Helena Laboratories). The CuZn-SOD is the smaller molecular weight enzyme and Mn-SOD the larger. Alternatively, immunoblotting can be performed by transfer from the gel onto nitrocellulose and application of commercially available antisera to CuZn or Mn-SOD conjugated to fluorescein isothiocyanate, or horseradish peroxidase is incubated with the sample overnight at room temperature. Visualization is either by indirect immunofluorescence, or by counterstaining for 5 min. with 3, 3'-diaminobenzidine in 0.06% hydrogen peroxide (77).

Antioxidant-Proxidant Status. The TEAC (Trolox Equivalent Antioxidant Capacity) assay reflects the total antioxidant capacity of a sample as compared to a water soluble analog of α-tocopherol (78). In addition to vitamin E, non-protein and protein compounds contributing to antioxidant protection include ascorbate, ß-carotene, flavenoids, glutathione, urate, bilirubin, transferrin, ceruloplasmin, albumin, glutathione peroxidase, selenium, superoxide dismutase and catalase. The assay is based on the reaction of free radicals with 2,2'-azinobis-(3-ethylbenzothiazoline-6-sulphonic acid or ABTS which gives rise to a peroxyl radical phenothiazine intermediate. In the presence of antioxidants, the control reaction between metmyoglobin (MetMb) and a peroxidase which produce the ABTS radical, is quenched in relation to the amount present. MetMb is synthesized and purified by reacting 400 μM type 3 myoglobin with 740 μM potassium ferricyanide followed by separation from other heme proteins on a Sephadex G15 column. The procedure mixes 8.4 μL of sample, 300 μL (500 μM) ABTS, 36 μL (70 μM) MetMb and 488 μL of normal saline, the reaction initiated by adding 167 μL (450 μM) of hydrogen peroxide, the sample incubated for 12 min. at room temperature and recorded at 734 nm. The degree of inhibition (%) by sample is determined from a Trolox standard curve.

The ORAC (Oxygen-Radical Absorbing Capacity) assay measures essentially the same antioxidant status as TEAC, but utilizes detection by fluorescence (79). This assay is based on the generation of peroxyl free radicals by 2,2'-azobis (2-amidinopropane) dihydrochloride or AAPH which oxidizes the bacterial protein porphyridium cruentum ß-phycoerythrin (ß-PE) to a fluorescent compound. The procedure mixes 20 μL each of ß-PE (1.67×10^{-8}M), and AAPH (3×10^{-3}M) with 1.96 mL of phosphate buffer (7.5×10^{-2}M) at pH 7, and the reaction incubated at 37°C with fluorescence measurements taken at 5 min. intervals at 540 nm excitation and 565 nm emission. One μL of serum is added to a duplicate tube and the areas under the two kinetic curves are compared ie; the presence of antioxidants reduces the relative fluorescence of ß-PE. One ORAC unit is equivalent to the protection provided by 1 μM Trolox.

Another test that appears useful has been developed which provides a measure of the balance between prooxidants and antioxidants in the body regardless of which factors are contributing to that balance.(80) The susceptibility of whole plasma to copper and hydrogen peroxide catalyzed oxidation is termed the Plasma Lipoprotein Peroxidation Potential. Duplicate samples (2 mL of plasma) are incubated for 2 and 24 hours at 60°C with 200 μL of prooxidant (10 μmol/L cupric acetate + 300 mL/L hydrogen peroxide). Thereafter, one sample is extracted into butyl alcohol and LHP analyzed by the TBARS method, while the second sample is saponified and the cholesterol oxides extracted into ethyl ether/petroleum ether. The latter samples are initially chromatographed by TLC, identified by spraying the side of the plate with phosphomolybdic acid or 2,6-dichlorofluorescein, the untreated bands are scraped, extracted into chloroform-methanol (2:1 v/v) and subjected to gas-liquid chromatography.

Commerical kits. A number of kits designed to measure different aspects of free radical reactions are already on the market and more can be expected in this expanding area of diagnositic medicine. Ready access to the products which are easily automated, will make it convenient and cost-effective for laboratories to offer assessments on oxidant and antioxidant status in disease. Richard et al (81) have reported on a kit marketed by Sobioda Grenoble that is basically the method of Yagi (44). Addition of butylhydroxytoluene (200 μg) does not improve sensitivitity, as we have found in our own studies. Their range of 2.5 ± 0.5 nmol/mL of plasma compares favorably with our results. These authors also state that TBARS is a measure of total lipid peroxidation. Several colorimetric and enzyme-linked immunoassay kits are available from Bioxytech S.A. France.* Their LHP kit is novel in that it uses a new synthetic substrate which reacts with MDA and 4-HNE and does not require boiling. To measure both aldehydes, 300 μL of plasma/serum, or standard is mixed with 225 uL of methanesulfonic acid + 975 μL of their chromogen, incubated for 40 min. at 45°C and absorbance measured at 586 nm. To measure only MDA, 37% HCl is substuted for the methanesulfonic acid and the reaction is incubated for 60 min. Their range of detection is around 2.5 mol/mL. A new chromogenic substrate has also been synthesized for the measurement of GSH which requires only 10 min of incubation at room temperature. Absorbance of the sample is measured at 400 nm, or at 356 nm if the total mercaptain level is required. An SOD assay on erythrocyte hemolysates is based on the antioxidation of another synthetic chromogen whose maximum absorbance is measured at 525 nm at alkaline pH (82). Glutathione peroxidase is measured in plasma using an enzyme-linked immunoassay. Samples are incubated in microplate wells coated with polyclonal antibodies specific for human GSH-P. Activity is detected by biotin-streptavidin coupling that is covalently linked to alkaline phosphatase. Hydrolysis of the p-nitrophenyl phosphate substrate is measured at 405 nm. Two other kits provide measures of neutrophil activation; myeloperoxidase and lactoferrin which are also EIA assays. Kamiya+ markets a kit for the colorimetric assay of LHP which utilizes a methylene blue derivative (MCDP) as substrate.

Calbiochem, Kallestad, Dako and The Binding Site to mention a few, market immunoassays for ceruloplasmin which require only 10 μL of plasma/serum, incubation at 37°C for 3 min. and is measured by random-access rate nephelometry. This protein has special significance because it 1) binds copper 2) acts as a ferroxidase by oxidizing ferrous iron to ferric and thereby 3) promotes incorportion of free iron into transferrin (83). These companies also offer transferrin analysis.

---

* Centre de Recherche Bioxytech, Z.A. des Petits Carreaux, 2 v. de coquelicots, 94385 Bonneuil-sur-Marne, France

+ Kamiya Biochemical Co., PO Box 6067, Thousand Oaks, CA 91359

Conclusion

Discussion. Assays presented in this chapter represent methods that are within the capability of clinical laboratories. It is up to the pathologist and clinical chemist to promote their use as valuable adjuncts to diagnosis and management of disease. At this time in the evolution of free radical-antioxidant research, there is an overwhelming body of data to support their implication in disease and to warrant further utilization by the physician and laboratory personnel. The availability of commercial kits and reference laboratories make it easy to incorporate free radical related testing into modern laboratory medicine.

Clinical chemistry was developed to identify end products of disease and provide markers for specific diagnosis. Diabetes mellitus can be used as a case in point. The level of glucose and glycosylated hemoglobin provide information on whether the patient is in glycemic control. These measures tell us little about the rate at which the disease is progressing, nor what effect they might have on development, severity or specificity of diabetic complications. LHP measurements do indeed correlate with some of these parameters (47,83) and clinical trials are encouraged to determine the effect of antioxidants (84). It is appropriate therefore, that the laboratory play an active role in providing analytic procedures that deal with the precursors of pathologic processes. The tests described here will complement existing methods and greatly expand our ability to identify disease in its earliest phase. With early diagnosis comes the potential for treatment, which should equate with improved cost-effectiveness and patient care.

Acknowledgements. We wish to thank LuAnn Kaite and Debbie Murello for assistance with the preparation of this manuscript, to Dr. Nicholas Miller for methodological information regarding automation, to Brian O'Connor for computer graphics and to the University at Buffalo for financial support to help establish and equip our research and reference laboratory.

REFERENCES

1. D. Armstrong, R. Sohol, R. Cutler and T. Slater. "Free Radicals in Molecular Biology, Aging and Disease," vol. 27, Raven Press, NY, pages 1-416 (1984).
2. C. Hooper, Free Radicals: Research on biochemical bad boys comes of age, *J NIH Res* 1:101-106 (1989).
3. B. Halliwell, J. Gutteridge and C. Cross, Free radicals, antioxidants and human disease: where are we now, *J. Lab. Clin. Med.* 119:598-620 (1992).
4. K. Cheeseman, Tissue injury by free radicals, *Toxicol. Industrial Health* 9:39-51 (1993).
5. K. Yagi. Role of lipid peroxides in aging and age-related diseases,*in* ""New Trends in Biological Chemistry," T. Ozawa, ed., Japan Sci. Soc. Press, Tokyo/Springer-Verlag, Berlin, pages 207-242 (1991).
6. E. Stadtman, Protein oxidation and aging, *Science* 257:1220-1224 (1992).
7. E. Harris, Regulation of antioxidant enzymes, *The FASEB Journal* 6:2675-2683 (1992).
8. A. Baouali, H. Aube, V. Maupoil, et al, Plasma lipid peroxidation in critically ill patients: importance of mechanical ventilation, *Free Rad. Biol. Med.* 16:223-227 (1994).
9. J. McCord and B. Omar, Sources of free radicals, *Toxicol. Industrial Health* 9:23-37 (1993).

10. J. Chaudiere, Some chemical and biochemical constraints of oxidative stress in living cells, *in* "Free Radical Damage and its Control," C. Rice-Evans and R. Burden eds., New Comprehensive Biochemistry, Elsevier Biomedical Press, Amsterdam, pages 24-64 (1994).
11. S. Liochev and J. Fridovich, The role of $O_2 \cdot^-$ in the production of HO· in vitro and in vivo, *Free Rad. Biol. Med.* 16:29-33 (1994).
12. K. Yagi, N. Ishida, S. Komura, et al, Generation of hydroxyl radical from linoleic acid hydroperoxide in the presence of epinephrine and iron, *Biochem. Biophys. Res. Comm.* 183:945-951 (1992).
13. M. Stampfer, C. Hennekens, J. Manson, et al, Vitamin E consumption and the risk of coronary disease in women, *NEJM* 328:1444-1449 (1993).
14. E. Rimm, M. Stampfer, A. Ascherio, et al, Vitamin E consumption and the risk of coronary disease in men, *NEJM* 328:1450-1456 (1993).
15. T. Emerson, Unique features of albumin: a brief review, *Crit. Care Med.* 17:690-694 (1989).
16. R. Pirisino, P. DiSimplicio, G. Ignesti, et al, Sulfhydryl group and peroxidase-like activity of albumin as scavenger of organic peroxides, *Pharmacol. Res. Commun.* 20:545-552 (1988).
17. T. Suematsu, T. Kamada, H. Abe, et al, Serum lipoperoxide level in patients suffering from liver disease, *Clin. Chem. Acta.* 79:267-270 (1977).
18. K. Satoh, Serum lipid peroxide in cerebrovascular disorders determined by a new colorimetric method, *Clin. Chem. Acta.* 90:37-43 (1987).
19. M. Santos, J. Vales, J. Angar, et al, Determination of plasma malondialdehyde-like material and its clinical application in stroke patients, *J. Clin. Pathol.* 33:973-976 (1980).
20. M. Maseki, J. Nishigaki, M. Hagihara, et al Lipid peroxide levels and lipid content of serum lipoprotein fractions of pregnant subjects with or without pre-eclampsia, *Clin. Chem. Acta.* 115:155-161 (1981).
21. J. Anzar, M. Santos, J. Valles, et al, Serum malondialdehyde-like material (MDA-LM) in acute myocardial infarction, *J. Clin. Pathol.* 36:312-715 (1983).
22. S. Wong, J. Knight, S. Hopfer, et al, Lipoperoxides in plasma as measured by liquid-chromatographic separation of malondialdehyde-thiobarbituric acid adduct, *Clin. Chem.* 33:214-220 (1987).
23. J. Knight, E. Smith, V. Kinder, et al, Reference intervals for plasma lipoperoxides: Age, sex and specimen-related variations, *Clin. Chem.* 33:2289-2291 (1987).
24. J. Knight, R. Pieper, S. Smith et al, Increased urinary lipoperoxides in drug abuses, *Ann. Clin. Lab. Sci.* 18:374-377 (1988).
25. J. Knight, R. Pieper and L. McClellan, Specificity of the thiobarbituric acid reaction: Its use in studies of lipid peroxidation, *Clin. Chem.* 34:2433-2438 (1988).
26. J. Knight, A. Cheung, R. Pieper et al, Increased urinary lipoperoxide levels in renal transplant patients, *Ann. Clin. Lab. Sci.*, 19:238-241 (1989).
27. E. Zarling and M. Clapper, Technique for gas-chromatographic measurement of volatile alkanes from single-breath samples, *Clin. Chem.* 33:140-141 (1987).
28. J. Knight, . McClellan and J. Stabell, Cerebrospinal fluid lipoperoxides quantified by liquid chromatography and determination of reference values, *Clin. Chem.* 36:139-142 (1990).

29. F. van Kuijk, D. Thomas, J. Konopelski et al, Transesterification of phospholipids or triglycerides to fatty acid benzyl esters with simultaneous methylation of free fatty acids for gas-liquid chromatographic analysis, *J. Lipid Res.* 27:452-456 (1986).

30. M. Selley, M. Bartlett, J. McGuiness, et al, Determination of the lipid peroxidation product trans-4-hydroxy-2-nonenal in biological samples by high performance liquid chromatography and combined capillary column gas chromatography-negative-ion chemical ionization mass spectroscopy, *J. Chromatogr.*, 488:329-340 (1989).

31. V. Sharov, V. Kazamanov, and Y. Vladimirov, Selective sensitization of chemiluminescence resulting from lipid and oxygen radical reactions, *Free Rad. Biol. Med* 7:237-242 (1989).

32. I. Minkenberg and E. Ferber, Lucigenin-dependent chemiluminescence as a new assay for NADPH oxidase activity in particulate practices of human polymorphonuclear leukocytes, *J. Immunol. Meth.*, 71:61-68 (1984).

33. M. Tosi and A. Hamedeni, A rapid, specific assay for superoxide release from phagocytes in small volumes of whole blood, *Am. J. Clin. Pathol.*, 97:566-573 (1992).

34. M. Nakano, Determination of superoxide radical and singlet oxygen based on chemiluminescence of Luciferin analogs, *in* "Oxygen Radicals in Biological Systems", Methods of Enzymology, vol. 186, L. Packer, and A. Glacer, eds., Academic Press, San Diego, CA, pages 585-591 (1990).

35. T. Miyazawa, Determination of phospholipid hydroperoxides in human blood plasma by a chemiluminescence-HPLC assay, *Free Rad. Biol. Med.* 7:209-217 (1989).

36. T. Nicotera, K. Thusa, and P. Dandona, Elevated production of active oxygen in Bloomus syndrome cell lines, *Cancer Res.* 53:5104-5107 (1993).

37. T. Mizawa, K. Fujimoto and S. Oikawa, Determination of lipid hydroperoxides in low density lipoprotein from human plasma using high performance liquid chromatography with chemiluminescence detection, *Biomed. Chromatogr.* 4:131-134 (1990).

38. W. Davis, B. Mohammad, D. Mays, et al, Hydroxylation of salicylate by activated neutrophils, *Biochem. Pharmacol.* 38:4013-4019 (1989).

39. B. Halliwell, H. Kaur and M. Ingelman-Sundberg, Hydroxylation of salicylate as an assay for hydroxyl radicals: a cautionary note, *Free Rad. Biol. Med.* 10:439-441 (1991).

40. A. Ghiselli, O. Laurenti, G. DeMattia, et al, Salicylate hydroxylation as an early marker of in vivo oxidative stress in diabetic patients, *Free Rad. Biol. Med.* 13:621-626 (1992).

41. F. Corongiu and A. Milia, An improved and simple method for determining diene conjugation in autoxidized polyunsaturated fatty acids, *Chem.-Biol. Interactions* 44:289-297 (1983).

42. D. Armstrong, T. Hiramitsu, J. Gutteridge, et al, Studies on experimentally induced retinal degeneration, I. Effect of lipid peroxides on electroretinographic activity in the albino rabbit, *Exp. Eye Res.* 35:157-171 (1982).

43. G. Metz, M. Gassull, A. Leeds, et al, A simple method of measuring hydrogen in carbohydrate malabsorption by end-expiratory sampling, *Clin. Sci. Mol. Med.* 50:237-240 (1976).

44. K. Yagi, A simple fluorometric assay for lipoperoxide in blood plasma, *Biochem. Med.* 15:212-216 (1976).

45. D. Janero, Malondialdehyde and thiobarbituric acid-reactivity as diagnostic indices of lipid peroxidation and peroxidative tissue injury, *Free Rad. Biol. Med.* 9:515-540 (1990).
46. A. Valenzuela, The biological significance of malondialdehyde determination in the assessment of tissue oxidative stress, *Life Sci.* 48:301-309 (1991).
47. D. Armstrong, N. Abdella, A. Salman, et al, Relationship of lipid peroxides to diabetic complications: comparison with conventional laboratory tests, *J. Diab. Comp.* 6:116-122 (1992).
48. D. Armstrong and F. Al-Awadi, Lipid peroxidation and retinopathy in streptozotorin-induced diabetes, *Free Rad. Biol. Med.* 11:433-436 (1991).
49. M. Conti, P. Morand, P. Levillain, et al, Improved fluorometric determination of malondialdehyde, *Clin. Chem.* 37:1273-1275 (1991).
50. H. Ikatsu, T. Nakajima, N. Murayama, et al, Flow-injection analysis for malondialdehyde in plasma with the thiobarbituric acid reaction, *Clin. Chem.* 38:2061-2065 (1992).
51. A. Miiki, N. Kayahara, Y. Yokote, et al, A simple assay for lipid peroxides using a leucomethylene blue derivative, *Clin. Chem.* 34:1228 (1988).
52. A. Yalcin, G. Haklar and K. Emerk, Simple colorimetric method for determination of peroxide, *Clin. Chem.* 39:2534-2535 (1993).
53. C. Smith, M. Mitchinson and B. Hallwell, Lipid peroxidation in hyperlipidaemic patients. A study of plasma using an HPLC-based thiobarbituric acid test, *Free Radic. Res. Comm.* 19:51-57 (1993).
54. H. Draper, E. Squires, H. Mahmoodi, et al, A comparison evaluation of thiobarbituric acid methods for the determination of malondialdehyde in biological materials, *Free Rad. Biol. Med.* 15:353-363 (1993).
55. A. Holley, M. Walker, K. Cheeseman , et al, Measurement of n-alkanals and hydroxyalkanals in biological samples, *Free Rad. Biol. Med.* 15:281-289 (1993).
56. V. Bowry, K. Stanley and R. Stocker, High density lipoprotein is the major carrier of lipid hydroperoxides in human blood plasma from fasting donors, *Proc. Natl. Acad. Sci.* 89:10316-10320 (1992).
57. J. Salonen, S. Yla-Herttula, R. Yamamota, et al, Autoantibody against oxidized LDL and progression of carotid atherosclerosis, *The Lancet* 339:883-887 (1992).
58. H. Hoff, J. O'Neil, G. Chisolm, et al, Modification of low density lipoprotein with 4-hydroxynonenal induces uptake by macrophages, *Arteriosclerosis* 9:538-549 (1989).
59. B. Kalyanaraman, W. Antholine and S. Parthasarathy, Oxidation of low-density lipoprotein by Cu^{2+} and lipoxygenase: an electron spin resonance study,*Biochem. Biophys. Acta* 1035:286-292 (1990).
60. H. Kleinveld, H. Hak-Lemmers, A. Stalenhoef, et al, Improved measurement of low-density-lipoprotein susceptibility to copper-induced oxidation: application of a short procedure for isolating low-density lipoprotein, *Clin. Chem.* 38:2066-2072 (1992).
61. W. Palinski, S. Yla-Herttuala, M. Rosenfeld, et al, Antisera and monoclonal antibodies specific for epitopes generated during oxidative modification of low density lipoprotein, *Arteriosclerosis* 10:325-335 (1990).

62. B. Frei, Y. Yamamota, D. Niclas, et al, Evaluation of an isoluminol chemiluminescence assay for the detection of hydroperoxides in human plasma, *Anal. Biochem.* 175:120-130 (1988).
63. H. Kleinveld, A. Naker, A. Stalenhoef, et al, Oxidation resistance, oxidation rate and extent of oxidation of human low-density lipoprotein depend on the ratio of oleic acid content to linoleic acid content: studies in vitamin E deficient subjects, *Free Rad. Biol. Med.* 15:273-280 (1993).
64. J. Teare, N. Punchard, J. Powell, et al, Automated spectrophotometric method for determining oxidized and reduced glutathione in liver, *Clin. Chem.* 39:686-689 (1993).
65. P. Hissin and R. Hilf, A fluorometric method for determination of oxidized and reduced glutathione in tissues, *Anal. Biochem.* 74:214-226 (1976).
66. J. Richie and C. Lang, The determination of glutathione, cyst(e)ine and other thiols and disulfides in biological samples using high-performance liquid chromatography with dual electrochemical detection, *Anal. Biochem.* 163:9-15 (1987).
67. J. Gutteridge, D. Rowley, E. Griffiths, et al, Low-molecular-weight iron complexes and oxygen radical reactions in idiopathic hemochromatosis, *Clinical Science* 68:463-467 (1985).
68. O. Aruma, A. Bomford, R.Polson, et al, Non-tranfonon-bound iron in plasma from hemochromatosis patients: effect of phlebotomy therapy, *Blood* 72:1416-1419 (1988).
69. B. Halliwell, O. Aruma, G. Maefi, et al, Bleomycin-detectable iron in serum from leukaemic patients before and after chemotherapy, *FEBS Letters* 241:202-204 (1988).
70. M. Grootveld, J Bell, b. Halliwell, et al, Non-transferrin-bound iron in plasma or serum from patients with idiopathic hemochromatosis. Characterization by high performance liquid chromatography and nuclear magnetic resonance spectroscopy, *J. Biol. Chem.* 264:4417-4422 (1989).
71. M. Richard, J. Arnaud, C. Jurkovitz, et al, Trace elements and lipid peroxidation abnormalities in patients with chronic renal failure, *Nephron* 57:10-15 (1991).
72. A. Alfthan, A micromethod for the determination of selenium in tissues and biological fluids by single-test-tube fluorimetry, *Anal. Chem. Acta* 165:187-194 (1984).
73. Z. Zaman, P. Fielden and P. Frost, Simultaneous determination of Vitamin A and E and carotenoids in plasma by reversed-phase HPLC in elderly and younger subjects, *Clin. Chem.* 39:2229-2234 (1993).
74. N. Miller, J. Shatti and C. Rice-Evans, Plasma vitamin C analysis by the ferrozine method using the Cobas Mira analyzer, (personal communication).
75. S. Umeki, M. Sumi, Y. Niki, et al, Concentrations of superoxide dismutase and superoxide anion in blood of patients with respiratory infection and compromised immune systems, *Clin. Chem.* 33:2230-2233 (1987).
76. J. Crapo, T. Oury, C. Rahouille, et al, Copper, zinc superoxide dismutase is primarily a cytosolic protein in human cells, *Proc. Natl. Acad. Sci.* 89:10405-10409 (1992).
77. D. Newsome, E. Dohard, M. Liles, et al, Human retinal pigment epithelium contains two distinct species of superoxide dismutase, *Invest. Ophthalmol. Vis. Sci.* 31:2508-2513 (1990).
78. N. Miller, C. Rice-Evans, M. Davies, et al, A novel method for measuring antioxidant capacity and its application to monitoring the antioxidant status in premature neonates, *Clin. Sci.* 84:407-412 (1993).

79. G. Cao, H. Alessio and R. Cutter, Oxygen-radical absorbance capacity assays for antioxidants, *Free Rad. Biol. Med.* 14:303-311 (1993).
80. M. Arshad, S. Bhadra, R. Cohen, et al, Plasma lipoprotein peroxidation potential: a test to evaluate individual susceptibility to peroxidation, *Clin. Chem.* 37:1756-1758 (1991).
81. M-J Richard, B. Portal, J. Meo, et al, Malondialdehyde kit evaluated for determining plasma and lipoprotein fractions that react with thiobarbituric acid, *Clin. Chem.* 38:704-709 (1992).
82. C. Nebot, M. Moutet, P. Huet, et al, Spectrophotometric assay of superoxide dismutase activity based on the activated autoxidation of a tetracyclic catechol, *Anal. Biochem.* 214:442-451 (1993).
83. P.E. Jennings, N.A. Scott, A.R. Saniabadi, et al, Effects of gliclazide on platelet reactivity and free radicals in type II diabetic patients: clinical assessment, *Metabolism* 41:36-39 (1992)
84. B. Halliwell and J. Gutteridge, The antioxidants of human extracellular fluids, *Arch. Biochem. Biophys.* 280:1-8 (1990).

FREE RADICALS AND THE PATHOGENESIS OF NEURONAL DEATH: COOPERATIVE ROLE OF EXCITATORY AMINO ACIDS

Domenico E. Pellegrini-Giampietro

Department of Preclinical and Clinical Pharmacology
University of Florence
Viale Morgagni 65
50134 Florence, Italy

INTRODUCTION

The potential role of oxygen-derived free radicals in the pathogenesis of neuropsychiatric diseases has been thoroughly discussed in the past few years[1, 2, 3]. The neurotoxic consequences of superoxide anion, hydrogen peroxide and hydroxyl radical formation have been described, as well as their relevance to abnormal conditions of the central nervous system, such as hyperoxia, hemorrhage, trauma, and aging. A separate line of investigation over the last ten years has established that excessive release of the excitatory neurotransmitter glutamate and sustained activation of glutamate receptors may also be responsible for neuronal degeneration associated with epilepsy, cerebral ischemia, hypoglycemia and other neurodegenerative diseases[4, 5, 6]. It is now emerging that free radical formation and glutamate receptor activation may act in concert, cooperating in the genesis and propagation of neuronal damage[7, 8, 9]. The goal of this report is to examine the potential relationship between these two pathogenic events in neurological disease, with particular stress on mechanisms underlying post-ischemic brain damage.

OXYGEN RADICALS AND THE CENTRAL NERVOUS SYSTEM

The brain is exceptionally vulnerable to the cytotoxic effects of oxygen-derived free radicals[1, 10, 11]. It is a highly oxygenated organ, consuming almost one fifth of the body's total oxygen, and it derives most of its energy from oxidative metabolism of the mitochondrial respiratory chain. The formation of superoxide anion and hydrogen peroxide along the electron transport chain can not be readily neutralized as in other tissues because of the poor catalase activity and the moderate amounts of superoxide dismutase (SOD) and glutathione peroxidase present in the brain. Moreover, brain membrane lipids are very rich in polyunsatured fatty acids, such as arachidonic acid, which are especially sensitive to free radical-induced peroxidation. Finally, several

Free Radicals in Diagnostic Medicine, Edited by
D. Armstrong, Plenum Press, New York, 1994

brain areas, including the globus pallidus and the substantia nigra, contain large amounts of iron, whereas the CSF content of transferrin is very low and close to iron saturation.

As in other tissues, accumulation of free radicals in the brain may result from increased formation of these compounds, as well as from reduction of naturally occurring free radical scavengers. For example, *hydrogen peroxide* may be formed as a normal byproduct of the activity of several enzymes including monoamine oxidase (MAO) and tyrosine hydroxylase or may result from auto-oxidation of endogenous ascorbic acid or catecholamines. *Superoxide anion* may be produced through the metabolism of arachidonic acid or, under conditions of energy failure, through the conversion of accumulated xanthine to uric acid by the Ca^{2+}-dependent enzyme xanthine oxidase (XOD). The *hydroxyl radical*, the most reactive species, is thought to be formed by the reaction of hydrogen peroxide with iron or copper ions (the Fenton reaction). Cell injury by any mechanism may accelerate free radical reactions, mainly because lysed cells release their intracellular iron pools into the extracellular space, hence generating hydroxyl radicals. In addition, conditions in which there is a relative decrease in antioxidants, such as vitamin E deficiency, are likely to be associated with increased radical formation.

As soon as they are formed, reactive free radicals directly attack proteins, nucleic acids and lipids[12]: damage to Ca^{2+} and other ion transport systems, altered gene expression, depletion of ATP and NAD(P)(H), and peroxidation of membrane lipids are among the mechanisms that lead to disruption of cellular functions and integrity. In particular, the formation of lipid peroxides in membranes (including the inner mitochondrial membrane, the site of the electron transport chain, and the plasma membrane) cause changes in membrane potential and fluidity and formation of peroxide pores, which allow Ca^{2+}, other ions and perhaps even neurotransmitter molecules to leak across the membrane. Radicals can also react with membrane-associated proteins, altering neurotransmitter uptake and receptor function.

EXCITATORY AMINO ACIDS AND EXCITOTOXICITY

L-glutamate, L-aspartate and related excitatory amino acids (EAAs), account for most of the excitatory synaptic activity in the central nervous system and are released via a Ca^{2+}-dependent mechanism by approximately 40% of all synapses[13]. EAAs produce their depolarizing action in post-synaptic neurons by activating three types of ionotropic receptors[14, 15], which can be distinguished by the selective agonists N-methyl-D-aspartate (NMDA), α-amino-3-hydroxy-5-methyl-4-isoxazole-propionate (AMPA), and kainate (Table 1). Whereas NMDA receptors are permeable to Na^{+}, K^{+} and Ca^{2+}, AMPA and kainate receptors (also known as "non-NMDA receptors") are normally permeable only to monovalent cations. NMDA receptors are blocked by extracellular Mg^{2+} in a voltage-dependent fashion and possess sites for allosteric potentiators, such as glycine (often considered a co-agonist, in that it is absolutely required for NMDA receptor activation) and spermine. A fourth class of glutamate receptors, the metabotropic receptors, are linked to G protein-mediated events, such as the production of inositol phosphates or the inhibition of adenylate cyclase (Table 1). As soon as they are released in the synaptic cleft, EAAs are immediately recaptured into presynaptic terminals or glial cells by means of high affinity and energy-dependent uptake mechanisms[16]. Hence, the concentration of glutamate can be very high in terminals (10 mM) or pre-synaptic vesicles (100 mM) but is normally much lower ($< 1\ \mu M$) in the extracellular space. However, if glutamate is released in excess, or if energy-dependent uptake systems fail to operate efficiently, the

Table 1. Excitatory amino acid receptors.

Receptors	Genes	Agonists	Competitive antagonists	Channel blockers	Allosteric potentiators	Allosteric antagonists	Effective pathways
NMDA	NMDAR1	NMDA (S)	D-AP5 (S)	Mg^{2+}	glycine	7-Cl-kynurenate	Na^+
	NMDAR2A-D	quinolinate	D-AP7 (S)	MK-801	D-serine		K^+
		ibotenate	CPP (S)	phencyclidine			Ca^{2+}
		glutamate		ketamine	spermine	arcaine	
		aspartate					
AMPA	GluR1-4	AMPA (S)	NBQX (S)				Na^+
		quisqualate	CNQX				K^+
		kainate	DNQX				
		glutamate	kynurenate				
KAINATE	GluR5-7	kainate	CNQX				Na^+
	KA1-2	domoate	DNQX				K^+
		glutamate	kynurenate				
METABOTROPIC	mGluR1-7	*t*-ACPD (M)	MCPG (M)				PLC
		quisqualate	AP3				AC
		AP4					PLA2
		CCG-I					

(S) selective
(M) active on most subtypes

concentration of extracellular glutamate may increase dramatically and become neurotoxic.

The concept of "excitotoxicity" (i.e. degeneration of neurons produced by activation of glutamate receptors, see for review Refs. 4, 5, 6) arose from studies principally by Olney[17], in which he demonstrated that glutamate, given systemically to immature animals, damages brain areas such as the retina and certain periventricular nuclei not protected by the blood brain barrier. The cytopathology of these lesions was striking, in that they were axon- and glia-sparing and they resembled pathology lesions seen in autoptic material of cerebral ischemia, hypoglycemia and epilepsy patients. A correlation between neurotoxic and excitatory potencies of various EAAs was also found. In the 1980s, the development of selective glutamate receptor antagonists confirmed that excitotoxicity is a consequence of excessive activation of glutamate receptors. Although ionotropic glutamate receptors appear to be primarily involved, recent evidence suggests that activation of subtypes of metabotropic receptors may either potentiate or attenuate excitotoxicity[18].

The activation of glutamate receptors is a fundamental step in excitotoxicity, but other post-synaptic mechanisms are thought to be directly responsible for neuronal degeneration. Cell culture studies have demonstrated that excitotoxic cell death can be separated into two distinct forms: an *acute* form, mediated by Na^+, K^+ and water influx, characterized by neuronal swelling and osmotic lysis, and a *delayed* form, mediated by an increase in cytosolic free Ca^{2+}, which more closely resembles EAA-mediated neurodegeneration *in vivo*. Glutamate may promote an increase in cytosolic free Ca^{2+} by at least three distinct mechanisms: *i)* activation of Ca^{2+}-permeable NMDA receptors (but also non-NMDA receptors, see below); *ii)* opening of voltage-dependent Ca^{2+} channels (indirectly, via membrane depolarization); or *iii)* stimulation of metabotropic receptor-mediated events (resulting in release of Ca^{2+} from intracellular stores). Elevated cytosolic Ca^{2+} then activates several enzymes (including protein kinase C, calpain I, phospholipase A2, XOD) capable of either directly or indirectly destroying cellular components.

To date, several genes encoding glutamate receptors of the ionotropic and metabotropic subtypes have been cloned in rodent and human brain[19, 20, 21] (Table 1). Ionotropic glutamate receptors, like other members of the ligand-gated channel superfamily, are thought to be membrane spanning proteins comprised of five subunits assembled around the pore of the channel. Of particular interest are observations that changes in subunit (GluR1-GluR4) composition of AMPA receptors can result in altered function: recombinant AMPA receptors assembled from GluR1 and/or GluR3 subunits in functional expression systems are permeable to Ca^{2+}, whereas the GluR2 subunit, expressed with GluR1 and/or GluR3, forms AMPA channels that are Ca^{2+}-impermeable (Figure 1). Since AMPA receptors, unlike NMDA receptors, normally *do not* gate Ca^{2+}, it is believed that in most adult neurons AMPA receptors are heterooligomers containing the dominant GluR2 subunit. However, the possibility exists that a switch in expression of GluR2 may result in altered Ca^{2+} permeability of newly synthesized AMPA receptors.

OXIDATIVE AND EXCITOTOXIC MECHANISMS IN NEUROPSYCHIATRIC DISORDERS

The existence of diseases of the central nervous system in which excitotoxic mechanisms are involved together with free radicals may have important clinical consequences, such as the possibility of more effective therapeutic intervention. Several *in vitro* studies have suggested the possibility that these two pathogenic events might

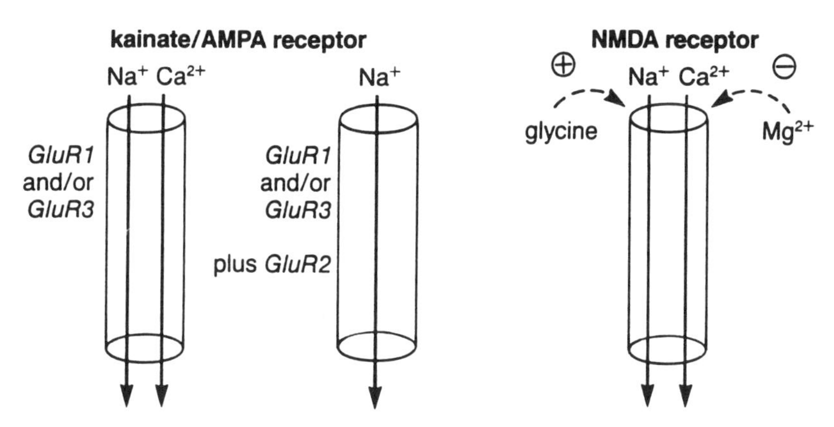

Figure 1. Ca^{2+} permeability of glutamate receptors. GluR1 and GluR3, expressed alone or in combination, form AMPA receptors that are permeable to Na^+ and Ca^{2+}, whereas GluR2, expressed with GluR1 and/or GluR3, forms channels that are Ca^{2+}-impermeable. NMDA receptors are permeable to Na^+ and Ca^{2+}, is potentiated by glycine, and blocked by Mg^{2+}. Adapted with permission from Ref. 23.

be linked. Dykens and co-workers[22] first demonstrated that kainate toxicity to cerebellar neurons in culture can be blocked by free radical scavengers. Many other papers have thereafter addressed the issue of free radical production in the brain as a consequence of glutamate receptor activation. For example, NMDA receptor agonists have been shown to promote the release of arachidonic acid from cultured striatal neurons[24] as well as the production of superoxide radicals in cerebellar granule cells[25]. In addition, free radical scavengers like 21-aminosteroids[26] or NO synthase inhibitors[27] have been demonstrated to be effective in attenuating excitotoxic injury in vitro. Since oxygen radicals promote the release of EAAs[28, 29] (see below) and arachidonic acid inhibits their re-uptake by glia and neurons[30, 31] it appears that glutamate may accumulate in the extracellular space as a consequence of free radical production, thus promoting a vicious cycle that could be responsible for the generation and propagation of neuronal death (Figure 2).

In a recent publication[2], Lohr has outlined the following criteria that, if fulfilled, would strongly suggest that a given disorder has a free radical-based pathophysiological component: *i)* evidence of increased radical production; *ii)* evidence of alteration in antioxidant enzymes or scavengers, or transition metals; *iii)* association of the measurements of radical activity with the severity of the disease; *iv)* ability to improve, or retard, the condition by use of antioxidants; and *v)* evidence that *in vivo* or *in vitro* models of the disease involve free radical production. No neuropsychiatric disorder meets *all* criteria, but there are some (tardive dyskinesia, schizophrenia, Down's syndrome, Alzheimer's disease among them) that meet many of them. We will discuss herein three disorders in which oxidative stress has been demonstrated to be implicated together with excitotoxic mechanisms: Parkinson's disease (PD), amyotrophic lateral sclerosis (ALS) and ischemic brain injury.

PD is a chronic, progressive disease characterized by rigidity, unintentional tremor and bradykinesia and caused by a selective degeneration of nigral dopaminergic neurons projecting to the caudate-putamen. Evidence for a free radical involvement in PD was initially centered around the autoxidation of dopamine and the production

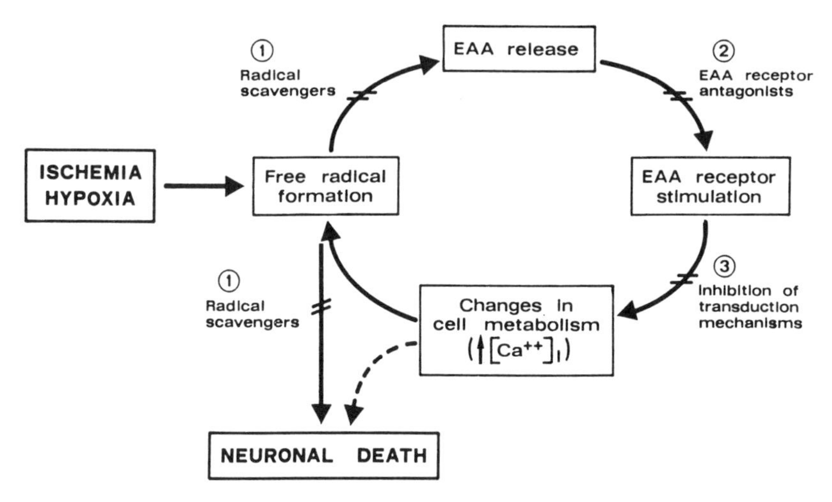

Figure 2. Hypothetical vicious cycle illustrating the main biochemical events leading to hypoxic-ischemic damage. Free radicals, formed during or following brain ischemia, promote the release of excitatory amino acids, which in turn may exert their toxic effects by interacting with specific receptors. Stimulation of these receptors increases cytosolic free Ca^{2+}, resulting in both neuronal damage and further production of free radicals. Adapted with permission from Ref. 29.

of hydrogen peroxide during its metabolism by MAO[32]. Subsequently, an increase in transition metals, such as copper and iron, as well as an increase in lipid peroxidation and SOD were reported in the substantia nigra of PD brains. Conclusive support to the free radical hypothesis has come from the discovery of 1-methyl-4-phenyl-1,2,3,6-tetrahydropyridine (MPTP) toxicity as a model of PD[33, 34]. MPTP, the agent responsible for sudden onset of PD symptoms in a group of heroin addicts in the United States, is converted to 1-methyl-4-phenylpyridinium (MPP+), the toxic metabolite, by MAO-B via a reaction that produces hydrogen peroxide. Moreover, MPP+-induced damage to nigral cells involves concentration in mitochondria and selective inhibition of complex I (nicotinamide adenine dinucleotide-ubiquinone oxireductase), thereby interfering with electron transport and generating oxygen radicals. According to this view, transgenic mice overexpressing CuZn-SOD are reported to be more resistant to the neurotoxic action of MPTP than control mice[35]. Interestingly, the findings that NMDA receptor antagonists protect against the dopaminergic degeneration induced by MPP+ in rats[36] and that AMPA antagonists possess antiparkinsonian effects in MPTP-treated monkeys[37] indicate that there is a link between oxidative stress and glutamate transmission in this model.

Similarly, a glutamate-mediated neurotoxic mechanism has been proposed for ALS, the late onset and progressive degeneration of motor neurons also known as Lou Gehrig's disease. Epidemiological studies have shown that the excitotoxin ß-N-methylaminoalanine (BMAA), implicated in a form of ALS and PD in Guam, causes selective degeneration of motor neurons, probably via activation of AMPA receptors[38]. More direct pieces of evidence are the findings that there is a loss of high affinity glutamate uptake sites in spinal cord of patients with ALS and that prolonged incubation of spinal cord explant cultures with agents that inhibit these sites (and therefore increase the extracellular concentration of EAAs) results in a delayed and selective degeneration of motor neurons that can be prevented by AMPA but not

NMDA receptor antagonists[39]. A free radical mechanism for ALS pathogenesis has also been suggested by the demonstration of structural defects in the gene encoding for CuZn-SOD in a familial variant of ALS[40]. A reasonable interpretation fusing the latter and the excitotoxic hypotheses is that the CuZn-SOD mutations sensitize motor neurons for AMPA-mediated glutamate transmission: the resulting "physiological" influx of Ca^{2+} through voltage-dependent Ca^{2+} channels activates the enzyme XOD generating a low level of superoxide anions, that accumulates and causes death of motor neurons. Sustained accumulation of glutamate in the extracellular space may trigger the oxidative stress for motor neurons in the sporadic forms of ALS, which are not associated with CuZn-SOD gene defects.

OXIDATIVE AND EXCITOTOXIC MECHANISMS IN ISCHEMIC BRAIN INJURY

Neurotoxic mechanisms mediated by both glutamate[41, 42, 43] and free radicals[44, 45, 46] have been repeatedly and independently reported to cause neuronal death following hypoxic or ischemic injury to the brain.

Glutamate is known to be released in excess during cerebral ischemia[47] and glutamate receptors are enriched in brain regions susceptible to ischemic injury[20]. Glutamate receptor antagonists under selected conditions protect against cerebral ischemia *in vivo*. The current idea envisions that, whereas both NMDA and AMPA receptor antagonists ameliorate focal neocortical damage produced by middle cerebral artery occlusion, the selective AMPA receptor antagonist NBQX[48] is considerably more effective than NMDA antagonists[49] in preventing delayed CA1 pyramidal cell death induced by transient but severe global or forebrain ischemia in various animal models. However, this perhaps oversimplified view has been challenged by recent studies, suggesting that selective antagonists at the glycine modulatory site of NMDA receptors like 7-Cl-kynurenate (7-Cl-KYNA)[50], as well as drugs acting at different sites like κ-opioid receptor agonists, K^+ channel openers, or N-type calcium channel antagonists, may also be beneficial in models of global ischemia.

Generation of toxic free radicals is also thought to play an important role in the pathogenesis of ischemia-induced neuronal death. The direct measurement of free radicals in brain tissue following ischemia/reperfusion has been achieved by using magnetic resonance techniques and spin trap agents[51, 52]. In addition, a number of studies have shown that free radical scavengers and/or lipid peroxidation inhibitors reduce brain damage following focal[53] and global[54, 55] ischemia. Similarly, the degree of cortical infarction induced by cerebral focal ischemia is reduced in rats treated with liposome-entrapped CuZn-superoxide dismutase (SOD)[56] or in transgenic mice overexpressing the CuZn-SOD1 gene[57].

In order to examine the relationship between the formation of free radicals and the release of glutamate in ischemic or post-ischemic conditions, we developed an *in vitro* model of cerebral ischemia by incubating hippocampal slices for 10 min in a medium lacking oxygen and glucose. Under these conditions the slices released, in a Ca^{2+}-independent fashion, a massive amount of pre-loaded D-[^{3}H]aspartate or of endogenous glutamate and aspartate, as detected by HPLC[29]. Since radical scavengers (D-mannitol), drugs reducing free radical formation (indomethacin, corticosterone), or enzymes able to metabolize them (catalase plus SOD) significantly reduced this outflow, it was supposed that free radicals caused release of EAAs. A direct demonstration of this concept was obtained by showing a significant release of EAAs after incubation of hippocampal slices with enzymes and substrates known to cause the formation of free radicals, such as xanthine plus XOD[28, 29] or arachidonic acid plus prostaglandin synthase[29]. It appears, therefore, that during ischemic states, brain production of reactive free radicals may lead to a Ca^{2+}-independent outflow of glutamate, presumably by reversal of the plasma membrane glutamate uptake carrier[16]. In such conditions of incomplete or almost complete ischemia (when mitochondria are in a maximally reduced state) the major site of oxygen-radical production appears to be the ubiquinone-cytochrome b region: in the presence of small amounts of O_2, these molecules can auto-oxidize to produce superoxide ions[46]. Since glutamate receptor activation is known to result in production of free radicals (see above) our data led us to hypothesize that free radical formation and glutamate receptor activation may affect each other as in a vicious cycle (Figure 2), acting in a sequential as well as in a reinforcing manner, ultimately leading to selective neuronal death.

Glutamate receptor activation and an increase in cytosolic free Ca^{2+} are important steps in the proposed cycle shown in Figure 2. To investigate the post-synaptic mechanisms underlying neuronal death following severe but transient forebrain ischemia, AMPA receptor subunit (GluR1, GluR2 and GluR3) gene expression was examined by *in situ* hybridization[58] in rats subjected to the four-vessel occlusion method of Pulsinelli. In this model, pyramidal cells in the CA1 subfield of the hippocampus are selectively damaged; however, histological signs of degeneration are not apparent until 48-72 hrs after circulation has been restored[59]. Figure 3 shows that at 24 hrs after ischemia, the expression of GluR2, the key subunit that controls Ca^{2+} fluxes through AMPA receptors (see Figure 1), was dramatically reduced in CA1, whereas GluR3 was reduced to a lesser extent and GluR1 was only slightly reduced. No significant change was seen in regions, such as CA3 and the dentate gyrus, that are known to be resistant to ischemic injury. The switch in expression of AMPA receptor subunits occurred at a time that clearly preceded CA1 pyramidal cell degeneration and coincided with the previously reported increase in Ca^{2+} influx into CA1 cells[60]. Timing of the switch indicates that it may play a causal role in postischemic cell death. As depicted in Figure 2, a modification in subunit composition of AMPA receptors and the resulting increase in cytosolic free Ca^{2+} may lead to free radical production. In turn, free radicals may be responsible for the observed switch because of their ability to damage DNA and alter gene expression[12, 45].

An important concept emerging from the proposed cycle (Figure 2) is that it may be possible to prevent cell death following ischemic injury to the brain by interfering with one or more of the following steps: *i)* free radical formation and reactivity, *ii)* EAA release, *iii)* glutamate receptor stimulation, *iv)* activation of transduction pathways leading to a rise in cytosolic free Ca^{2+}. For example, Oh and Betz have shown that pretreatment with the radical scavenger dimethylthiourea and the NMDA receptor blocker MK-801 (alone or in combination) can reduce brain edema during the early stages of cerebral ischemia in rats[61]. We have tested the protective effect of a compound, 7-Cl-thio-kynurenic acid (7-Cl-thioKYNA), which is a potent antagonist at the glycine site of the NMDA receptor[62] and, in addition, a free radical scavenger[63]. As such, 7-Cl-thioKYNA proved to be a more effective inhibitor of glutamate toxicity *in vitro* than 7-Cl-KYNA, which is equally potent as a glycine antagonist (Table 1) but fails to inhibit lipid peroxidation[63]. In a permanent middle cerebral artery occlusion stroke model in the rat, 7-Cl-thioKYNA was effective and 5-Cl-thioKYNA (a good lipid peroxidation inhibitor but a poor glycine antagonist) was inactive in attenuating infarct volume[64]. Figure 4 shows that in gerbils subjected to 5 min of severe global ischemia, 7-Cl-thio-KYNA dramatically attenuated ischemia-induced CA1 cell loss; the protection was associated with a delayed and marked reduction in the animals' temperature. However, when the gerbils were maintained normothermic for at least 360 min, 7-Cl-thio-KYNA still provided partial

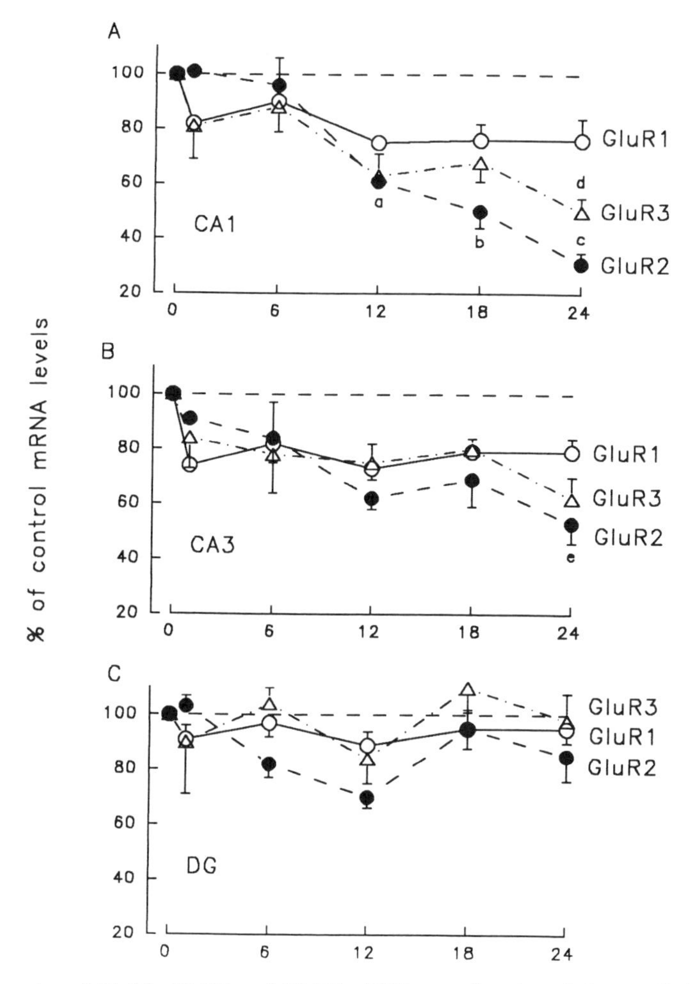

Figure 3. Expression of GluR1, GluR2, and GluR3 mRNAs as a function of time postischemia in CA1, CA3 and dentate gyrus (DG) of rats subjected to 10 min of severe forebrain ischemia. Values are expressed as percent of mean control mRNA levels ± SEM. a = $p < 0.05$ vs. 1 hr; b = $p < 0.05$ vs. 1 hr or 6 hrs; c = $p < 0.05$ vs. 18 hrs and vs. GluR3 at 24 hrs and $p < 0.01$ vs. 12 hrs and vs. GluR1 at 24 hrs; d = $p < 0.05$ vs. 1 hr and $p < 0.01$ vs. 6 hrs, ANOVA followed by Tukey's *w*-test. Reproduced with permission from Ref. 58.

but significant protection, indicating that the latter could not be ascribed to hypothermia alone. Moreover, no protection was observed when a reduction in temperature with a time course similar to that caused by 7-Cl-thio-KYNA was experimentally induced in saline-treated ischemic animals (Figure 4).

CONCLUSIONS

It has been established that activation of glutamate receptors is an important source of free radicals in neurons and that, in turn, free radicals may promote an

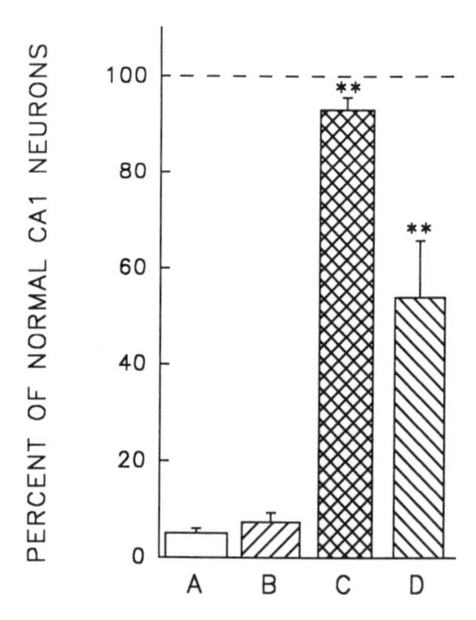

Figure 4. CA1 neuronal damage in ischemic gerbils. **A:** Saline-treated ischemic gerbils. **B:** Saline-treated ischemic gerbils, in which temperature was experimentally decreased starting at 60 min after reperfusion. **C:** Ischemic gerbils treated with 7-Cl-thioKYNA (100 mg/kg X 5, i.p.). **D:** 7-Cl-ThioKYNA ischemic gerbils, in which temperature was maintained at 37°C until 360 min post-ischemia. ** $p<0.01$ vs. A, ANOVA followed by Dunnett's test.

increase in the extracellular concentration of glutamate. The following hypothesis has thus been proposed: that free radical formation and glutamate receptor activation may affect each other as in a vicious cycle, cooperating in the genesis and propagation of neuronal damage. Several recent reviews address the link between glutamate and free radicals as an important pathogenic event in neuropsychiatric diseases[7, 8, 9, 43].

Elucidation of the relation between oxidative stress and glutamate neurotransmission could result in the development of more selective and effective therapeutic agents that interfere with the pathological events of the proposed cycle. We propose antagonists at the glycine site of NMDA receptors possessing radical scavenger properties as prototypes of new anti-ischemic drugs.

REFERENCES

1. H.A. Kontos, Oxygen radicals in CNS damage, *Chem.-Biol. Interact.* 72:229 (1989).
2. J.B. Lohr, Oxygen radicals and neuropsychiatric illness. Some speculations, *Arch. Gen. Psychiatry* 48:1097 (1991).
3. C.W. Olanow, A radical hypothesis for neurodegeneration, *Trends Neurosci.* 16:439 (1993).
4. S.M. Rothman and J.W. Olney, Excitotoxicity and the NMDA receptor, *Trends Neurosci.* 10:299 (1987).
5. D.W. Choi, Glutamate neurotoxicity and diseases of the nervous system, *Neuron* 1:623 (1988).
6. B. Meldrum and J. Garthwaite, Excitatory amino acid neurotoxicity and neurodegenerative disease, *Trends Pharmacol. Sci.* 11:379 (1990).
7. S.C. Bondy and C.P. LeBel, The relationship between excitotoxicity and oxidative stress in the central nervous system, *Free Radical Biol. Med.* 14:633 (1993).

8. J.T. Coyle and P. Puttfarcken, Oxidative stress, glutamate and neurodegenerative disorders, *Science* 262:689 (1993).
9. G.J. Lees, Contributory mechanisms in the causation of neurodegenerative disorders, *Neuroscience* 54:287 (1993).
10. H.B. Demopoulos, E.S. Flamm, D.D. Pietronigro, and M.L. Seligman, The free radical pathology and the microcirculation in the major central nervous system disorders, *Acta Physiol. Scand.* Suppl. 492:91 (1980).
11. B. Halliwell, Reactive oxygen species and the central nervous system, *J. Neurochem.* 59:1609 (1992).
12. J.P. Kehrer, Free radicals as mediators of tissue injury and disease, *Crit. Rev. Toxicol.* 23:21 (1993).
13. F. Fonnum, Glutamate: a neurotransmitter in mammalian brain, *J. Neurochem.* 42:1 (1984).
14. G.L. Collingridge and R.A.J. Lester, Excitatory amino acid receptors in the vertebrate central nervous system, *Pharmacol. Rev.* 40:143 (1989).
15. D.T. Monaghan, R.J. Bridges, and C.W. Cotman, The excitatory amino acid receptors: their classes, pharmacology and distinct properties in the function of the central nervous system, *Ann. Rev. Pharmacol. Toxicol.* 29:365 (1989).
16. D. Nicholls and D. Attwell, The release and uptake of excitatory amino acids, *Trends Pharmacol. Sci.* 11:462 (1990).
17. J.W. Olney, Excitotoxins: an overview, *in:* "Excitotoxins," K. Fuxe, P. Roberts, and R. Schwarcz, ed., Macmillan, London (1983).
18. D.D. Schoepp and P.J. Conn, Metabotropic glutamate receptors in brain function and pathology, *Trends Pharmacol. Sci.* 14:13 (1993).
19. G.P. Gasic and M. Hollmann, Molecular neurobiology of glutamate receptors, *Ann. Rev. Physiol.* 54:507 (1992).
20. S. Nakanishi, Molecular diversity of glutamate receptors and implications for brain function, *Science* 258:597 (1992).
21. B. Sommer and P.H. Seeburg, Glutamate receptor channels: novel properties and new clones, *Trends Pharmacol. Sci.* 13:291 (1992).
22. J.A. Dykens, A. Stern, and E. Trenkner, Mechanism of kainate toxicity to cerebellar neurons in vitro is analogous to reperfusion tissue injury, *J. Neurochem.* 49:1222 (1987).
23. R.J. Miller, Metabotropic excitatory amino acid receptors reveal their true colors, *Trends Pharmacol. Sci.* 12:365 (1991).
24. A. Dumuis, M. Sebben, L. Haynes, J.-P. Pin, and J. Bockaert, NMDA receptors activate the arachidonic acid cascade system in striatal neurons, *Nature* 336:68 (1988).
25. M. Lafon-Cazal, S. Pietri, M. Culcasi, and J. Bockaert, NMDA-dependent superoxide production and neurotoxicity, *Nature* 364:535 (1993).
26. H. Monyer, M. Hartley, and D.W. Choi, 21-Aminosteroids attenuate exitotoxic neuronal injury in cortical cell cultures, *Neuron* 5:121 (1990).
27. V.L. Dawson, T.M. Dawson, E.D. London, D.S. Bredt, and S.H. Snyder, Nitric oxide mediates glutamate neurotoxicity in primary cortical neurons, *Proc. Natl. Acad. Sci. USA* 88:6368 (1991).
28. D.E. Pellegrini-Giampietro, G. Cherici, M. Alesiani, V. Carlà, and F. Moroni, Excitatory amino acid release from rat hippocampal slices as a consequence of free radical formation, *J. Neurochem.* 51:1960 (1988).
29. D.E. Pellegrini-Giampietro, G. Cherici, M. Alesiani, V. Carlà, and F. Moroni, Excitatory amino acid release and free radical formation may cooperate in the genesis of ischemia-induced neuronal damage, *J. Neurosci.* 10:1035 (1990).
30. B. Barbour, M. Szatkowski, N. Ingledew, and D. Attwell, Arachidonic acid induces a prolonged inhibition of glutamate uptake into glial cells, *Nature* 342:918 (1989).
31. A. Volterra, D. Trotti, P. Cassutti, C. Tromba, A. Salvaggio, R.C. Melcangi, and G. Racagni, High sensitivity of glutamate uptake to extracellular free arachidonic acid levels in rat cortical synaptosomes and astrocytes, *J. Neurochem.* 59:600 (1992).

32. P. Jenner, Oxidative stress as a cause of Parkinson's disease, *Acta Neurol. Scand.* 84 (suppl. 136):6 (1991).
33. J.D. Adams and I.N. Odunze, Biochemical mechanisms of 1-methyl-4-phenyl-1,2,3,6-tetrahydropyridine toxicity, *Biochem. Pharmacol.* 41:1099 (1991).
34. K.F. Tipton and T.P. Singer, Advances in our understanding of the mechanisms of the neurotoxicity of MPTP and related compounds, *J. Neurochem.* 61:1191 (1993).
35. S. Przedborski, V. Kostic, V. Jackson-Lewis, et al., Transgenic mice with increased Cu/Zn-superoxide dismutase activity are resistant to N-methyl-4-phenyl-1,2,3,6-tetrahydropyridine-induced neurotoxicity, *J. Neurosci.* 12:1658 (1992).
36. L. Turski, K. Bressler, K.-J. Rettig, P.-A. Loschmann, and H. Wachtel, Protection of substantia nigra from MPP+ neurotoxicity by N-methyl-D-aspartate antagonists, *Nature* 349:414 (1991).
37. T. Klockgether, L. Turski, T. Honoré, Z. Zhang, D.M. Gash, R. Kurlan, and J.T. Greenamyre, The AMPA receptor antagonist NBQX has antiparkinsonian effects in monoamine-depleted rats and MPTP-treated monkeys, *Ann. Neurol.* 30:717 (1991).
38. P.S. Spencer, P.B. Nunn, J. Hugon, A.C. Ludolph, S.M. Ross, D.N. Roy, and R.C. Robertson, Guam amyotrophic lateral sclerosis-parkinsonism-dementia linked to a plant excitant neurotoxin, *Science* 237:517 (1987).
39. J.D. Rothstein, L. Jin, M. Dykes-Hoberg, and R.W. Kuncl, Chronic inhibition of glutamate uptake produces a model of slow neurotoxicity, *Proc. Natl. Acad. Sci. USA* 90:6591 (1993).
40. D.R. Rosen, T. Siddique, D. Patterson, et al., Mutations in Cu/Zn superoxide dismutase genes are associated with familial amyotrophic lateral sclerosis, *Nature* 362:59 (1993).
41. D. Choi, Cerebral hypoxia: some new approaches and unanswered questions, *J. Neurosci.* 10:2493 (1990).
42. W. Pulsinelli, Pathophysiology of acute ischaemic stroke, *Lancet* 339:533 (1992).
43. B. Peruche and J. Krieglstein, Mechanisms of drug actions against neuronal damage caused by ischemia - An overview, *Prog. Neuropsychopharmacol. Biol. Psychiat.* 17:21 (1993).
44. J.M. Braughler and E.D. Hall, Central nervous system trauma and stroke. I. Biochemical considerations for oxygen radical formation and lipid peroxidation, *Free Radical Biol. Med.* 6:289 (1989).
45. R.A. Floyd, Role of oxygen free radicals in carcinogenesis and brain ischemia, *Faseb J.* 4:2587 (1990).
46. J.R. Traystman, J.R. Kirsch, and R.C. Koehler, Oxygen radical mechanisms of brain injury following ischemia and reperfusion, *J. Appl. Physiol.* 71:1185 (1991).
47. H. Benveniste, J. Drejer, A. Schousboe, and N. Diemer, Elevation of extracellular concentrations of glutamate and aspartate in rat hippocampus during transient cerebral ischemia monitored by intracerebral microdialysis, *J. Neurochem.* 43:1369 (1984).
48. M.J. Sheardown, E.O. Nielsen, A.J. Hansen, P. Jacobsen, and T. Honoré, 2,3-Dihydroxy-6-nitro-7-sulfamoyl-benzo(F)quinoxaline: a neuroprotectant for cerebral ischemia, *Science* 247:571 (1990).
49. A.M. Buchan, H. Li, and W. Pulsinelli, The N-methyl-D-aspartate antagonist, MK-801, fails to protect against neuronal damage caused by transient, severe forebrain ischemia in adult rats, *J. Neurosci.* 11:1049 (1991).
50. D.K.J.E. Von Lubitz, R.C.-S. Lin, R.J. McKenzie, T.M. Devlin, R.T. McCabe, and P. Skolnick, A novel treatment of global cerebral ischemia with a glycine partial agonist, *Europ. J. Pharmacol.* 219:153 (1992).
51. C.N. Oliver, P.E. Starke-Reed, E.R. Stadtman, G.J. Liu, J.M. Carney, and R.A. Floyd, Oxidative damage to brain proteins, loss of glutamine synthetase activity, and production of free radicals during ischemia/reperfusion-induced injury to gerbil brain, *Proc. Natl. Acad. Sci. USA* 87:5144 (1990).
52. A. Sakamoto, S.T. Ohnishi, T. Ohnishi, and R. Ogawa, Relationship between free radical production and lipis peroxidation during ischemia-reperfusion injury in the rat brain, *Brain Res.* 554:186 (1991).

53. H. Hara, K. Kogure, H. Kato, A. Ozaki, and T. Sukamoto, Amelioration of brain damage after focal ischemia in the rat by a novel inhibitor of lipid peroxidation, *Europ. J. Pharmacol.* 197:75 (1991).
54. E.D. Hall, K.E. Pazara, and J.M. Braughler, 21-Aminosteroid lipid peroxidation inhibitor U74006F protects against cerebral ischemia in gerbils, *Stroke* 19:997 (1988).
55. C. Clough-Helfman and J.W. Phillis, The free radical trapping agent *N-tert*-butyl-α-phenylnitrone (PBN) attenuates cerebral ischaemic injury in gerbils, *Free Radical Res. Commun.* 15:177 (1991).
56. S. Imaizumi, V. Woolworth, R.A. Fishman, and P.H. Chan, Liposome-entrapped superoxide dismutase reduces cerebral infarction in cerebral ischemia in rats, *Stroke* 21:1312 (1990).
57. H. Kinouchi, C.J. Epstein, T. Mizui, E. Carlson, S.F. Chen, and P.H. Chan, Attenuation of focal cerebral ischemic injury in transgenic mice overexpressing CuZn superoxide dismutase, *Proc. Natl. Acad. Sci. USA* 88:11158 (1991).
58. D.E. Pellegrini-Giampietro, R.S. Zukin, M.V.L. Bennett, S. Cho, and W.A. Pulsinelli, Switch in glutamate receptor subunit gene expression in CA1 subfield of hippocampus following global ischemia in rats, *Proc. Natl. Acad. Sci. USA* 89:10499 (1992).
59. W.A. Pulsinelli, J.B. Brierley, and F. Plum, Temporal profile of neuronal damage in a model of transient forebrain ischemia, *Ann. Neurol.* 11:491 (1982).
60. J.K. Deshpande, B.K. Siesjö, and T. Wieloch, Calcium accumulation and neuronal damage in the rat hippocampus following cerebral ischemia, *J. Cereb. Blood Flow Metabol.* 7:89 (1987).
61. S.M. Oh and A.L. Betz, Interaction between free radicals and excitatory amino acids in the formation of ischemic brain edema in rats, *Stroke* 22:915 (1991).
62. F. Moroni, M. Alesiani, A. Galli, et al., Thiokynurenates: a new group of antagonists of the glycine modulatory site of the NMDA receptor, *Europ. J. Pharmacol.* 199:227 (1991).
63. F. Moroni, M. Alesiani, L. Facci, et al., Thiokynurenates prevent excitotoxic neuronal death in vitro and in vivo by acting as glycine antagonists and as inhibitors of lipid peroxidation, *Europ. J. Pharmacol.* 218:145 (1992).
64. J. Chen, S. Graham, F. Moroni, and R. Simon, A study of the dose-dependency of a glycine receptor antagonist in focal ischemia, *J. Pharmacol. Exp. Ther.* 267:937 (1993).

FREE RADICALS AND OCULAR DISEASE

Robert E. Anderson, Frank L. Kretzer, and Laurence M. Rapp
Cullen Eye Institute
Baylor College of Medicine
Houston, TX 77030

INTRODUCTION

The eye is a unique organ in that its very reason for existence, to capture photons, exposes it to the harmful effects of a number of light-induced processes. The high levels of long chain polyunsaturated fatty acids (PUFA) and prodigious utilization of oxygen by the retina make this tissue especially vulnerable to the consequences of free radical induced lipid peroxidation. This chapter will briefly review evidence for lipid peroxidation in several ocular disorders.

RETINOPATHY OF PREMATURITY

Retinopathy of prematurity (ROP) is a potentially devastating disease that can result in retinal detachment (blindness),[1] profound macular ectopia and vessel traction (low visual acuity),[2] and prolonged retinal traction (late retinal tears and detachments).[3] Despite judicious curtailment of oxygen and new developments in neonatology technology, ROP nevertheless still plagues those preterms who are ≤ 1250 grams birth weight.[4,5] With the increasing survival rate of preterms ≤ 750 grams birth weight, the number of infants developing ROP is increasing.[6] Fortunately, there is now an international classification for retinopathy of prematurity,[7] a protocol for screening neonatal intensive care units,[8] and a recommended surgical intervention to halt threshold disease.[9] However, the challenge is how ROP can be prevented so that surgical intervention is unnecessary.

The pathogenesis of ROP appears to be a cascade of events starting with lipid peroxidation and oxygen perturbation of spindle cells in the avascular retina during the first hours after birth and terminating with invasion of myofibroblasts into the vitreous 8 - 12 weeks later. When these myofibroblast sheets contract, they induce retinal detachments, cause macular dragging, or create long term traction. From 278 pairs of whole eye donations from preterm infants (320 - 1500 grams birth weight) who survived from 20 minutes to 1 year, a model has evolved concerning the pathogenesis of ROP.[10,11]

Spindle cells are perturbed by oxygen-related postnatal insults: they form extensive gap junctions; the rough endoplasmic reticulum proliferates within their cytoplasm; they secrete angiogenic factors; angiogenic factors trigger abnormal neovascularization at the shunt; myofibroblasts invade the vitreous; and the myofibroblast sheets contract. In the preterm retina, spindle cells are transretinal to immature photoreceptors; yet, the choroidal vasculature is fully

Free Radicals in Diagnostic Medicine, Edited by
D. Armstrong, Plenum Press, New York, 1994

developed, has high vascular flow, and nourishes the maturing, thin retina in its hypoxic intrauterine environment (25mm Hg). At term birth, all spindle cells have become inner retinal vessels, and the inner retinal vasculature approaches the temporal ora serrata.

With premature birth, the situation is completely different. Spindle cells are now exposed to an enormous oxygen flux across the retina. The working hypothesis is that these vulnerable spindle cells in an environment low in endogenous antioxidants experience oxidative damage which triggers an 8 - 12 week cascade of events which eventually can lead to blindness. The unifying principle of this hypothesis is that damage to spindle cell plasma membranes is initiated by multiple clinical parameters. Each parameter increases the probability that free radicals generated by enhanced cellular oxygenation can overwhelm the sequestering ability of the scarce endogenous antioxidant systems operant in the premature retina.

The most consistent risk factors for ROP are oxygen administration, blood transfusions, confirmed sepsis, reperfusion, and high ambient light intensity.[11,12] Each of these ROP risk factors supports the hypothesis that the initial insult to spindle cells might be oxidative damage.

One consistent ROP risk factor is the cumulative amount of administered oxygen. The pathologic response of spindle cells is increased by oxygen tension greater than the hypoxic, intrauterine environment. Risk factors such as hyaline membrane disease, pneumothorax, patent ductus arteriosus, apnea, and bronchopulmonary dysplasia are associated with the therapeutic use of supplemental oxygen. They enhance the degree of free radical damage which potentiates gap junction formation between adjacent spindle cells. The amount of oxygen administered is not the sole cause of ROP, but it is one unavoidable factor if these infants are to survive with intact central nervous systems. There is no safe, minimal level of oxygen that can be administered. All extrauterine levels are hyperoxic challenges to the peripheral, avascular retina with its low level of endogenous antioxidants.

Transfusions with adult, packed red blood cells change the fetal oxygen dissociation curve. Adult hemoglobin has less affinity for oxygen than fetal hemoglobin. As a result, more free radical damage is imposed on the developing retina. Very sick preterms receive relatively large quantities of adult blood in replacement transfusions. Transfusions are necessary due to blood loss related to multiple blood tests, to intraventricular and pulmonary hemorrhage, and to exchange transfusions related to hyperbilirubinemia.

Sepsis is also identified repeatedly as a risk factor for ROP. Macrophages release a large quantity of free radicals concomitant with the ingestion of bacteria. The burst of free radicals overloads the endogenous antioxidant systems of the preterm retina and potentiates gap junction formation between adjacent spindle cells.

Tissue reperfusion occurs repeatedly in the difficult to stabilize preterm. Bouts of perinatal asphyxia cause low pH. Postnatal transport often results in hypothermia. These acidic and hypothermic insults increase tissue utilization of oxygen during moments of reperfusion; this results in increased free radical formation.

Prolonged high intensity light has been suggested to increase the risk of ROP. Light is known to damage photoreceptors (see next section), resulting in decreased utilization of oxygen by even the most immature photoreceptors. Thus, the minimal oxygen barrier created by photoreceptors is further reduced and permits a greater oxygen flux across the retina from choroidal vessels. This increases the degree of free radical damage to spindle cell plasma membranes.

Three masked, controlled clinical trials suggested the efficacy of vitamin E supplementation in suppressing the development of severe ROP.[12-14] In preterms ≥ 1000 grams birth weight (≥ 27 weeks gestational age), such vitamin E supplementation suppressed the development of severe ROP. However, vitamin E supplementation was not an equal panacea for those preterms ≤ 1000 grams birth weight. There was an age dependent response to the efficacy of vitamin E supplementation. Univariant statistical analysis of these data bases showed no significance; multivariant analysis suggested efficacy. The rationale for these clinical results were as follows: the preterm infant is deficient in vitamin E because the vitamin has low permeability across the placenta and the preterm infant has no adipose tissue to store reserves of vitamin E. With daily and continuous vitamin E supplementation, spindle cells in preterms ≥ 1000 grams birth weight are protected against oxidative damage. Their spindle cells never develop gap junctions, they never secrete angiogenic factors, abnormal neovascularization never occurs, normal inner retinal vasoformation continues towards the ora serrata, spindle cells disappear from the retina, myofibroblasts never invade the vitreous, and ROP does not develop to any severe stage. Apparently, despite vitamin E supplementation, significant retinal uptake does not occur in those preterms ≤ 1000 grams birth weight.

The clinical data from these three centers are controversial. The age-dependent efficacy is troublesome. The three centers utilized different routes of vitamin E administration, targeted different plasma vitamin E levels, and documented the resultant ROP before the adoption of the international classification of ROP in 1984. Furthermore, the literature abounds with avoidable vitamin E toxicities.[11] The bottom line is that the American Academy of Pediatrics has never sanctioned vitamin E supplementation as an antioxidant mode of suppressing the development of severe ROP. But, the clinical efficacy supports the basic research fact that spindle cells can be protected by a favorable retinal oxidant/antioxidant environment.

Nielsen et al.[15] published vitamin E levels in the avascular retinas of 15 preterm infants. Figure 1 plots their tabular data and reveals a very important concept. The nine preterms (22 - 33 weeks gestations age) not supplemented with vitamin E show a slow increase in the endogenous vitamin E levels in the avascular retina as gestational age increases; however, this level does not approach that of the peripheral retina in adults (n = 9). Also, 10 days of vitamin E supplementation in a 23 week gestational age preterm, 11 days in a 25 week preterm, and 20 days in a 25 week preterm resulted in minimal retinal antioxidant elevation above the unsupplemented baseline. The spindle cells in these three retinas were extensively gap junction linked, the cytoplasm of the spindle cells contained a massive proliferation of rough endoplasmic reticulum, and angiogenic factors were assayable in extracts of their avascular retinas. Thus, vitamin E supplementation was not preventing the cascade of events that most likely would have led to severe ROP had these infants survived for 8 - 12 weeks. In contrast, two preterms of 28 weeks gestational age who were vitamin E supplemented for 14 and 43 days, respectively, and one preterm of 31 weeks gestational age given vitamin E for 4 days

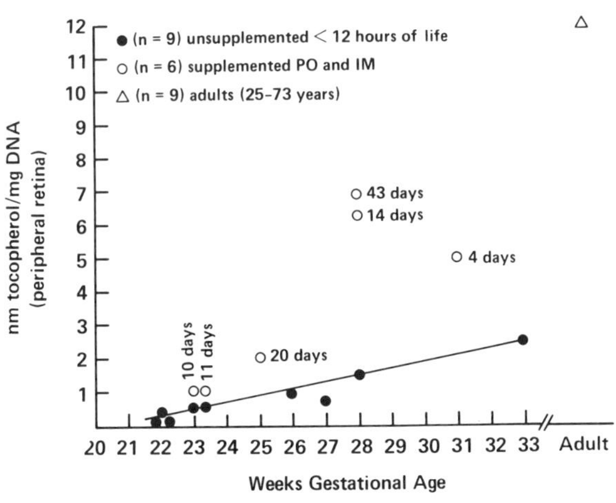

Figure 1. Replotted data of Nielsen et al. (1988) concerning vitamin E (nm tocopherol/mg DNA) in the avascular retina of 15 preterm infants (circles) and the peripheral retina of nine adults (triangles). Vitamin E supplemented preterms (open circles) received 25mg/kg x 4 oral dl-alpha-tocopherol acetate in MCT oil per day and intramuscular injections of 15,10,10 and 10mg/kg of dl-alpha-tocopherol (Ephynal, Hoffman-LaRoche) on days 1,2,3 and 6 after birth.

demonstrated significant antioxidant elevation in their avascular retinas. Such vitamin E retina levels protected spindle cells from oxidative damage. Their spindle cells demonstrated no gap junctions, the spindle cells had no proliferation of rough endoplasmic reticulum within their cytoplasm, and there were no assayable levels of angiogenic factors in extracts of their avascular retinas. These combined biochemical and ultrastructural data show that vitamin E supplementation results in an age dependent uptake of plasma vitamin E into the avascular retina. In those preterms < 27 weeks gestational age, plasma elevation does not result in significant retinal uptake of vitamin E to protect spindle cells. In contrast, in those preterms ≥ 27 weeks gestational age (≥ 1000 grams birth weight), plasma vitamin E elevation to physiologic levels (1.2 - 3.5 mg%) results in elevated antioxidant levels in the avascular retina where the spindle cells reside. When the spindle cells are protected by the antioxidant, severe ROP does not develop clinically.

Lastly, the recent national multicenter trial to appraise the efficacy of cryotherapy to halt the progression of threshold ROP has resulted in a cohort of 4099 preterms who have been analyzed in depth. One of the surprising findings is that Black preterms of matched birth weight, gestational age, and clinical risk factors are at less risk of developing ROP than their Caucasian matched counterparts.[4] The concept of free radical damage to spindle cells as the initial pathologic event of ROP is supported by this clinical reality. Perhaps the increased melanin in the retinal pigment epithelium and the increased melanocytes among the choroidal vessels of highly pigmented persons absorb oxygen free radicals and diminish the probability that spindle cells will be perturbed.

The final solution to the saga of ROP lies in understanding the oxygen free radical attack on spindle cells.

LIGHT DAMAGE

The mature retina provides an ideal environment for the generation of oxygen free radicals and the subsequent formation of lipid peroxides. The levels of long chain PUFA, especially docosahexaenioc acid (DHA, 22:6n-3), are higher in the retina than in any other body tissue.[16] Retinal oxygen consumption is greater than even for the brain and there is a daily bombardment of the retina with light. However, in order for oxidation to occur there must also be an initiator of the reaction. Two features of the retinal environment favor free radical generation. First is the presence of light which, of course, is focused on the retina for the purpose of vision. Second is the presence of a number of molecular photosensitizers which are capable of absorbing light and generating reactive oxygen species. As has been shown for other biological systems, these species can react with polyunsaturated lipids to form lipid hydroperoxides, which in turn can damage the cellular membranes in which they reside and lead to the destruction of the cell.

Since the work of Noell et al.,[17] it has been known that light can damage the retina by a photochemical process. In albino and pigmented rats exposed to constant light, this damage was characterized at early stages by photoreceptor cell disruption seen as disorganization and vesiculation of the outer segment disk membranes.[18-21] Under favorable conditions the retina could recover, but if the damage was severe, the photoreceptors and adjacent RPE were permanently destroyed. In contrast to rodents, the primary site of light damage in primates most often was the RPE which underwent swelling and hypopigmentation.[22] Studies measuring the action spectrum of retinal light damage indicated that more than one, if not several, endogenous photosensitizers may be mediating the damage. Visual pigments are likely candidates, but other chromophores absorbing in the shorter wavelength blue and ultraviolet are undoubtedly involved with exposures to these wavelengths. Potential mediators of short wavelength light damage include cytochromes, flavins, melanin, and retinoids.[23]

Perhaps the most direct evidence to date that oxidative mechanisms play a role in retinal light damage comes from studies that have observed the formation of end products of lipid peroxidation in retinas exposed to constant light. An early study by Kagan et al.[24] demonstrated lipid hydroperoxide formation, indicated by increased 232 nm absorbance (i.e. conjugated diene formation) in extracted lipids following exposure of the frog retina to bright light for 30 minutes. Subsequently, Kagan et al.[25] found that in albino rats exposed to bright light for 210 minutes, lipid hydroperoxide formation was correlated with reductions in electroretinogram amplitude, a measure of retinal function. In albino rats exposed to constant light for 1 and 3 days, Wiegand et al.[26] reported an increase in retinal lipid hydroperoxides and a selective loss of 22:6n-3 from rod outer segment membranes. This finding was latter replicated in pigmented rats and biochemical evidence for lipid peroxidation was associated with damage to outer segment membranes as observed by morphological examination.[27]

Studies manipulating retinal oxygen levels provide indirect evidence for an oxidative mechanism of light damage. In the presence of elevated blood oxygen, the radiant dose required to produce ophthalmoscopically-visible blue light lesions to the monkey retina was reduced from 30 to 11 J/cm^2 (ref 28). A similar effect was observed in cultured bovine RPE cells, where the radiant dose required to cause blue light damage was reduced by a factor of ten in 95 versus 20% oxygen[29]. Because of the enhancement of light damage susceptibility in the presence of elevated oxygen, these studies are consistent with the idea that oxidation may be an important mechanism in retinal damage.

Other experiments have used various means to alter retinal fatty acid composition and then examine the susceptibility of these retinas to light damage. The hypothesis to be tested was that light damage susceptibility would decrease with lower, and increase with higher amounts of long-chain polyunsaturated fatty acids in the retina. Retinas of rats raised in bright cyclic light had lower levels of 22:6n-3 and were less susceptible to acute light stress than animals raised in dim cyclic light. The bright-light reared animals also had higher levels of antioxidant vitamins and enzymes which could have been a factor in the protection against light damage (reviewed in ref. 30). Dietary deprivation of N-3 fatty acids has also been used to lower retinal 22:6n-3 levels in albino rats and this treatment protected against acute structural alterations in photoreceptor cells.[31] Albino rats fed linseed oil which is enriched in N-3 fatty acids were more susceptible to acute and chronic light damage than rats given safflower oil (only N-6 PUFA) or hydrogenated coconut oil (no PUFA).[32] However, when albino rats were fed fish oil, acute photoreceptor cell damage was diminished rather than enhanced.[33] Reme et al.[33] suggested that increased retinal levels of eicosapentaenoic acid (20:5n-3) and arachidonic acid (20:4n-6) provided some protection through the formation of eicosaniods.

Biological tissues, in particular those susceptible to oxidative attack, contain a high concentration of endogenous antioxidants. These molecules are thought to protect against oxidation either by scavenging reactive species or by converting reactive intermediaries into non-reactive products. The high level of antioxidants found in the retina may be indicative of a role in protecting against photo-oxidative damage. Molecules found in high concentration in the retina that are thought to act as antioxidants include vitamins E and C, glutathione, glutathione-coupled enzymes, and superoxide dismutase.[34-36] Perhaps the most widely studied of the endogenous antioxidants is vitamin E. Early studies provided conflicting results concerning the capability of vitamin E to protect against retinal light damage. Kagan et al.[25] reported that albino rats maintained on vitamin E deficient diets had a significantly increased production of lipid hydroperoxides and electroretinogram deficits when exposed to 210 minutes of constant light. In contrast, Stone et al.[37] found that vitamin E deficient albino rats incurred a lesser degree of functional damage (measured by reduction in electroretinogram amplitude) than supplemented controls when exposed to constant light for 12 hours. In fact, the majority of studies have shown that light-induced photoreceptor cell death was not exacerbated in albino rats with vitamin E levels reduced by vitamin E deficiency.[37-39] However, these findings do not necessarily rule out a protective role for vitamin E against photo-oxidation since enhanced activity of other endogenous antioxidants could possibly counterbalance the effects of vitamin E deficiency. Other studies have examined changes in vitamin E levels in response to constant light exposure. Albino rats exposed to bright light constantly for 2 days were found to have lower vitamin E levels in the retina.[40] However, when using lower light intensities to expose pigmented rats to a longer (5 day) period of constant light exposure, Wiegand et al.[27] found an increase in the level of vitamin E in rod outer segments. An increase in vitamin E during constant light exposures of relatively low intensity may reflect the ability of the retina to mobilize this antioxidant to protect against light stress.

A considerable amount of work in recent years has focused on the role of vitamin C as an antioxidant protecting against light damage. In contrast to the vitamin E studies described above, the role of vitamin C has mostly been examined in animals supplemented with this vitamin. Intraperitoneal injection of albino rats with either L-ascorbic acid, sodium ascorbate, or dehydroascorbate significantly increased vitamin C levels in the retina and decreased the amount of photoreceptor cell damage caused by constant or intermittent light.[41-43] Evidence that vitamin C may be acting as an antioxidant comes partly from the finding that light-induced loss of polyunsaturated fatty acids was reduced in retinas with elevated levels of vitamin C.[41]

Glutathione peroxidase (and other associated glutathione enzymes) and superoxide dismutase have been demonstrated to have high activities in the retina and are effective in detoxifying lipid hydroperoxides and superoxide radical, respectively. Contrary to expected results, reduction in glutathione peroxidase activity by dietary deprivation of its required metal selenium, in combination with vitamin E deficiency, did not enhance retinal light damage as measured by electroretinographic evaluation.[37] However, as discussed above for vitamin E, this finding cannot be taken as conclusive evidence that glutathione peroxidase is not involved in protecting against light damage. The potential role of superoxide dismutase in protecting against light-induced retinal damage has not been studied, probably because of the difficulty of specifically altering the activity of this enzyme in the retina.

Although not ordinarily found in measurable amounts in the retina, beta-carotene is a potent antioxidant found in other tissues.[44,45] Following intraperitoneal injection of albino rats with beta-carotene, its level in the retina was found to be 0.4 ug/ml wet tissue, which effectively protected against photoreceptor and RPE cell death following a 24 hour exposure to green-filtered fluorescent light.[46] However, no protection against ultraviolet-A induced photoreceptor cell loss was afforded to pigmented rats similarly treated with beta-carotene injections in which retina levels of this antioxidant were found to be 12.4 ng/retina.[45] Human macula contains the two caroteniods zeaxanthin and lutein[47] which have been suggested to provide antioxidant protection.[48]

Treatments with exogenous compounds thought to protect against oxidation have been examined as a means of ameliorating retinal light damage. Intraperitoneal injection of albino rats with the synthetic antioxidant dimethylthiourea greatly reduced the severity of retinal light damage as indicated by rhodopsin measurements and histological findings.[49,50] Protection against acute damage to rod outer segments, observed 1 hour following a 30 minute exposure to fluorescent light, was afforded by intraperitoneal injections of the radioprotective agent and free radical scavenger WR-77913 (ref. 51).

The pathway by which oxidation leads to retinal damage is not clearly understood. Lipid hydroperoxides eventually combine to form malondialdehyde, a bifunctional compound that reacts with primary amines of lipids and proteins to form cross-links in biological membranes. Extensive crosslinking would undoubtedly affect enzyme activity, which could directly alter the structure and ion permeability of the photoreceptor membranes. Studies using *in vitro* preparations derived from tissues other than retina have shown that light can induce the inactivation of cytochrome oxidase, succinate dehydrogenase, catalase, and lysosomal enzymes.[52-53] Enzyme inactivation has also been demonstrated in light-damaged retina,[54] and very recently, cytochrome oxidase activity in the retina was shown to be inhibited by blue light exposure.[55] A direct link between photo-oxidation and enzyme inactivation is, however, lacking at this point.

AGE-RELATED MACULAR DEGENERATION

The macular degenerations comprise a group of severely debilitating disorders that result in the loss of central vision. Some forms develop in the first decade and are clearly familial, whereas others, such as age-related macular degeneration (ARMD) develop in older individuals and have been attributed primarily to age. However, recent studies have also suggested a familial component in ARMD. ARMD is a significant vision problem for the elderly, as these disorders are the leading causes of blindness in persons over 65 years of age in the United States. Clinically, there are two stages and two distinctively different classes of ARMD: 1) a background stage, which rarely is accompanied with any visual loss and 2) a degenerative stage, which has been classified as atrophic (dry) or neovascular (wet). The latter, which comprises about 10% of ARMD, has the worst visual prognosis.[56]

Several epidemiological studies involving large populations have provided valuable information on the risk factors associated with ARMD. Among these are The Health and Nutrition Examination Survey (HANES),[57,58] Beaver Dam Eye Study,[59] Framingham Eye Study,[60] and Eye Disease Case-Control Study Group (EDC-CSG).[61,62] Although each study looked at different parameters of eye disease, some interesting information regarding the role of lipid peroxidation has emerged. In the HANES study, there was a negative association with prevalence of ARMD and the consumption of vegetables and fruits rich in vitamins C and A.[58] The Eye Disease Case-Control Study Group[61] report of 421 patients with neovascular ARMD and 615 controls showed strong negative association between ARMD and serum carotenoids in both men and women and with post-menopausal estrogens in women. An antioxidant index that combined vitamin C, vitamin E, selenium, and carotenoids showed highly significant negative association with ARMD.[62] In a small clinical trial, Newsome et al.[63] reported that patients with drusen or ARMD given oral zinc for 12 to 24 months had significantly less visual loss than controls. However, the EDC-CSG study[61] did not find a significant correlation between serum zinc levels and ARMD. If light-induced free radical formation is important in ARMD, ocular pigmentation could possibly be a risk factor. While some studies have suggested that blacks have a lower rate of ARMD,[64,65] the HANES study did not find this correlation.[57] Hyman et al.[66] reported that people with blue or medium iris pigmentation were at greater risk for ARMD. However, iris pigmentation was not found to be significant in the EDC-CSG study.[62] Also, the latter study did not find any relation between sunlight exposure and ARMD.

An interesting correlation was found in several studies between cigarette smoking and macular degeneration. Hyman et al.[66] reported increased risk in males. The Beaver Dam Study found increased risk for exudative macular degeneration in both sexes[59] as did the EDC-CSG report.[61] As discussed by Klein et al.,[59] the effects of cigarette smoking could be due to changes in choroidal blood flow dynamics or reduction in antioxidants levels. Chow et al.[67] found significantly lower levels of vitamin C and total carotenes, but not of vitamin A, selenium, and vitamin E in 125 male cigarette smokers compared to 125 age- and race-matched non-smokers. In another study, smokers had lower plasma levels of beta-carotene, but no of vitamin E, compared to non-smokers.[68] Since cigarette smoking generates free radicals,[69] the reduction in plasma antioxidants in smokers may reflect increased antioxidant usage due to oxidant stress. One factor in the increased incidence of macular degeneration and cataracts (discussed later) in smokers could be increased oxidant stress in retina and lens, respectively.

Limited experimental data support a role for lipid peroxidation in macular degeneration. Prasher et al.[70] reported highly significant decreases in superoxide dismutase, glutathione peroxidase, and catalase activities in red blood cells from persons with ARMD compared to age-matched controls. De La Paz and Anderson[71] compared the susceptibility of macular and peripheral punches of human retinas to iron-induced lipid peroxidation. There was an age-related increase in susceptibility of macula, but not of peripheral retina to oxidation.

While epidemiological and experimental results support a role for lipid peroxidation in macular degeneration, no recommendation for antioxidant supplementation has been made by any ophthalmological association.[62,72] The National Eye Institute of the National Institutes of Health is currently conducting a randomized clinical trial to evaluate the role of antioxidants and mineral supplements in the development of macular degeneration.

CATARACTS

Cataracts are an inevitable consequence of the ageing process and are a major cause of blindness in the world. In the United States alone, over a million cataract operations are performed each year at a cost of billions of dollars. Although much effort and many resources have been directed towards understanding the molecular etiology of cataract formation, amazingly little is still known that could lead to development of an effective therapy or treatment.

For some time, evidence has accumulated that oxidative stress leads to the formation of cataracts in ageing human lenses. Over forty years ago, Dische and Zil[73] reported the oxidation of the amino acid cysteine to cystine in lens proteins during the formation of cataracts. Many investigators have reported similar changes that result in the accumulation with age of insoluble material in the lens, particularly in the nucleus. The oxidation of sulfhydryl groups and polymerization of lenticular proteins as a function of age is now well-established. Since the lens operates under reduced oxygen tension and has very low PUFA levels, oxidative processes will out of necessity involve redox reactions of small molecules. Ascorbic acid and glutathione, two small molecules with great reductive power, are present in very high concentrations in the lens (millimolar). In human lens, ascorbic acid levels may be 60 times those in the blood. Although it is quite clear from studies with animal lenses in organ culture that pro-oxidant conditions promote opacification and that anti-oxidant treatment can prevent or reverse these effects,[74] there is no direct evidence that oxidation is causal in development of cataracts in older human lenses. In fact, this has been a most difficult hypothesis to test. However, there is sufficient indirect evidence to make the hypothesis attractive.

The concentration of ascorbic acid[75] and glutathione[76] are lower in cataractous lens than in clear lens. Older lenses have less of these antioxidants than young lens. The activities of some antioxidant enzymes in the lens can decrease with age and with development of cataract.[77] Rathbun et al.[78] recently reported an inverse relationship between the activities of two enzymes involved in glutathione synthesis and the degree of opacification in human subcapsular cataracts. Other groups have made a similar observation.[79,80]

Several studies have linked cataract risk and antioxidant intake. In general, the studies show a reduced risk of cataract in those persons who have taken antioxidants or eaten food high in antioxidants. In a study of 50,828 female registered nurses between 45 and 67 years of age, Hankinson et al.[81] found carotene and vitamin A intake was inversely associated with cataract. In multivariate analysis, neither dietary vitamins E or C were associated with cataract, although the risk for cataract was 45% lower in women who had used vitamin C for 10 years. Knekt et al.[82] examined the blood levels of alpha-tocopherol, beta-carotene, retinol, and selenium in 47 patients with senile cataracts and 94 age-, sex-, and municipality-matched controls and concluded that alpha-tocopherol and beta-carotene were risk factors for end stage senile cataract. A study of 112 subjects, 77 with cataract, concluded that consumption of at least two of three antioxidant vitamins (E, C, or carotenoids) reduced the risk for cataracts, although no correlation was found with superoxide dismutase, glutathione peroxidase, and glucose-6-phosphate dehydrogenase.[83]

The Lens Opacities Case-control Study evaluated risk factors for several types of cataracts in 1380 participants.[84] Pertinent to lipid peroxidation, they found that regular intake of multivitamin supplements, vitamin E, vitamin C, and carotenoids were associated with decreased risk for cataract. In a case-control of 175 cataract patients and an equal number of matched controls, Robertson et al.[85] concluded that the consumption of supplemental vitamins C and E reduce the risk of senile cataracts by st least 50%. The role of nutrients in cataracts was recently reviewed by Taylor.[86]

A strong association between cigarette smoking and cataracts has recently been reported. In a study of 22,071 US male physicians aged 40 to 84 years.[87] compared to persons who had never smoked, current smokers had a significantly increased risk of nuclear sclerosis and posterior subcapsular cataract ($P < 0.001$). A similar study in 50,828 registered nurses found a significant increase for posterior subcapsular cataracts among smokers.[88] This is not surprising, given the relationship discussed above between antioxidants and cataracts, and the several reports that cigarette smokers have lower blood levels of carotenoids and vitamin C.[67,68] Whether the effect is due to reduced blood levels or the direct interaction of tobacco products with lens proteins or antioxidants remains to be determined.

SUMMARY AND CONCLUSIONS

Ames, Shigenaga, and Hagen[89] recently published a thorough review of the relatiohship between oxidants, antioxidants, and degenerative diseases of ageing. They point out that only 9% of Americans daily consume the two fruits and three vegetables recommended by the National Cancer Institute and the National Research Council/National Academy of Science. In addition to antioxidants, these foodstuffs contain many essential micronutrients. To date, specific recommendations for antioxidant supplementation have not been made by any governmental agency or professional association.

A number of clinical, basic, and epidemiological studies have implicated free radical induced lipid peroxidation in various ocular disorders. It would seem prudent that those persons at greatest risk for these disorders take some precautions, which could include sunglasses that filter ultraviolet light; hats that shield the eyes from direct sunlight; and the ingestion of fruits, vegetables, and antioxidants.

ACKNOWLEDGMENTS

The author's research is supported by grants from The National Eye Institute (EY04149, EY00871, EY07001, EY02520, EY04554), The Retina Research Foundation, The RP Foundation Fighting Blindness, and Research to Prevent Blindness, Inc. We thank Ms. Ann Koval for assistance in preparing and editing the manuscript.

REFERENCES

1. G.E. Quinn, V. Dobson, C.C. Barr, B.R. Davis, J.T. Flynn, E.A. Palmer, J. Robertson, and M.T. Trese, Visual acuity in infants after vitrectomy for severe retinopathy of prematurity, *Ophthalmol.* 98:5 (1991).
2. H.A. Mintz-Hittner, T.C. Prager, and F.L. Kretzer, Visual acuity correlates with seveity of retinopathy of prematurity in untreated infants weighing 750 g or less at bith, *Arch. Ophthalmol.* 110:1087 (1992).
3. H.A. Mintz-Hittner, and F.L. Kretzer, Postnatal retinal vascularization in former pretrm infants with retinopathy of prematurity, *Ophthalmol.* 101:548 (1994).
4. E.A. Palmer, J.T. Flynn, R.J. Hardy, D.L. Phelps, C.L. Phillips, D.B. Schaffer, and B. Tun, Incidence and early course of retinopathy of prematurity, *Ophthalmol.* 98:1628 (1991).
5. D.B. Schaffer, E.A. Palmer, D.F. Plotsky **et al.**, Prognostic factors in the natural course of reinopathy of prematurity, *Ophthalmol.* 100:230 (1993).
6. D.L. Phelps, D.R. Brown, B. Tung, G. Cassady, R.E. McClead, D.M. Purohit, and E.A. Paler, 28-day survival rates of 6676 neonates with birth weights of 1250 grams or less, *Pediatrics.* 87:7 (1991).
7. Committee for the Classification of Retinopathy of Prematurity, An international classification of retinopathy of prematurity, *Arch. Ophthalmol.* 102:1130 (1984).
8. Cryotherapy for Retionpathy of Prematurity Cooperative Group, Multicenter trial of cryotherapy for retinopathy of prematurity - preliminary results, *Arch. Ophthalmol.* 106:471 (1988).
9. Cryotherapy for Retionpathy of Prematurity Cooperative Group, Multicenter trial of cryotherapy for retinopathy of prematurity: 3½ year outcome - structure and function., *Arch. Ophthalmol.* 111:339 (1993).
10. F.L. Kretzer, R.S. Mehta, A.T. Johnson, D.G. Hunter, E.S. Brown, and H.M. Hittner, Vitamin E protects against retinopathy of prematurity through action on spindle cells, *Nature.* 309:793 (1984).
11. F.L. Kretzer, and H.M. Hittner, H.M., Retinopathy of prematurity: clinical implications of retinal development, *Arch. of Dis. Child.* 63:1151 (1988).
12. H.M. Hittner, L.B. Godio, A.J. Rudolph, J.M. Adams, J.A. Garcia-Prats, Z. Friedman, J.A. Kautz, and W.A. Monaco, Retrolental fibroplasia: efficacy of vitamin E in a double-blind clinical study of preterm intants, *N. Engl. J. Med.* 305:1365 (1981).
13. N.N. Finer, G. Grant, R.F. Schindler, G.B. Hill, K.L. and Peters, Effect of intramuscular vitamin E on frequency and severity of retrolental fibroplasia; a controlled trial, *Lancet.* 1:1087 (1982).
14. L. Johnson, G.E. Quinn, S. Abbasi, C. Otis, D. Goldstein, L. Sacks, R. Porat, E. Fong, M. Delivoria-Papadopoulos, G. Peckham, D. B. Schaffer, and F. W. Bowen, Effect of sustained pharmacologic vitmin E levels on incidence and severity of retinopathy of prematurity: A controlled clinical trial, *J. Pediatr.* 114:827 (1989).

15. J.C. Nielson, M.I. Naash, and R.E. Anderson, R.E., The regional distribution of vitamins E and C in mature and premature human retinas, *Invest. Ophthalmol Vis. Sci.* 29:22 (1988).
16. S.J. Fliesler and R.E. Anderson, Chemistry and metabolism of lipids in the vertebrate retina. *Prog. Lipid Res.* 22:79 (1983).
17. W.K. Noell, V.S. Walker, B.S. Kang, and S. Berman, Retinal damage by light in rats. *Invest. Ophthalmol. Vis. Sci.* 5:450 (1966).
18. T. Kuwabara, and R.A. Gorn, Retinal damage by visible light: An electron microscope study, Arch. Ophthalmol. 79:69 (1968).
19. A. Grignolo, N. Orzalesi, R. Castellazzo, and P. Vittone, Retinal damage by visible light in albino rats: An electron microscope study, *Ophthalmologica.* 157:43 (1969).
20. W.K. O'steen, C.R. Shear, and K.V. Anderson, Retinal damage after prolonged exposure to visible light: A light and electron microscopic study. *Am. J. A. Nat.* 134:5 (1972).
21. L.M. Rapp, and S.C. Smith, Morphological comparisons between rhodopsin-mediated and short-wavelength classes of retinal light damage, *Invest. Ophthalmol. Vis. Sci.* 33:3367 (1992).
22. W.T. Ham, Jr., J.J. Ruffolo, Jr., H.A. Mueller, A.M. Clarke, and M.E. Moon, Histologic analysis of photochemical lesions produced in rhesus retina by short-wavelength light, *Invest. Ophthalmol. Vis. Sci.* 17:1029 (1978).
23. J.D. Spikes, Photosensitization, *in:* "The Science of Photobiology," K.C. Smith, ed., Plenum Publishing Corp., New York (1977).
24. V.E. Kagan, A.A. Shvedova, K.N. Novikov, and Y.P Kozlov, Light-induced free radical oxidation of membrane lipids in photoreceptors of frog retina, *Biochimica et Biophysica Acta.* 330:76 (1973).
25. V.E. Kagan, I.Y. Kuliev, V.B. Spirichev, A.A. Shvedova, and Y.P. Kozlov, Accumulation of lipid peroxidation products and depression of retinal electrical activity in vitamin E deficient rats exposed to high-intensity light, *Bull. Exp. Biol. Med.* 91:144 (1981).
26. R.D. Wiegand, N.M. Guisto, L.M. Rapp, and R.E. Anderson, Evidence for rod outer segment lipid peroxidation following constant illlumination of the rat retina, *Invest. Ophthalmol. Vis. Sci.* 24:1433 (1983).
27. R.D. Wiegand, C.D. Joel, L.M. Rapp, J.C. Nielsen, M.B. Maude, and R.E. Anderson, Polyunsaturated fatty acids and vitamin E in rat rod outer segments during light damage, *Invest. Ophthalmol. Vis. Sci.* 27:727 (1986).
28. W.T.J. Ham, H.A. Mueller, J.J.J. Ruffolo, J.E. Millen, S.F. Cleary, R.K. Guerry, and D.I. Guerry, Basic mechanisms underlying the production of photochemical lesions in the mammalian retina, *Curr. Eye Res.* 3:165 (1984).
29. R.S. Crockett, and T. Lawhill, Oxygen dependence of damage by 435 nm light in cultured retinal epithelium, *Curr. Eye Res.* 3:209 (1984).
30. J.S. Penn, and R.E. Anderson, Effects of light history on the rat retina, *in:* "Progress in Retinal Research," N. Osborne and G. Chader, ed., Pergamon Press, New York (1992).
31. R.A. Bush, C.E. Reme, and A. Malnoe, Light damage in the rat retina: the effect of dietary deprivation of n-3 fatty acids on acute structural alterations, *Exp. Eye Res.* 53:741 (1991).
32. C.A. Koutz, R.D. Wiegand, H. Chen, and R.E. Anderson., Effect of dietary fat and environmental lighting on the susceptibility of rat photoreceptors to light damage, *Invest. Ophthalmol. Vis. Sci. supp.* 32:1096 (1992).

33. C.E. Reme, A. Malnoe, H.H. Jung, Q. Wei, and K. Munz, Effect of dietary fish oil on acute light-induced photoreceptor damage in the rat retina, *Invest. Ophthalmol. Vis. Sci*. 35:78 (1994).
34. H. Heath, A.C. Rutter, and T.C. Beck, Changes in the ascorbic acid and glutathione content of the retinae and adrenals from alloxandiabetic rats, *Vis. Res*. 2:431 (1962).
35. M. Hall and D. Hall, Superoxide dismutase of bovine and frog rod outer segments. *Biochem. Biophys. Res. Com*. 67:1199 (1975).
36. L.M. Rapp, M.I. Naash, R.D. Weigand, C.D. Joel, J.D. Nielsen, and R.E. Anderson, Morphological and biochemical comparisons between retinal regions having differing susceptibility to photoreceptor degeneration, *in:* "Retinal Degeneration," M.M. LaVail, ed., Alan R. Liss, Inc., New York (1985).
37. W.L. Stone, M.L. Katz, M. Lurie, M.F. Marmor, and E. A. Dratz, Effects of dietary vitamin E and selenium on light damage to the rat retina, *Photochem. Photobiol.* 29:725 (1979).
38. S.M. Sykes, W.G.J. Robinson, and J.G. Bieri, Retinal damage by cyclic light and the effect of vitamin E, *DHHS Publication FDA* 81:8156 (1981).
39. M.L. Katz, and G.E. Eldred, Failure of vitamin E to protect the retina against damage resulting from bright cyclic light exposure, *Invest. Ophthalmol. Vis. Sci*. 30:29 (1989).
40. C.D. Joel, S. Briggs, D. Gall, J. Hannan, M. Kahlow, M. Stein, A. Tarver, and A. Yip, Light causes early loss of retinal tocopherol *in vivo*, *Invest. Ophthalmol. Vis. Sci.* 20:166 (1981).
41. D.T. Organisciak, H.M. Wang, Z.Y. Li, and M.O.M. Tso, The protective efect of ascorbate in retinal light damage of rats. *Invest. Ophthalmol. Vis. Sci*. 26:1580 (1985).
42. Z.Y. Li, M.O.M. Tso, H.M. Wang, and D.T. Organisciak, Amelioration of photic injury in rat retina by ascorbic acid: a histopathologic study, *Invest. Ophthalmol. Vis. Sci.* 26:1589 (1985).
43. D.T. Organisciak, Y.L. Jiang, H.M. Wang, and I. Bicknell, The protective effect of ascorbic acid in retinal light damage of rats exposed to intermittent light, *Invest. Ophthalmol. Vis. Sci.* 31:1195 (1990).
44. G.J. Handleman, E.A. Dratz, C.C. Reay, and F.J. G.M. Van Kuijk, Carotenoids in the human macula and whole retina. *Invest. Ophthalmol. Vis. Sci*. 29:850 (1988).
45. L.M. Rapp, P.L. Fisher, and D.W. Suh, Beta-carotene supplementation and light-induced retinal degeneration. *Invest. Ophthalmol. Vis. Sci. suppl.* 35:1391 (1994).
46. M.O.M. Tso, Experiments on visuals cell by nature and man: In search of treatment for photoreceptor degeneration. *Invest. Ophthalmol. Vis. Sci*. 30:2430 (1989).
47. R.A. Bone, J.T. Landrum, and S.L. Tarsis, Preliminary identirication of the human macular pigment, *Vision Res*. 25:1531 (1985).
48. W. Schalch, Carotenoids in the retina - A review of their possible role to prevent or limit damage caused by light and oxygen, *in:* "Free Radicals and Aging," B.I. Emerit, and B. Chance eds., Birkhäuser Verlag, Basel (1992).
49. S. Lam, M.O.M. Tso, and D.H. Gurne, Amelioration of retinal photic injury in albino rats by dimethylthiourea, *Arch. Ophthalmol*. 108:1751 (1990).
50. D.T. Organisciak, R.M. Darrow, Y.L. Jiang, G.E. Marak, and J.C. Blanks, Protection by dimethylthiourea against retinal light damage in rats, *Invest. Ophthalmo. Vis. Sci*. 33:1599 (1992).

51. C.E. Reme, U.F. Braschler, J. Roberts, and J. Dillon, Light damage in the rat retina: Effect of a radioprotective agent (WR-77913) on acute rod outer segment disk disruptions, *Photochem. Photobiol.* 54:137 (1991).
52. B.B. Aggarwal, A.T. Quintanilha, R. Cammack, and L. Packer, Damage to mitochondrial electron transport and energy coupling by visible light. *Biochimica et Biophysica Acta.* 502:367 (1978).
53. H. Ninnemann, W.L. Butler, and B.L. Epel, Inhibition of respiration and destruction of cytochrome A_3 in mitochondria by light in mitochondria and cytochrome oxidase from beef heart, *Biochem. Biophys. Acta.* 205:507 (1970).
54. H.A. Hansson, A histochemical study of oxidative enzymes in rat retina damaged by visible light, *Exp. Eye Res.* 9:285 (1970).
55. E. Chen, P.G. Soderberg, and B. Lindstrom, Cytochrome oxidase activity in rat retina after exposure to 404 nm blue light, *Curr. Eye Res.* 11:825 (1992).
56. F.L. Ferris, III, S.L. Fine, and L. Hyman, Age-related macular degeneration and blindness due to neovascular maculopathy, *Arch. Ophthalmol.* 102:1640 (1984).
57. B.E. Klein, and R. Klein, Cataracts and macular degeneration in older Americans, *Arch. Ophthalmol.* 100:571 (1982).
58. J. Goldberg, G. Flowerdew, E. Smith, J.A. Brody, and M.O.M. Tso, Factors associated with age-related macular degeneration, *Am. J. Epidemiology.* 128:700 (1988).
59. R. Klein, B.E. Klein, K.L.P. Linton, and D.L. DeMets, The Beaver Dam eye study: the relation of age-related maculopathy to smoking, *Am. J. Epidemiology.* 137:190 (1993).
60. H.A. Kahn, H.M. Leibowitz, J.P. Ganley, M.M. Kini, T. Colton, R.S. Nickerson, and T.R. Dawber, The Framingham Eye Study II. Association of ophthalmic pathology with single variables previously measured in the Framingham Heart Study, *Am. J. Epidemiology.* 106:33 (1977).
61. Eye Disease Case-Control Study Group, Risk factors for neovascular age-related macular degeneration, *Arch. Ophthalmol.* 110:1701 (1992).
62. Eye Disease Case-Control Study Group, Antioxidant status and neovascular age-related macular degeneration, *Arch. Ophthalmol.* 111:104 (1993).
63. D.A. Newsom, M. Swartz, N.C. Leone, R.C. Elston, and E. Miller, *Arch. Ophthalmol.* 106:192 (1988).
64. L. C. Chambley, Impressions of eye diseases among Rhodesian blacks in Marshonland, *S. Africa Med. J.* 52:316 (1977).
65. I. Mann. "Culture, Race, Climate and Eye Disease," Charles C. Thomas, Springfield, IL (1966).
66. L.G. Hyman, A.M. Lillienfeld, F.L. Ferris, and S.L. Fine, Senile macular degeneration: a case-control study, *Am. J. Epidemiology.* 118:213 (1983).
67. C.K. Chow, R.R. Thacker, C. Changchit, R.B. Bridges, S.R. Rehm, J. Humble, and J. Turbek, Lower levels of vitamin C and carotenes in plasma of cigarette smokers, *J. Am. Coll. Nutr.* 5:305 (1986).
68. W.S. Stryker, L.A. Kaplan, E.A. Stein, M.J. Stampfer, A. Sober, and W.C. Willett, The relation of diet, cigarette smoking, and alcohol consumption to plasma beta-carotene and alpha-tocopherol levels, *Am. J. Epidemiology.* 127:283 (1988).
69. W.A. Pryor, K.I. Terauchi, and W.H. Davies, Jr., Electron spin resonance (ESR) study of cigarette smoke by use of spin trapping techniques, *Environ. Health Perspect.* 16:161 (1976).

70. S. Prashar, S.S. Pandav, A. Gupta, and R. Nath, R., Antioxidant enzymes in RBCs as a biological index of age related macular degeneration, *Acta Ophthalmol.* 71:214 (1993).
71. M. De La Paz and R.E. Anderson, Region and age-dependent variation in susceptibility of the human retina to lipid peroxidation. *Invest. Ophthalmol. Vis. Sci.* 33:3497 (1992).
72. R.D. Sperduto, F.L. Ferris, and N. Kurinij, Do we have a nutritional treatment for age-related cataract or macular degeneration, *Arch. Ophthalmol.* 108:1403 (1990).
73. Z. Dische and H. Zil, Studies on the oxidation of cysteine to cystine in lens proteins during cataract formation, *Am. J. Ophthalmol.* 34:104 (1951).
74. A. Spector, G.M. Wang, R.R. Wang, W.H. Garner, and H. Moll, The prevention of cataract by oxidative stress in cultured rat lenses. I. H_2O_2 and photochemically induced cataract, *Cur. Eye Res.* 12:163 (1993).
75. M. Wilczek and H. Zygulska-Machowa, Zawartosc witaminy C W. roznych typack zaem, *J. Klin. Oczna* 38:477 (1968).
76. V.N. Reddy, Glutathione and its function in the lens—an overview, *Exp. Eye Res.* 150:771 (1990).
77. E.R. Berman, "Biochemistry of the Eye," Plenum Press, New York (1991).
78. W.B. Rathbun, A.J. Schmidt, and A.M. Holleschau, Activity loss of glutathione synthesis enzymes associated with human subcapsular cataract, *Invest. Ophthalmol. Vis. Sci.* 34:2049 (1993).
79. P.Y. Xie, A. Kanai, A. Nakajima, S. Kitahara, A. Ohtsu, and K. Fujii, Glutathione and glutathione-related enzymes in human cataractous lenses, *Ophthalmol. Res.* 23:133 (1991).
80. H. Pau, P. Graf, and H. Sies, Glutathione levels in human lens: regional distribution in different forms of cataract, *Exp. Eye Res.* 50:17 (1990).
81. S.E. Hankinson, M.J. Stampfer, J.M. Seddon, G.A. Colditz, B. Rosner, F.E. Speizer, and W.C. Willett, Nutrient intake and cataract extraction in women: a prospective study, *Brit. Med. J.* 305:335 (1992).
82. P. Knekt, M. Heliövaara, A. Rissanen, A. Aromaa, R.K. Aaran, Serum antioxidant vitamins and risk of cataract, *BMJ.* 305:1392 (1992).
83. P.F. Jacques, L.T. Chylack Jr., R.B. McGandy, and S.C. Hartz, Antioxidant status in persons with and without senile cataract, *Arch. Ophthalmol.* 106:337 (1988).
84. M.C. Leske, L.T. Chylack Jr., and S.Y. Wu, The lens opacities case-control study, risk factors for cataract, *Arch. Ophthalmol.* 109:244 (1991).
85. J.M.C.D. Robertson, A.P. Donner, and J.R. Trevithick, Vitamin E intake and riske of cataracts in humans, *Ann. N.Y. Acad. Sci.* 570:372 (1989).
86. A. Taylor, Role of nutrients in delaying cataracts, *Ann. N.Y. Acad. Sci.* 669:111 (1992).
87. W.G. Christen, J.E. Manson, J.M. Seddon, R.J. Glynn, J.E. Buring, B. Rosner, and C.H. Hennekens, A prospective study of cigarette smiking and risk of cataract in men, *JAMA*. 268:989.
88. S.E. Hankinson, W.C. Willett, G.A. Colditz, J.M. Seddon, B. Rosner, F.E. Speizer, and M.J. Stampfer, A prospective study of cigarette smoking and risk of cataract surgery in women, *JAMA*. 268:994 (1992).
89. B.N. Ames, M.K. Shigenaga, T.M. Hagen, Oxidants, antioxidants, and the degenraive diseases of aging, *Proc. Natl. Acad. Sci.* 90:7915 (1993).

ACTIVE OXYGEN MECHANISMS OF UV INFLAMMATION

Alice P. Pentland

Division of Dermatology, Washington University
School of Medicine
St. Louis, Missouri 63110

SUMMARY

Active oxygen radicals are important in the pathogenesis of UV irradiation injury. The initiating mechanisms involve the generation of hydroxyl radicals, superoxide, and organic hydroperoxides due to photochemical reactions. These active oxygen species lead to DNA strand breakage, mutation and the generation of inflammatory mediators such as cytokines and arachidonic acid metabolites which amplify the irradiation-induced inflammation. Several compounds have recently been utilized to successfully decrease these effects. Improved understanding of the mechanisms by which active oxygen species induce injury in skin now promises improved treatment.

INTRODUCTION

The sun emits electromagnetic radiation over a broad range, with wavelengths between 290 nm and 4000 nm reaching the earths surface through the atmosphere[1]. The portion of this emission spectrum which is most important in the production of human disease are the shorter, ultraviolet wavelengths. Exposure of the skin to a sufficient dose of 200 nm to 400 nm wavelength light produces characteristic inflammation commonly known as sunburn[2-4]. Distinct patterns of inflammation are produced by exposure to specific wavelengths of light, with shorter wavelengths being the most potent. Because different patterns of inflammation are produced by these wavelengths, investigators generally divide the ultraviolet radiation into 3 main groups (Figure 1). The shorter wavelengths (200-290 nm) are called UVC, and are used for germicidal radiation. These wavelengths do not penetrate the ozone layer in the atmosphere, and so are not responsible for human disease. Wavelengths between 290-320 nm are termed mid-UV or UVB radiation. Although UVB wavelengths comprise $< 0.1\%$ of the solar spectrum, they are responsible for UV irradiation-induced inflammation and carcinogenesis, particularly squamous cell and basal cell carcinoma[5,6]. Those wavelengths between 320 and 400 nm are designated UVA or long-wave UV[1,7]. The difference in the patterns of inflammation

Free Radicals in Diagnostic Medicine, Edited by
D. Armstrong, Plenum Press, New York, 1994

produced by 290 - 320 nm wavelengths and those produced by wavelengths from 320 - 400 nm wavelengths and those produced by wavelengths from 320 - 400 nm are probably due to the specific photochemistry which these wavelengths initiate, resulting in different injury produced by 290 - 320 mechanisms. Because of the importance of UVB radiation to human pathophysiology, the active oxygen-mediated mechanisms which are implicated in UVB-induced inflammation will be discussed.

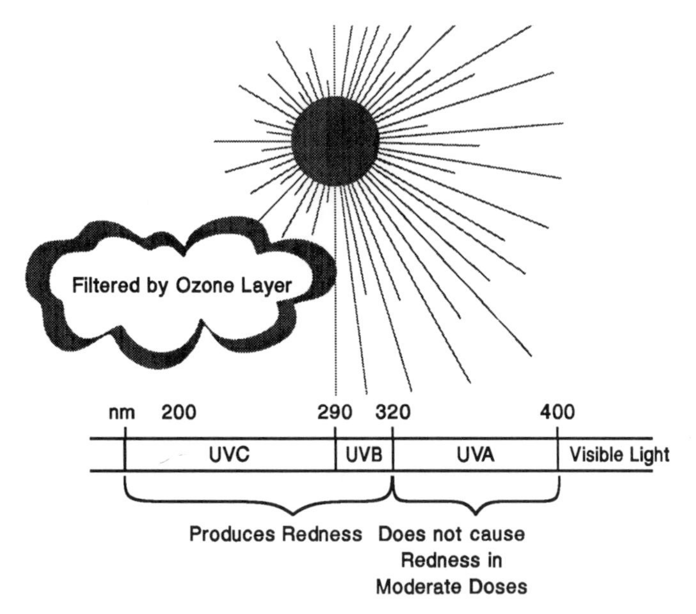

Figure 1. Wavelengths of light in the ultraviolet spectrum of the sun. Wavelengths less than 290 nm are filtered by the ozone layer, but are erythemogenic (UVC). Wavelengths between 290 nm and 320 nm are termed UVB, and are primarily responsible for sun-induced skin cancers. Wavelengths longer than 320 nm are only erythemogenic in large doses.

ACTIVE OXYGEN SPECIES PARTICIPATE IN UV INFLAMMATION

The physiologic response to excess UVB exposure is familiar to all. Acute exposure results in erythema, heat, edema, pain and pruritus, followed later by tanning and epidermal thickening[8,9].The clinical pattern of erythema produced after UVB exposure generally begins 3-5 hours after exposure, is maximal between 12-24 hours and is generally gone by 72 hours. This time course may be manipulated by the exposure dose; small doses produce short-lived erythema, while larger doses produce erythema which is faster in onset, more intense and persistent[9].

The specific chemistry responsible for UV-induced inflammation is unclear, but

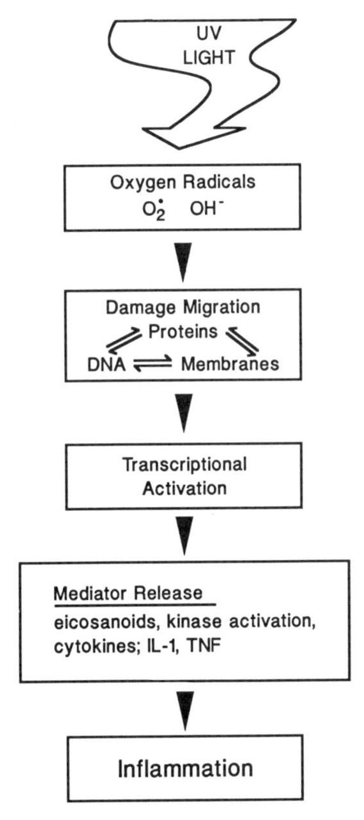

Figure 2. Putative pathway by which UVB-induced inflammation occurs.

must result from photochemical reactions in the skin initiated by the absorption of incident light. The chromophores which absorb the light are not fully defined, but UV wavelengths less than 300 nm are thought to act on nucleic acids, amino acids, urocanic acid, melanin and water as the major chromophores[10]. Subsequent to the absorption of UV light, it is clear that active oxygen intermediates are formed[11-13]. Recent observations have allowed a putative pathway mediating UV inflammation to be constructed (Figure 2). At very early time points after UV exposure, the production of superoxide anion and hydroxyl radicals in irradiated tissue has been documented by several investigators[11,12,14]. Because of their extreme chemical reactivity, these oxygen radicals can disrupt cellular processes by a number of mechanisms, including lipid peroxidation, cleavage of DNA, altered enzyme activity, polymerization of polysaccharides and cell death[13,15,16]. In combination, these reactions are likely to contribute to the clinical pattern seen as UVB inflammation.

Many UV effects on DNA are mediated by light absorption by DNA. Such absorption by nucleic acids leads to formation of cis, syn-cyclobutane pyrimidine dimers and pyrimidine-pyrimidine (6-4) dimers, including the Dewar valance isomer of the latter compound. The 6-4 Dewar photoproduct is highly mutagenic and likely to underlie UV carcinogenesis[17,18]. However, in addition to these photoproducts, singlet oxygen causes alteration of guanine bases, forming several polar photooxidation products. These have been characterized as desoxyribosyl-cyanuric acid and desoxyribosyl-4,8-dihydro-4-hydroxy-8-oxoguanine. The desoxyguanosine derivatives are formed by the addition of oxygen across the 4,5 or 1,4 double bond of the purine ring[19]. Such alterations to DNA, either direct photoproducts or due to free radicals, are known to induce activation of genes important in the response to UV[16,19,20]. However, recent work addressing the relative importance of such DNA damage versus active oxygen-mediated effects on membrane or cytoplasmic signalling in the UV inflammatory response suggests that early events may be primarily mediated by active oxygen species interacting with membrane based signalling systems, while later effects (including carcinogenesis) may be more related to DNA damage and repair[21,22].

Among the initial effects of UV exposure, lipid peroxidation induced by active oxygen species is likely to play an important role. This process has been linked circumstantially to increased formation of lipid mediators important in UV inflammation. UV irradiation can produce lipid peroxidation when superoxide anion and hydrogen peroxide form hydroxyl radicals in the presence of iron[15,23-25]. The hydroxyl radicals can then react with polyunsaturated fatty acids to abstract hydrogen, forming lipid hydroperoxyl radicals which have half-lives of seconds, and are thus capable of diffusing significant distances before detoxification. Propagation and diffusion allow a peroxidative chain reaction to spread through the membrane, generating new radical species as the reaction proceeds and greatly amplifying the damage produced[15,23,26]. In addition to the demonstration of hydroxyl radicals and superoxide anion in irradiated epidermis, evidence showing that the natural antioxidants glutathione, ubiquinone and vitamin E are drastically depleted supports the role of oxygen radical species and resultant lipid peroxidation in mediating UV injury[27]. It seems likely that UVB-induced inflammation occurs when the antioxidant capacity of epidermis is exceeded.

After the immediate events during which UV light interacts with epidermal chromophores to produce active oxygen species and the resultant damage to membranes, proteins and DNA occurs, transcriptional activation is induced (Figure 2). Whether it is direct damage to DNA or the effects of oxygen radicals which predominate in producing this activation is not clear[16,20-22]. The production of many of these gene products is probably directed at downregulating cellular pathways which are potentially injurious while the cell is under oxidative stress, as well as repairing UV-induced damage. This period of transcriptional activation, also termed the UV response, is thought to be related to the bacterial SOS response, and begins within minutes of exposure to light. Several of the gene products induced appear very rapidly, including c-fos, c-jun, jun-B, c-myc, v-src, Ha-ras, and c-raf[16,20]. Heme oxygenase is also likely to be induced, since it has been documented to be increased after exposure to UVA light. Presumably, the induction of heme oxygenase occurs to the decrease iron available for production of hydroxyl radical, and attenuate hydroxyl radical-mediated damage[28,29]. At somewhat later times, synthesis of IL-1 and TNF-alpha are also induced[30-32]. The synthesis of these cytokines is likely to be responsible for many of the later inflammatory events, including increased prostaglandin synthesis at time points 12-24 hours after irradiation[33]. Maintenance of inflammation at later time points may also be related to the synthesis of nitric oxide[34]. Increased blood flow in UVB irradiated rats has been linked to nitric oxide production, which may also contribute to increased PGE_2 synthesis by irradiated cells[35].

MECHANISMS OF INCREASED EICOSANOID SYNTHESIS AFTER UV INJURY

To understand how the initial photochemistry produced by UV results in inflammation, it is useful to examine UV effects on cellular synthesis of arachidonic acid metabolites. Prostaglandins and lipoxygenase products, termed eicosanoids, are synthesized from arachidonic acid present in membrane phospholipid[36]. Their synthesis is markedly induced by UV exposure, and they play an important role as mediators of the delayed erythema which occurs after UVB irradiation. As evidence for this, the synthesis of eicosanoids occurs in parallel with the onset of erythema, and inhibitors of eicosanoid formation decrease UVB inflammation[37-40]. The effects of UVB irradiation on eicosanoid synthesis in skin have been measured in several different model systems: suction blister fluid exudates, animals and cells in culture[38-47]. In all of these instances, arachidonic acid metabolism is enhanced. Prostaglandins E_2, D_2, $F_{2\alpha}$ and 12-hydroxyeicosatetraenoic acid are all increased in suction blister aspirates shortly after irradiation by UVB, and remain increased for 48 hours, with peak concentrations occurring approximately 24 hours after exposure[38,40-42]. The prominence of this aspect of UV inflammation has prompted several investigators to undertake detailed study of the underlying stimulatory mechanisms to provide insight into the underlying mechanisms by which the immediate free radical-mediated mechanisms produce the later changes observed in UV inflammation.

UV Irradiation Increases Phosopholipase A_2 Activity by Oxidant-Mediated Activation of the Epidermal Growth Factor Receptor

The key event in the increased formation of eicosanoids after UVB irradiation is likely to be an increase in the release of free arachidonic acid from membrane phospholipids at time points less than 12 hours. Studies of cultured cells support the contribution of enhanced phospholipase activity[41,44-46]. In human epidermal cultures, release of arachidonic acid increases dose-dependently after UVB exposure. This change is confirmed by associated increases in glycerophosphorylcholine release from membrane phosphatidylcholine[44]. Recent work suggests this increase in phospholipase activity is likely to occur through oxidant stress-mediated activation of the EGF receptor at early time points[47]. Type IV high molecular weight cytosolic phospholipases ($cPLA_2$) have recently been described, and are currently thought to be the class of phospholipase responsible for hormone-mediated prostaglandin synthesis[48-52]. In studies examining mechanisms of $cPLA_2$ activation, phosphorylation of $cPLA_2$ was shown to increase its activity 2-6 fold[53,54]. EGF receptor can stimulate $cPLA_2$ phosphorylation through the action of MAP kinase[53]. EGF receptor activity is sensitive to the redox status of the cell, and the receptor phosphorylated when cells are exposed to oxidants such as tert-butyl hydrogen peroxide[47]. UV irradiation produces a similar phosphorylation of the EGF receptor, an effect which is blocked by use of the anti-oxidant N-acetylcysteine[47]. Collectively, this evidence suggests that the pathway by which oxidant stress stimulates prostaglandin synthesis is as follows: UV produces oxygen radicals, which result in phosphorylation and activation of the epidermal growth factor receptor. The receptor's intrinsic tyrosine kinase activity can then phosphorylate and activate downstream MAP kinase resulting in activation of $cPLA_2$, and producing increased synthesis of PGE_2 (Fig.3)[47]. The initial activation of EGFR by oxidation may be subsequent to (although not necessarily caused directly by) the action v-src or Ha-ras tyrosine kinases which are membrane associated[21]. Src tyrosine kinases have been linked to the activation of MAP kinase, thus strengthening their potential role in initiating the UV response[55-57]. In addition, activation of these kinases has been documented in UVC-irradiated HELA cell cultures[21]. Although UVC is filtered out by the earths ozone layer and thus is not responsible for human disease, mechanisms producing inflammation due to UVC wavelengths (200-290) appear similar in many instances to those found in UVB irradiation, and are also thought to occur via active oxygen species[6,7,9].

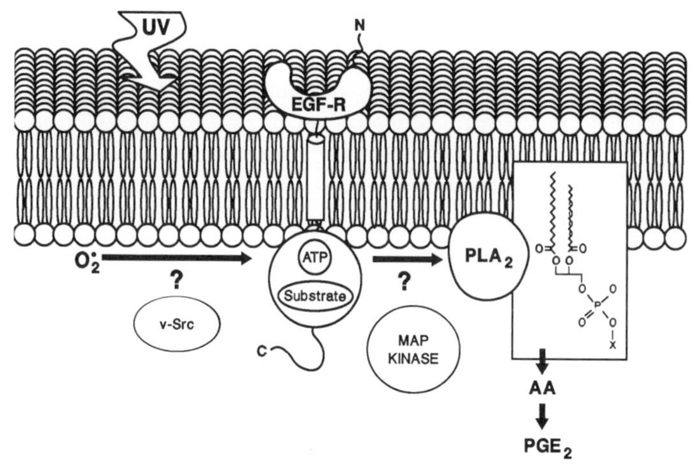

Figure 3. Proposed mechanism by which UV exposure increases the synthesis of prostaglandins. Oxidative stress may activate membrane based tyrosine kinases such as v-src, which phosphorylate the epidermal growth factor receptor. This subsequently results in the activation of MAP kinase, which phosphorylates and activates $cPLA_2$, resulting in increased PGE_2 synthesis.

In addition to effects on the state of phosphorylation of $cPLA_2$, UVB irradiation also increases $cPLA_2$ synthesis. In studies done using normal adult human keratinocytes, synthesis of $cPLA_2$ was found to increase 3-5 fold 3 to 9 hours after UV exposure. This increase in synthesis is accompanied by increases in the quantities of $cPLA_2$ detected by immunohistochemical stains of irradiated human skin[58]. The increase in $cPLA_2$ may be due to increases in TNF produced by UV irradiation, as TNF has been documented to increase the synthesis of $cPLA_2$ in other systems[59].

EPIDERMAL ANTIOXIDANTS OFFER PHARMACOLOGIC STRATEGIES FOR PROTECTION FROM UV INJURY

Epidermis covers the surface of the body, adjacent to the oxygen-rich atmosphere. This location dictates the need for well developed defenses against oxygen free radical damage. Vitamin E is the only lipid, membrane based antioxidant, and as such is the only defense against UV-mediated lipid peroxidation[59,60]. Peroxidation of lipid results in increased activation of phospholipase[61]. The documented increases in $cPLA_2$ activation occurring in irradiated cells suggest that vitamin E or its more water-soluble analogues could help attenuate free radical-induced lipid peroxidation produced by UV irradiation. 2,2,5,7,8- Pentamethyl-6-hydroxychromane (PMC), an analogue of vitamin E in which the phyto side chain is replaced by a methyl group. PMC has been studied for its capacity to decrease prostaglandin synthesis. It is capable of profoundly suppressing synthesis of prostaglandin by decreasing the activity of phospholipase[62]. The capacity of PMC to suppress phospholipase activity was evident even in simple vesicle preparations, suggesting that the mechanism by which activity was decreased was through its ability to decrease vesicle peroxidation or through direct interaction with the enzyme. Although the mechanism remains unclear, analogues of vitamin E or other membrane-based antioxidants may be useful in providing protection against UV damage.

Glutathione is a cysteine-containing tripeptide present in high concentrations in cells which serves an important role as an endogenous antioxidant[64]. In addition to functioning as an antioxidant, it is a cofactor for the synthesis of prostaglandins by cyclooxygenase. Glutathione helps preserve the normal reduced cellular environment by being oxidized itself in irradiation injury[65-67]. The importance of glutathione in attenuating UV injury has been demonstrated by its ability to decrease the cytotoxic effects of UVB coupled with the observation that cellular glutathione content is greatly depleted by ultraviolet light exposure[65-67]. Furthermore evidence of the utility of enhancing cellular reducing capacity to decrease irradiation-induced inflammation is provided by studies utilizing N-acetylcysteine (NAC). NAC is readily taken up by cells then rapidly converted to glutathione[68]. Pretreatment of Hela cells with NAC prior to irradiation by UVC prevented induction of c-jun mRNA, and prevented the activation of c-src[21]. NAC also eliminates increased prostaglandin synthesis by irradiated cells, an effect mediated by decreased activation of $cPLA_2$[47]. Agents which can provide reducing capacity without toxicity are therefore likely to be helpful in irradiation injury.

While much remains to be discovered about the mechanisms by which ultraviolet light produces inflammation, it is clear that active oxygen species play a major role. A better understanding of these processes should improve the treatment of irradiation injury, and may potentially decrease the carcinogenic potential of recurrent UV exposure.

References

1. L.C. Harber and D.R. Bickers, and A. Lamola, Principles of light absorption and photochemistry, *in:* Photosensitivity Diseases. W.B. Saunders Co.; Second ed., Philadelphia (1989).
2. R.S. Cotran and M.A. Pathak, The pattern of vascular leakage induced by monochromatic UV irradiation in rats, guinea pigs and hairless mice, *J Invest Dermatol.* 51:155-164 (1968).
3. G. Logan and D.L. Wilhelm, Vascular permeability changes in inflammation: I. The role of endogenous permeability factors in ultraviolet injury. *Br J Exp Pathol.* 47:300-314 (1966).
4. N.A. Soter, Acute effects of ultraviolet radiation on the skin. *Semin Dermatol.* 9:11-5 (1990).

5. D.E. Brash, J.A. Rudolph, J.A. Simon, A. Lin, G.J. McKenna, H.P. Baden, A.J. Halperin and J Ponten, A role for sunlight in skin cancer: UV-induced p53 mutations in squamous cell carcinoma. *Proc Natl Acad Sci.* 88:10124-10128 (1991).
6. D.R. Bickers and L.C. Harber, Non-melanonia skin cancer and melanomas *in:* Photosensitivity Diseases. Saunders Co., Second ed., Philadelphia: W.B. (1989).
7. I.E. Kochevar, M.A. Pathak and J.A. Parrish, Photophysics, photochemistry, and photobiology. *in:* Dermatology in General Medicine. T.B. Fitzpatrick, A.Z. Eisen, K. Wolff, I.M. Freedberg and K.F. Austen eds., Fourth ed. McGraw-Hill, New York (1993).
8. A. Bachem, Time factors of erythema and pigmentation produced by ultraviolet rays of difference wavelengths. *J Invest Dermatol.* 25:215-218 (1955).
9. J.L.M. Hawk and J.A. Parrish, Responses of normal skin to ultraviolet radiation. *in:* Photoimmunology. J.A. Parrish, M.L. Kriple and W.L. Morrison eds., Plenum Medical Book, New York (1983).
10. R.R. Anderson and J.A. Parrish, The optics of human skin. *J Invest Dermatol.* 77:13-19 (1981).
11. M.A. Pathak and K. Stratton, Free radicals in human skin before and after exposure to light. *Arch Derm and Biophysics.* 123:468-476 (1968).
12. E. Pelle, D. Maes, G.A. Padulo, E-K Kim, and W.P. Smith, An in vitro model to test relative antioxidant potential: Ultraviolet-induced lipid peroxidation in liposomes. *Arch of Biochem and Biophys.* 283:234-240 (1990).
13. I. Fridovich, The biology of oxygen radicals. *Science* 201:875-880 (1978).
14. R. Dixit, H. Mukhtar and D.R. Bickers, Studies on the role of reactive oxygen species in mediating lipid peroxide formation in epidermal microsomes of rat skin. *J Invest Dermatol.* 81:369-375 (1983).
15. A. Petkau, Protection and repair of irradiated membranes, *in:* Free Radicals, Aging, and Degenerative Disease, Alan R. Liss, Inc. (1986).
16. N.J. Holbrook and A.J. Fornace, Jr., Response to adversity: molecular control of gene activation following genotoxic stress. *New Biol.* 3:825-833 (1991).
17. J.E. LeClerc, A. Borden and C.W. Lawrence, The thymine-thymine pyrimidine-pyrimidine (6-4) ultraviolet light photoproduct is highly mutagenic and specifically induces 3'thymine-to-cytosine transitions in *Escherichia coli. Proc Natl Acad Sci. USA* 88:9685-9689 (1991).
18. D. L. Svoboda, C.A. Smith, J-S A. Taylor and A. Sancar, Effect of sequence, adduct type, and opposing lesions on the binding and repair of ultraviolet photodamage by DNA photolyase and (A)BC excinuclease. *J Biol Chem.* 268:10694-10700 (1993).
19. J. Piette, M. Paule, M. Louis and J. Decuyper, Damages induced in nucleic acids by photosensitization. *Photochem and Photobiol.* 44:793-802 (1986).
20. Z.A. Ronai, M.E. Lambert and I.B. Weinstein, Inducible cellular responses to ultraviolet light irradiation and other mediators of DNA damage in mammalian cells. *Cell Biol and Toxicology.* 6:105-126 (1990).
21. Y. Devary, R.A. Gottlieb, T. Smeal and M. Karin, The mammalian ultraviolet response is triggered by activation of src tyrosine kinases. *Cell.* 71:1081-1091.
22. Y. Devary, C. Rosette, J.A., DiDonato and M. Karin, NF-kB activation by ultraviolet light not dependent on a nuclear signal. *Science.* 261:1442-1445 (1993).
23. Y.A. Vladimirov, Free radical lipid peroxidation in biomembranes: mechanism, regulation, and biological consequences, *in:* Free Radicals, Aging, and Degenerative Diseases, Plenum Press, New York (1986).

24. E.D. Wills, Mechanisms of lipid peroxide formation in animal tissues. *Biochem J.* 99:667-676, (1966).
25. M.K. Logani and R.E. Davies, Lipid oxidation: biologic effects and antioxidants -- A review. *Lipid.* 15:485-495, (1980).
26. J. Nishi, R. Ogura, M. Sugiyama, T. Hidaka and M. Kohno, Involvement of active oxygen in lipid peroxide radical reaction of epidermal homogenate following ultraviolet light exposure. *J Invest Dermatol.* 97:115-119 (1990).
27. J. Fuchs, M.E. Huflejt, A.B. Rothfuss, A.B. Wilson, G. Carcamo and L. Packer, Impairment of enzymic and nonenzymic antioxidants in skin by UVB irradiation. *J Invest Dermatol.* 93:769-773 (1989).
28. S.M. Keyse and R. M. Tyrrell, Induction of the heme oxygenase gene in human skin fibroblasts by hydrogen peroxide and UVA (365 nm) radiation: evidence for the involvement of the hydroxyl radical. *Carcinogenesis.* 11:787-791 (1990).
29. D. Lautier, P. Luscher and R.M. Tyrrell, Endogenous glutathione levels modulated both constitutive and UVA radiation/hydrogen peroxide inducible expression of the human heme oxygenase gene. *Carcinogenesis.* 13:227-232 (1992).
30. A.P. Pentland and M.G. Mahoney, Keratinocyte prostaglandin synthesis is enhanced by IL-1. *J Invest Dermatol.* 94:43-46 (1990).
31. T.S. Kupper, A.O. Chua, P. Flood, J. McGuire and U. Gubler, Interleukin 1 gene expression in cultured human keratinocytes is augmented by ultraviolet irradiation. *J Clin Invest.* 80:430-436 (1987).
32. A. Kock, T. Schwarz, R. Kirnbauer, A. Urbanski, P. Perry, J.C. Ansel and T.A. Luger, Human keratinocytes are a source for tumor necrosis factor α: evidence for synthesis and release upon stimulation with endotoxin or ultraviolet light. *J Exper Med.* 172:1609-1614 (1990).
33. M. Grewe, U. Trefzer, A. Ballhorn, K. Gyufko, H. Henninger and J. Krutmann, Analysis of the mechanism of ultraviolet (UV) B radiation-induced prostaglandin E_2 synthesis by human epidermoid carcinoma cells. *J Invest Derm.* 4:528-531 (1993).
34. J.B. Warren, R.K. Loi and M.L. Coughlan, Involvement of nitric oxide synthase in the delayed vasodilator response to ultraviolet light irradiation of rat skin in vivo. *Br. J Pharmacol.* 109:802-806 (1993).
35. D. Salvemini, T.P. Misko, J.L. Masferrer, K. Seibert, M.G. Currie and P. Needleman, Nitric oxide activates cyclooxygenase enzymes. *Proc Natl Acad Sci.* 90:7240-7244 (1993).
36. W.L. Smith, Prostanoid biosynthesis and mechanisms of action. *Am J Phys.* 11:F181-F191 (1992).
37. W.S. Miller, F.R. Ruderman, and J.G. Smith Jr., Aspirin and ultraviolet light-induced erythema in man. *Arch Dermatol.* 95:357-358 (1967).
38. A.K. Black, N. Fincham, M.W. Greaves, and C.N. Hensby, Time course changes in levels of arachidonic acid and prostaglandins D_2 E_2 F_2 in human skin following ultraviolet irradiation. *Br J Clin Pharmacol.* 10:453-457 (1980).
39. D.S. Snyder and W.H. Eaglstein, Intradermal anti-prostaglandin agents and sunburn. *J Invest Dermatol.* 62:47-50 (1974).
40. A.K. Black, M.W. Greaves, C.N. Hensby and N.A. Plummer, Increased prostaglandins E_2 and $F_{2\alpha}$ in human skin at 6 and 24 h after ultraviolet B irradiation (290-320 nm). *Br J Clin Pharmacol.* 5:431-436 (1978).
41. V.A. DeLeo, H. Horlick, D. Hanson, M. Eisinger and L.C. Harber, Ultraviolet radiation stimulates the release of arachidonic acid from mammalian cells in culture. *Photchem Photobiol.* 41:51-56 (1985).

42. A.P. Pentland, M. Mahoney, S.C. Jacobs and M.J. Holtzman, Enhanced prostaglandin synthesis after ultraviolet injury is mediated by endogenous histamine stimulation. *J Clin Invest.* 86:566-574 (1990).
43. B.A. Gilchrest, N.A. Soter, J.S. Stoff and M.C. Mihm, The human sunburn reaction: histologic and biochemical studies. *J Am Acad Dermatol.* 5:411-422 (1981).
44. V. DeLeo, S. Scheide, J. Meshulam, D. Hanson and A. Cardullo, Ultraviolet radiation alters choline phospholipid metabolism in human keratinocytes. *J Invest Dermatol.* 91:303-308 (1988).
45. A.P. Pentland and S.C. Jacobs, Bradykinin-induced prostaglandin synthesis is enhanced in keratinocytes and fibroblasts by UV injury. *Am J Physiol.* 261:R543-R547 (1991).
46. C.H. Kang-Rotondo, C.C. Miller, A.R. Morrison and A.P. Pentland. Enhanced keratinocyte prostaglandin synthesis after UV injury is due to increased phospholipase activity. *Am J Physiol.* 264:C396-C401 (1993).
47. C.C. Miller, P. Hale and A.P. Pentland, Ultraviolet B injury increases prostaglandin synthesis through a tyrosine kinase-dependent pathway. Evidence for UVB-induced epidermal growth factor receptor activation. *J Biol Chem.* 269:3529-3533 (1994).
48. W.L. Smith, Prostanoid biosynthesis and mechanisms of action. *Am J Physiol.* 263:F181-F191 (1992).
49. J.D. Clark, L.L. Lin, R.W. Kris, C.S. Ramesha, L.A. Sutzman, X.L. Lin, N. Milona, and J.L. Knopf, A novel arachidonic acid-selective cytosolic phospholipase A_2 contains a Ca^{++}-dependent translocation domain with homology to PKC and GAP. *Cell.* 65:1043-1051 (1991).
50. J.H. Gronich, J.V. Bonventre and R.A. Nemenoff, Purification of a high-molecular mass form of phospholipase A_2 from rat kidney is activated at physiological Ca^{++} concentrations. *Biochem J.* 271:37-43 (1990).
51. R.J. Mayer and L.A. Marshall, New insights on mammalian phospholipase A_2(s); comparison of arachidonoyl-selective and nonselective enzymes. *FASEB J.* 7:339-348 (1993).
52. L.L. Lin, A.Y. Lin and J.L. Knopf, Cytosolic phospholipase A_2 is coupled to hormonally regulated release of arachidonic acid. *Proc Natl Acad Sci. USA* 89:6146-6151 (1992).
53. L.L. Lin, M. Wartman, A.Y. Lin, J.L. Knopf, A. Seth, and R. J. Davies, $cPLA_2$ is phosphorylated and activated by MAP kinase. *Cell.* 72:69-278 (1993).
54. R.A. Nemenoff, S. Winitz, N.X. Quian, V. Van Putten, G.L. Johnson and L.E. Heasley, Phosphorylation and activation of a high molecular weight form of phospholipase A_2 by p42 microtubule-associated protein 2 kinase and protein kinase C. *J Biol Chem.* 268:1640-1663 (1993).
55. B.J. Pulverer, J.M. Kyriakis, J. Avruch, E. Nikolakaki and J.R. Woodgett, Phosphorylation of c-jun mediated by MAP kinases. *Nature.* 353:670-674 (1991).
56. S.A. Moodie, B.M. Willumsen, M.J. Weber and A. Wolfman, Complexes of Ras GTP with Raf-1 and mitogen-activated protein kinase. *Science.* 260:1658-1661 (1993).
57. R. Muller, D. Mumberg and F.C. Lucibello, Signals and genes in the control of cell cycle progression. *Biochem et Biophys Acta.* 1155:151-179 (1993).
58. A. Gresham, J. Masferrar, A. Morrison and A.P. Pentland, Cytosolic phospholipase A_2 synthesis is enhanced in human keratinocytes by acute UVB irradiation. *J Invest Dermatol.* 100:595A (1993).

59. W.G. Hoeck, C.S. Ramesha, D.J. Chang, N. Fan and R.A. Heller, Cytoplasmic phospholipase A_2 activity and gene expression are stimulated by tumor necrosis factor: Dexamtheansone blocks the induced synthesis. *Proc Natl Acad Sci. USA* 90:4475-4479 (1993).
60. V. Kagan, E. Witt, R. Goldman, G. Scita and L. Packer, Ultraviolet light-induced generation of vitamin E radicals and their recycling. A possible photosensitizing effect of vitamin E in skin. *Free Rad Res Comm.* 16:51-64 (1992).
61. H.J. Forman and A.B. Fisher, Antioxidant defenses. *in:* Gilbert DL eds., Oxygen and Living Processes-An interdisciplinary Approach. Springer-Verlag, New York, 1981.
62. D.A. Gamache, A.A. Fawzy and R.C. Franson, Preferential hydrolysis of peroxidized phospholipid by lysosomal phospholipase C. *Biochim Biophys Acta.* 958:116-124 (1988).
63. A.P. Pentland, A.R. Morrison, S.C. Jacobs, L. Hruza, J.S. Hebert and L. Packer, Tocopherol analogs suppress arachidonic acid metabolism via phospholipase inhibition. *J. Biol Chem.* 267:15578-15584 (1992).
64. A. Meister and M.E. Anderson, Glutathione. *Ann Rev Biochem.* 52:711-760 (1983).
65. R.M. Tyrell and M. Pidoux, Endogenous glutathione protects human skin fibroblasts against the cytotoxic action of UVB, UVA and near-visible radiations. *Photochem Photobiol.* 44:561-564 (1986).
66. M.J. Connor and L.A. Wheeler, Depletion of cutaneous glutathione by ultraviolet radiation. *Photochem Photobiol.* 46:239-245 (1987).
67. K. Hanada, R.W. Gange and M.J. Connor, Effect of glutathione depletion on sunburn cell formation in the hairless mouse. *J Invest Dermatol.* 96:838-840 (1991).
68. C.A. Rice-Evans and A. T. Diplock, Current status of antioxidant therapy. *Free Rad Biol and Med.* 15:77-96 (1993).

OXYGEN-FREE RADICALS AT MYOCARDIAL LEVEL: EFFECTS OF ISCHAEMIA AND REPERFUSION

Roberto Ferrari

Chair of Cardiology, University of Brescia, Italy
and Fondazione Clinica del Lavoro di Pavia
Centro di Fisiopatologia Cardiovascolare "S. Maugeri"
Gussago (Brescia), Italy

INTRODUCTION

In the past several years, much interest has arisen over the involvement of free radical metabolism in the biochemical events associated with ischaemia and reperfusion injury to the heart. A decade ago, it would have been difficult to propose a single plausible mechanism to implicate oxygen in injury since it was generally held that the determinants of the degree of damage after a coronary occlusion were simply the oxygen supply vs the oxygen demand of the heart. However, it has been recently recognized that the myocardium cannot recover without the restitution of adequate coronary flow, or in other words: without being reperfused[1]. However, reperfusion may result in numerous negative consequences and it has been shown that reperfusion of heart muscle after > 60 min of ischaemia is associated with release of enzymes, transient rise of diastolic pressure, persistent reduction of contractility, influx of calcium, alteration of mitochondrial function, disruption of cell membranes, and eventual necrosis of at least a proportion of tissue[2-8].

This phenomenon has been called reperfusion damage since much of the damage is believed to be the consequence of events occurring at the moment of reperfusion rather than a result of biochemical changes during the period of ischaemia. Many hypotheses have been put forward to account for the events on reperfusion and have been recently reviewed[1,5,9-11]. They are by no means exhaustive.

A rather popular hypothesis takes into consideration alteration of the permeability of the cell membrane due to the effect of lipases or of oxygen or other radicals, generated at the moment of reperfusion. This hypothesis has gained interest because reperfusion damage can be mimicked by peroxides, and several antioxidants have been proven to be protective against ischaemic and reperfusion damage[12-15]. It follows that the concept of an oxygen-free radicals-mediated cardiotoxicity has important clinical implications in situations of myocardial ischaemia followed by reperfusion.

Free Radicals in Diagnostic Medicine, Edited by
D. Armstrong, Plenum Press, New York, 1994

Interventions such as streptokinase, tissue plasminogen activator, and percutaneous transluminal angioplasty are commonly used to re-establish coronary flow in patients with myocardial infarction. In addition, ischaemia and reperfusion sequences occur in patients with vasospastic angina or during coronary angioplasty or cardiopulmonary bypass. However, the relative importance of free radical production to irreversible injury of the animal, and particularly of the human heart is far from clear. Equally unclear, and certainly contradictory, is the value of therapeutic approaches for limiting the free-radical mediated component of reperfusion injury.

Thus, several controversies still exist regarding the role of oxygen radicals in the setting of myocardial ischaemia and reperfusion. The objective of this chapter is to summarize the current knowledge which we accumulated in our laboratories and the evidence in favour of a role for oxygen-free radicals in the pathogenesis of myocardial reperfusion injury. Some of the data presented here have already been published[16-21]. We will not examine the biochemistry of oxygen-free radicals which has been extensively reviewed in other chapters of this book.

SOURCES OF OXYGEN RADICALS DURING ISCHAEMIA AND REPERFUSION OF THE HEART

By definition, myocardial ischaemia is a condition that exists when oxygen delivery to the myocardium is not sufficient to meet the needs for mitochondrial oxidation [22]. However, the fact that during ischaemia there is a lack of oxygen availability does not necessarily mean that oxygen-free radicals cannot be formed. On the contrary, the metabolic disarrangements that occur during ischaemia might predispose the formation of free radicals from the residual molecular oxygen.

There are many potential sources of free radicals in the myocardium. This obviously makes it difficult to determine which is the most important site of production. This problem is even more complex because oxygen-derived free radicals may be produced from different sources after different durations of ischaemia and reperfusion, such that, upon reperfusion after a certain period of ischaemia, only one or two sources may be involved; however, if reperfusion is further delayed, different other sources could become more important.

Altered Mitochondrial Electron-Transport System

It is well accepted that during ischaemia the adenine nucleotide pool is partially degraded, thus leaving the mitochondrial carriers in a more fully reduced state [23]. This condition will result in a higher increase of electron leakage from the respiratory chain that, in turn, will react with the residual molecular oxygen entrapped within the inner mitochondrial membrane, thus leading to the formation of superoxide radicals. Most likely during the early phase of ischaemia, the increased oxygen-free radical production from the mitochondria is neutralized by the normal functioning of the defense mechanisms against oxygen toxicity, i.e. mainly the mitochondrial SOD. Increasing the duration or the severity of ischaemia leads to a progressive decline of mitochondrial SOD [16,18], leaving the mitochondria less equipped to deal with the increased flux of radicals.

Re-introduction of oxygen with reperfusion will re-energise the mitochondria, but electron egress through cytochrome oxidase will be reduced because of the lack of ADP. As a result, the percentage of electron leakage will further increase and they will have ample availability of molecular oxygen to react. Interestingly, after prolonged periods of ischaemia, reperfusion does not restore mitochondrial SOD activity [18].

If ischaemia and reperfusion lead to an increase in mitochondrial radical production with a decrease in radical scavenging capability, it is possible to suppose a self-induced progression of mitochondrial injury with further deterioration of normal electron flow causing production of oxygen-free radicals. If the above postulated sequence of events is correct, then it should be possible to measure an increased production of oxygen-free radicals of mitochondria isolated from ischaemic and reperfused myocardium. Surprisingly, only few studies have been undertaken in this direction; they have shown that an increased production of reduced oxygen intermediates after ischaemia and particularly after reperfusion, from sub-mitochondrial particles [24,25].

Otani et al[26] studied the effects of oxygen on ATP production by mitochondria isolated from globally ischaemic rabbit heart. Incubation of the mitochondria with SOD and catalase enhanced ATP production and suppressed the appearance of hydroxyl radical signal detected with electron spin resonance. Guarnieri et al[27] found that in vitro formation of free radicals (generated by xanthine xanthine-oxidase system) resulted in an increase of malondialdehyde formation and reduced mitochondrial respiration, whereas Ceconi et al[28] showed that iron-induced lipid peroxidation of myocardial mitochondria results in impaired calcium transporting capacities and oxidative phosphorylation: effects that could be inhibited by addition of SOD and catalase.

Xanthine Oxidase

Saugstad and Aasen[29] recognized that the increase of plasma hypoxanthine during hypoxia and/or ischaemia could, during re-oxygenation and/or reperfusion, support the production of superoxide by xanthine oxidase. At the same time, the drug allopurinol, an inhibitor of xanthine dehydrogenase and oxidase, has been shown to exert protective effects against ischaemia and reperfusion[30] or hemorrhagic shock[31]. For this reason, xanthine oxidase was a prime candidate for the production of oxygen-free radicals[32].

The enzyme is not localized in the myocytes, but rather in the vascular endothelial cells. During ischaemia, xanthine dehydrogenase is converted to the oxidase form. At the same time, ATP is degraded to hypoxanthine which accumulates in ischaemic tissue. On reperfusion, with the readmission of large quantities of molecular oxygen in the presence of high concentrations of hypoxanthine which is the other substrate for xanthine oxidase, there may be a burst of O^2 production. Following this hypothesis, many investigators have provided evidence that inactivation of xanthine oxidase, either with allopurinol or tungsten, results in a protection of the ischaemic and reperfused myocardium, suggesting that generation of superoxide by xanthine oxidase may play an important pathogenetic role in reperfusion injury[33,34,35].

However, although this conclusion, is very challenging it is not completely justified. First of all, there is the possibility that allopurinol may reduce the severity of myocardial ischaemic injury by alternative mechanisms to an inhibition of xanthine oxidase activity[36]. Secondly, not all the studies are positive. Using very similar experimental models, some authors provided entirely negative results[37-40], suggesting that a careful analysis of the interpretation of the positive results and of the experimental condition is warranted. Additionly, as reported earlier, the distribution of xanthine oxidase among tissues and species varies widely; the rabbit, pig, and human myocardium have essentially no activity and yet these species are not immune to reperfusion injury. Finally, it is surprising to realize that one of the first studies[32] suggesting that production of superoxide by xanthine oxidase might be a problem for re-oxygenated myocardium was carried out in pigs which later were shown to have no xanthine oxidase activity.

Activated Neutrophils

The other potential extracellular sources of oxygen-free radicals are the activated neutrophils. They possess a membrane-bound NADPH oxidase as an important part of their antibacterial armament. This enzyme produces superoxide as a broa- spectrum antibiotic that aids in the killing of any micro-organism engulfed by these cells. The superoxide released by metabolically activated neutrophils amplifies the inflammatory response by activation of a latent chemo-attractant present in extracellular fluids. The presence of this chemotactic factor would cause neutrophils to attach to the endothelium and to enter into the interstices. These activated neutrophils would begin to injure the tissue further by release of additional oxidative enzymes (myeloperoxidase) and hydrolytic enzymes (elastase, collagenase, cathepsins. hyaluronidase). In addition, the phenomenon of neutrophils plugging capillary beds will further reduce the circulation of blood, thereby exacerbating ischaemia. The increased capillary permeability due to the superoxide anions will create edema and increase interstitial pressure impairing the local circulation of blood.

Whether or not neutrophils represent a major source of free radicals during ischaemia and reperfusion depends critically on how soon they arrive in the tissue. It has been shown that after 24 h in infarcted myocardium, the neutrophil content is increased by 17-fold[41]. However, 3 h after the onset of ischaemia, the neutrophils are only slightly increased and appear to be marginated in the veins and not in tissue[42]. Finally, Engler et al.[43] found that no neutrophil had entered the tissue even as late as 5 h after the onset of ischaemia; this is a period in which most of the tissues had become irreversibly injured.

Mullane et al[44] documented the attraction of neutrophils toward a developing infarct in dogs and showed that anti-inflammatory compounds could reduce infarct size after occlusion of a coronary artery. Several groups have attempted to prevent myocardial damage by adding SOD, with or without catalase, to the perfusate on different timing from the onset of ischaemia so as to destroy radicals as soon as they are formed. However, the results are still controversial[45].

Metabolism of Arachidonic Acid

Arachidonic acid is released and subsequently metabolized to prostaglandins and leukotrienes during ischaemia. These metabolic pathways involve electron transfers that can initiate the formation of free radicals. Kontos et al[46] found that application of arachidonate to the brain resulted in vascular endothelial lesions that were similar to those seen in ischaemic, traumatic, or hypertensive injuries: these lesions were reduced by SOD, a catalase, or indomethacin added simultaneously with arachidonate. In addition, it has been suggested that the ischaemia and reperfusion-induced calcium overload would activate phospholipases[47], that in turn may degrade cell membrane phospholipids, releasing arachidonic acid.

Studies with isolated myocytes[48] have shown that lipid peroxidation can be activated by increasing calcium concentration. Furthermore, even the phenomenon of "calcium paradox" has been associated with an enhanced production of free radicals[49], although the occurrence of oxidative damage during the event of the calcium paradox has been called into doubt[50]. However, at present, the role of arachidonic acid metabolism as a source of free radicals in the setting of myocardial ischaemia and reperfusion is still not completely understood.

Auto-Oxidation of Catecholamines

The auto-oxidation of catecholamines could provide oxygen-free radicals through the formation of adrenochrome[51]. Vitamin E has been shown to protect against isoprenaline-induced myocardial damage, whereas depletion of the vitamin exacerbates the damage[52]. However, as in the case of arachidonic acid metabolism, the precise role of catecholamines in the production of oxygen-free radicals is, at the moment, still not known.

MYOCARDIAL ANTIOXIDANT MECHANISMS

The aerobic myocardium is able to handle and survive the continuous production of oxygen-free radicals because of the existence of a delicate balance between cellular systems that generate the various oxidants and those that maintain the antioxidant defense mechanism. In the heart, these defense mechanisms include the enzymes SOD, catalase, and glutathione peroxidase together with other endogenous antioxidants such as vitamin E, ascorbic acid, and cysteine.

The primary mechanism responsible for clearance of superoxide anions is the SOD. This enzyme, which was isolated in 1969 by McCord and Fridovich[53], catalyzes the dismutation of superoxide anions to H_2O_2 and O_2. The reaction can also proceed spontaneously, but SOD is able to increase the rate of intracellular dismutation by a factor of 10^9.

Two enzyme systems are important in the metabolism of H_2O_2 produced by the univalent reduction of superoxide anion[54,55]. The first is catalase, an enzyme mainly present in cytosol and which catalyzes the reduction of H_2O_2 to water. Catalase, however, is present only at very low concentrations in the myocardium, whereas the second enzyme, glutathione peroxidase (a selenium-dependent enzyme) is present in significant concentration in the cytosol of the heart. Through glucose-6-phosphate oxidation, the hexose monophosphate shunt produces the reducing equivalents (NADPH) for the action of glutathione reductase. Reduced glutathione (GSH) is then used by GSH peroxidase to form oxidized glutathione (GSSG), but it is also in dynamic equilibrium with all cellular sulphydryl groups. In fact, glutathione-mixed disulfide with proteins constitutes an important part of total cellular glutathione pool and the entire equilibrium is regulated by thiol transferases.

There are several indications proving that glutathione plays an important role in myocardial metabolism[56]. In the heart, glutathione is predominantly intracellular in concentrations of 1.1 µmol/L. More than 95% of cardiac glutathione is in the form of GSH, the GSH/GSSG ratio of aerobic myocardium being over 50[57]. Among other functions, glutathione is a key factor in the detoxification of electrophilic metabolites and reactive oxygen intermediates. As the determinant of the sulphydryl/disulfide ratio[58], glutathione modulates the activity of a number of enzymes and is also involved in the transport of amino acids across cell membranes[57]. Furthermore, as a co-substrate of glutathione peroxidase, GSH plays an essential protective role against oxygen-free radicals and prevents peroxidation of membrane lipids, the activity of SOD in the heart being nearly four time less than in liver and catalase activity being extremely low[57]. This protective mechanism results in an increased formation of intracellular GSSG. It follows that the changes of glutathione status provide important information in the cellular oxidative events and tissue accumulation and/or release of GSSG in the coronary effluent is a sensitive and accurate index of oxidative stress[56-59].

In addition, an antioxidant that has been long known in biological systems is vitamin E. It has been identified in significant concentrations in both myocardial cytosolic and mitochondrial membranes[36-64]. In vitro studies have shown vitamin E to function as a free-radical scavenger which protects heart membrane from lipid peroxidation by free radicals[63,34]. It functions synergistically with ascorbic acid (vitamin C), which can react with vitamin E radicals to regenerate vitamin E. Vitamin C radicals, in turn, can be reduced by NADH reductase.

In light of its lipophilic nature, vitamin E is likely to serve as antioxidant within membranes, whereas vitamin C serves as a water-soluble electron-transport system in the cytosol or in extracellular fluid. It should be emphasized, however, that although there is sufficient in vitro and in vivo evidence to support vitamin E as an important antioxidant, a protective role for this compound in physiological concentration in humans has not been well documented.

ISCHAEMIA AND REPERFUSION INJURY

With this information as background, we shall now consider the potential role of the myocardial oxygen-free radicals in the pathogenesis of ischaemic and reperfusion injury.

The subject is quite complex, as there is still a fair amount of debate concerning the existence of additive, reflow-induced damage. It is, in fact, known that reperfusion is essential for the mechanical recovery of the ischaemic myocardium since without reperfusion, no recovery is possible at all. There is experimental and clinical evidence that early coronary reperfusion reduces acute myocardial infarct size and mortality[65,66] and, with the exception of rhythm distrubances, has no negative effects. With increasing duration of ischaemia prior to reperfusion, the degree of recovery is less pronounced and, in many conditions, there is no recovery at all with a further exacerbation of the biochemical, mechanical and ultrastructural disarrangements induced by ischaemia[67].

The existence of reperfusion damage has, however, been questioned, and it has been argued that, with the exception of induction of arrhythmias, it is difficult to believe that reperfusion causes further injury[68].

Recently, the concept has been promoted that reperfusion injury is a syndrome that includes four distinct forms: reperfusion-induced arrhythmias, myocardial stunning, induction of lethal injury in tissue that was potentially viable before reperfusion, and accelerated necrosis[68]. These conditions should however, be kept separate as they describe two opposite outcomes of ischaemia. The first two forms of reperfusion injury refer to a fully reversible and transient abnormality with complete, although delayed, recovery in contractility. The others describe an almost irreversible condition that results in further depression of contractility.

Recently, it has been shown that interventions given at the time of reperfusion result in substantial improvement in the extent of cardiac function and in the quality of cardiac metabolism. These interventions are aimed at reducing calcium entry and damage induced by oxygen-free radicals. The finding that the outcome of postischaemic reperfusion can be manipulated by interventions given concurrent with, or even after reperfusion, at least suggests that reperfusion itself may be deleterious. In addition, reperfusion damage is often, if not always, associated with the occurrence of oxidative stress, suggesting that re-introduction of oxygen may also be deleterious for the heart[25,56,57].

However, not all studies assessing myocardial infarct size in occlusion reperfusion models in which agents against oxygen radicals were given, have been positive[69,70].

Thus several controversies exist regarding the role of oxygen-free radicals in mediating myocardial damage during ischaema and, particularly, during reperfusion.

To date, the majority of evidence of toxic effects of oxygen-free radicals is indirect and derived from studies in which substances known to eliminate or reduce these toxic species have resulted in evidence of less severe myocardial injury in the setting of ischaemia and reperfusion. However, we cannot ignore that some of these studies have been negative, that oxygen-radical scavengers might have complementary, pharmacological or hemodynamic effects, other than that of reducing oxygen toxicity. Furthermore, it is not clear if their positive effects are permanent or merely a delay in the development of necrosis.

There are, then, several fundamental questions which must be addressed before accepting the thesis that oxygen radicals are important mediators of ischaemia and reperfusion injury. First, what evidence is there that they are produced in ischaemic and reperfusion injury above the neutralizing capacity of the myocardium, and what are the potential sources of production? What are the effects of ischaemia and reperfusion on the defence mechanisms against oxygen-free radicals? In addition, what are the critical intracellular targets leading to irreversible myocardial injury and how might this process be interrupted?

DEMONSTRATION OF OXYGEN-FREE RADICAL PRODUCTION DURING ISCHAEMIA AND REPERFUSION

Direct Detection

Direct measurements of free-radical species has been limited primarily by the instability of these oxygen metabolites. Essentially one technique has been used: electron paramagnetic resonance (EPR) spectroscopy, a system employed for many years to identify and characterize free radicals in simple chemical systems with or without the use of spin-trap agents. Nevertheless, conflicting results have been obtained. Zweier et al[71] identified a spectral signal similar to superoxide oxygen-centered free radicals that increased significantly during 10 min of ischaemia and even more significantly during the first 10 s of reperfusion. On the other hand, using the same EPR technique in a very similar model of ischaemia and reperfusion, Luber and associates[72] failed to confirm the presence of such oxygen-derived free radical species and doubted the importance of free radicals generating during postischaemic reperfusion.

In addition to the report of Zweier et al[71], many other authors using different spin-trap agents have reported that superoxide or its derivative radicals can be demonstrated in the reperfused isolated hearts[73,74] or in vivo models even 3 h after reperfusion[75]. On the other hand, electron spin resonance spectra reported by other investigators did not allow the same conclusion[76,77]. Additionally, Nakazawa et al.[78], performing experiments with either isolated perfused rat and rabbit hearts, or open-chest canine hearts subjected to ischaemia and reperfusion, have pointed out that electron-spin resonance results need to be analyzed with caution since artifactual radicals are misleading problems common to this method. In particular, the superoxide and nitrogen-centered radicals commonly detected have been shown to be artifactually produced by pulverization of the frozen samples. Interestingly, these authors showed that the radicals native to the myocardium were identified as derived from coenzyme Q_{10}, suggesting that the mitochondria might be an important site of production for these radicals. Therefore, the studies employing EPR technique seem to suggest that free radicals are produced in significant amount during aerobic reperfusion.

Indirect Detection

Specific tissue damage of lipids caused by free radicals may be easier to measure. It is known that free radical activity and the extent of tissue damage are related to the amount of lipid peroxidation. Malondialdehyde (MDA) formation is a final product of lipid peroxidation and can be used to indirectly prove occurrence of damage caused by oxygen-free radicals.

Several methods have been developed to quantify MDA formation in vitro and in vivo. In the past, most of the detection methods were based on the reaction of MDA with thiobarbituric acid (TBA), in which one molecule of MDA reacts with two molecules of TBA to form a stable red chromogen. The MDA-TBA adduct was measured by spectrophotometry or fluorometry. Several positive results have been obtained by employing this method[78,79,80]. However, the TBA-test is often criticized due to its lack of specificity in complex biological systems because: 1) TBA can react with various other compounds, and 2) the strong acidic or heating conditions might induce artificial MDA formation during determination[81].

Lipid peroxidation does not necessarily prove that peroxidation itself is a primary mechanism of damage. Lipid peroxidation may be a secondary phenomenon in tissue already damaged by other means[82]. In addition, when the more accurate high pressure liquid chromatography (HPLC) technology employed for MDA measurements was involved, no sign of lipid peroxidation was detected either after early or late reperfusion of the ischaemic hearts[83].

More precise evidence of production of oxygen-free radical comes from experiments in which the occurrence of oxidative stress was measured. Oxidative stress is a condition in which oxidant metabolites can exert their toxic effects because of increased production or of altered cellular mechanism of protection[58,84].

In previous studies, we demonstrated that in the isolated and perfused rabbit hearts, ischaemia induces a progressive reduction of tissue content of reduced GSH with a concomitant decline of the GSH-GSSG ratio[85,86]. After a short period of ischaemia, reperfusion normalizes the myocardial glutathione status, whereas after a prolonged period of ischaemia, reperfusion results in a release of GSH from the myocardium with a further depauperation of the tissue content[58]. At the same time, an important accumulation and release of GSSG occurs causing a further reduction of GSH-GSSG ratio and a shift of cellular thiol redox state toward oxidation. Thus, during reperfusion after prolonged ischaemia, the glutathione system which represents an important mechanism of defense against oxygen toxicity, is under stress. The reduced availability of cellular GSH becomes a rate-limiting factor for detoxification of oxygen metabolites, which are most likely hydrogen peroxide and lipid hydroperoxides.

Interestingly, the occurrence of an oxidative stress has been demonstrated also in patients with coronary artery disease subjected to different periods of global ischaemia followed by reperfusion during coronary artery bypass grafting[87,88]. Oxidative stress was measured by determining the arterial and coronary sinus difference of GSH and GSSG. When the period of clamping ischaemia was contained within 30 min, reperfusion resulted in only a small and transient rise of GSH and GSSG in the coronary sinus which probably represents a wash-out process. Reperfusion reinstated after 60 min of ischaemia led to important releases of GSH and GSSG in the coronary sinus which was still continuing at the end of the procedure, suggesting the occurrence of oxidative damage. Furthermore, there was an inverse correlation of hemodynamic function after surgery[87,88,89]. It is relevant to recall here that these studies, together with that of Ferreira et al[89], are the only evidence that oxygen-free radicals might be involved in reperfusion damage in humans.

In another study in CAD patients subjected to coronary artery bypass, we also showed that neutrophils do accumulate within the myocardium but are not activated during early reperfusion when oxidative stress is already manifested[90].

This suggest that the likely source of oxygen-free radicals during the early phases (first minutes) of reperfusion in man, is the mitochondria. Neutrophils probably play an additional role in the later phases of reperfusion.

Further indirect evidence of the role of oxygen-free radicals in myocardial ischaemia and reperfusion comes from all the positive results obtained with antioxidants to protect the myocardium against these two conditions. Several experimental models have been used and several different end points have been employed. We cannot review all the literature here, but it is fair to says that, in general, antioxidant and agents able to donate SH group to the myocardium have been found protective, particularly when administered prior or at the time of reperfusion. When the same agents are given 1 minute after reperfusion they have little effect. This very short time lapse probably explains the controversial data. In addition, it is possible that antioxidant act through different mechanisms, and more importantly there are, at the present, no clinical data to prove their effectiveness, although several multicentric trials have been started or are in preparation.

CONCLUSIONS

In this chapter, the results, conclusions and criticisms from different experiments have been brought together to consider the possibility that oxygen-derived free radicals are generated from the myocardium during ischaemia and reperfusion above the neutralizing capacity of the myocytes.

This possibility is by no means proven. There is, however, evidence that prolonged ischaemia reduces the defence mechanism against oxygen toxicity, thus making the myocardium more vulnerable to the damaging effects of oxygen-free radicals. On the other hand, there is also evidence, albeit preliminary, that the formation of these toxic species is enhanced after ischaemia and, particularly, after reperfusion. At the moment, it is not possible to discriminate which radicals are produced under these circumstances and from which sources they come. It seems that the presence of blood is not a prerequisite for the formation of oxygen-free radicals, and that mitochondria are likely to be an important source of production.

Finally, several cotnroversies exist regarding the meaning of the studies in which agents known to eliminate or reduce oxygen-free radical species have provided protection against ischaemia and repefusion. The pathophysiological conclusion derived from such studies should be taken into account with some caution.

Acknowledgements

This work was supported by the Italian Research Council (C.N.R.) targeted project "Prevention and Control Disease Factors" and by the C.N.R. target Project on Biotechnology and Bioinstrumentation. We thank Dr. Bill Dotsoon Smith for his editing of the manuscript.

REFERENCES

1. J.M. Turner and A. Boveris, Gencration of superoxide anion by NADH dehydrogenasc of bovine heart mitochondria, *Biochem J* 1291: 421 (1980).
2. R. Ferrari, S. Bongrani and F. Cucchini, Effects of molecular oxygen and calcium on heart metabolism during reperfusion, *in*: M.E. Bertrand ed., *Coronary Arterial Spasm*, 46 (1982).
3. N. Nohl, The biochemical mechanism of the formation of reactive oxygen species in heart mitochondria, in: C.M. Caldarera and P. Harris, eds., "Advances in Studies on Heart Metabolism," Cooperativa Libraria Universitaria Editrice, Bologna, 413 (1982).
4. H. Otani, H. Tanaka, T Inove et al., In vitro studies on contribution of oxidative metabolism of isolated rabbit heart mitochondria to myocardial reperfusion injury, *Circ Res* 55:168 (1984).
5. R.W. Egan, J. Paxton and F.A. Kuehl Jr., Mechanisms for the irreversible self deactivation of prostaglandin synthetase, *J Biol Chem* 251: 7329 (1976).
6. D.E. Charnbers, D.A. Parks, G. Patterson, et al., Xanthine oxidase as a source of free radical in myocardial ischaemia, *J Mol Cell Cardiol* 17:145 (1985).
7. S.W. Werns, M.J Shea, S.E. Mitsos, et al., Reduction of the size of infarction by allopurinol in the ischaemic reperfused canine heart, *Circulation* 73: 518 (1986).
8. S. Akizuki, S. Yoshida, D.E. Chambers, et al., Infarct size limitation by the xanthine oxidase inhibitor, allopurinol, in closed chest dogs with small infarcts, *Cardiovasc Res* 19: 686 (1985).
9. W.L. Arnold, R.H. de Wall, P. Keydi, et al., The effect of allopurinol on the degree of early myocardial ischaemia, *Am Heart J* 99: 614 (1985).
10. J.R. Parratt and C.L. Wainwright, Failure of allopurinol and a spin-trapping agent N-t-alpha-phenyl nitrone to modify significantly ischaemia and reperfusion-induced arrhythmias, *Br J Pharmacol* 91: 49 (1987).
11. J.P. Kehrer, H. Piper and H. Sies, Xanthine oxidase is not responsible for reoxygenation injury in isolated perfused rat heart, *Free Rad Res Commun* 3: 69 (1987).
12. U.A.S. Al-Khalidi and T.H. Chaglassian, The species distribution of xanthine oxidase, *Biochem J* 97: 318 (1965).
13. J.M. Downey, D.J. Hearse and D.M. Yellon, The role of xanthine oxidase during myocardial ischaemia in several species including man, *J Mol Cell Cardiol* 20(suppl.II): 55 (1988).
14. T.D. Enhgerson, T.G. McKelvey, D.B. Rhynie, et al., Conversion of xanthine dehydrogenase to oxidase in ischaernic rat tissue, *J Clin Invest* 79: 2564 (1987).
15. S.J. Weiss, M.B. Lampert, Test ST., Long-lived oxidants generated by human neutrophils: characterization and bioactivity, *Science* 222: 625 (1983).
16. R. Ferrari, C. Ceconi, S. Curello, C. Guarnieri, C.M. Caldarera, A. Albertini and O. Visioli, Oxygen mediated myocardial damage during ischaemia and reperfusion: role of the cellular defences against oxygen toxicity, *J. Moll. Cell. Cardiol.* 17:937 (1985).
17. R. Ferrari, C. Ceconi, S. Curello, A. Cargnoni, G. Agnoletti, G.M. Boffa and O. Visioli, Intracellular effects of myocardial ischaemia and reperfusion: role of calcium and oxyge, *Eur. Heart J.* 7:3 (1986).
18. R. Ferrari, C. Ceconi, S. Curello, A. Cargnoni and D. Medici, Oxygen free radicals and reperfusion injury: effect of ischaemia and reperfusion on the cellular ability to neutralise oxygen toxicity,*J. Mol. Cell. Cardiol.* 18: 67 (1986).
19. S. Curello, C. Bigoli, R. Ferrari, A. Albertini and C. Guarnieri, Changes in the cardiac glutathione status after ischaemia and reperfusion, *Experientia* 41: 42 (1985).
20. R. Ferrari, C. Ceconi, S. Curello, A. Cargnoni, E. Pasini, F. De Giuli and A. Albertini, Role of oxygen free radicals in ischaemic and reperfused myocardium, *Am. J. Clin. Nutr.* 53: 215 (1991).
21. R. Ferrari, The role of free radicals in ischaemic myocardium, *British Journal of Clinical Practice* 44: 301 (1990).
22. R.B. Jennings, Myocardial ischaemia. Observations, definitions and speculation, *J Mol Cell Cardiol* 1: 345 (1970).
23. J.M McCord, Free radicals and myocardial ischaemia: overview and outlook, *Free Radic Biol Med* 4: 9 (1988).
24. Nohl H. The biochemical mechanism of the formation of reactive oxygen species in heart mitochondria *in*: C.M Caldarera and P. Harris, eds., "Advances in Studies on Heart Metabolism," Cooperativa Libraria Universitaria Editrice, Bologna, 413 (1982).
25. R. Ferrari, S. Bongrani F. Cucchini, F. Di Lisa, C. Guarnieri and O. Visioli, Effect of molecular oxygen and calcium on heart metabolism during reperfusion, *in*: Bertrand ME, ed., "Coronary arterial spasm," Lille, France, 46 (1982).

26. H. Otani, H. Tanaka, T. Inove, et al., In vitro studies on contribution of oxidative metabolism of isolated rabbit heart mitochondria to myocardial reperfusion injury, *Circ Res* 55:168 (1984).
27. C. Guarnieri C. Ceconi, C. Muscari and F. Flamigni, Influence of oxygen radicals on heart metabolism, *in*: C.M. Caldarera and P. Harris, eds., "Advances in studies on heart metabolism," Clueb, Bologna (1982).
28. C. Ceconi, S. Curello, A. Albertini and R. Ferrari, Effect of lipid peroxidation on heart mitochondria oxygen consuming and calcium transporting capacities, *Mol Cell Biochem* 81:131 (1988).
29. O.D. Saugstad and A.O. Aasen, Plasma hypoxanthine concentrations in pigs a prognostic aid in hypoxia, *Eur Surg Res* 12:123 (1980).
30. R. A. de Wall, K.A. Vasko, E.L. Stanley and P. Kezdi, Responses of the ischaemic myocardium to allopurinol, *Am Heart J* 82:362 1971).
31. C.E. Jones, J. W. Crowell and E.E. Smith, Significance of increased blood uric acid following extensive hemmorrhage, *Am J Physiol* 214:1374 (1968).
32. D.N. Granger, G. Rutilio and J.M. Mc Cord, Superoxide radicals in feline intestinal ischaemia,*Gastroenterology* 81:22 (1981).
33. A.S. Manning, D.J. Coltart and D.J. Hearse, Ischaemia and reperfusion induced arrhythmias in the rat. Eflect of xanthine oxidase inhibition with allopurinol,*Circ Res* 55:545 (1984).
34. R.A. Kloner, Introduction to the role of oxygen radicals in myocardial ischaemia and infarction.,*Free Radic Biol Med* 4:5 (1988).
35. J.M. Downey, D.J. Hearse and D.M.Yellon, The role of xanthine oxidase dunng myocardial ischaemia in several species including man, *J Mol Cell Cardiol* 20(suppl II):55 (1988).
36. W.L Arnold , R.H. de Wall, P. Keydi and H.H. Eward, The effect of allopurinol on the degree of early myocardial ischaemia, *Am Heart J* 99:614 (1985).
37. J.R. Parrtt and C.L. Wainwright, Failure of allopurinol and a spin-trapping agent N-t-alpha-phenyl nitrone to modify significantly ischaemia and reperfusion-induced arrhythmias, *Br J Pharmacol* 91:49 (1987).
38. J. Podzuweit, W. Braun, A. Muller and W. Schaper, Arrhythmias and infarction in the ischaemic pig heart are not mediated by xanthine-derived free oxygen radicals, *Circulation* 74(suppl II):311 (1986).
39. J.P. Kehrer, H. Piper and H. Sies, Xanthine oxidase is not responsible for reoxygenation injury in isolated-perfused rat heart, *Free Radic Res Commun* 3:69 (1987).
40. K.A. Reimer and R.B. Jennings, Failure of xanthine oxidase inhibitor allopurinol to limit infarct size after ischaemia and reperfusion in dogs, *Circulation* 71:1069 (1985).
41. J.L Romson, B.G. Hook, S.L. Kunkel, G.R. Abrams, M.A. Schork and B.R. Lucchesi, Reduction of the extent of ischaemic injury by neutrophil deplelion in the dog, *Circulation* 62:1016 (1983).
42. A. Manning, Reperfusion induced arrhythmias: do free radicals play a critical role? *Free Radic Biol Med* 4:305 (1988).
43. R.L Engler, G.W. Schmid Schonbein and R.S. Pavelec, Leucocyte capillary plugging in myocardial ischaemia and reperfusion in the dog, *Am J Pathol* 111:98 (1983).
44. K.M. Mullane, N. Read, J.A. Salmon and S. Moncada, Role of leukocytes in acute myocardial infarction in anesthetized dogs: relationship to myocardial salvage by anti-inflammatory drugs, *J. Pharmacol Exp Ther* 228:510 (1984).
45. Y. Qui, M. Galinanes, R. Ferrari, A. Cargnoni, A. Erzin and D.J. Hearse, PEG-SOD improves post-ischemic functional recovery and anti-oxidant status in the isolated blood-perfused rabbit heart, *Americam Journal of Physiology* 263:H1243 (1992).
46. H.A. Kontos, E.P. Wei and J.T. Povlishock, Cerebral arteriolar damage by arachidonic acid and prostaglandin G2, *Science* 209:156 (1983).
47. A. Schomig, D.M. Dart, R. Dietz, E. Mayer and W. Kubler, Release of endogenous catecholamines in the ischaemic myocardium of the rat, *Circ Res* 55: 689 (1984).
48. P.K. Singal, R.E. Beamish and N.S. Dhalla, Potential oxidative pathways of catecholamines in the formation of lipid peroxides and genesis of heart disease, *Adv Exp Med Biol* 161:391 (1983).
49. R. Julicher, L. Sterrenberg, J. Koomen, A. Bast and J. Noordhoek, Evidence for lipid peroxidation during the calcium paradox in vitamin E-deficient rat heart, *Arch Pharmacol* 326:87 (1984).
50. R. Ferrari, C. Ceconi, A. Cargnoni, S. Curello and T. Ruigrok, No evidence of oxidative stress during calcium paradox, *Basic Res Cardiol* 84: 396 (1989).
51. R.A. Wolf and R.W. Gross, Identification of a neutral phospholipase C which hydrolyzes choline giycerophospholipids and plasmalogen selective phospholipase A2 in canine myocardium,*J Biol Chem* 260:7295 (1985).

52. V.E. Kagan, V.M. Savov, V.V. Didenko, Yu V. Arkhypenko and F.Z. Meerson, Calcium and lipid peroxidation in mitochondrial and microsomal membranes ofthe heart, *Bull Exp Biol Med* 95:458 (1983).
53. I.M. McCord and I. Fridovich, Superoxide dismutase: An enzymatic function for erythrocuprein(hemocuprein), *J Biol Chem* 244: 6049 (1969).
54. D. Roos, R.S. Weening and S.R. Wyss, Protection of human neutrophils by endogenous catalase: studies with cells from catalase-deficient individuals. *J Clin Invest* 65:1515 (1980).
55. B. Chance, H. Sies and A. Boveris, Hydroperoxide metabolism in mammalian organs. *Physiol Rev* 59:527 (1979).
56. R. Ferrari, C. Ceconi, S. Curello, A. Cargnoni, A. Albertini and O. Visioli, Oxygen utilization and toxicity at myocardial level, in: G. Benzi, L. Packer and N. Siliprandi, eds., "Biochemical aspects of physical exercise," Elsevier, Amsterdam, 145 (1986).
57. S. Curello, C. Ceconi, C. Bigoli, R. Ferrari, A. Albertini and C. Guarnieri, Change in the cardiac glutathione status aner ischaemia and reperfusion, *Experientia* 41:42 (1985).
58. S. Curello, C. Ceconi, A. Cargnoni, R. Ferrari and A. Albertini, Improved procedure for determining glutathione plasma as an index of myocardial oxidative stress, *Clin Chem* 33:1448 (1987).
59. S. Curello, C. Ceconi, D. Medici and R. Ferrari, Oxidative stress during myocardial ischaemia and reperfusion: experimental and clinical evidences, *J Appl Cardiol* 1: 311 (1986).
60. R. Ferrari, O. Visioli, C.M. Caldarera and W.G. Nayler, Vitamin E and the heart possible role as antioxidant, *Acta Vitaminol Enzymol* 5: 11 (1982).
61. C. Guarnieri, R. Ferrari, O. Visioli, C.M. Caldarera and W.G. Nayler, Effect of alpha-tocopherol on hypoxic reperfused and reoxygenated rabbit heart, *J Mol Cell Cardiol* 10:893 (1978).
62. R. Ferrari, A. Cargnoni, C. Ceconi, S. Curello, A. Albertini and O. Visioli, Role of oxygen in myocardial ischaemic and reperfusion damage: protective effects of vitamin E, *in*: Hayaishi O, Mino M, eds., "Clinical and nutritional aspects of vitamin E," Elsevier, Amsterdam 209 (1987).
63. C. Guarnieri, F. Flamigni, C. Rossoni-Caldarera and R. Ferrari, Myocardial mitochondrial function in alpha-tocopherol deficient and refed rabbits, *in*: "Advances in myocardiology," Vol. 3, Plenum Press, New York, 621 (1982).
64. R. Ferrari, A. Cargnoni, C. Ceconi, S. Curello, A. Albertini and O. Visioli, Role of oxygen in the myocardial ischaemic and reperfusion damage: protective effects of vitamin E, *in*: O. Hayaishi, M. Mino, eds., "Clinical and biochemical aspects of vitamin E," Elsevier, Amsterdam, 209 (1987).
65. Gruppo Italiano per lo Studio della streptochinasi nell'infarto miocardico(GISSI). Effectiveness of intravenous thrombolytic treatment in acute myocardial infarction, *Lancet* 22: 397.
66. TIMI Study Group. The thrombolysis in myocardial infarction(TIMI) trial: phase 1 findings, *N Engl J Med* 312: 932 (1985).
67. R. Ferrari S. Curello A. Cargnoni, E. Condorelli, S. Belloli, A. Albertini and O. Visioli, Metabolic changes durng post-ischaemic reperfusion, *J Mol Cell Cardiol* 20: 119 (1988).
68. D.J. Hearse, Ischaemia, reperfusion, and the determinants of tissue injury, *Cardiovasc Drugs Ther* 4: 767 (1990).
69. K.A. Reimer and R.B. Jennings, Failure xanthine oxidase inhibitor allopurinol to limit infarct size aftr ischaemia and reperfusion in dogs, *Circulation* 71: 1069 (1985).
70. K.P. Gallagher, A.J. Buda, D. Pace, R.A. Gerren and M. Shlafer, Failure of superoxide dismutase reduces oxygen-free radical concentrations in reperfused myocardium, *J Clin Invest* 73: 1065 (1986).
71. J.L Zweier, J.T. Flaherty and M.L. Weisfeldt, Direct measurement of free radical generation following reperfusion of ischemic myocardium, *Proc Natl Acad Sci USA* 84:1404 (1987).
72. J.M. Luber, P.S. Rao, and M.S. Crowder, Identification of free radicals produced during myocardial ischemia and reperfusion using electron paramagnetic resonance spectroscopy and high precision liquid chromatography, *J. Thorac Cardiovasc Surg* (in press).
73. C.M. Arroyo, J.H. Kramer, B.F. Dickens and W.B. Weglicki, Identification of free radicals in mypcardial ischemia/reperfusion by spin trapping with nitron DMPO. FEBS *Lett* 221:101 (1987)
74. C.M. Arroyo, J.H. Kramer, R.H. Leiboff, G.W. Mergner, B.F. Diskens and W.B. Weglicki, Spin trapping of oxygen and carbon-centered free radicals in ischaemic canine myocardium, *Free Radic Biol Med* 3:313 (1987).
75. P.B. Garlick, M.J. Davies, D.J. Hearse and T.F. Slater, Direct detection of free radicals in the reperfused rat heart using electron spin resonance spectroscopy,*Circ Res* 61:757 (1987).
76. R. Bolli, B.S. Patel, M.O. Jeroudi, E.K. Lai and P.B. McCay, Demonstration of free radicals generation in "stunned" myocardium of intact dogs with use of the spin trap Alpha-phenyl n-ter-butyl nitrone, *J Clin Invest* 82:476 (1988).

77. W. Limm, M. Mugiishi, L.H. Piette and J.J. Mc Namara. Quantitative assessment of free radical generation during ischemia and reperfusion in the isolated rabbit heart, "Proc Fourth Int Cong Oxygen Radicals", La Jolla CA (1987).
78. Y. Gauduel and M.A. Duvelleroy, Role of oxygen radicals in cardiac injury due to reoxygenation, *J Mol Cell Cardiol* C. 16: 459 (1984).
79. C. Guarnieri, F. Flamigni and C.M. Caldarera, Role of oxygen in the cellular damage induced by re-oxygenation of hypoxic heart, *J Mol Cell Cardiol* 12: 797 (1980).
80. F.Z. Meerson, V.E. Kagan, YuP. Kozlov, et al., The role of lipid peroxidation in pathogenesis of ischaemic damage and the antioxidant protection of the heart, *Basic Res Cardiol* 77: 465 (1982).
81. J.M.C. Gutteridge and G.J. Quinlan, Malondialdehyde formation from lipid peroxides in the thiobarbituric acid test: the role of lipid radicals, iron salts and metal chelators, *J Appl Biochem* 5: 293 (1983).
82. T.L. Dormandy, Free radical oxidation and anti-oxidants, *Lancet* 1: 647 (1978).
83. C. Ceconi, A. Cargnoni, E. Pasini, E. Condorelli, S. Curello and R. Ferrari, Evaluation of phospholipid peroxidation as malondialdehyde during myocardial ischaemia and reperfusion injury, *Am J Physiol* 260: 105 (1991).
84. J.D. Adams, B.M. Lauterburg and J.R. Mitchell, Plasma glutathione and glutathione disulfide in the rat: regulation and response to oxidative stress, *J Pharmacol Exp Ther* 227 749 (1983).
85. R. Ferrari, C. Ceconi, S. Curello, A. Cargnoni, A. Albertini and O. Visioli, Molecular events occurring during post-ischaemic reperfusion, *in*: N.S. Dhalla, I.R. Innes and R.E. Beanish, eds., "Myocardial ischaemia," Nijhoff Publishing, Boston, 67 (1987).
86. R. Ferrari, S. Curello, C. Ceconi, A. Cargnoni, E. Condorelli and A. Albenini, Alterations of glutathione status during myocardial ischaemia and reperfusion, *in*: P.K. Singal, ed., "Oxygen radicals in the pathophysiology of heart disease," Vol 10. Kluwer Academic Publisher, Amsterdam, 145 (1988).
87. R. Ferrari, O. Alfieri, S. Curello, C. Ceconi, A. Cargnoni, P. Marzollo, A. Pardini, E. Caradonna and O. Visioli, Occurrence of oxidative stress during reperfusion of the human heart, *Circulation* 81:201 (1990).
88. R. Ferrari, C. Ceconi, S. Curello, A. Cargnoni, F. De Giuli and O. Visioli, Occurrence of oxidative stress during myocardial reperfusion, *Mol. Cell. Biochem*, 111:61 (1992).
89. R. Ferrari, Importance of oxygen-free radicals during ischaemia and reperfusion in the experimental and clinical setting, *The American Journal of Cardiovascular Pathology* 4 (1992).
90. S. Curello, C. Ceconi, F. De Giuli, A.F. Panzali, B. Milanesi, M. Calarco, Pardini, P. Marzollo, O. Alfieri, R. Ferrari and O. Visioli, Oxidative stress during reperfusion of human hearts: potential sources of oxygen-free radicals, Manuscript submitted to *Circulation.*

MECHANISMS OF ATHEROSCLEROSIS: ROLE OF LDL OXIDATION

Peter D. Reaven

Division of Endocrinology and Metabolism
Department of Medicine
9500 Gilman Drive, 0682
University of California, San Diego
La Jolla, CA 92093

There is now extensive evidence suggesting that atherosclerosis begins with the formation of foam cells (the initial stage of the fatty streak) underneath an intact endothelial layer (1). An early step in foam cell formation is the adherence of monocytes to the endothelium overlying an initial accumulation of cholesterol. Subsequently, the monocytes enter the artery wall through cell gap junctions, presumably attracted by a variety of chemoattractants. Within the subendothelial space, monocytes differentiate into macrophages which may then take up lipoproteins (smooth muscle cells may also take up lipoproteins, although to a lesser extent) forming foam cells. The fatty streak, through a series of poorly defined steps, may develop into complex atherosclerotic lesions, called fibrous plaques, which may eventually lead to clinically apparent coronary artery disease (CAD). The fibrous plaques are covered by a thick cap of connective tissue and smooth muscle cells and overlay a core of necrotic cellular debris and lipid. Plaques may eventually grow large enough that they can project into the lumen of the artery, reducing blood flow. Most clinical events, such as myocardial infarctions, appear to be due to ruptures, in the margins of the fibrous plaques, which are macrophage-enriched, leading to hemorrhage into the plaque with subsequent thromboses and acute occlusion of the vessel.

There are many lines of evidence that support the above proposed sequence of events (1-3), although only during the last decade has a tenable explanation for the uptake of lipoproteins by macrophages been proposed. Low density lipoproteins (LDL) are believed to be the primary source of cholesterol present in atherosclerotic lesions. However, there is good evidence that macrophage uptake of LDL does not occur primarily via the LDL receptor. Macrophages have very few LDL receptors, and those present on the cell surface are down regulated by a rise in intracellular cholesterol content. Thus, *in vitro*, macrophages do not form foam cells, even when incubated with high concentrations of LDL. Furthermore, individuals with homozygous Familial Hypercholesterolemia and the

Free Radicals in Diagnostic Medicine, Edited by
D. Armstrong, Plenum Press, New York, 1994

WHHL rabbit, both of whom have absent or dysfunctional LDL receptors, still develop atherosclerosis at accelerated rates. The fact that animal and human atherosclerotic lesions contained macrophages loaded with cholesterol ester, yet *in vitro* macrophages could not take up LDL in sufficient amounts to become foam cells was a significant stumbling block in our understanding of the early events of atherosclerosis. Subsequent work by Brown, Goldstein and colleagues suggested a possible solution to this paradox. They proposed that since a role of the macrophage is to scavenge damaged material, lipoproteins may need to become "damaged" (or modified) prior to their rapid uptake by the macrophage. They observed that if LDL was modified by acetylation, then rapid uptake of the modified LDL particle occurred by a novel macrophage receptor, referred to as the acetyl or "scavenger receptor" (4). The acetyl receptor has subsequently been cloned and sequenced and is apparently expressed in two slightly different forms (5). Both forms of the acetyl receptor are capable of recognizing and binding derivatized lysine amine groups of LDL's apoprotein B-100, as well as that of other modified proteins. Important to the concept of atherosclerosis, the expression of the scavenger receptor is not down-regulated by high intracellular LDL content. Thus, macrophages will continue to take up LDL cholesterol even as they become foam cells. Although there is no evidence to suggest that acetylation of LDL occurs *in vivo*, studies by several groups of investigators have demonstrated forms of LDL modification that do occur *in vivo* (6-10). Incubation of LDL with macrophages, endothelial cells or smooth muscle cells leads to LDL modification and rapid uptake of the LDL by way of the acetyl LDL receptor (11-13). Subsequent studies have clearly demonstrated that these cell-mediated modifications involve LDL oxidation. Although this chapter will focus primarily on oxidation of LDL, other forms of modification such as enhanced non-enzymatic glycation (as occurs in diabetes), lipoprotein aggregation and the formation of immune complexes also occur *in vivo* and may play an important role in the development of atherosclerosis (2, 3).

THE PROCESS OF LDL OXIDATION

The chemical events of LDL oxidation are very complex and have been excellently reviewed elsewhere (14, 15). Although the exact sequence of events leading to LDL oxidation *in vivo* is unknown, initiation begins with either abstraction of hydrogen atoms from polyunsaturated fatty acids within LDL by various reactive oxygen species or by direct enrichment of the LDL with lipoperoxides from cells. Either mechanism leads to a seeding of LDL with lipoperoxides which decompose into more reactive intermediates that can initiate oxidation in neighboring polyunsaturated fatty acids or polyunsaturated fatty acids in nearby LDL particles. *In vitro*, the decomposition of lipoperoxides to more reactive peroxyl radicals is dependent on the presence of transition metals (e.g., copper or iron) and can be inhibited by metal chelators. Importantly, the oxidation process is autocatalytic such that theoretically, a single hydrogen abstraction could lead to oxidation of the entire LDL particle as well as neighboring LDL particles. This process may be facilitated by LDL's intrinsic phospholipase A_2 activity, which cleaves off oxidized fatty acids from lecithin, generating lysolecithin (16). The oxidized fatty acids released by phospholipase A_2 are more mobile and presumably may help spread the oxidation process to other areas of the LDL particle. Initially, the oxidation process proceeds slowly, but eventually the antioxidant content within LDL is depleted and the number of fatty acid lipoperoxides amplify such that the oxidation process rapidly accelerates (the propagation phase). Eventually, the polyunsaturated fatty acids are cleaved into a variety of reactive aldehydes, ketones and other short chain fragments (the decomposition phase). These in turn may bind to apoprotein B-100 in LDL which leads to decreased recognition and binding by the LDL receptor. In addition, new epitopes are formed that lead to recognition and enhanced uptake of modified LDL by the scavenger receptor of macrophages. There may in fact be a family of "scavenger" receptors whose function is to remove modified or altered proteins, including LDL (17-19). Additionally, oxidation mediated decomposition of LDL's polyunsaturated fatty acids and cholesterol generates a variety of substances with atherogenic properties. These include chemoattractants, cytotoxic agents, and a variety of products which stimulate cytokine release and which are described below in more detail. It is important to realize that this is not an uniformly consistent process and that oxidized LDL is not a homogenous collection of particles. The extent to which LDL is oxidized and the chemical changes which occur during oxidization as well as the substances released during this process may vary. Some of the factors which may influence this process include the duration or mode of oxidation (cell type or transition metals, etc.), factors intrinsic to the LDL (fatty acid content, antioxidants, etc.,) or the surrounding milieu (media types, the presence of serum or other proteins or lipoproteins, etc.). It is not surprising therefore that “oxidized LDL” may not be identical when prepared in different laboratories or that each preparation does not yield identical results in any given assay even within the same laboratory.

Oxidation of LDL may be initiated by a number of different mechanisms. However, cell-mediated oxidation of LDL may be the most relevant to our understanding of LDL oxidation *in vivo*. In culture, all the cells normally present in the artery wall, including endothelial cells, smooth muscle cells, macrophages and lymphocytes can oxidize LDL (11-13, 20). Additionally, macrophage derived foam cells extracted from arteries of WHHL rabbits can also oxidize lipoproteins (21). The mechanisms by which cells initiate oxidation of LDL are poorly defined. Cells can produce reactive oxygen species and can secrete thiols into the medium (which facilitate formation of free radicals) and thus may participate, directly or indirectly, in the initiation of LDL oxidation (15, 22, 24). In addition, cellular lipoxygenases, which can also initiate hydrogen abstraction, may contribute to this process and in some cells, such as macrophages, oxidation can be inhibited by addition of lipoxygenase inhibitors (25, 26). 15-lipoxygenase mRNA and protein have also been demonstrated in macrophage-rich lesions of humans and WHHL rabbits (27). Furthermore, transfection of cells with 15 lipoxygenase appears to accelerate LDL oxidation (28). Cell mediated LDL oxidation *in vitro* in most cases requires the presence of transition metals to facilitate decomposition of preformed lipoperoxides to more reactive peroxy radicals. Removal of transition metals, such as iron or copper, by addition of chelators such as EDTA will inhibit LDL oxidation. It has not yet been demonstrated whether there is normally sufficient unbound copper or iron in the artery wall to allow oxidation to proceed. However, in areas of inflammation or injury, when tissue pH is lowered, release of iron from various binding proteins can undoubtedly occur. Which of the above mechanisms of oxidation are most relevant to *in vivo* oxidation is unknown and is under active investigation.

ATHEROGENIC PROPERTIES OF OXIDIZED LDL

When oxidized, LDL takes on a variety of properties that make it more atherogenic than unmodified LDL (14). These are outlined in Table 1. With our current appreciation of the many atherogenic properties of oxidized LDL, we now can propose a new sequence of events leading to the development of atherosclerosis (Figure). A very early step in the initial stages of atherosclerosis is the focal adherence of monocytes to the artery wall. This may result from expression on endothelial cells of specific leukocyte adherence molecules such as VCAM-1 (29). These adherence proteins are expressed prior to monocyte binding to the endothelium and accumulate over areas of the artery wall which later become sites of foam cell formation (1, 29). Endothelial cell expression of adherence proteins may result from hemodynamically altered cell function or perhaps are stimulated by focal subendothelial collections of modified LDL (30, 31). LDL that has been minimally oxidized (mm-LDL) stimulates the expression and secretion of many different cytokines. *In vitro*, adding minimally modified LDL to endothelial and smooth muscle cells stimulates their secretion of monocyte chemoattractant protein (MCP-1) (31). Presumably, this cytokine will attract monocytes to the endothelium where they bind to specific adherence proteins. During LDL oxidation phosphatidylcholine is formed, and can stimulate expression of the adherence molecule VCAM-1 (30), and is one example of the many products of modified lipoproteins that likely influence monocyte recruitment. Phosphatidylcholine is also chemotactic for monocytes, and thus can help attract additional monocytes into the subendothelial space (32). After adhering to the endothelial cells, monocytes then migrate into the artery wall where they differentiate into macrophages. Normally, after a period of residence within the artery wall, the macrophages will return to the circulation. However, oxidized LDL inhibits the mobility of monocyte derived macrophages (33), in essence, trapping them in the intima of the artery wall. This in turn propagates the oxidation process further, as retained macrophages can generate newly oxidized LDL as well as also further oxidizing minimally modified LDL. Thus, the longer macrophages remain in the intima the more lipoproteins they may take in and the greater their contribution to oxidation of LDL. As noted above, oxidation of LDL leads to rapid uptake of LDL via the scavenger(s) receptor leading to cholesteryl ester enrichment in the macrophage and foam cell formation.

Table 1. Mechanisms By Which Oxidized LDL May Be Atherogenic

1.	It has enhanced uptake by macrophages leading to cholesteryl ester enrichment and foam cell formation.
2.	It is chemotactic for circulating monocytes and T-lymphocytes.
3.	It inhibits the motility of tissue macrophages.
4.	It is cytotoxic.
5.	It renders LDL more susceptible to aggregation, which leads to enhanced macrophage uptake.
6.	It can alter gene expression of neighboring arterial cells such as induction of MCP-1, colony-stimulating factors, IL-1 and endothelial expression of adhesion molecules.
7.	It can adversely alter coagulation pathways, such as by induction of tissue factor.
8.	It can adversely alter vasomotor properties of coronary arteries.
9.	It is immunogenic and can elicit autoantibody formation and reactive T-cells.

Investigators have also shown *in vitro* that oxidized LDL is toxic to a variety of cell types (34). Several soluble polar sterols formed during oxidation have been implicated in this cytotoxicity (35). As macrophages and other artery wall cells oxidize LDL and in turn internalize or are exposed to these toxic oxidation products they may induce their own death. These dying macrophages may release toxic products as well as proteolytically active enzymes. In this milieu, it is easy to imagine how accumulated toxic products may damage and perhaps contribute to the disruption of the artery wall structure and rupture of the overlying intact endothelium. Damage to the endothelium may also have several direct atherogenic and thrombogenic effects. For example, it may result in greater inflow of

LDL into the arterial wall and it may also stimulate the adhesion of leukocytes and platelets.

The effect of oxidized LDL on the cellular environment may be more subtle than inducing cell death. Oxidized LDL has been demonstrated to alter gene expression in many of the cells present in the artery wall (36, 37). Minimally modified LDL stimulates the expression and secretion of M-CSF and GM-CSF by human aortic endothelial cells (38). These cytokines may contribute to cellular proliferation of a variety of cell types in the artery wall and may be crucial in the progression of the early foam cell lesions to the more complicated atherosclerotic plaques.

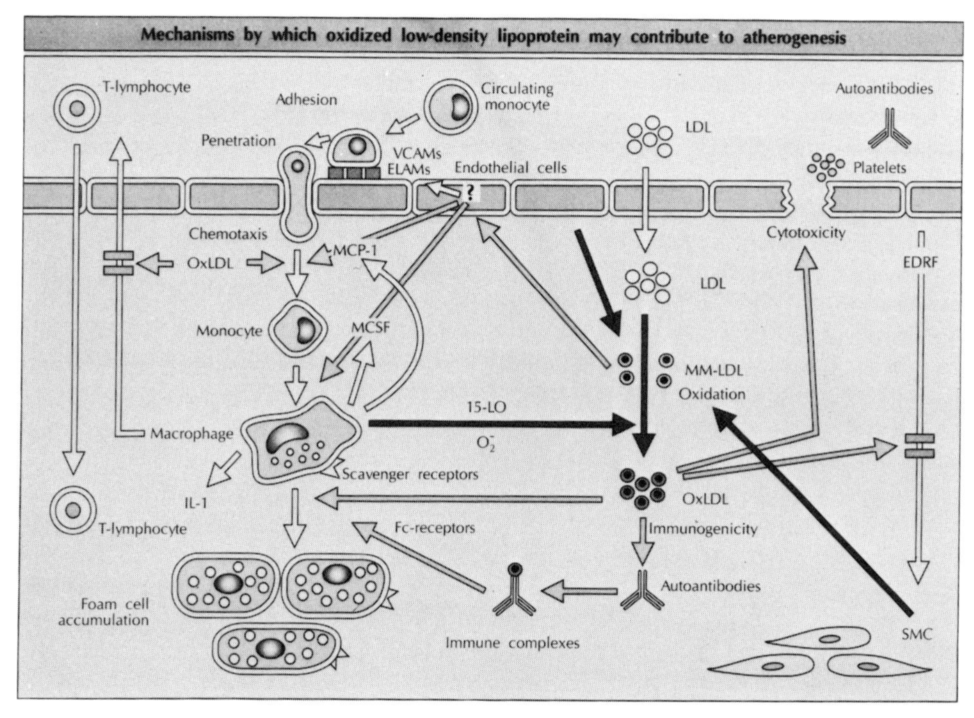

Figure 1. Hypothetical mechanisms by which oxidized low density lipoprotein (OxLDL) contributes to atherogenesis. Pathways enhancing the oxidation of LDL are indicated by dark grey arrows. The light grey arrows indicated potential atherogenic effects of OxLDL, for details see text (IL-1 = interleukin 1; VCAMs = vascular cell adhesion molecules; ELAMs = endothelial leukocyte adhesion molecules; EDRF = endothelial derived relaxing factor; SMC = smooth muscle cell; MM-LDL = minimally modified LDL: 15-LO = 15-lipoxygenase; MCSF = macrophage colony stimulating factor; MCP-1 = macrophage chemotactic protein-1). Reprinted with permission from Palinski, *In*: New Horizons in Coronary Heart Disease. Born GVR, Schwartz CD, eds. Current Science, London, 1993; pp 13.1-13.12.

Oxidized LDL may also interfere with vascular vasomotor tone. For example, it may inhibit vascular smooth muscle cell relaxation response to endothelial-derived relaxation factor. This may have important clinical implications for coronary artery flow. *In vitro*, oxidized LDL appears to decrease levels of EDRF and *in vivo*, treatment with antioxidants appears to reverse this process and improves coronary artery flow (39).

Lastly, there is increasing evidence for the involvement of immune mechanisms in the development of atherosclerosis. Atherosclerotic lesions from humans contain immune competent cells such as T-lymphocytes and macrophages (1). There is evidence that these cells are in fact activated, with expression of IL-2 receptors on lymphocytes and Class II histocompatibility antigens on smooth muscle cells. The presence of C5b-9 complement complexes have also been detected. Oxidized LDL is immunogenic and may contribute to this immune activation within the arterial wall in several ways. Oxidized LDL may stimulate the formation of circulating antibodies which can in turn form immune complexes with oxidized LDL. These immune complexes may lead to more rapid uptake of oxidized LDL by macrophages either through the Fc receptors or via phagocytosis. Additionally, oxidized LDL is chemotactic for lymphocytes which will further increase the number of lymphocytes in the intima (40). Lymphocytes and other immune competent cells can also produce a variety of inflammatory and immune mediators such as interleukins, tumor necrosis factor, leukotrienes and gamma interferon which undoubtedly affect the atherogenic process.

EVIDENCE THAT OXIDIZED LDL OCCURS *IN VIVO*

Occurrence of oxidized lipoproteins *in vivo* has been demonstrated by a number of lines of evidence. This has recently been extensively reviewed (3, 14) and I will only briefly summarize this information. The first line of evidence consists of immunocytochemical demonstration of specific epitopes in atherosclerotic lesions that are recognized by antibodies developed against modifications of LDL. These antibodies were produced by immunizing guinea pigs and mice with homologous LDL bound to reactive short-chain fragments (MDA and 4-hydroxynonenal) or by immunizing with copper-oxidized LDL. The antibodies developed recognize these epitopes on a variety of modified proteins, such as MDA-albumin or oxidized LDL. A number of immunocytochemical studies were performed with these antibodies and demonstrated that oxidation-specific epitopes were present in atherosclerotic lesions in both animals and humans (6-10). Although antibodies to different forms of modified LDL were used, the staining in atherosclerotic tissue was quite consistent. The most intense staining of these oxidation-specific epitopes was near or within the macrophages, consistent with the concept that oxidized LDL is taken up by the scavenger receptor(s) on cells such as macrophages. The predominance of macrophage-associated staining suggests that the oxidation process may be occurring in microdomains created in the vicinity of macrophages. This may explain how oxidation can occur in the intima while the plethora of antioxidants in the plasma may prevent this process from occurring.

A second line of evidence supporting the occurrence of oxidized LDL *in vivo* is related to the demonstration that gentle extraction of LDL from atherosclerotic lesions contains oxidation-specific epitopes. LDL extracted from artery wall lesions also shares many of the compositional and functional properties of LDL that is oxidized *in vitro* (7, 41, 42). Moreover, LDL extracted from atherosclerotic lesions is recognized by antibodies generated against oxidatively modified LDL and is also recognized and taken up by the macrophage scavenger receptor.

A third line of evidence is the demonstration that there are circulating autoantibodies present in serum from both humans and rabbits, which recognize the MDA lysine epitope of LDL modified with MDA (6). Since we know that modifications of LDL make it highly immunogenic, it is conceivable that oxidized LDL generated *in vivo* could lead to the formation of these autoantibodies. The role that these autoantibodies may play in atherosclerosis is unknown although autoantibodies against MDA lysine were elevated in a group of people who had accelerated carotid atherosclerosis compared to controls. Additionally, autoantibody titers were a relatively good predictor of progression of the carotid lesions (43). Recent work has demonstrated that antibodies that recognize oxidized LDL are present in the artery wall. In an elegant series of experiments, investigators used Western blots to demonstrate that immunoglobulins extracted from the atherosclerotic lesions of Watanabe rabbits can bind to LDL oxidized *in vitro* (44). This documentation of immune complexes between specific autoantibodies and oxidized LDL in lesions is an additional line of evidence for the formation of oxidized LDL *in vivo*.

A final line of evidence for the presence of oxidized LDL *in vivo* is that treatment of a number of different animal models with antioxidants can reduce atherosclerosis by mechanisms which appear independent of lipid lowering. This will be reviewed in more detail later in this chapter. Although there is now substantial evidence that oxidized LDL does occur *in vivo*, an important remaining question is whether oxidized LDL *in vivo* is related to the development of atherosclerosis or perhaps is only a consequence of atherosclerosis.

EVIDENCE SUPPORTING THE CONCEPT THAT OXIDIZED LDL IS IMPORTANT IN ATHEROSCLEROSIS

There are currently no good animal models of accelerated oxidation, and no adequate methods to measure oxidation status *in vivo*, particularly in the artery wall. Therefore, much of the evidence supporting the role of LDL oxidation in atherosclerosis comes from studies in animals and man of the relationship between atherosclerosis and antioxidant intake and/or plasma levels.

Table 2. Studies of antioxidants and atherosclerosis in animals

Antioxidant	Reference	Animal Model	Lesion Area	LDL Oxidation
Probucol	45	WHHL Rabbit	decreased 50%	decreased
Probucol	67	WHHL Rabbit	decreased 80%	decreased
Probucol	68	Cholesterol-fed Rabbit	decreased 70%	decreased
Probucol	69	Cholesterol-fed Rabbit	no effect	N.D.
Probucol	70	WHHL Rabbit	decreased 74% decreased 54% (1)	N.D. N.D.
Probucol Analog	71	WHHL Rabbit	decreased 35%	decreased
Probucol	72	Cholesterol-fed Primate	decreased 50%	decreased
DPPD	73	Cholesterol-fed Rabbit	decreased 71%	decreased
BHT	74	Cholesterol-fed Rabbit	decreased 68%	decreased
Vit E	75	Cholesterol-fed Rabbit	decreased (in thoracic aorta only) 46%	N.D.
Vit E	76	Cholesterol-fed Primate	decreased (carotid) progression 79%	N.D.

(1) Probucol started after 8 months of lesion formation
N.D. Not done

To directly evaluate the effect of antioxidants on atherosclerosis, intervention studies have been performed in animals. In an early study by Carew et al. probucol, a potent lipophilic antioxidant (and lipid lowering agent) was fed to WHHL rabbits for 30 weeks and compared to two control groups (45). One control group was given small doses of lovastatin to control for the lipid lowering effects of probucol while the third group was fed only standard rabbit chow. Throughout the study duration lipid levels were highest in the standard chow fed group, with the lovastatin group having slightly lower values than those of the probucol fed group. Despite cholesterol levels higher on average than occurred in the lovastatin group, the extent of aortic surface area covered with lesion in the probucol group was significantly lower. Thus, probucol decreased the extent of atherosclerosis beyond that which was obtained by cholesterol lowering only. Subsequently, there have been a number of other studies in animals using a variety of different antioxidants that have yielded in general similar results (Table 2). Although these agents may have other properties, in addition to their antioxidant activity, that might contribute to their antiatherosclerotic effects, the fact that very different antioxidants successfully inhibit atherosclerosis strongly supports the hypothesis that lipoprotein oxidation is crucial to the development of CAD. Of note, there have been no convincing intervention studies in animals using naturally occurring antioxidants, such as β-carotene, vitamin E and vitamin C.

Epidemiologic studies do however suggest a relationship between dietary intake and/or serum levels of natural antioxidants such as vitamin C, vitamin E, and β-carotene and CAD. This was noted by Gey et al. in a cross sectional study of heart disease risk factors among European countries with diverse rates of coronary heart disease. It was observed that there was a strong inverse relationship between vitamin C and vitamin E intake and rate of CAD (within each country's sample population) (46). Vitamin A, β-carotene and selenium intake were not different in countries with different rates of heart disease. A small prospective study in a region of Switzerland with relatively high levels of vitamin E intake in the normal diet showed no relationship between development of CAD and plasma vitamin E levels at entry to the study (47). There was however, a trend for increased relative risk for CAD in the subjects with the lowest vitamin C intake. Subsequently, there have been a number of other case control and cohort studies (Table 3). In general, the results of these studies have been inconsistent. More recently, however, several large cohort studies have demonstrated in both men and women that high vitamin E intake is associated with a reduced risk for CAD (48, 49). Those in the highest quintiles of vitamin E intake (including use of supplements) had a reduction in CAD events by 30-50%. No protective effect was seen for increased vitamin C intake in either study, and increased β-carotene intake was protective only in a subset of men, who were current or former smokers. In the only long-term intervention study in humans to date, a preliminary presentation from the Harvard Physicians Health Study, was reported on a subset of 333 men with chronic angina or prior coronary revascularization who were followed for seven years while taking β-carotene 50 mg every other day. Those taking β-carotene had a 50% reduction in combined cardiovascular endpoints (revascularization, myocardial infarction, or coronary death) compared to those receiving a β-carotene placebo (50).

MECHANISMS OF INHIBITION OF LDL OXIDATION

With increasing evidence that lipoprotein oxidation is fundamental to the development of atherosclerosis there is a need to understand the factors which influence the susceptibility of LDL to oxidation. Factors that may influence LDL oxidation *in vivo* can be classified into factors extrinsic or intrinsic to the lipoprotein particle.

Extrinsic factors refer primarily to cellular prooxidant activity, plasma and extracellular fluid pro-oxidants (trace metals) and antioxidants (e.g., ascorbate, bilirubin, urate, HDL), as well as a variety of other factors which may increase residence time of LDL in the intima. In general, our limited understanding of which factors extrinsic to the LDL particle significantly influence lipoprotein oxidation, as well as concerns about the

Table 3. Studies Relating Antioxidants and Pro-oxidants to Ischemic Heart Disease

Study	Reference	# of Participants/ Study Type	Variables Evaluated	Association with Ischemic Heart Disease
		Case Control Studies		
Netherlands (EPOZ - study)	(77)	84 cases with CAD vs. 168 controls/ Prospective Nested	Serum, Vitamin E, A and Selenium	No association
Eastern Finland Heart Survey	(78)	92 cases with MI vs. 92 matched controls	Serum Selenium, retinol and tocopherol levels	No association
Scottish Heart Health Study	(79)	9,527 men & women	Dietary β-carotene, Vitamin C, and Vitamin E	O.R. reduced in highest quintiles of Vit C, β-carotene, in newly diagnosed male CAD pts
EURAMIC Study	(80)	683 cases with CAD 727 controls	Adipose β-carotene, Adipose Vitamin E	O.R.* = 1.78 No association
Edinburgh Study	(81)	125 Angina cases 430 Controls	Vitamin E, β-carotene Vitamin C	O.R.* = 2.68 (lowest quintile) No association
		Cohort Studies		
Massachusetts Elderly Cohort Study	(82)	1,299 elderly men & women/ Prospective observation	Dietary β-carotene/ CVD Mortality	RR* = .54 (highest quartile)
World Health Organization Study	(46)	1,796 men/ Cross-Cultural Comparison of CVD Mortality Rates	Plasma Vitamin E, Vitamin C, Carotenoids, Vitamin A, Selenium	r^2 = .62 for Vit E No association for others
Basle Study (12 year follow-up)	(47)	2,974 men/ Prospective observation	Plasma Vitamin E, A, β-carotene/ CAD Mortality	No association RR = 1.53 (lowest quartile)
First National Health and Nutrition Examination Study	(83)	11,348 men & women/ Prospective observation	Dietary Vitamin C/ CVD death	males $SMR^{2,3}$ = .58 females $SMR^{2,3}$ = .75
Kuopio Ischemic Heart Disease Risk Factor Study I	(84)	1,132 men, cohort/ Cross Sectional	Serum Selenium Serum Vitamin C Serum Vitamin E	Lower levels in CAD pts No association No association
Kuopio Ischemic Heart Disease Risk Factor Study II	(85)	1,666 men/ Prospective observation	Serum $Copper^{1}$ /MI	3.5-4.0 fold excess risk (highest tertiles)
Kuopio Ischemic Heart Disease Risk Factor Study III	(86)	1,931 men/ Prospective observation	Serum $Ferritin^{1}$ /MI	RR* = 2.2 (Ferritin 7200 mg/L)
Nurses' Health Study	(48)	87,245 women/ Prospective observation	Use of Vitamin E supplements/CAD	RR* = .59
Health Professionals Follow-up Study	(49)	45,770 men Prospective observation	Current Vitamin E intake (>250 IU/d)/CAD	RR* = .74
		Intervention Studies		
U.S. Physicians' Health Study (interim report)	(50)	333 men with stable angina/ Prospective intervention	β-carotene 50 mg Alternate days/ CAD events	RR* = .49

* - After multivariate adjustment; 1 - Pro-oxidant; 2 - SMR = Standardized monthly ratio (expected/observed defined as 1.0 for US Caucasians); 3 - All cardiovascular disease. Not adjusted for other vitamin intake; O.R. - Odds ratio; RR - Relative Risk; MI - Myocardial infarction; CVD - Cardiovascular Disease; CAD - Coronary Artery Disease. Reprinted with permission from Reaven PD, Witztum JL, *In*: The Endocrinologist; 1994; in press.

safety of manipulating these factors *in vivo*, currently make it difficult to test interventions to alter factors extrinsic to LDL. Supplementation with the water soluble vitamin C may be an exception. It is well characterized as an antioxidant, and long term studies have documented its safety (51, 52). Further studies with this agent are clearly needed. The inherent susceptibility of LDL particles to oxidation undoubtedly also plays an important role in its overall rate of oxidation. Factors intrinsic to LDL include fatty acid composition, antioxidant content, and possibly phospholipase A_2 activity. As the polyunsaturated fatty acids of LDL are the primary substrates for lipid peroxidation their contribution to the susceptibility of LDL to oxidation is substantial. The antioxidant content of the LDL particle itself obviously also plays an important role in protecting it from oxidation. Presumably, if one could develop effective and safe ways to reduce the intrinsic susceptibility of the LDL particle to oxidation by modifying dietary fatty acids or antioxidants, one may succeed in inhibiting all of the subsequent events (outlined above in Figure) which occur as a result of the oxidation of LDL and as a consequence, inhibit atherosclerosis. Thus, much of the current clinical research in this area is directed towards this goal.

Dietary interventions that could reduce the content of polyunsaturated fatty acids in LDL by replacing them with fatty acids that are less susceptible to oxidation, while not raising plasma cholesterol levels may be one approach to reducing LDL oxidation *in vivo*. The feasibility of this approach in humans has now been demonstrated by several investigators, who compared the effects of feeding diets enriched in oleic acid (18:1) to diets enriched in linoleic acid (18:2) on LDL susceptibility to oxidation (53-56). LDL isolated from subjects fed linoleate-enriched diets were more susceptible to oxidation *ex vivo* than were LDL from subjects fed oleate-enriched diets. Diets enriched in oleate may be particularly effective in reducing the susceptibility of small, dense LDL to oxidation (57).

There are a number of antioxidant compounds, both natural and synthetic, that have also been demonstrated *in vitro* to inhibit LDL oxidation. Most supplementation studies, however, have focused on the lipid-soluble antioxidants (vitamin E, β-carotene, and ubiquinol) naturally present in LDL. Supplementation with all three of these antioxidants has been shown to be safe and relatively free of side effects. Supplementation with vitamin E (400-1600 mg/day) has been repeatedly shown to increase vitamin E levels in LDL and reduce LDL susceptibility to oxidation (58-62). Similarly, short-term feeding of ubiquinol to humans also appears to protect LDL from oxidation (63), although data on this antioxidant is limited. In contrast, supplementation studies with β-carotene have demonstrated that this agent does not appear to protect LDL from oxidation, despite reaching quite high levels within LDL (59, 61, 64). Although studies of antioxidant supplementation have in general been promising, very little is known about their antiatherosclerotic effects *in vivo*, particularly in humans.

Antioxidants may have effects on atherosclerosis not related to their free radical scavenging activity within LDL. For example, all of the above antioxidants, including β-carotene, may significantly influence the overall rate of LDL oxidation *in vivo*, not because of protection of LDL, but because of a direct effect on cellular pro-oxidant activity. Parthasarathy has shown that pre-incubation of macrophages with a water-soluble form of probucol reduces their ability to oxidize LDL (65). Navab et al., using a coculture of endothelial and smooth muscle cells that can oxidatively modify LDL, have demonstrated that pretreatment of their coculture with vitamin E or β-carotene can inhibit the ability of these cells to modify LDL (66). Thus, *in vivo*, antioxidants may have numerous effects, and their overall usefulness can only be evaluated in clinical trials.

ACKNOWLEDGMENTS

I would like to thank Joseph L. Witztum for his invaluable advice and comments, and Lisa Gallo for her excellent assistance in preparing this chapter.

REFERENCES

1. R. Ross, The pathogenesis of atherosclerosis: a perspective for the 1990s, *Nature* 362(6423):801-9 (1993).
2. D. Steinberg and J.L.Witztum, Lipoproteins and atherogenesis: Current concepts, *JAMA* 264(23):3047-52 (1990).
3. D. Steinberg, S. Parthasarathy, T.E. Carew, J.C. Khoo, and J.L. Witztum, Beyond cholesterol: Modifications of low density lipoprotein that increase its atherogenicity, *New Engl J Med* 320:915-924 (1989).
4. J.L. Goldstein, Y.K. Ho, S.K. Basu, and M.S. Brown, Binding site on macrophage that mediates uptake and degradation of acetylated low density lipoprotein, producing massive cholesterol deposition, *Proc Natl Acad Sci USA* 76:333-7 (1979).
5. M. Freeman, Y. Ekkel, L. Rohrer, M. Penman, N.J. Freedman, G.M. Chisolm, and M. Krieger, Expression of type I and type II bovine scavenger receptors in Chinese hamster ovary cells: lipid droplet accumulation and non-reciprocal cross competition by acetylated and oxidized low density lipoprotein, *Proc Natl Acad Sci USA* 88(11):4931-5 (1991).
6. W. Palinski, M.E. Rosenfeld, S. Ylä-Herttuala, G.C. Gurtner, S.A. Socher, S.W. Butler, S. Parthasarathy, T.E. Carew, D. Steinberg, and J.L. Witztum, Low density lipoprotein undergoes oxidative modification *in vivo*, *Proc Natl Acad Sci USA* 86: 1372-6 (1989).
7. Ylä-Herttuala S, Palinski W, Rosenfeld ME, S. Parthasarathy, T.E. Carew, S. Butler, J.L. Witztum, and D. Steinberg, Evidence for the presence of oxidatively modified low density lipoprotein in atherosclerotic lesions of rabbit and man, *J Clin Invest* 84(4):1086-1095 (1989).
8. M.E. Rosenfeld, W. Palinski, S. Ylä-Herttuala, S. Butler, and J.L. Witztum, Distribution of oxidation-specific lipid-protein adducts and apolipoprotein B in atherosclerotic lesions of varying severity from WHHL rabbits, *Arteriosclerosis* 10 (3):336-349 (1990).
9. M.E. Haberland, D. Fong, and L. Cheng, Malondialdehyde-altered protein occurs in atheroma of Watanabe heritable hyperlipidemic rabbits, *Science* 241(4862):215-8 (1988).
10. H.C. Boyd, A.M. Gown, G. Wolfbauer, and A. Chait, Direct evidence for a protein recognized by a monoclonal antibody against oxidatively modified LDL in atherosclerotic lesions from a Watanabe heritable hyperlipidemic rabbit, *Am J of Path* 135(5):815-25 (1989).
11. U.P Steinbrecher, S. Parthasarathy, D.S. Leake, J.L. Witztum, and D. Steinberg, Modification of low density lipoprotein by endothelial cells involves lipid peroxidation and degradation of low density lipoprotein phospholipids, *Proc Natl Acad Sci USA* 83:3883-7 (1984).
12. S. Parthasarathy, D.J. Printz, D. Boyd, L. Joy, and D. Steinberg, Macrophage oxidation of low density lipoprotein generates a modified form recognized by the scavenger receptor, *Arteriosclerosis* 6:505-10 (1986).
13. J.W Heinecke, H. Rosen, and A. Chait, Iron and copper promote modification of low density lipoprotein by human arterial smooth muscle cells in culture, *J Clin Invest* 74(5):1890-4 (1984).

14. J.L. Witztum and D. Steinberg, Role of oxidized low density lipoprotein in atherogenesis, *J Clin Invest* 88:1785-1792 (1991).
15. H. Esterbauer, O. Quehenberger, and G. Jürgens, Oxidation of human low density lipoprotein with special attention to aldehydic lipid peroxidation products. *in*: "Free Radicals: Methodology and Concepts," C. Rice-Evans and B. Halliwell, eds., The Richelieu Press, London (1988), pp 243-268.
16. S. Parthasarathy, U.P. Steinbrecher, J. Barnett, J.L Witztum, and D. Steinberg, Essential role of phospholipase A2 activity in endothelial cell-induced modification of low density lipoprotein, *Proc Natl Acad Sci USA* 82(9):3000-4 (1985).
17. C.P. Sparrow, S. Parthasarathy, and D. Steinberg, A macrophage receptor that recognizes oxidized low density lipoprotein but not acetylated low density lipoprotein, *J Biol Chem* 264:2599-2604 (1989).
18. H. Arai, T. Kita, M. Yokode, S. Narumiya and C. Kawai, Multiple receptors for modified low density lipoproteins in mouse peritoneal macrophages: different uptake mechanisms for acetylated and oxidized low density lipoproteins, *Biochem Biophys Res Comm* 159(3):1375-82 (1989).
19. Y.B. de Rijke and T.J.C. Van Berkel, Rat liver Kupffer and endothelial cells express different binding proteins for modified low density lipoproteins, *J Biol Chem* 269:824-27 (1994).
20. D.J. Lamb and D.S. Leake, CD4-positive T-lymphocytes can oxidatively modify low density lipoprotein, *Biochem Soc Transactions*. 21(2):1328 (1993).
21. M.E Rosenfeld, J.C. Khoo, E. Miller, S. Parthasarathy, W. Palinski, and J.L. Witztum, Macrophage-derived foam cells freshly isolated from rabbit atherosclerotic lesions degrade modified lipoproteins, promote oxidation of LDL, and contain oxidation specific lipid-protein adducts, *J Clin Invest* 87:90-99 (1991).
22. S. Parthasarathy, Oxidation of low-density lipoprotein by thiol compounds leads to its recognition by the acetyl LDL receptor, *Biochim Biophys Acta*. 917(2):1328 (1987).
23. J.W. Heinecke, H. Rosen, L.A. Suzuki, and A. Chait, The role of sulfur-containing amino acids in superoxide production and modification of low density lipoprotein by arterial smooth muscle cells, *J Biol Chem* 262(21):10098-103 (1987).
24. K. Hiramatsu, H. Rosen, J.W. Heinecke, G. Wolfbauer, and A. Chait, Superoxide initiates oxidation of low density lipoprotein by human monocytes. *Arterioscler* 7:55-60 (1987).
25. C.P. Sparrow, S. Parthasarathy, and D. Steinberg, Enzymatic modification of low density lipoprotein by purified lipoxygenase plus phospholipase A_2 mimics cell-mediated oxidative modification, *J Lipid Res* 29:745-53 (1988).
26. S. Parthasarathy, E. Wieland, and D. Steinberg, A role for endothelial cells lipoxygenase in the oxidative modification of low density lipoprotein, *Proc Natl Acad Sci USA* 86:1046-50 (1989).
27. S. Ylä-Herttuala, M.E. Rosenfeld, S. Parthasarathy, E. Sigal, T. Sarkioja, J.L. Witztum, and D. Steinberg, Gene expression in macrophage-rich human atherosclerotic lesions. 15-lipoxygenase and acetyl low density lipoprotein receptor messenger RNA co-localize with oxidation specific lipid-protein adducts, *J. Clin Invest* 87(4):1146-52 (1991).
28. D.J. Benz, N. Mori-Ito, J.L. Witztum, A. Miyanohara, T. Friedmann, D. Steinberg, and S. Parthasarathy, Expression of 15-lipoxygenase in fibroblasts confers an enhanced capacity to oxidatively modify LDL (Presented Abstract), *Circulation* 86:I-209 (1992), (Submitted).

29. L. Hongmei, M.I. Cybulsky, M.A. Gimbrone, Jr., and P. Libby, An atherogenic diet rapidly induces VCAM-1, a cytokin-regulatable mononuclear leukocyte adhesion molecule, in rabbit aortic endothelium, *Arterioscler Thromb* 13:197-204 (1993).
30. N. Kume, M.I. Cybulsky, and M.A. Gimbrone, Jr., Lysophosphatidylcholine, a component of atherogenic lipoproteins, induces mononuclear leukocyte adhesion molecules in cultured human and rabbit arterial endothelial cells. *J Clin Invest* 90:1138-1144 (1992).
31. J.A. Berliner, M.C Territo, A. Sevanian, et al., Minimally modified low density lipoprotein stimulates monocyte endothelial interactions, *J Clin Invest* 85:1260-1266 (1990).
32. M.T. Quinn, S. Parthasarathy, and D. Steinberg, Lysophosphatidylcholine: a chemotactic factor for human monocytes and its potential role in atherogenesis, *Proc Natl Acad Sci USA* 85:2805-9 (1988).
33. M.T. Quinn, S. Parthasarathy, L.G. Fong, and D. Steinberg, Oxidatively modified low density lipoproteins: a potential role in recruitment and retention of monocyte/macrophages during atherogenesis, *Proc Natl Acad Sci USA* 84:2995-8 (1987).
34. M.K. Cathcart, D.W. Morel, and G.M. Chisolm, III, Monocytes and neutrophils oxidized low density lipoproteins making it cytotoxic, *J Leukocyte Biol* 38:341-50 (1985).
35. D.W. Morel, J.R. Hessler, and G.M. Chisolm, Low density lipoprotein cytotoxicity induced by free radical peroxidation of lipid. *J Lipid Res* 24:1070-1076 (1983).
36. S.D. Cushing, J.A. Berliner, A.J. Valente, M.C. Territo, M. Navab, F. Parhami, R. Gerrity, C.J. Schwartz, and A.M. Fogelman, Minimally modified low density lipoprotein induces monocyte chemotactic protein 1 in human endothelial cells and smooth muscle cells, *Proc Natl Acad Sci USA* 87(13):5134-5138 (1990).
37. F. Liao, A. Andalibi, F.C. deBeer, A.M. Fogelman, and A.J. Lusis, Genetic control of inflammatory gene induction and NF-Kappa B-like transcription factor activation in response to an atherogenic diet in mice, *J Clin Invest* 91(6):2572-9 (1993).
38. T.B. Rajavashisth, A. Andalibi, M.C. Territo, J.A. Berliner, M. Navab, A.M. Fogelman, and A.J. Lusis, Induction of endothelial cell expression of granulocyte and macrophage colony-stimulating factors by modified low-density lipoproteins, *Nature* 344(6263):254-7 (1990).
39. K. Kugiyama, S.A. Kerns, J.D. Morrisett, R. Roberts, and P.D. Henry, Impairment of endothelium-dependent arterial relaxation by lysolecithin in modified low density lipoproteins, *Nature* 344:160-162 (1990).
40. H.F. McMurray, S. Parthasarathy, and D. Steinberg, Oxidatively modified low density lipoprotein is a chemoattractant for human T Lymphocytes, *J Clin Invest* 92:1004-1008 (1993).
41. M. Shaikh, S. Martini, J.R. Quiney, P. Baskerville, A.E. LaVille, N.L. Browse, R. Duffield, P.R. Turner, and B. Lewis, Modified plasma-derived lipoproteins in human atherosclerotic plaques, *Atherosclerosis* 69:165-172 (1988).
42. A. Daugherty, B.S. Zweifel, B.E. Sobel, and G. Schonfeld, Isolation of low density lipoprotein from atherosclerotic vascular tissue of Watanabe heritable hyperlipidemic rabbits, *Arteriosclerosis* 8:768-777 (1988).
43. J.T. Salonen, S. Ylä-Herttuala, R. Yamamoto, S. Butler, H. Korpela, R. Salonen, K. Nyyssonen, W. Palinski, and J.L. Witztum, Autoantibody against oxidized LDL and progression of carotid atherosclerosis. *Lancet* 339(8798):883-7 (1992).
44. S. Ylä-Herttuala, W. Palinski, S.W. Butler, S. Picard, and D. Steinberg, Rabbit and human atherosclerotic lesions contain IgG that recognizes epitopes of oxidized low density lipoprotein, *Arterioscler Thromb* 14:32-40 (1994).

45. T.E. Carew, D.C. Schwenke, and D. Steinberg, Antiatherogenic effect of probucol unrelated to its hypercholesterolemic effect: evidence that antioxidants *in vivo* can selectively inhibit low density lipoprotein degradation in macrophage-rich fatty streaks and slow the progression of atherosclerosis in the Watanabe heritable hyperlipidemic rabbit, *Proc Natl Acad Sci USA* 84:7725-9 (1987).
46. F.K. Gey, Lipids, lipoproteins and antioxidants in cardiovascular dysfunction, *Biochemical Society Transactions* 18:1041-1045 (1990).
47. M. Eichholzer, H.B. Stähelin, and K.F. Gey, Inverse correlation between essential antioxidants in plasma and subsequent risk to develop cancer, ischemic heart disease and stroke respectively: 12-year follow-up of the Prospective Basel Study, *in*: "Free Radicals and Aging," I. Emerit and B. Chance, eds., Birkhäyser /verlag. Basel (1992), pp 398-410.
48. M.J. Stampfer, C.H. Hennekens, J.E. Manson, G.A. Colditz, B. Rosner, and W.C. Willet, Vitamin E consumption and the risk of coronary disease in women, *NEJM* 328(20): 1444-1449 (1993).
49. E.B. Rimm, M.J. Stampfer, A. Ascherio, E. Giovannucci, G.A. Colditz, B. Rosner, and W.C. Willet, Vitamin E consumption and the risk of coronary heart disease in men, *NEJM* 328(20):1450-1456 (1993).
50. J.M. Gaziano, J.E. Manson, P.M. Ridker, J.E. Buring, C.H. and Hennekens, Beta carotene therapy for chronic stable angina, *Circulation* 82:Suppl III:III-201 (1990).
51. D.H. Hornig and U. Moser, The safety of high vitamin C intakes in man, *in*: "Vitamin C (Ascorbic Acid)," J.N. Counsell and D.H. Hornig, eds., Applied Science Publishers Ltd., London. 1st ed. (1981), pp 225-247.
52. S. Fahn, A pilot trial of high-dose alpha-tocopherol and ascorbate in early Parkinson's Disease, *Ann Neurol* 32:S128-S132 (1992).
53. P.D. Reaven, S. Parthasarathy, B.J. Grasse, E. Miller, F. Almazan, F.H. Mattson, J.C. Khoo, D. Steinberg, and J.L. Witztum, Feasibility of using an oleate-enriched diet to reduce the susceptibility of low density lipoprotein to oxidative modification in humans, *Am J Clin Nutr* 54:701-706 (1991).
54. P. Reaven, S. Parthasarathy, B.J. Grasse, E. Miller, D. Steinberg, and J.L. Witztum, Effects of oleate-rich and linoleate-rich diets on the susceptibility of low density lipoprotein to oxidative modification in hypercholesterolemic subjects, *J Clin Invest* 91:668-676 (1993).
55. A. Bonanome, A. Pagnan, S. Biffanti, A. Opportuno, F. Sorgato, M. Dorella, M. Maiorino, and F. Ursini, Effect of dietary monounsaturated and polyunsaturated fatty acids on the susceptibility of plasma low density lipoproteins to oxidative modification, *Arterioscler. Thromb.* 12:529-533 (1992).
56. M. Abbey, G.B. Belling, M. Noakes, F. Hirata, and P.J. Nestel, Oxidation of low-density lipoproteins: intra-individual variability and the effect of dietary linoleate supplementation, *Am. J. Clin. Nutr.* 57:391-398 (1993).
57. P.D. Reaven, B.J. Grasse, and D.L. Tribble, Effects of linoleate-rich and oleate-rich diets in combination with α-tocopherol on the susceptibility of low-density lipoproteins (LDL) and LDL subfractions to oxidative modification in humans, *Arterioscler. Thromb.* 14:557-566 (1994).
58. H. Esterbauer, M. Dieber-Rotheneder, G. Striegl, and G. Waeg, Role of vitamin E in preventing the oxidation of low-density lipoprotein, *Am. J. Clin. Nutr* 53:314S-321S (1991).
59. P.D. Reaven, A. Khouw, W. Beltz, S. Parthasarathy, and J.L.Witztum, Effect of dietary antioxidant combinations in humans: Protection of LDL by vitamin E, but not by β-carotene, *Arterioscler. Thromb.* 13:590-600 (1993).

60. P.D. Reaven and J.L Witztum, Comparison of supplementation of RRR-α-tocopherol and racemic-α-tocopherol in humans: Effects on lipid levels and lipoprotein susceptibility to oxidation, *Arterioscler. Thromb.* 13:601-608 (1993).
61. H. Princen, G. van Poppel, C. Vogelezang C, R. Buytenhek, F.J. Kok, Supplementation with vitamin E but not β-carotene in vivo protects low density lipoprotein from lipid peroxidation *in vitro:* Effect of cigarette smoking, *Arterioscler Thromb* 12:554-562 (1992).
62. I. Jialal and S.M. Grundy, Effect of dietary supplementation with alpha-tocopherol on the oxidative modification of low density lipoprotein, *J Lipid Res* 33:899-906 (1992).
63. D. Mohr, V.W. Bowry, and R. Stocker, Dietary supplementation with coenzyme Q_{10} results in increased levels of ubiquinol-10 within circulating lipoproteins and increased resistance of human low-density lipoprotein to the initiation of lipid peroxidation, *Biochim Biophys Acta* 1126:247-254 (1992).
64. P. Reaven, E. Ferguson, M. Navab, and F. Powell, Susceptibility of human low density lipoprotein to oxidative modification: Effects of variations in β-carotene concentration and oxygen tension, *Arterioscler Thromb* In press (1994).
65. S. Parthasarathy, Evidence for an additional intracellular site of action of probucol in the prevention of oxidative modification of low density lipoproteins: Use of a new water-soluble probucol derivative, *J Clin Invest* 89:1618-1621 (1992).
66. M. Navab, S.S. Imes, S.Y. Hama, G.P. Hough, L.A. Ross, R.W. Bork, A.J. Valente, J.A. Berliner, D.C. Drinkwater, H. Laks, and A.M. Fogelman, Monocyte transmigration induced by modification of low density lipoprotein in cocultures of human aortic wall cells is due to induction of monocyte chemotactic protein 1 synthesis and is abolished by high density lipoprotein, *J Clin Invest* 88:2039-2046 (1991).
67. T. Kita, Y. Nagano, M. Yokode, K. Ishii, N. Kume, A. Ooshima, H. Yoshica, and C Kawai, Probucol prevents the progression of atherosclerosis in Watanabe heritable hyperlipidemic rabbit, an animal model for familial hypercholesterolemia, *Proc Natl Acad Sci USA* 84:5928-5931 (1987).
68. A. Daugherty, B.S. Zweifel, and G. Schonfeld, Probucol attenuates the development of aortic atherosclerosis in cholesterol-fed rabbits, *Br J Pharmacol* 98:612-618 (1989).
69. Y. Stein, O. Stein, B. Delplanque, J.D. Fesmire, D.M. Lee, and P. Alaupovic, Lack of effect of probucol on atheroma formation in cholesterol-fed rabbits kept at comparable cholesterol levels, *Atherosclerosis* 75:145-155 (1989).
70. Y. Nagano, T. Nakamura, Y. Matsuzawa, M. Cho, Y. Ueda, and T. Kita, Probucol and atherosclerosis in the Watanabe heritable hyperlipidemic rabbit -- long-term antiatherogenic effect and effects on established plaques, *Atherosclerosis* 92:131-140 (1992).
71. S.J.T. Mao, M.T. Yates, R.A. Parker, E.M. Chi, and R.L. Jackson, Attenuation of atherosclerosis in a modified strain of hypercholesterolemic Watanabe rabbits using a probucol analog (MDL 29,311) that does not lower serum cholesterol, *Arterioscler Thromb* 11:1266-1275 (1991).
72. M. Sasahara, E.W. Raines, T.E. Carew, D. Steinberg, P.W. Wahl, A. Chait, and R. Ross, Inhibition of hypercholesterolemia-induced atherosclerosis in Macaca Nemestrina by probucol: I. Intimal lesion area correlates inversely with resistance of lipoproteins oxidation, *J Clin Invest* In press (1994).
73. C.P. Sparrow, T.W. Doebber, J. Olszewski, M.S. Wu, J. Ventre, K.A. Stevens, and Y.S. Chao, Low density lipoprotein is protected from oxidation and the progression

of atherosclerosis is slowed in cholesterol-fed rabbits by the antioxidant N,N'-diphenyl-phenylenediamine, *J Clin Invest* 89:1885-1891 (1992).

74. I. Björkhem, A. Henriksson-Freyschuss, O. Breuer, U. Diczfalusy, L. Berglund, and P. Henriksson, The antioxidant butylated hydroxytoluene protects against atherosclerosis, *Arterioscler Thromb* 11:15-22 (1991).
75. T.M.A. Bocan, S.B. Mueller, E.Q. Brown, P.D. Uhlendorf, M.J. Mazur, R.S. Newton, Antiatherosclerotic effects of antioxidants are lesion-specific when evaluated in hypercholesterolemic New Zealand White rabbits, *Exp Molec Pathol* 57:70-83 (1992).
76. A.J. Verlangieri and M.J. Bush, Effects of d-α-tocopherol supplementation on experimentally induced primate atherosclerosis. *J Am Coll Nutr* 11(2):131-138 (1992).
77. F.J. Kok, A.M. de Bruijn, R. Vermeeren, A. Hofman, A. van Laar, M. de Bruin, R.J.J. Hermus, and H.A.Valkenburg, Serum selenium, vitamin antioxidants, and cardiovascular mortality: a 9-year follow-up study in the Netherlands, *Am J Clin Nutr* 45:462-8 (1987).
78. J.T. Salonen, R. Salonen, I. Penttilä, J. Herranen, M. Jauhiainen, M. Kantola, R. Lappeteläinen, P.H. Mäenpää, G. Alfthan, and P. Puska, Serum fatty acids, apolipoproteins, selenium and vitamin antioxidants and the risk of death from coronary artery disease, *Am J Cardiol* 56:226-231 (1985).
79. C. Bolton-Smith, M. Woodward, and H. Tunstall-Pedoe, The Scottish Heart Health Study. Dietary intake by food frequency questionnaire and odds ratios for coronary heart disease risk. II. The antioxidant vitamins and fibre, *Europ J Clin Nutr* 46:85-93 (1992).
80. A.F.M. Kardinaal, F.J. Kok, J. Ringstad, J. Gomez-Aracena, V.P. Mazaev, L. Kohlmeier, B.C. Martin, A. Aro, J.D. Kark, M. Delgado-Rodriguez, R.A. Riemersma, P. van't Veer, J.K. Huttunen, and J.M. Martin-Moreno, Antioxidants in adipose tissue and risk of myocardial infarction: the EURAMIC study, *Lancet* 342:1379-1384 (1993).
81. R.A. Riemersma, D.A. Wood, C.C.A. MacIntyre, R.A. Elton, K.F. Gey, and M.F. Oliver, Risk of angina pectoris and plasma concentrations of vitamins A, C and E and carotene, *Lancet* 337:1-5 (1991).
82. J.M. Gaziano, J.E. Manson, L.G. Branch, F. LaMotte, G.A. Colditz, J.E. Buring, and C.H. Hennekens, Dietary beta carotene and decreased cardiovascular mortality in an elderly cohort, Abstract # 982-14, *JACC* 19(3):377A (1992).
83. J.E. Enstrom, L.E. Kanim, and M.A Klein, Vitamin C intake and mortality among a sample of the United States population, *Epidemiology* 3(3):194-202 (1992).
84. J.T. Salonen, R. Salonen, R. Seppänen, M. Kantola, M. Parviainen, G. Alfthan, P.H. Mäenpää, E. Taskinen, and R. Rauramaa, Relationship of serum selenium and antioxidants to plasma lipoproteins, platelet aggregability and prevalent ischaemic heart disease in Eastern Finnish men, *Atherosclerosis* 70:155-160 (1988).
85. J.T. Salonen, R. Salonen, H. Korpela, S. Suntioinen, and J. Tuomilehto, Serum copper and the risk of acute myocardial infarction: A prospective population study in men in Eastern Finland, *Am J Epidemiol* 134(3):268-276 (1991).
86. J.T. Salonen, K. Nyyssönen, H. Korpela, J. Tuomilehto, R. Seppänen, and R. Salonen, High stored iron levels are associated with excess risk of myocardial infarction in Eastern Finnish men, *Circulation* 86(3): 803-811 (1992).

TISSUE IRON OVERLOAD AND MECHANISMS OF IRON-CATALYZED OXIDATIVE INJURY

Edward J. Lesnefsky

Division of Cardiology
Case Western Reserve University
Cleveland VA Medical Center
Cleveland, OH 44106

INTRODUCTION

Tissue iron overload causes cell damage and organ dysfunction. The mechanisms of iron uptake by tissues and the probable biochemical pathways of iron-derived tissue injury will be reviewed. The iron-overload states are the initial clinical setting where the contribution of iron-catalyzed oxidative injury to the pathogenesis of a clinical disease has been appreciated. When viewed from the perspective of oxidative injury, the iron-overload syndromes also provide a model of tissue injury that is applicable to other disease states, including inflammation, ischemia-reperfusion injury, and anthracycline-induced cardiac toxicity. The goal of this chapter will be to provide a brief clinical overview of the iron-overload states, review the mechanisms of tissue iron uptake in normal and pathologic situations, the likely cellular targets and reactions of iron-catalyzed oxidative injury, and the clinical therapy of the iron-overload syndromes. The potential role of iron-catalyzed oxidative injury in the myocardial damage resulting from ischemia and reperfusion, and from anthracycline administration will also be discussed.

CLINICAL SYNDROMES

Primary hemochromatosis is a recessive disorder of increased gastrointestinal absorption of iron that results in increased iron deposition in the heart, skin, liver, and pancreas. The specific biochemical defect that causes increased iron absorption has

Free Radicals in Diagnostic Medicine, Edited by
D. Armstrong, Plenum Press, New York, 1994

not been defined. The genetic defect is located on the short arm of chromosome 6 in close proximity to the HLA loci. An autosomal recessive pattern of inheritance best describes the mode of transmission.[1] Approximately 8 to 10 percent of Caucasian populations are heterozygous, and 1 in 333 individuals in such populations are homozygotes.[1] The phenotype has variable penetrance, depending on other clinical factors.

Transfusion-related tissue iron overload occurs following the transfusion of approximately 100 units of blood without significant bleeding. Chronic transfusion therapy of syndromes including pure red cell aplasia, aplastic anemia, and thalassemia result in transfusion-related iron overload.

Tissue iron overload affects several organs. The incidence of neoplasia is increased, suggesting that somatic mutations are increased in the presence of elevated tissue iron levels. Cardiac iron overload initially results in increased perinuclear iron deposits, then more extensive iron deposition throughout the cell. Cardiac iron overload is associated with systolic and diastolic dysfunction, cardiac hypertrophy, ventricular dilatation, and congestive heart failure.[2] A restrictive cardiomyopathy can also occur. Supraventricular arrhythmias are common, and atrioventricular block may occur. Most investigators feel that the heart is the most susceptible organ to damage in iron-overload states, sustaining injury at lower tissue iron levels (1000 ug iron/ gm tissue) than the liver (>3000 ug iron/ gm tissue). Prior to the use of therapies to reduce tissue iron levels, approximately one-third of patients with hemochromatosis died from congestive heart failure. Increased liver iron stores result in hepatomegaly, fibrosis, and cirrhosis. Liver failure and portal hypertension can result. Approximately half of the patients with hemochromatosis suffer diabetes mellitus due to islet failure, which may be the presenting symptom.

MECHANISMS OF TISSUE IRON OVERLOAD

Tissue iron uptake occurs via both transferrin-mediated and non-transferrin mechanisms. Transferrin (Tf), an 80,000 dalton protein, has two binding sites for iron. Cell uptake of Tf-bound iron depends on the number of membrane transferrin receptors (TR). Increases in intracellular iron down-regulate TR number.[3,4] Following binding of the Tf-iron complex by the TR, the entire receptor-Tf-iron complex is internalized, and Fe^{2+} released to intracellular pools following the reduction of bound Fe^{3+}. Reduction is accomplished in hepatocytes by NADH diferric reductase, and in endothelial cells by superoxide anion generated by xanthine oxidase. TR number is modulated by post-transcriptional regulation of TR mRNA (Figure 1). In the setting of low cellular and cytosolic iron, the iron responsive element binding protein (IRE-BP) binds to a 3′ regulatory untranslated region in TR mRNA, increasing mRNA stability, and leading to increased transcription and greater receptor number.[3] Conversely, when intracellular iron is elevated, iron interacts with the IRE-BP, markedly reducing the affinity of IRE-BP for TR mRNA. Without bound IRE-BP, the stability of TR mRNA is greatly decreased, leading to a reduction in receptor number.

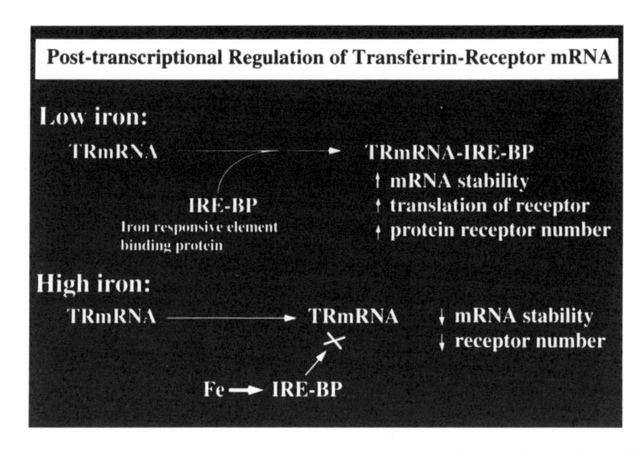

Figure 1. Transferrin receptor number is regulated by tissue iron levels via the iron-dependent binding of the iron responsive element binding protein (IRE-BP) to transferrin receptor mRNA that in turn regulates transferrin receptor number in a negative feedback mechanism.

The mechanism by which cytosolic iron modulates the binding affinity of IRE-BP to TR mRNA remains unclear.[4] Possible mechanisms of interaction include direct binding of iron to IRE-BP, or iron-mediated oxidation of protein sulfhydryl groups on IRE-BP.[4] During tissue iron-overload, the uptake of iron by TR-mediated mechanisms is very low due to appropriate down regulation of the TR receptor.

Serum ferritin concentration is markedly increased in iron overload. Ferritin uptake contributes to iron overload in hepatocytes,[5] but provides an uncertain contribution to increased cardiac iron uptake. The hepatic uptake of ferritin-iron remains unchanged in the setting of iron overload.[5] Ferritin is taken up by endocytosis, degraded in lysosomes, and the iron is released to mitochondria and cytosolic ferritin.[6]

Iron is essentially insoluble in physiologic buffers without an appropriate ligand to enhance the solubility of iron. Iron bound to negatively charged ligands including citrate, ammonium-citrate, nitriloacetic acid, or adenosine diphosphate (ADP) has increased solubility. These ligands enhance the redox cycling of iron and allow iron to catalyze oxidative damage.[7,8] Iron bound to these ligands is non-transferrin bound iron (NTBI). The serum NTBI in normal individuals is very low, while in patients with primary hemochromatosis NTBI as iron-citrate ranges between 8-20 μM[9].

NTBI is taken up by the cell. In fact, the uptake of elevated NTBI iron by rat cardiac myocytes greatly exceeds iron uptake from transferrin.[10] Viable cells with intact membranes can accumulate NTBI, probably via a specific receptor or channel. It appears that iron may share a channel with other transition metals, such as Cu^{2+} or Zn^{2+}.[11] Stimulated NTBI uptake in cultured cells involves an increase in the maximal rate of iron uptake, while the K_m for iron is unchanged.[11] Uptake is rapid, occurring within one hour of increased NTBI, and is reversible within several hours of removal of the stimulus for NTBI uptake.[12] NTBI uptake increases as extracellular NTBI levels increase.[12] More importantly, elevated intracellular iron also <u>increases</u> NTBI uptake in a <u>positive</u> feedback manner (Figure 2),[12] in contrast to the down regulation of TR number and Tf-bound-iron uptake in the presence of increased intracellular iron.

Conversely, reducing cytosolic iron levels by a chelator intervention reverses the increase in NTBI uptake.

Cellular iron uptake, whether via Tf-mediated or NTBI Tf-independent mechanisms, leads to an increase in iron in the low molecular weight cytosolic pool.[13] The iron in this pool is bound to low molecular weight chelates (Figure 3), and is able to catalyze oxidative injury to the cell. Iron in the low molecular weight pool is rapidly removed and stored in a redox-inactive form in ferritin, or is incorporated into hemeproteins. Iron is also deposited into hemosiderin, which increases as cells become iron overloaded.[12] Thus, under normal conditions, iron in the low molecular weight cytosolic pool is tightly controlled at very low concentrations in order to avoid oxidative tissue injury. However, tissue iron overload, in addition to increasing ferritin and hemosiderin levels, also increases iron levels in the cytosolic pool and iron associated with hemeproteins, predisposing to oxidative injury.[12] In addition, the superoxide anion ($\cdot O_2^-$)[14] or ischemia[15,16] can reductively release iron from ferritin into the redox-active cytosolic pool. Iron is present in ferritin as Fe^{3+}, and reduction to Fe^{2+}, either by $\cdot O_2^-$ascorbate, or other mechanisms, results in the release of iron into pools that can catalyze oxidative injury. Similarly, during oxidative stress, H_2O_2 can release iron from hemeproteins, exacerbating oxidative damage.[17,18] Ischemia causes approximately a 10-20 nM increase in cytosolic iron.[15]

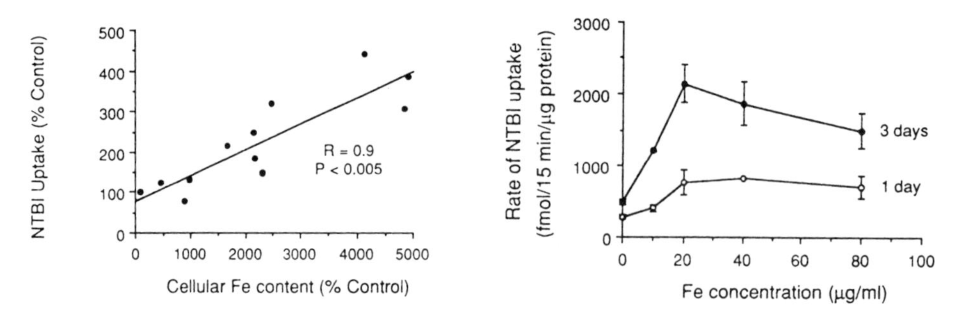

Figure 2. The uptake of nontransferrin bound iron (NTBI) is increased by elevated levels of extracellular and intracellular iron in a positive feedback mechanism. Reprinted with permission from Parkes, et. al, Journal of Laboratory and Clinical Medicine 122:36-47 (1993).

In the extracellular space, pathologic states release iron from transferrin (Tf). Tf-bound iron does not catalyze oxidative injury. Tf binding of iron is pH dependent, since iron is carried as part of a Tf-iron-bicarbonate complex. Decreases in pH, especially below 5.6, release iron from Tf to redox-active pools.[19] Acidic environments caused by ischemia or inflammation can result in the release of iron from Tf, even at physiologic levels of iron saturation.[20] Thus, while in normal homeostasis extracellular and intracellular iron is carefully sequestered to minimize oxidative damage, in iron-overload states or following tissue injury, available redox-active iron is increased, predisposing to tissue injury via oxidative mechanisms.

MECHANISMS AND TARGETS OF IRON-CATALYZED OXIDATIVE INJURY

Redox-cycling by iron between Fe^{2+} and Fe^{3+} damages biomolecules, causing cell injury. $Fe^{2+/3+}$ reacts with superoxide anion ($\cdot O_2^-$) and hydrogen peroxide (H_2O_2) via the Haber-Weiss mechanism to generate the very reactive hydroxyl radical ($\cdot OH$).[21] The second reaction (see below) between Fe^{2+} and H_2O_2 that generates $\cdot OH$ is the Fenton reaction. The generation of "free" $\cdot OH$ may not actually occur. Instead, a highly-reactive "crypto" $\cdot OH$ in the iron coordination complex, perhaps the perferryl [$(Fe^{3+}\text{---}\cdot O_2^-)^+$] ion, may be the oxidizing species.[22] Iron-catalyzed $\cdot OH$ formation requires at least one aqueous coordination site in the iron coordination sphere.[23] Thus, deferoxamine, which simultaneously complexes all six coordination sites of iron and excludes water from the coordination complex, prevents iron-catalyzed $\cdot OH$ formation.[23] Iron ligands such as EDTA, sugars, citrate, and nucleotides (including ATP and ADP)[7] allow water in the iron coordination complex and catalyze subsequent iron-catalyzed $\cdot OH$ formation. Thus, $\cdot O_2$ and H_2O_2 present during tissue oxidative stress, though poorly reactive toward most cellular biomolecules[22], release iron from ferritin and hemeproteins,[14,17] increasing the available iron in the cytosol to catalyze $\cdot OH$ formation and oxidative damage.

Haber-Weiss Reactions:

$$Fe^{3+} + \cdot O_2^- \dashrightarrow Fe^{2+} + H_2O_2$$

Fenton reaction:

$$Fe^{2+} + H_2O_2 \dashrightarrow Fe^{3+} + \cdot OH + OH^-$$

In the cell, $\cdot OH$ causes oxidative damage to DNA,[24] membrane lipids,[25] and proteins.[26] Protein damage can involve protein sulfhydryl oxidation, carbonyl group formation, or oxidative bond cleavage. Protein damage leads to enzyme inactivation, structural protein alteration, and often accelerated proteolysis.[26] Oxidative DNA damage causes strand breaks and cross-linking, as well as the hydroxylation of bases, the latter providing a marker of oxidative damage to DNA. By any of these mechanisms, oxidative damage to DNA increases the incidence of mutations.[24] Iron can both initiate and propagate lipid peroxidation, leading to altered membrane fluidity, inactivation of membrane-bound enzyme complexes, and eventual membrane disruption.[25,27] All of these processes can disrupt cell homeostasis and lead to cell injury.

The potential role of the iron-catalyzed Fenton reaction in tissue injury deserves additional comment. Experiments with electron paramagnetic resonance technique have shown that Fe^{2+} complexed to EDTA, ADP, or ATP is a catalyst of $\cdot OH$ formation in the presence of H_2O_2.[7,23] Fe^{2+} bound to ADP and ATP is an especially good catalyst for $\cdot OH$ generation, and of potential physiologic importance as well.[7] The Fenton reaction between Fe^{2+}-ADP and H_2O_2 clearly damages tissues.[28,29] ADP, present in the cytosol, enhances iron-catalyzed formation of $\cdot OH$ from H_2O_2 in part by delaying the auto-oxidation of Fe^{2+} to Fe^{3+}, so that Fe^{2+} has a greater likelihood of reacting with H_2O_2. H_2O_2 is generated by mitochondria and neutrophils. H_2O_2 crosses membranes, and thus can diffuse into the cytosol from sites of generation in mitochondria or extracellular sites of inflammation to react with cytosolic Fe^{2+}-ADP complexes. Increased cytosolic pool iron present in iron-overload states or following ischemia is likely to be bound in part by ADP, and therefore available to react with H_2O_2.

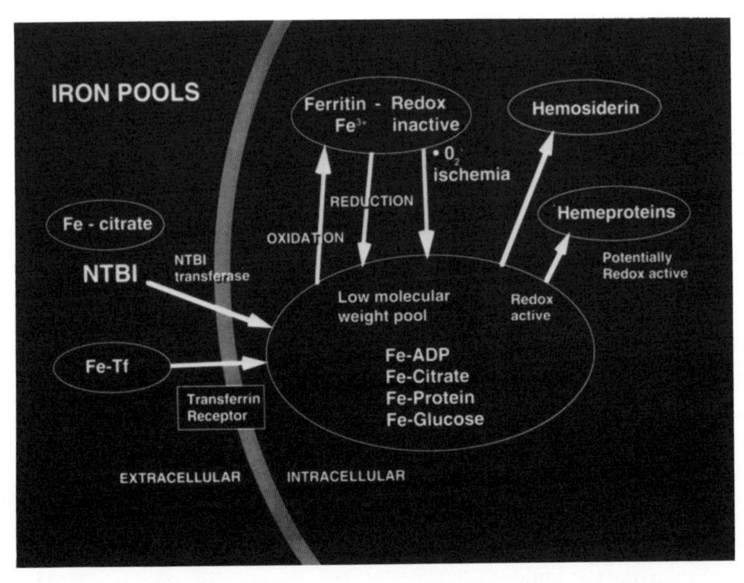

Figure 3. The cellular uptake of iron occurs via both transferrin-dependent and independent mechanisms. Intracellular iron is present initially in the cytosolic pool, and is stored in a redox inactive form as ferritin-bound iron. Iron can be reductively released from ferritin into the redox-active cytosolic pool.

Iron-generated $\cdot OH$ abstracts hydrogen atoms from unsaturated membrane lipids, initiating lipid peroxidation by generating lipid radicals in the membrane.[27] However, while the Haber-Weiss and Fenton reactions have been widely considered as potential causes of iron-catalyzed lipid peroxidation, the ability of iron to cause oxidative damage by redox-cycling independent of oxygen metabolites has also been described. Fe^{2+}-ADP chelates alone cause oxidative damage to myocardial lipids, and the lag phase prior to the initiation of oxidative damage is reduced by the presence of small amounts of Fe^{3+}.[31] Fe^{3+}/Fe^{2+} ratios of 1:1 to 7:1 result in the highest rates of lipid peroxidation.[30] The lipid peroxidation caused by these systems cannot be blocked by scavengers of $\cdot OH$, $\cdot O_2^-$, or H_2O_2, suggesting that these reactive oxygen species are not involved.[30,31] Complexes of iron with molecular oxygen ($Fe^{2+}\text{--}O_2 \leftrightarrow Fe^{3+}\text{--}\cdot O_2^-$) may the active species.[32] The potential for similar direct non-oxyradical mediated iron-redox cycling to damage proteins or DNA has not been adequately explored.

Iron-catalyzed lipid peroxidation is a free radical chain reaction, involving initiation and propagation steps.[27] During initiation, lipid radical generation leading to lipid peroxide formation occurs (Fig 4). Initiation might involve iron-generated $\cdot OH$.[33] Lipid peroxides are also decomposed by iron salts with the formation of lipid radicals.[34] (Figure 4, Reaction B). Fe^{2+}-ADP is an efficient initiator, but does not result in propagation.[35] However, Fe^{2+}-EDTA is an efficient propagator of lipid

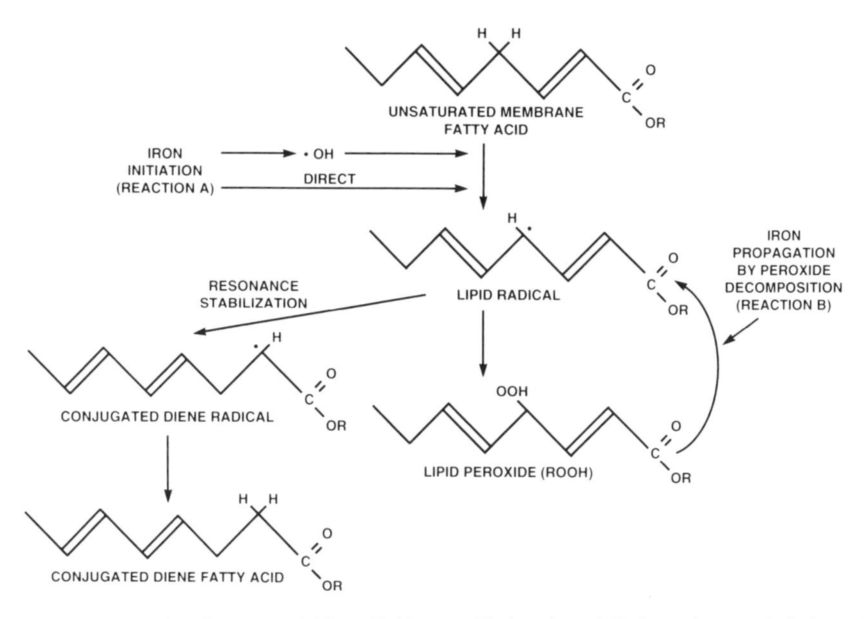

Figure 4. Redox-active iron can initiate lipid peroxidation by ·OH-dependent and independent mechanisms (Reaction A), and propagate lipid peroxidation via the iron-catalyzed decomposition of membrane lipid peroxides (Reaction B).

peroxidation.[35] Thus, iron could be involved in damage to cell membrane lipids in at least three ways. Iron-generated ·OH could initiate lipid peroxidation[33]; iron could initiate lipid peroxidation independent of ·OH[30,31,34]; or iron could propagate lipid-radical chain reactions via lipid peroxide decompositions independent of ·OH[32-34] (Figure 4, Reaction B).

The deleterious reactions caused by iron can also vary depending upon the chemical nature of the ligand, with Fe^{2+}chelated to EDTA vs ADP exhibiting different behavior toward lipids.[35] Iron associated with oxidized hemeproteins also catalyzes lipid peroxidation independent of ·OH.[18]. Also, the mechanism of injury of iron may vary depending on the molecular target. Studies suggest that iron mediated decomposition of deoxyribose is via ·OH, while participation in lipid peroxidation may be independent of ·OH.[36] Thus, in tissue injury, elevated intracellular iron could participate in several distinct chemical mechanisms of injury, and different ligands of iron might accentuate tissue injury to differing degrees, depending on the iron reactions that they alter.[36,37]

Iron-catalyzed oxidative damage to lipids and proteins results in cellular organelle damage and dysfunction, involving lysosomes, mitochondria, and the plasma membrane. Exposure of rat liver mitochondria in vitro to 1-3 mM Fe^{2+}-chelates increases lipid peroxidation and progressively inhibits electron transport.[38] The probable mechanism involves iron-catalyzed oxidative damage to mitochondrial membrane lipids, with alteration of the inner membrane environment, and subsequent inhibition of the membrane associated enzymes of the electron transport chain. Direct oxidative damage to electron transport enzyme proteins could also have occurred. Similar processes appear to occur in vivo, since mitochondria isolated from iron-overloaded hepatocytes have decreased rates of oxidative phosphorylation.[39] Even prior to the onset of electron transport defects, subtle iron overload increases mitochondrial lipid peroxidation and calcium release.[40] Mitochondrial calcium release initiates futile cycles of calcium uptake and release from mitochondria that consume energy and impair cellular calcium homeostasis, predisposing to cell damage. The mitochondrial changes occur with tissue iron levels in the range described in hemochromatosis.

The ferritin and hemosiderin in iron-overloaded cells is largely contained in lysosomes. An elevation in iron stores has been proposed to increase lysosomal fragility and enzyme release. Consistent with this potential mechanism of iron-mediated damage, iron-loaded cardiac myocytes have an increased release of lysosomal enzymes into the cytosol, indicating greater lysosomal fragility.[41] Iron-catalyzed lipid peroxidation of lysosomal membranes causes increased lysosomal fragility and enzyme release.[42] The release of lysosomal enzymes, in turn, can accentuate cell damage.

Sarcolemmal membranes are susceptible to iron-catalyzed lipid peroxidation in vitro.[30,31] If similar lipid peroxidation occurred in the intact cell, alterations in membrane fluidity and structure, potentially impairing membrane function and integrity, could occur.[27] Increased peroxidation of cardiac membranes has been observed in the presence of excess iron. Iron-loaded cardiac myocytes sustain a depletion of polyunsaturated fatty acids and a decrease in total protein sulfhydryl groups, suggesting that oxidative damage to membrane lipids and membrane-associated proteins occurs.[10,41] Aldehyde products of iron-mediated lipid peroxidation bind to amino groups on lysine residues, inactivating membrane bound proteins and enzymes. Lipid peroxidation products bound to proteins are increased in iron-overloaded animals.[43]

Thus, iron-catalyzed oxidative damage to biomolecules can lead to cell organelle damage and dysfunction, and these mechanisms are thought to be the causes of the organ dysfunction clinically observed in the iron-overload states. Evidence of increased lipid peroxidation and lysosomal damage has been observed in tissues from iron-overload patients.[44-46]

CLINICAL THERAPY OF IRON-OVERLOAD SYNDROMES

The goal of clinical therapy in the iron-overload syndromes is the reduction of tissue iron stores. In primary hemochromatosis caused by increased gastrointestinal iron uptake, this is achieved by intermittent phlebotomy. In secondary hemochromatosis caused by chronic transfusion therapy, iron removal must be achieved by iron chelator therapy. The successful reduction of tissue iron stores can delay the onset of clinical abnormalities or stabilize the extent of current disease. In some situations, iron chelation can even reverse clinical abnormalities, especially in the heart.[47,48] Removal of iron stores can decrease hepatic iron stores and stabilize cirrhosis, improve liver and pancreatic islet function, and allow for the recovery of cardiac function and reduction in arrhythmias.[47-49]

Chronic iron removal has been achieved by the use of deferoxamine therapy.[48] Deferoxamine is a specific, high affinity chelator of Fe^{3+}. Deferoxamine-chelated iron does not catalyze $\cdot OH$ formation[23], and it prevents iron-catalyzed lipid peroxidation in vitro.[36] Deferoxamine is weak scavenger of $\cdot O_2^-$ and $\cdot OH$, but only has important scavenging activity at concentrations greater than 500 uM.[50] Deferoxamine does not scavenge H_2O_2 and does not inhibit neutrophil generation of $\cdot O_2^-$.[51] Thus, deferoxamine can be used to bind iron, and to implicate iron-catalyzed processes in tissue injury. However, while useful as an experimental tool, deferoxamine has limitations in clinical use. Due to poor oral absorption, deferoxamine must be administered by repeated injection or chronic subcutaneous infusion.[48] The difficulty in deferoxamine administration often reduces patient compliance.[48] The concurrent administration of ascorbate with deferoxamine can enhance iron mobilization and removal, presumably by reductive release from tissue ferritin stores into the cytosolic pool, where iron is then accessible to deferoxamine. Of interest, consistent with the redox-active nature of the cytosolic iron pool discussed above,, adjunctive ascorbate therapy is occasionally associated with an increase in organ toxicity.

Deferoxamine, while a high affinity iron-chelator, has significant limitations. Following acute administration, deferoxamine has relatively poor intracellular penetration, requiring relatively large extracellular levels (5-20 mM) and prolonged exposure to achieve significant tissue penetration.[52,53] The ability to achieve these extracellular levels by the administration of large deferoxamine doses in vivo is limited by toxicity, including dose-related hypotension caused by histamine release from mast cells. Deferoxamine achieves tissue iron reduction in part by chelating extracellular iron, with secondary removal of iron from tissues. Deferoxamine that does enter cells forms an intracellular, slowly released ferrioxamine complex.[53] In cell culture studies, the half-life of release of the ferrioxamine complex was approximately 100 hours.[53] One potential mechanism of deferoxamine effect in the iron-overload states is achieved by reducing extracellular and free intracellular iron to extremely low levels, thereby causing a marked reduction of cellular uptake of NTBI.[12] As discussed above, cellular uptake of the 8-20 uM extracellular Fe-citrate is probably the major mechanism of cellular iron uptake in the iron-overload states. Cessation of NTBI uptake would at least stabilize intracellular iron levels. Despite these difficulties, when long-term deferoxamine therapy is adhered to, survival in the iron-overload states is enhanced and organ toxicity can be markedly reduced and even avoided.[47-49]

The limitations of deferoxamine, including difficulties with oral administration and reduced cell penetration, have stimulated interest in the development and clinical use of orally active iron chelators. The pyridones are a group of orally active iron chelators that have a high affinity for iron as well as enhanced lipid solubility and cell penetration.[54] Pyridones chelate iron in a redox active form via a ratio of three pyridone ligands to one iron. These compounds not only have adequate bioavailability following administration, but have an enhanced access to intracellular iron stores. Increased cell permeability may increase iron removal, while the ability to administer drugs orally enhances patient compliance. Based on in vitro studies, cell permeability due to lipid solubility is the critical determinant for effective cellular iron mobilization.[55] Pyridone chelators with an approximately equal partition into aqueous and lipid phases are the most efficient iron chelators.[55] Pyridone chelators with greater lipid solubility increase cell toxicity.[55] The iron binding constant is a secondary consideration, but is important at low chelator concentrations, and is thus probably important in clinical settings. Pyridone chelators, including L1, have superior iron affinity and cell permeability properties compared to deferoxamine. In cultured hepatocytes, iron mobilization by these compounds is superior to that achieved by deferoxamine.[55] In initial clinical studies, the pyridone compounds have increased iron excretion in patients suffering from chronic transfusion-induced iron overload, and are the subject of continuing study.[56] Iron excretion with the pyridone chelators equals that achieved with deferoxamine, with an acceptable side effect profile.[57] Certainly they have the potential to supplement or perhaps even supplant deferoxamine as the therapeutic agent for the mobilization and removal of tissue iron in the iron-overload states.

It is interesting to note that iron bound to both deferoxamine and the pyridones cannot participate in redox-cycling reactions or catalyze oxidative injury. Other cell-permeable iron ligands, such as tropolone and omadine, are able to enter the cell and access intracellular iron.[58] However, these ligands bind iron in a redox-active form, and in cell culture, can enhance iron toxicity.[58] Thus, clinically safe iron-binding requires that the iron be removed in a redox-inactive form, underscoring the importance of iron-catalyzed oxidative injury from a clinical perspective as well.

INSIGHT INTO IRON-CATALYZED OXIDATIVE INJURY PROVIDED BY THE IRON OVERLOAD STATES

Myocardial injury associated with iron occurs in conditions other than the tissue iron-overload states, including ischemia-reperfusion injury and anthracycline-induced cardiotoxicity. A large body of experimental work implicates redox-active intracellular iron as a contributor to myocardial injury during ischemia and reperfusion. From a purely clinical perspective, emerging work suggests that cardiac iron stores may also be important in the outcome of atherosclerotic coronary artery disease. Women have a decreased morbidity and mortality related to coronary artery disease, and this has been hypothesized to be due to their lower tissue iron stores caused by menstrual bleeding.[59] Conversely, the incidence of acute myocardial infarction in a population of middle-aged Finnish men was strongly predicted by the serum ferritin level.[60] Perhaps increased tissue iron stores contribute to the oxidation of low-density lipoproteins, predisposing to increased atherosclerosis, to endothelial cell damage that may contribute to the genesis of acute coronary syndromes, or to enhanced myocardial oxidative injury leading to greater myocyte necrosis during acute myocardial infarction.

Based on both in vitro and in vivo experimental studies, intracellular redox-active iron contributes to myocardial injury during ischemia and reperfusion. The use of large doses of deferoxamine in isolated hearts in vitro, either prior to ischemia or during early reperfusion, reduces myocardial injury, free radical generation, and lipid peroxidation.[61] Consistent with their greater access to the intracellular space, pyridone chelators were also successful, but at lower concentrations than deferoxamine.[62] In vivo, treatment prior to ischemia with deferoxamine reduces experimental canine infarct size.[51,63] However, therapy around the time of reperfusion with deferoxamine was ineffective.[51] Perhaps the lack of efficacy was related to the inability of deferoxamine to access critical sites of iron-catalyzed damage. To more directly determine the sites of iron-catalyzed injury during reperfusion, isolated buffer-perfused hearts were exposed to catalytic (5 uM) levels of redox-active iron bound to cell-permeable ligands during early reperfusion. Subtle elevation of cytosolic iron, but not of extracellular iron alone, increased lipid peroxidation, contractile dysfunction, and myocardial necrosis.[64] Thus, iron-catalyzed injury during reperfusion appears to occur predominantly at intracellular sites.[64] Since iron-catalyzed oxidative injury is "site-specific" due to the very reactive nature of iron-catalyzed ·OH or "·OH-like" species,[21] this also suggests that the predominant targets of iron-catalyzed oxidative injury during reperfusion are intracellular in location (Figure 5). In fact, the extent of oxidative injury during reperfusion, including lipid peroxidation and contractile dysfunction, increases in proportion to the tissue iron level present during the reperfusion period[65] (Figure 6).

The primary importance of intracellular iron stores as the catalytic site of oxidative damage during reperfusion explains why deferoxamine and the large enzyme scavengers superoxide dismutase and catalase, with their limited intracellular penetration, have not consistently reduced experimental infarct size when administered during reperfusion.[66] In contrast, therapy around the time of reperfusion with cell-permeable scavengers of ·OH, including dimethylthiourea (DMTU)[67] and N-2-mercaptopropionyl glycine (MPG),[68] reduce experimental infarct size, including at 48 hours post-infarction,[67] when definitive infarct size is manifest. Thus, if a scavenger of iron-catalyzed ·OH can gain access to the appropriate compartment where iron-catalyzed injury occurs, oxidative damage can be reduced in vivo. The $\cdot O_2^-$ and H_2O_2 precursors of iron-catalyzed injury can be generated either internal or external to the cell, yet still participate in intracellular iron-catalyzed oxidative damage (Figure 5). H_2O_2 can diffuse across cell membranes, while superoxide can traverse anion channels. This finding is consistent with results in cell culture that target cells provide the intracellular iron for their own destruction, even when $\cdot O_2^-$ and H_2O_2 are generated in the extracellular space.[53] Thus, a cell-permeable ·OH scavenger or iron-chelator might emerge as adjunctive therapy to reduce the component of oxidative injury associated with the reperfusion of acute myocardial infarction.[69]

Anthracycline chemotherapy agents cause cardiotoxicity in approximately 20% of patients when the cumulative dose exceeds 700 mg/m^2.[70] The cardiac damage is thought to involve anthracycline-induced redox-cycling of iron that leads to oxidative damage. In contrast, the mechanism of the chemotherapeutic effect may not involve iron-catalyzed oxidative mechanisms.[71] Anthracycline toxicity increases lipid peroxidation and impairs membrane function and cardiac contractility.[72] A further increase in tissue iron content by approximately 25% markedly increased anthracycline toxicity to cardiac myocytes, while exposure to iron alone or anthracycline alone had only minimal toxic effects in this cell culture model.[73] The anthracycline toxicity was reduced by pre-treatment with deferoxamine.[73] Since many patients receiving anthracycline chemotherapy may have increased tissue iron stores resulting from multiple transfusions required to treat anemia and bone marrow suppression associated with malignancy, this synergistic toxicity deserves additional consideration. Underscoring the importance of intracellular iron in anthracycline cardiac toxicity, a cell permeable iron-chelator reduced anthracycline toxicity in a clinical trial.[74] In the future, chelation of cardiac iron may become routine adjunctive therapy that will allow the continued use of anthracycline agents.

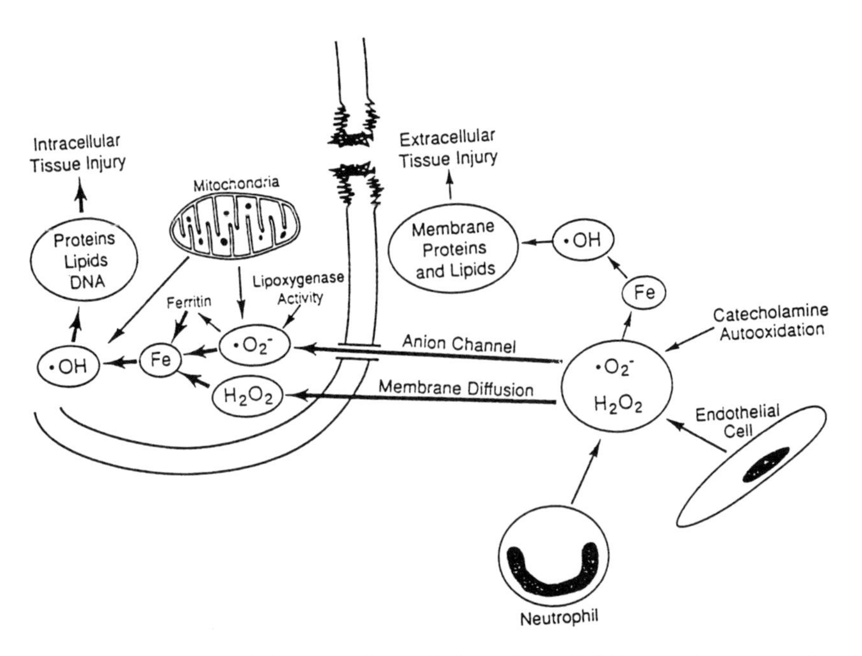

Figure 5. During myocardial ischemia and reperfusion, $\cdot O_2^-$ and H_2O_2 can be generated at both intracellular and extracellular sites, which can then react with intracellular iron to direct oxidative injury to intracellular locations in a "site-specific" fashion. Reprinted with permission from Lesnefsky, Free Radicals Biology and Medicine 12:429-446 (1992).

Cytosolic iron, whether dramatically increased in iron-overload states, interacting with anthracyclines that enhance iron redox cycling, or reacting with H_2O_2 and $\cdot O_2^-$ during reperfusion, can contribute to cardiac toxicity. The presence of additional factors that enhance iron-catalyzed oxidative damage during reperfusion ($\cdot O_2^-$ and H_2O_2 generation) or with anthracyclines (enhanced iron redox-cycling) produce cell injury at tissue iron levels lower than in the iron-overload states, where such factors are presumably absent.

However, it is likely that the mechanisms of iron-catalyzed tissue injury in all three states are similar. Intervention with iron chelating therapy to lower tissue iron, or to chelate iron in a redox-inactive state, holds the promise of reduced cardiac dysfunction and myocyte necrosis. A reduction in cardiac injury with lowering of tissue iron has been achieved in the iron-overload states, and may also be successful in managing anthracycline toxicity. While these two diseases allow for the opportunity for chronic iron chelator therapy, the reduction of reperfusion injury requires the prompt access of an iron chelator that binds iron in a redox-inactive form to the intracellular space. A pyridone-type chelator may be hold promise in this regard. Administration of a cell-permeable hydroxyl radical scavenger that reduces iron-catalyzed oxidative injury could also be successful. Conversely, in chronic iron-overload states or anthracycline-induced cardiotoxicity, one would speculate that chronic antioxidant therapy might be beneficial.

SUMMARY

Tissue iron overload causes clinical syndromes that involve the heart, liver, and pancreas. While tissue iron uptake occurs by both transferrin-dependent and independent processes, tissue uptake in the iron overload syndromes occurs

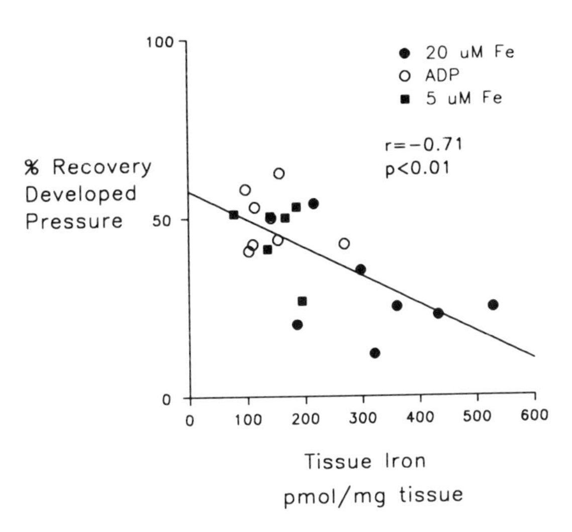

Figure 6. The recovery of myocardial systolic function during reperfusion following 30 minutes of ischemia in the isolated, buffer perfused rabbit heart is reduced in proportion to the increase in tissue iron present during reperfusion. Tissue iron was increased by exposure to 20 μM Fe^{2}-100μM ADP during the initial four minutes of reperfusion, while controls received ADP alone or 5μM Fe^{2+}-25μM ADP during early reperfusion, regimens that did not increase tissue iron levels.[65]

predominantly via transferrin-independent mechanisms. Increased redox-active iron present in hemeproteins and the cytosolic iron pool can catalyze oxidative damage to lipids, proteins, and nucleic acids, either by oxyradical dependent or independent mechanisms.

Iron-catalyzed injury results in damage to cell constituents, including mitochondria, lysosomes, and the sarcolemmal membrane. These mechanisms of iron-mediated damage are involved in the pathogenesis of organ dysfunction in primary hemochromatosis, transfusion-related iron overload, ischemia-reperfusion injury, and cardiac anthracycline toxicity.

ACKNOWLEDGEMENTS

The author wishes to thank Charles L. Hoppel, M.D. for his thoughtful review. The secretarial assistance of Ms. Sandra Vincent and the graphics assistance of the Cleveland VAMC Medical Media Department, especially Mr. James Suchy, is greatly appreciated.

REFERENCES

1. J.L. Goldstein, M.S. Brown. Genetics and cardiovascular disease, in: Heart Disease A Textbook of Cardiovascular Medicine, E. Braunwald, ed., W.B. Saunders Company Philadelphia, PA. (1988).
2. D.S. Rosenthal, E. Braunwald. Hematological-oncological disorders and heart disease, in: Heart Disease A Textbook of Cardiovascular Medicine, E. Braunwald, ed., W.B. Saunders Company Philadelphia, PA. (1988).
3. D.M. Koeller, J.A. Horowitz, J.L. Casey, R.D. Klausner, J.B. Harford, Translation and the stability of mRNAs encoding the transferrin receptor and c-fos, *Proc Natl Acad Sci USA* 88:7778-7782 (1991).

4. R.D. Klausner, J.B. Harford, Cis-trans models for post-transcriptional gene regulation. *Science* 246:870-872 (1989).
5. P.C. Adams, L.A. Chau, Hepatic ferritin uptake and hepatic iron, *Hepatology* 11:805-808 (1990).
6. J.C. Sibille, M. Ciriolo, H. Kondo, R.R. Crichton, P. Aisen, Subcellular localization of ferritin and iron taken up by rat hepatocytes, *Biochem J* 262:685-688 (1989).
7. R.A. Floyd, Direct demonstration that ferrous ion complexes of di- and triphosphate nucleotides catalyze hydroxyl free radical formation from hydrogen peroxide. *Arch Biochem Biophys* 225;263-270 (1983).
8. R.A. Floyd, C.A. Lewis, Hydroxyl free radical formation from hydrogen peroxide by ferrous iron-nucleotide complexes, *Biochemistry* 22:2645-2649 (1983).
9. M. Grootveld, J.D. Bell, B. Halliwell, O.I. Aruoma, A. Bomford, P.J. Sadler, Non-transferrin-bound iron in plasma or serum from patients with idiopathic hemochromatosis. Characterization by high performance liquid chromatography and nuclear magnetic resonance spectroscopy, *J Biol Chem*, 264:4417-4422 (1989).
10. G. Link, A. Pinson, C. Hershko, Heart cells in culture: a model of myocardial iron overload and chelation, *J Lab Clin Med*, 106:147-153 (1985).
11. J. Kaplan, I. Jordan, A. Sturrock, Regulation of the transferrin-independent iron transport system in cultured cells, *J Biol Chem*, 266:2997-3004 (1991).
12. J.G. Parkes, R.A. Hussain, N.F. Olivieri, D.M. Templeton, Effects of iron loading on uptake, speciation, and chelation of iron in cultured myocardial cells, *J Lab Clin Med* 122:36-47 (1993).
13. A. Jacobs, Low molecular weight intracellular iron transport compounds. *Blood* 50:433-439 (1977).
14. P. Biemond, A.J.G. Swaak, C.M. Beindorff, J.F. Koster JF, Superoxide-dependent and -independent mechanisms of iron mobilization from ferritin by xanthine oxidase, *Biochem J*, 239:169-173 (1986).
15. G. Healing, J. Gower, B. Fuller, C. Green, Intracellular iron redistribution, an important determinant of reperfusion damage to rabbit kidneys, *Biochem Pharm* 7:1239-1245 (1990).
16. S. Holt, M. Gunderson, K. Joyce, N.R. Nayini, G.F. Eyster, A.M Garitano, C. Zonia, G.S. Krause, S.D. Aust, B.C. White, Myocardial tissue iron delocalization and evidence for lipid peroxidation after two hours of ischemia, *Ann Emerg Med* 15:1155-1159 (1986).
17. S. Harel, M.A. Salan, J. Kanner, Iron release from methyoclobin, methemoglobin and cytochrome c by a system generating hydrogen peroxide, *Free Rad Res Comms* 5:11-19 (1988).
18. S. Harel, J. Kanner, The generation of ferryl or hydroxyl radicals during interaction of haemproteins with hydrogen peroxide, *Free Rad Res Comms* 5:21-33 (1988).
19. J.K. Brieland, J.C. Fantone, Ferrous iron release from transferrin by hemin neutrophil-derived superoxide anion: effect of pH and iron saturation. *Arch Biochem Biophys* 284:78-83 (1991).
20. J.K. Brieland, S.J. Clarke, S. Karminol, S.H. Phan, J.C. Fantone, Transferrin: a potential source of iron for oxygen free radical-mediated endothelial cell injury, *Arch Biochem Biophys* 294:265-270 (1992).

21. B. Halliwell, J.M.C. Gutteridge, Oxygen toxicity, oxygen radicals, transition metals and disease, *Biochem J* 219:1-14 (1984).
22. B. Halliwell, J.M.C. Gutteridge, Oxygen free radical and iron in relation to biology and medicine: Some problems and concepts, *Arch Biochem Biophys* 246:501-514 (1986).
23. E. Graf, J.R. Mahoney, R.G. Bryant, J.W. Eaton, Iron-catalyzed hydroxyl radical formation, *J Biol Chem* 259:3620-3624 (1984).
24. L.A. Loeb, E.A. James, A.M. Waltersdorph, S.J. Klebanoff, Mutagenesis by the autoxidation of iron with isolated DNA, *Proc Natl Acad Sci USA* 85:3918-3922 (1988).
25. H.W. Gardner, Oxygen radical chemistry of polyunsaturated fatty acids, *Free Rad Biol Med* 7:65-86 (1989).
26. E.R. Stadtman, Metal ion-catalyzed oxidation of proteins: Biochemical mechanism and biological consequences *Free Rad. Biol. Med* 9:315-325 (1990).
27. F.Z. Meerson, V.E. Kagan, N.P. Kozlov, L.M. Belkina, Y.V. Arkhipenko, The role of lipid peroxidation in pathogenesis of ischemic damage and the antioxidant protection of heart, *Basic Res Cardiol* 77:465-485 (1982).
28. E.J. Lesnefsky, K.G.D. Allen, F.P. Carrea, L.D. Horwitz, Iron-Catalyzed Lipid Peroxidation Occurs in the Intact Heart: Detection Using a New Lipid Peroxide Assay, *J Mol Cell Cardiol* 24:1031-1038 (1992).
29. K.P. Burton, J.M. McCord, G. Ghai, Myocardial alteration due to free radical generation. *Am J Physiol* 246:H776-H783 (1984).
30. J.M. Braughler, L.A. Duncan, R.L. Chase, The involvement of iron in lipid peroxidation, *J Biol Chem* 261:10282-10289 (1986).
31. J.M. Gutteridge, The role of superoxide and hydroxyl radicals in phospholipid peroxidation catalyzed by iron salts, *FEBS Lett* 150:454-458 (1982).
32. J.R. Bucher, M. Tien, S.D. Aust, The requirement for ferric in the initiation of lipid peroxidation by chelated ferrous iron, *Biochem Biophys Res Comm* 111:777-784 (1983).
33. J.M.C. Gutteridge, Lipid peroxidation initiated by superoxide-dependent hydroxyl radicals using complexed iron and hydrogen peroxide. *FEBS Lett* 172:245-249 (1984).
34. P.J. O'Brien, Intracellular mechanisms for the decomposition of a lipid peroxide. I. Decomposition of a lipid peroxide by metal ions, heme compounds, and nucleophiles, *Can J Biochem* 47:485-499 (1969).
35. B.A. Svinson, J.A. Buege, F.O. O'Neal, S.D. Aust, The mechanism of NADPH-dependent lipid peroxidation, *J Biol Chem* 254:5892-5899 (1979).
36. J.M. Gutteridge, R. Richmond, B. Halliwell, Inhibition of the iron-catalyzed formation of hydroxyl radicals from superoxide and of lipid peroxidation by desferrioxamine, *Biochem J* 184:469-472 (1979).
37. G. Cohen, P.M. Sinet, The Fenton reaction between ferrous-diethylene triamepenta-acetic acid and hydrogen peroxide. *FEBS Lett* 138:258-260 (1982).
38. B.R. Bacon, R. O'Neill, Park C.H, Iron-induced peroxidative injury to isolated rat hepatic mitochondria. *Free Rad Biol Med* 2:339-347 (1986).
39. B.R. Bacon, C.H. Park, G.M. Brittenham, R. O'Neil, A.S. Tavill, Hepatic mitochondrial oxidative metabolism in rats with chronic dietary iron overload, *Hepatology* 5:789-797 (1985).

40. A. Masini, D. Ceccareli, T. Trenti, F.P. Corongiu, U. Muscatello, Perturbation in liver mitochondrial Ca^{2+}homeostasis in experimental iron overload: a possible factor in cell injury, *Biochim Biophys Acta* 1014:133-140 (1989).
41. G. Link, A. Pinson, Hershko C, Iron loading of cultured cardiac myocytes modifies sarcolemmal structure and increases lysosomal fragility. *J Lab Clin Med* 121:127-134 (1993).
42. I.T. Mak, W.B. Weglicki, Characterization of iron-mediated peroxidative injury in isolated hepatic lysosomes, *J Clin Invest* 75:58-63 (1985).
43. K. Houglum, M. Filip, J.L. Witzutum, M. Choikier, Malondialdehyde and 4-hydroxynonenal protein adducts in plasma and liver of rats with iron overload, *J Clin Invest* 86:1991-1998 (1990).
44. C.A. Seymour, T.J. Peters, Organelle pathology in primary and secondary haemochromatosis with special reference to lysosomal changes, *Br J Haematol* 40:239-253 (1978).
45. M.P. Weir, J.F. Gibson, T.J. Peters, Haemodiserin and tissue damage, *Cell Biochem Funct* 2:186-194 (1984).
46. A.D. Heys, T.L. Dormandy, Lipid peroxidation in iron loaded spleens, *Clin Sci* 60:295-301 (1981).
47. L. Wolfe, N. Olivieri, D. Sallan, Prevention of cardiac disease by subcutaneous deferoxamine in patients with thalassemia major, *New Engl J Med* 312:1600-1603 (1985).
48. A. Cohen, Current status of iron chelation therapy with deferoxamine, *Semin Hematol* 27:86-90 (1990).
49. R. Marcus, S. Davies, H. Bantock, S. Underwood, S. Walton, E. Huehns, Desferrioxamine to improve cardiac function in iron-overloaded patients with thalassemia major, *Lancet* 1:392-393 (1984).
50. S. Hoe, D.A. Rowley, B. Halliwell, Reactions of ferrioxamine and desferrioxamine with the hydroxyl radical, *Chem Biol Inter* 41:75-81 (1982).
51. E.J. Lesnefsky, J.E. Repine, L.D. Horwitz, Deferoxamine pretreatment reduces canine infarct size and oxidative injury, *J Pharm Exp Ther* 253:1103-1109 (1990).
52. R. Laub, Y.J. Schneider, J.N. Octave, A. Trouet, R.R. Crichton, Cellular pharmacology of deferoxamine B and derivatives in cultured rat hepatocytes in relation to iron mobilization, *Biochem Pharm* 34:1175-1183 (1985).
53. D.E. Gannon, J. Varani, S.H. Phan, J.H. Ward, J. Kaplan, G.O. Till, R.H. Simon, U.S. Ryan, P.A. Ward, Source of iron in neutrophil-mediated killing of endothelial cells, *Lab Invest* 57:37-44 (1987).
54. G.J. Kontoghiorghes, The study of iron mobilisation from transferrin using a-ketohydroxy heteroaromatic chelators, *Biochim Biophys Acta* 869:141-146 (1986).
55. J.B. Porter, M. Gyparaki, L.C. Burke, E.R. Huehns, P. Sarpong, V. Saez, R.C. Hider, Iron mobilization from hepatocyte monolayer cultures by chelators: the importance of membrane permeability and the iron-binding constant, *Blood* 72:1497-1503 (1988).
56. P. Tondury, G.J. Kontoghiorghes, A. Ridolfi-Luthy, A. Hirt, A.V. Hoffbrand, A.M. Lottenbach, T. Sonderegger, H.P. Wagner, L1 (1,2-dimethyl-3-hydroxypyrid-4-one) for oral iron chelation in patients with bets-thalassemia major, *Br J Haematol* 76:550-553 (1990).

57. N.R. Olivieri, G. Koren, C. Hermann, Y. Bentur, D. Chung, J. Lkein, P. St. Louis, M.H. Freedman, R.A. McClelland, D.M. Templeton, Comparison of oral iron chelator L1 and desferrioxamine in iron-loaded patients, *Lancet* 336:1275-1279 (1990).
58. G.J. Kontoghorgies, A. Piga, A.V. Hoffbrand, Cytotoxic and DNA-inhibitory effects of iron chelators on human leukaemic cell lines. *Hematol Oncol* 4:195-204 (1986).
59. J.L. Sullivan, The iron paradigm of ischemic heart disease, *Am Heart J* 117:1177-1188 (1989).
60. J.T. Salonen, K. Nyyssonen, H. Koppela, J. Tuomilehto, R. Seppanen, R. Salonen, High stored iron levels are associated with excess risk of myocardial infarction in Finnish men. *Circulation* 86:803-811 (1992).
61. R.E. Williams, J.L. Zweier, J.T. Flaherty, Treatment with deferoxamine during ischemia improves functional and metabolic recovery and reduces reperfusion-induced oxygen radical generation in rabbit hearts, *Circulation* 83:1006-1014 (1991).
62. A.M.M. van der Kraaij, H.G. van Eijk, J.F. Koster, Prevention of postischemic cardiac injury by the orally active iron chelator 1,2-dimethyl-3-hydroxy-4-pyridone (L1) and the antioxidant (+)-cyanidanol-3, *Circulation* 80:158-164 (1989).
63. B.R. Reddy, R.A. Kloner, K. Przyklenk, Early treatment with deferoxamine limits myocardial ischemic/reperfusion injury, *Free Rad Biol Med* 7:45-52 (1989).
64. E.J. Lesnefsky, J. Ye, Exogenous Intracellular, But Not Extracellular, Iron Augments Myocardial Oxidative Injury During Ischemia and Reperfusion, *Am J Physiol* (1994).
65. E.J. Lesnefsky, J. Ye, Excess Intracellular, But Not Extracellular Iron Augments Myocardial Lipid Peroxidation and Injury During Reperfusion. *Clin Res* 39:691A (1991).
66. R. Engler, E. Gilpin, Can superoxide dismutase alter myocardial infarct size? *Circulation* 79:1137-1142 (1989).
67. F.P. Carrea, E.J. Lesnefsky, J.E. Repine, R.H. Shikes, L.D. Horwitz, Reduction of Canine Myocardial Infarct Size by a Diffusible Reactive Oxygen Metabolite Scavenger: Efficacy of Dimethylthiourea Given at the Onset of Reperfusion, *Circ Res* 68:1652-59 (1991).
68. S.E. Mitsos, T.E. Askew, J.C. Fantone, S.L. Kunkel, G.D. Abrams, A. Schork, B.R. Lucchesi, Protective effects of N-2-mercaptopropionyl glycine against myocardial reperfusion injury after neutrophil depletion in the dog: evidence for the role of intracellular-derived free radicals, *Circulation* 73:1077-1086 (1986).
69. E.J. Lesnefsky, Reduction of Infarct Size by Cell Permeable Oxygen Metabolite Scavengers. *Free Rad Biol and Med* 12:429-446 (1992).
70. R.C. Young, R.F. Ozols, C.E. Myers, The anthracycline antineoplastic drugs. *N Engl J Med* 305:139-153 (1981).
71. C.E. Myers, W.P. McGuire, R.H. Liss, Adriamycin: the role of lipid peroxidation in cardiac toxicity and tumor response. *Science* 197:165-167 (1977).
72. P.K. Singal, G.N. Pierce, Adriamycin stimulates low affinity Ca^{2+} binding and lipid peroxidation but depresses myocardial function. *Am J Physiol* 250:H419-H425 (1986).

73. C. Hershko, G. Link, M. Tzahor, J.P. Kaltwasser, P. Athias, A. Grynberg, A. Pinson. Anthracycline toxicity is potentiated by iron and inhibited deferoxamine: Studies in rat heart cells in culture. *J Lab Clin Med* 122:245-251 (1993).

74. J.L. Speyer, M.D. Green, A. Zeleneiuch-Jacquotte, J.C. Wernz, M. Rey, J. Sanger, E. Kramer, V. Ferrans, H. Hochster, M. Meyers, et al, ICRF-187 permits longer treatment with doxorubicin in women with breast cancer. *J Clin Oncol* 10:117-127 (1992).

FREE RADICALS IN THE PATHOPHYSIOLOGY OF PULMONARY INJURY AND DISEASE

Daniel J. Brackett[1,2,3] and Paul B. McCay[1,3]

[1]Departments of Surgery and Anesthesiology
University of Oklahoma Health Sciences Center
[2]Veterans Affairs Medical Center
[3]Free Radical Biology and Aging Program
Oklahoma Medical Research Foundation
Oklahoma City, Oklahoma 73190

INTRODUCTION

There is a substantial volume of literature indicating that free radicals are significantly involved in lung injury and disease in a wide variety of clinically relevant situations. Evidence to support this relationship is derived from both experimental animal studies and clinical studies of human patients. Data is convincing that pulmonary injury can be induced by a myriad of stimuli that have the capacity to activate the inflammatory system. These stimuli initiate a cascade of events that appear to involve the inflammatory mediators complement, tumor necrosis factor, thromboxane, xanthine oxidase, neutrophils, and the subject of this chapter, free radicals. The conclusions of reports which have utilized manipulation of these mediators in experimental studies and measurements of indices of excessive free radical generation in patients suffering from pulmonary disorders strongly suggest that free radicals may be essential and perhaps final mediators of lung injury in many of these experimental and clinical situations. These data support further evidence that measurements detecting the presence of excessive free radical generation or cellular injury known to be associated with free radical reactions may be useful tools in revealing the potential for the development of pulmonary damage and as predictors of the intensity of the injury. Clinical measurements of this type may result in the identification of patients who are at high risk for pulmonary injury or disease and who are candidates for antioxidant therapy. This chapter will review a selection of reports from the literature documenting the significant role of free radicals in lung injury, their presence in pulmonary disease, and the potential for clinically relevant measurements of free radical generation and activity to aid in the detection, prediction, and therapy of damage to the pulmonary system.

Free Radicals in Diagnostic Medicine, Edited by
D. Armstrong, Plenum Press, New York, 1994

EVIDENCE OF FREE RADICAL INVOLVEMENT IN PULMONARY INJURY

Pulmonary injury and dysfunction occurs following multiple types of insults to peripheral tissue and organs. Mediators, documented to be associated with free radical generation and produced as a result of these systemic cellular insults, are activated or released into the venous system and the first organ encountered is the lung. The pulmonary capillary bed endothelium becomes the primary target of any toxic or potentially injurious mediator(s) since it is the first capillary bed encountered by blood leaving the affected tissue. Even though there is wide variation in the types of insults, the mediators involved in the resulting pulmonary injury are essentially the same. The insults and the mediators and the sources of their release are discussed below.

Reperfusion of Ischemic Tissue

Reperfusion of ischemic tissues and organs (I/R) in patients is an important and not uncommon clinical event, occurring during revascularization of compromised vessels, organ transplantation, cardiac surgery, aneurysm repair, extremity replantation, and many types of reconstructive surgery. In many of these cases re-establishment of blood flow is followed by pulmonary distress.[1,2,3,4] Experimental animal models have generated evidence that free radicals are an important factor in reperfusion-evoked pulmonary disturbances following ischemia of diverse tissues and organs.

Intestinal. Reperfusion of ischemic intestinal tissue is used frequently as a model to study and elucidate the mechanisms of acute respiratory distress since the majority of the pulmonary responses are the same. The inflammatory mediator that appears to be common to essentially all reports concerning pulmonary distress following intestinal reperfusion, as well as reperfusion of other systemic ischemic tissue, is the activated neutrophil and the associated oxidative burst which gives rise to powerful oxidizing agents. The source of these neutrophil-derived oxidants is the superoxide radical.[5] Intestinal I/R evokes increased pulmonary microvascular permeability, alveolar capillary endothelial cell injury, and pulmonary neutrophil accumulation suggesting an injurious role for the neutrophil.[6] The injurious role for neutrophil-derived cytotoxins including oxygen species was confirmed when intestinal I/R-induced pulmonary injury was eliminated in neutropenic animals.[7] Phospholipase A_2 appears to be a key mediator in this injury produced by products of activated neutrophils, since inhibition of this enzyme prevented enhanced neutrophil superoxide production and pulmonary injury; however, neutrophil to endothelial cell adhesion was not prevented.[8] The dissociation between neutrophil-endothelial cell adhesion and pulmonary injury and the critical role of neutrophil extravasation and the oxidative burst has been noted in similar intestinal I/R studies involving the mediators xanthine oxidase and tumor necrosis factor (TNF) and the adhesion molecules Mac-1, CD18, and P-selectin. Depletion of xanthine oxidase eliminated intestinal I/R-induced lung injury[9] and was associated with attenuated neutrophil priming as measured by superoxide radical production.[10] However, there was no effect on neutrophil to endothelial cell adhesion. Similarly, an anti-TNF antibody did not reduce neutrophil to endothelial cell adhesion in the lung, but did attenuate pulmonary microvascular injury.[11] Differences found using monoclonal antibodies to neutrophil and endothelial cell adhesion molecules in this model may help explain the attenuation of injury in the presence of neutrophil adhesion and emphasize the significance of the oxidative burst and free radical production. Monoclonal antibodies to the neutrophil integrin CD18 significantly reduced neutrophil to endothelial cell adhesion and pulmonary injury produced by intestinal I/R.[12] However, antibodies to the integrin Mac-1, which is required for neutrophil adhesion before the oxidative burst can occur, prevented lung permeability but had no effect on neutrophil adhesion to the endothelium.[13] These data indicate that Mac-1 is required for the oxidative burst and pulmonary injury, but not neutrophil adhesion, while CD18 is required for both. The expression of both of these adhesion molecules appears to be essential for pulmonary injury to occur following reperfusion of the ischemic intestine. Antibodies to the endothelial cell adhesion molecule P-selectin inhibit the lung injury elicited by intestinal I/R, but have no effect on the accumulation of neutrophils demonstrating a functional similarity with the Mac-1 integrin.[14] These studies provide evidence that stimulation of the neutrophil to adhere to the endothelium is not necessarily sufficient to provoke injury, but extravasation and the oxidative burst releasing oxidative and proteolytic products appear to be required.

Limb / Torso. The pulmonary distress and injury induced by reperfusion of ischemic limb or lower torso tissue and the mediators involved are similar to those found in intestinal I/R and the injury appears to be dependent on neutrophils and neutrophil-derived oxidative products. Neutropenia has been shown to effective in normalizing lung histology and pulmonary function during reperfusion of limb ischemia.[15] During reperfusion in the same ischemia model, treatment with either the free oxygen radical scavengers, superoxide dismutase and catalase, or the xanthine oxidase inhibitor, allopurinol, prevented lung injury and neutrophil accumulation.[16] Further evidence of significant free radical involvement was demonstrated using the oxygen radical scavengers (DMSO, DMTU) or inhibitors of processes that generate oxygen radicals (allopurinol, superoxide dismutase, and catalase). These inhibitors act on the hydroxyl radical, xanthine oxidase, the superoxide radical, and hydrogen peroxide respectively. Each of these antioxidants blocked pulmonary injury and significantly diminished pulmonary neutrophil sequestration that occurred following limb I/R.[17] The administration of a monoclonal antibody to the CD18 neutrophil adhesion molecule prevented pulmonary neutrophil sequestration and reduced pulmonary microvascular permeability induced by I/R of the hind limbs or torso.[18] These studies provide convincing evidence of the role of neutrophil-derived free radicals as significant mediators in the final stages of distal organ cellular damage initiated by reperfusion of ischemic tissue. Data also exists that implicate thromboxane and TNF as mediators in the cascade of inflammatory events evoked during reperfusion of limb or torso ischemia. Thromboxane synthesis is stimulated during ischemia. Thromboxane inhibition using OK-046 during limb[19] or torso[20] I/R prevented increased pulmonary vascular permeability and neutrophil sequestration. Both of these pulmonary responses were also prevented by inhibition of lipoxygenase using diethylcarbamazine[19,21] and both inhibitors blocked the increase in circulating thromboxane.[19,20,21] TNF has been detected in the plasma following reperfusion of torso ischemia accompanied by sequestration of neutrophils in the lungs, microvascular permeability, pulmonary edema, and generation of leukotriene B_4 in the bronchoalveolar lavage fluid.[22] When a TNF antiserum was applied in this study the lung neutrophil sequestration was reduced, permeability and edema were limited, and the rise in leukotriene B_4 was prevented. During limb I/R, TNF and interleukin-1 are both elevated and are associated with accumulation of neutrophils in the lung, expression of the endothelial cell adhesion molecule E-selectin in the pulmonary vasculature, vascular permeability, and pulmonary hemorrhage.[23] Polyclonal antibodies to either TNF or interleukin-1 significantly attenuated or prevented each of these responses. Each of these inflammatory mediators, thromboxane, TNF, and interleukin, appears to be important in the pulmonary injury related to reperfusion of limb or torso ischemia and the final injurious mediators seem to be neutrophil-derived oxidative products.

Liver. The development of adult respiratory distress syndrome and pulmonary dysfunction and injury are associated with liver transplantation and end stage liver disease.[3] Experimental studies have demonstrated that reperfusion of ischemic hepatic tissue induced measurable circulating concentrations of TNF, pulmonary alterations including neutrophil infiltration, edema, and intra-alveolar hemorrhage. Anti-TNF antiserum caused a reduction in neutrophil accumulation and complete inhibition of injury.[24] In another study TNF concentrations were also documented to be elevated and pulmonary capillary permeability was significantly increased in response to reperfusion to ischemic hepatic tissue. Anti-TNF antiserum also prevented the damage to the pulmonary microvasculature in this study.[25] These data suggest that the process of pulmonary injury evoked by hepatic I/R is comparable to that of other ischemic tissue and that the damage is dependent on neutrophil activation, adhesion to the endothelium, migration into the interstitial tissue, and subsequently, the oxidative burst.

Whole Body (Circulatory Shock). Resuscitation following an episode of circulatory shock (inadequate tissue perfusion) is a form of ischemia-reperfusion involving total body tissue ischemia. Multiple organ failure is one of the complications of circulatory shock and either pulmonary dysfunction or the adult respiratory distress syndrome is one of the first signs that organ integrity is being threatened and is associated with a high mortality rate.[26,27] During circulatory shock marked accumulation of neutrophils in the vasculature[28,29] and damage to pulmonary vascular endothelium[30,31] occurs. In this situation the lung is particularly vulnerable, since while it is compromised with its own reperfusion injury, it is also the target of the devastating consequences of ischemia followed by reperfusion of the rest of the body, which have been documented above. There is a substantial amount of information implicating the generation of free radicals as a significant deleterious event during circulatory shock[32,33] and as with other tissue ischemia, the role of neutrophil-derived oxidative products seems to be significant. Lung damage and increased neutrophil adhesiveness are characteristic occurrences during lethal, experimental circulatory shock. When a monoclonal antibody to the neutrophil adherence molecule CD18 was given under these study conditions lethality was prevented and histologic evidence of lung injury was absent or markedly attenuated.[34] Further studies have demonstrated that during circulatory shock there is pulmonary sequestration of neutrophils and that the activation of neutrophils is dependent on xanthine oxidase.[35] When a deficiency of the antioxidant, glutathione, is imposed on circulatory shock, organ failure and lethality are enhanced.[36] These data suggest that neutrophil activation and the generation of oxidative products are significantly involved in the respiratory distress observed during circulatory shock.

Tumor Necrosis Factor

Tumor necrosis factor (TNF) concentrations are increased in response to reperfusion of ischemic tissue and with the use of monoclonal antibodies to TNF it has been demonstrated that TNF is one of the important mediators in the development of pulmonary injury during reperfusion conditions (documented in the discussion above). Further convincing evidence that TNF can induce pathologic pulmonary responses and that these responses are mediated by neutrophil-derived free radicals has been generated from studies using intravenous infusion of TNF. Tumor necrosis factor infusion elicited significant pulmonary damage that on histopathologic examination included acute inflammation of the interstitium and thickening of the alveolar walls[37] and diffuse hyperemia and punctate areas of hemorrhage.[38] Evaluation of the areas of hemorrhage revealed occlusion of arteries by thrombi containing large numbers of neutrophils, intense neutrophil margination, and migration of neutrophils through the vessel walls. This data implies an extraordinary stimulation and activation of neutrophils. *In vitro* studies have shown that neutrophil stimulation by TNF evokes the oxidative burst (measured by chemiluminescence and H_2O_2 production) and degranulation[39] and adherence to human umbilical vein endothelium.[40] These studies provide evidence that pulmonary injury is a threat during any physiological conditions of elevated circulating TNF concentrations and most importantly, that free radicals are the final mediator of pulmonary injury in situations of accelerated TNF synthesis.

Thermal / Burn

The primary cause of death in burn or thermal injury patients is infection followed by sepsis and pulmonary failure.[41,42] Activation of the complement system is an early response occurring in most burn patients[43,44] and experimental animals[45]. The magnitude of complement activation appears to be correlated with mortality.[43] The role of complement in the development of thermal injury induced progressive lung injury has been clearly demonstrated by the significant pulmonary protection provided by complement depletion.[46] In this same model of burn-induced lung injury, neutrophil depletion resulted in the same extent of pulmonary protection as complement depletion; however, when antioxidant therapy was applied utilizing the combination of superoxide dismutase and catalase, virtually complete protection was afforded.[46] Similar protection was found when the potent hydroxyl radical scavenger, dimethyl sulfoxide, or iron chelators (iron is a catalyst for hydroxyl radical generation) were used.[47] Evidence that the occurrence of lipid peroxidation (an ultimate process of cellular membrane damage) following burn injury is dependent on neutrophils and can be blocked by treatment with hydroxyl radical scavengers and iron chelators strongly implies that hydroxyl radical-induced lipid peroxidation may be the final process involved in pulmonary damage following systemic thermal injury.[48] Additional studies have demonstrated that lipid peroxidation occurs in lung tissue during burn-elicited pulmonary dysfunction and that thromboxane may be involved.[49] Increased serum catalase activity is also present associated with burn-induced lung injury and the injury is decreased by administration of the oxygen radical scavenger, dimethylthiourea, with no effect on the serum catalase activity.[50] Taken together these studies provide evidence that pulmonary distress evoked by thermal injury is initiated by complement activation followed by neutrophil activation, margination, migration, and degranulation; the generation of oxidative products; and finally, hydroxyl radical-induced lipid peroxidation in the pulmonary vasculature and interstitial tissue (for review, see ref. 51).

Complement

As documented repeatedly above, the activation of endogenous complement activity and the subsequent generation of free radicals is a significant event that occurs in the development of pulmonary injury and dysfunction in response to an insult that is confined to distal tissue. There is also more direct evidence derived from studies involving infusion of complement-activated plasma and cobra venom factor activation of the complement system to support these conclusions concerning the free radical-mediated deleterious effects of complement on the pulmonary system. Infusion of complement activated plasma resulted in pulmonary vascular thrombi composed of degenerating neutrophils and endothelial injury, as well as interstitial edema,[52] and pulmonary hypertension and increased vascular permeability[53]. These morphologic and functional changes were prevented or markedly inhibited by either neutrophil depletion[52] or treatment with superoxide dismutase to metabolize superoxide radicals.[53] There is also data establishing that thromboxane inhibition partially attenuates some of these pathologic effects inferring that thromboxane may have a mediator role in the cascade of injurious events initiated by activation of the complement system.[53] Intravascular activation of the complement system with cobra venom factor produced very similar pulmonary damage with neutrophils being prominent on histologic evaluation.[46,54] Complement associated lung injury was profoundly inhibited by depletion of neutrophils or administration of superoxide dismutase or catalase.[54] Treatment with the hydroxyl radical scavenger, dimethyl sulfoxide, or iron chelators resulted in protection from tissue injury without affecting the aggregation or margination of neutrophils which supports the concept that complement and neutrophil-induced pulmonary injury are dependent on the capacity of the local biochemical environment to produce hydroxyl radicals from neutrophil-derived hydrogen peroxide.[46] These data are supportive of studies implicating complement system activation as a triggering event in the induction of neutrophil and hydroxyl radical-dependent pulmonary injury.

Pancreatitis

It has been reported that 45 to 58% of patients with acute pancreatitis developed respiratory insufficiency[55,56] and the pulmonary dysfunction in one of every three progressed to adult respiratory distress syndrome[56]. A characteristic morphologic pattern appears to be present in these cases that features a loss of integrity of alveolar-capillary membrane resulting in pulmonary edema[57] and histologic findings reveal injury similar to that found in the "shock lung syndrome" [58]. However, this pulmonary damage is usually reversible with appropriate care.[59] In a large retrospective autopsy study in which 405 deaths from pancreatitis were identified, it was found that sixty percent died within seven days of admission.[60] Pulmonary edema and congestion were the prevalent causes of death suggesting that early detection leading to treatment should result in increased survival. Experimental studies designed to determine mechanisms, using a cerulein-induced pancreatitis that produced lung injury similar to that found in humans with pancreatitis-induced lung injury, revealed a profound, early neutrophil infiltration in the pulmonary tissue.[61] Either neutrophil or complement depletion or the combination of superoxide dismutase plus catalase (implicating the superoxide radical and hydrogen peroxide) protected against this injury.[62] However, contrary to other neutrophil-dependent, free radical-dependent pulmonary injury, the injury associated with pancreatitis does not appear to be mediated by the hydroxyl radical, since dimethyl sulfoxide and iron chelators were not effective in attenuating this injury. Nevertheless this pulmonary injury is similar to other injuries discussed thus far and seems to be initiated by complement system activation resulting in pulmonary damage caused by neutrophil-derived oxidative products, but is independent of iron and the hydroxyl radical.

Sepsis / Endotoxemia

There are approximately 350,000 cases of septic shock each year in the United States resulting in 100,000 to 150,000 deaths.[63] Multiple organ failure or dysfunction is usually the ultimate cause of death in these patients and one of the first organs to be effected is the lung and the development of adult respiratory distress syndrome frequently occurs. The most common predisposing causes of adult respiratory distress syndrome include gram-negative sepsis and gram-negative bacterial pneumonia.[64] It is generally agreed that many of the peripheral responses characteristic of septic shock are triggered by increases in circulating endotoxin, a component of the outer membrane of gram-negative bacteria.[65-71] These responses include increased synthesis of TNF and activation of the complement system. The detrimental effects of these two inflammatory agents on pulmonary function and morphology, as mediated by neutrophils and free radicals, have been documented above. A disturbance in the oxidant-antioxidant balance (as established by increased serum catalase and manganese superoxide dismutase concentrations) has been demonstrated in septic patients that eventually developed adult respiratory distress syndrome.[72,73] Direct detection of *in vivo* free radical generation in the lung induced by endotoxin has been achieved using electron spin resonance spectroscopy.[74] The free radical scavenger N-acetylcysteine significantly attenuated endotoxin-induced pathophysiologic alterations.[75] The induction of sepsis by infusion of live *Pseudomonas aeruginosa* caused significantly enhanced neutrophil superoxide production and increased airspace bronchoalveolar lavage protein and extravascular lung water[76] and elevated TNF plasma concentrations[77]. Inhibition of cyclooxygenase attenuated all of these responses,[76,77] even when inhibition was delayed until after the bacterial infusion[78]. Endotoxin induces substantial pulmonary neutrophil infiltration and these neutrophils are the predominant source of TNF-α

messenger RNA in the lung tissue 6-12 hours after the endotoxin challenge.[79] This additional source of TNF from neutrophils sequestered in the lung may perpetuate the potential for TNF-induced free radical generation. The evidence for free radical involvement in pulmonary injury in response to a bacterial or endotoxin challenge is convincing. Free radicals have been detected in pulmonary tissue in response to endotoxin, and interventions that block bacteria-induced neutrophil superoxide production and elevated circulating TNF concentrations prevent the associated pulmonary damage.

Glucose-Glucose Oxidase

Acute lung injury can be induced in intact animals or isolated perfused lungs by intratracheal instillation of glucose-glucose oxidase. This injury was shown to be inhibited by superoxide dismutase, catalase, and hydroxyl radical scavengers or iron chelators implicating superoxide radicals, hydrogen peroxide, and the hydroxyl radical as significant mediators in the development of this injury.[80,81] Administration of antioxidants in this model essentially prevented injury when given simultaneously with the insult, but afforded only minimal protection when delayed twenty minutes suggesting that the free radical mediated injury occurs very quickly.[82] Injury in this *in vivo* model was also blocked by administration of the manganese ion and was enhanced by EDTA, a manganese chelator, providing further evidence of free radical involvement.[83] In addition to inhibition of lipid peroxidation,[84] manganese causes disproportionation of hydrogen peroxide,[85-87] can catalyze superoxide anion dismutation,[88] and is a component of superoxide dismutase[89] all of which antagonize two important ingredients (superoxide anion and hydrogen peroxide) in the reactions generating the highly toxic hydroxyl radical.

Paraquat

The toxicity of paraquat is mediated initially by the conversion of molecular oxygen to the superoxide anion when an electron is transferred from paraquat.[90] Pulmonary tissue appears to be a target of paraquat[91] and paraquat-induced lipid peroxidation;[92] thereafter, pulmonary edema, intraalveolar hemorrhage, extensive fibrosis, and epithelial swelling subsequently occur.[93] Paraquat has the capacity to induce the release of chemoattractants for neutrophils in bronchoalveolar fluid, increased infiltration of inflammatory cells into the alveolae, and impairment of the ability of alveolar macrophages to release superoxide radicals in response to PMA stimulation.[94] Treatment with modest doses of the antioxidant N-acetylcysteine, a precursor of glutathione,[95] significantly reduced all of these injurious responses to paraquat suggesting an important role for free radicals in the pulmonary toxicity of this compound.

Air Pollutants (Ozone and Nitrous Dioxide)

Ozone and nitrous dioxide are major components of air pollutants that have toxic effects on the respiratory system.[96-98] There is evidence that lipid peroxidation is the injurious process mediating pulmonary damage induced by both ozone[99,100] and nitrous dioxide.[99,101] Dietary alpha-tocopherol significantly increased survival to acute toxic exposure to either ozone or nitrous dioxide.[102] Animals that died showed marked signs of pulmonary edema and hemorrhage. Animals fed the highest α-tocopherol diet with the supplemental antioxidants methionine, ascorbic acid, and butylated hydroxy-acetate were the least effected. The thiobarbituric acid assay for pulmonary malondialdehyde concentrations revealed an inverse relationship between malondialdehyde content and the amount of α-tocopherol in the diet in animals exposed to toxic levels of ozone. Both acute and chronic exposure to ozone significantly decreased tracheal mucous velocity.[103] The antioxidant N-acetylcysteine prevented the mucociliary dysfunction that occurred in both exposure groups. The efficacious effect of antioxidants on lung injury and pulmonary dysfunction induced by ozone and nitrous dioxide implies that free radicals and reactive oxygen species participate in the toxicity of these air pollutants.

Formyl-Norleucyl-Leucyl-Phenylalanine (FNLP)

FNLP is a neutrophil chemotactic and activator component of the cell wall of gram negative bacteria.[104] When FNLP was instilled in the trachea of experimental animals there was an increase in permeability and neutrophil alveolitis was evoked.[105] Dimethyl sulfoxide prevented these *in vivo* FNLP-induced responses. *In vitro* tests demonstrated that dimethyl sulfoxide inhibited neutrophil chemotaxis, but did not affect neutrophil superoxide generation or adherence to cultured endothelial cells or nylon fiber. The *in vivo* action of dimethyl sulfoxide may have been due to the effect on chemotaxis, but since the neutrophils retained the capacity to adhere and to produce superoxide, the potent free radical scavenging properties of dimethyl sulfoxide[106] may have been the component responsible for preventing the injury. The marginated, activated neutrophil and its ability to initiate the reactions necessary to produce the hydroxyl radical are probably the mediators of tissue injury in this model.[107,108]

Tobacco Smoke

Cigarette smoke contains approximately 10^{16} oxidant molecules per puff.[109] Neutrophils traveling through the capillary bed of the lungs may be exposed to this oxidant insult. Cigarette smoke has been shown to induce oxidative changes in the cell membrane and cytoskeleton of neutrophils, to reduce their deformability, and to diminish their ability to release oxidative products.[110] These responses to cigarette smoke increase the transit time of the neutrophil through the pulmonary capillary bed increasing the potential for activation and adherence. The endogenous antioxidant, glutathione, significantly diminishes these smoke-induced effects in an *in vitro* situation implicating oxidative products as having a role in these neutrophil responses. The antioxidant, N-acetylcysteine, a precursor to glutathione, was shown to increase both plasma and bronchoalveolar glutathione raising the possibility of a therapeutic intervention to enhance the antioxidant balance to resist pulmonary oxidant injury. Platelet activating factor stimulates increased amounts of superoxide radicals from the alveolar macrophages from human smokers[111] and spontaneous release of both superoxide and hydrogen peroxide is increased.[112] These observations concerning a potential pulmonary oxidant-antioxidant imbalance induced by cigarette smoke suggest that free radicals may be involved in the lung injury in smokers.

Immunoglobulin

Mouse immunoglobulins IgA and IgG instilled into the trachea of rats caused similar pulmonary injury.[113] Morphologic evaluation revealed blebbing of endothelial cells, intra-alveolar hemorrhage, and fibrin deposition. In addition significant increases in lung permeability were induced by both immunoglobulins. IgA immune complexes have the capacity to stimulate the activation of the complement system and activation of neutrophils, as measured by superoxide radical production, in a dose related manner. Complement depletion significantly inhibited the lung injury induced by both IgA and IgG; however, even though neutrophil depletion significantly inhibited pulmonary damage induced by IgG there was no effect of neutrophil depletion on the injury in response to IgA. Therefore, both complement and neutrophils are significant mediators of IgG-induced injury, apparently through the production and release of toxic oxidative products.[114] Information concerning the final pathway of IgA induced complement-dependent injury is not currently available.

HUMAN PULMONARY DISORDERS RELATED TO FREE RADICAL ACTIVITY

There are numerous pulmonary diseases and disorders that result in or are involved in processes of cellular injury to pulmonary tissue. Among these diseases are several that have been associated with the presence of free radicals and this has initiated the concept that free radical processes may be involved as mediators of the injury. Most of these particular diseases are also among the most well known and the evidence of their involvement with free radicals is presented below.

Adult Respiratory Distress Syndrome

The adult respiratory distress syndrome (ARDS) identifies a situation of increased pulmonary capillary permeability resulting in respiratory failure that is characterized by edema, hypoxemia, and mechanical abnormalities.[115-117] ARDS can be initiated by a diverse variety of insults including those discussed in the section above.[30] A subpopulation of primed neutrophils has been demonstrated to exist in the blood of patients with ARDS.[118] These neutrophils possessed the capacity to generate significantly elevated amounts of hydrogen peroxide and the hydrogen peroxide generation was correlated with circulating TNF concentrations as well as the ability of circulating monocytes to produce TNF. Spontaneous increased oxygen radical production by neutrophils, as demonstrated by chemiluminescence measurements, has been found to be significantly increased in patients with confirmed ARDS and the chemiluminescence value for patients identified to be at risk for ARDS was between the confirmed ARDS patients and the controls.[119] Both the confirmed and at risk ARDS groups were found to have evidence of increased complement consumption in their plasma. This set of studies suggests a role for free radicals, neutrophils, and complement in the development of ARDS. Further evidence of free radical involvement is acquired from data demonstrating that glutathione levels are decreased in plasma and red blood cells from patients with ARDS.[120] Glutathione is a major scavenger of hydrogen peroxide, a precursor of the hydroxyl radical. When N-acetylcysteine, a precursor of glutathione, was administered to these patients their red cell and plasma glutathione concentrations increased concurrently with improvement of multiple cardiopulmonary parameters. Another study has established that plasma concentrations of xanthine oxidase (an enzyme that catalyzes a reaction that produces superoxide anions) are increased in patients with ARDS compared to other critically ill patients even though hypoxanthine (one of the substrates for xanthine oxidase) was the same between both groups of patients.[121] These data indicate the existence of the potential for intravascular oxygen metabolite production. The collective data from these studies taking diverse experimental approaches have arrived at the conclusion that oxidative products are prime candidates as mediators of cellular injury during ARDS.

Respiratory Distress Syndrome in Premature Infants

Studies in very low birth weight infants have used ethane and pentane in expired air to elucidate the mechanisms of pulmonary injury in these patients. Ethane and pentane are volatile products of lipid peroxidation and therefore indicators of excessive free radical generation. Measurement of these lipid peroxidation products in small preterm infants revealed that maximal amounts were significantly higher in infants with a poor outcome (death or permanent pulmonary disorder) than in infants that survived without damage.[122] Interestingly, an association has been reported between parenteral lipids nutrition and chronic lung disease in preterm infants.[123] It has been shown that a marked increase in excretion of ethane and pentane occurred in preterm infants that were given parenteral lipids than in those given regular parenteral nutrition.[124] This suggests that parenteral lipid infusion should be used with caution in preterm infants due to the established relationship between lipid peroxidation and chronic lung disease in these patients.

Idiopathic Pulmonary Fibrosis

Distinguishing features of chronic idiopathic fibrosis (IPF) include alveolar wall fibrosis, injury of the parenchymal cells, and accumulation of alveolar macrophages and neutrophils.[125-128] Inflammatory cells from the lungs of patients with IPF spontaneously released increased amounts of superoxide radicals and hydrogen peroxide.[129] When AKD alveolar epithelial cells were incubated with IPF epithelial lining fluid in the presence of hydrogen peroxide there was increased cellular injury which was suppressed by methionine, a myeloperoxidase system scavenger. IPF epithelial lining fluid has high concentrations of myeloperoxidase;[129] however, there are dramatically decreased concentrations of the important antioxidant glutathione.[130] This glutathione deficiency creates an oxidant-antioxidant imbalance at the surface of the alveolar cells of patients with IPF. Glutathione given by aerosol to patients with IPF significantly increased epithelial lining fluid glutathione concentrations with associated increased oxidized glutathione concentrations and decreased release of spontaneous superoxide anions by alveolar macrophages.[131] These data suggest that glutathione administration may favorably effect the oxidant-antioxidant balance in these patients and provides strong evidence supporting the important role of free radicals in IPF and the possibility of a therapeutic intervention.

Sarcoidosis

One of the dominating features of sarcoid alveolitis is the presence of large numbers of macrophages.[132] Bronchoalveolar lavage fluid macrophages from patients with high intensity alveolitis have a decreased oxidative burst as measured by superoxide production.[133] This may be due to increased *in vivo* activation of these cells and has raised the possibility that chronic low grade release of macrophage-derived oxygen products may contribute to the pathology of sarcoidosis. The broncoalveolar lavage of patients with sarcoidosis or IPF contained a greater number of alveolar macrophages expressing the adhesion molecules CD11a, CD11b, CD11c, and CD18 and the superoxide production of these macrophages was greater than macrophages from control patients.[134] The increased expression of adhesion molecules of these cells appears to be involved with the greater superoxide production in patients with sarcoidosis or IPF.

Cystic Fibrosis

A central feature of cystic fibrosis in the lung is progressive epithelial injury or degeneration which may be linked to free radical activity. There is an accumulation of alginate, a linear co-polymer of uronic acids, in the cystic fibrosis lung that appears to be due to poor degradation by alveolar phagocytes. Phagocytic cells representative of those from the cystic fibrosis lung stimulated to generate free radicals, did not degrade alginate compared to a control situation.[135] The plasma from cystic fibrosis patients were found to have elevated ascorbic acid concentrations and reduced total radical-trapping antioxidant potential with a strong negative correlation between the two.[136] The high ascorbic acid concentrations in children with cystic fibrosis may result in a prooxidant effect, induce an oxidant/antioxidant imbalance, and increase susceptibility to oxidative stress. Other reported evidence suggesting an increased susceptibility to oxidative stress, demonstrated that glutathione concentrations are significantly reduced in the pulmonary epithelial lining fluid of cystic fibrosis patients.[137] An alteration in free radical activity has also been reported in cystic fibrosis patients based on evidence of increased free radical interaction with polyunsaturated fatty acids in nasal epithelial cells (a characteristic target of cystic fibrosis).[138] These data infer that a treatment regimen to reestablish the oxidant/antioxidant balance may be beneficial in these patients.

Emphysema

Pulmonary emphysema is characterized by derangements of lung connective tissue which *in vitro* studies have shown to be directly degradable by oxidants (for review, see ref. 139) and by increased interstitial proteolytic activity[140,141]. The activity of elastase, a neutrophil-derived protease, is controlled by α_1-antitrypsin[142,143] which has been shown to be susceptible to oxidant reactions at its methionine residues.[144] This data suggests that during episodes of increased free radical activity in the lung such as during cigarette smoking (documented above), there may be compromised opposition to elastase activity (oxidation of α_1-antitrypsin) and its deleterious reactions with pulmonary proteins. Additional data offers evidence that neutrophils obtained from the plasma of emphysema patients generate increased amounts of both superoxide anion and elastase activity and importantly, that the superoxide release was correlated with the duration of the disease.[145] These significant data should stimulate interest in further elucidation of this relationship in emphysemic patients.

CLINICAL MEASUREMENT OF FREE RADICAL ACTIVITY AS A PREDICTOR OF DISEASE, INJURY, AND OUTCOME

There is overwhelming evidence derived from experimental and clinical studies to support the concept that oxidative products are not simply involved in injury to pulmonary tissue, but that they are the principal mediators of this damage in a large number of circumstances associated with pulmonary dysfunction. Neutrophils are associated with this damage in almost every situation and appear to be a principal source of free radicals in this injury process. The involvement of these cells is subsequent to activation by one of several inflammatory agents (complement, TNF) released into the circulatory system in response to a diverse array of insults, most of which are documented above. The documentation that processes involved with free radical activity are important in pulmonary injury and disease opens the possibility that measurements establishing the presence or absence of these processes may be of use clinically. The ability to establish and quantitate the imbalance of an intrinsic oxidant-antioxidant relationship, whether due to *excessive* generation of oxidative products or a *deficiency* of antioxidant agents will provide the means to detect the presence and to predict the potential for cellular damage to pulmonary tissue. Additionally, with this information it may be possible to adjust the oxidant-antioxidant balance by administering the appropriate antioxidant or substrate-limiting compounds to correct, minimize, or prevent free radical-induced pulmonary injury. Examples of potentially beneficial measurements can be taken from the literature presented in this discussion. Increased concentrations of serum ***catalase*** and ***manganese superoxide dismutase*** from septic patients are predictive of the development of the adult respiratory distress syndrome (ARDS).[72,73] The measurements of these two antioxidant enzymes were as sensitive, specific, and efficient as measurements of serum lactate dehydrogenase activity and factor VIII concentrations for the prediction of ARDS in septic patients. The development of this syndrome in critically ill patients is highly correlated with increased concentrations of ***xanthine oxidase*** which in the presence of elevated hypoxanthine levels creates a situation conducive for intravascular free radical generation.[120] Quantitation of spontaneous oxygen radical generation from neutrophils by evaluation of ***chemiluminescence*** measurements of intensive care patients distinguished patients at risk of developing ARDS from control patients as well as patients with established ARDS.[118] Patients "at risk" were also distinguished from control patients by chemiluminescence measurements from neutrophils stimulated with opsonized zymosan. Patients with established ARDS have reduced plasma and red cell ***glutathione*** concentrations;[119] it would be useful to know if these conditions also exist in patients at risk for ARDS and have the potential to be used as predictors. A subpopulation of neutrophils with an increased capacity to generate ***hydrogen peroxide*** also exists in patients with confirmed ARDS and correlated better with lung injury scores than with the general clinical severity scores.[117] ***Ethane*** and ***pentane*** are products of free radical induced lipid peroxidation that can be measured in expired air[121,123] These indicators of free radical activity and cellular damage have been shown to be significantly higher in low birth weight infants that have a poor outcome and predict life-threatening complications in these patients.[121] Lipid peroxidation in broncoalveolar lavage cells and fluid has been shown to be detectable by measurement of ***malondialdehyde.***[146] This methodology using bronchoalveolar lavage samples has been proposed for use in monitoring the development and extent of pulmonary damage due to lipid peroxidation after inhalation of a toxic substance. These well-developed assays indicating the activity of biochemical interactions involving free radicals are potential predictive parameters that may assist the physician in selecting and applying specific therapeutic interventions designed to control excessive free radical activity before clinical symptoms are expressed. Antioxidant interventions have been developed and used successfully to inhibit pulmonary damage; however, documentation of these interventions, as well as potential therapeutic protocols, are not within the scope of this discussion. The ability to identify a population of patients who are at risk for free radical-induced pulmonary dysfunction and injury using clinically relevant assays has substantial potential to aid in the prevention or attenuation of cellular damage with a resultant improvement in patient health care and outcome.

Acknowledgments

The authors wish to express their gratitude to Ms. Megan R. Lerner for editing and critical review of the manuscript and to Ms. Theresa J. Lander and Mr. Son Van Do for the retrieval, collection, and collation of the material used in this chapter. The completion of this work would not have been possible without their support.

REFERENCES

1. R.J. Stallone, R.C. Lim, Jr., and F.W. Blaisdell, Pathogenesis of the pulmonary changes following ischemia of the lower extremities, Ann. Thor. Surg. 7:539 (1969).
2. H. Haimovici, Metabolic complications of acute arterial occlusions, J. Cardiovas. Surg. 20:349 (1979).
3. G.M. Matuschak, J.E. Rinaldo, M.R. Pinsky, J.S. Gavaler, and D.H. Van Thiel, Effect of end-stage liver failure on the incidence and resolution of the adult respiratory distress syndrome, J. Crit. Care 2:162 (1987).
4. I.S. Paterson, J.M. Klausner, G. Goldman, R. Pugatch, H. Feingold, P. Allen, J.A. Mannick, C.R. Valeri, D. Shepro, and H.B. Hechtman, Pulmonary edema after aneurysm surgery is modified by mannitol, Ann. Surg. 210:796 (1989).
5. B.M. Babior, J.T. Curnutte, and N. Okamura, The respiratory burst oxidase of the human neutrophil, *in*: "Oxygen Radicals and Tissue Injury," B. Halliwell, ed., Federation of American Societies for Experimental Biology, Bethesda (1988).
6. D.J. Schmeling, M.G. Caty, K.T. Oldham, K.S. Guice, and D.B. Hinshaw, Evidence for neutrophil-related acute lung injury after intestinal ischemia-reperfusion, Surgery 106:195 (1989).
7. R.S. Poggetti, F.A. Moore, E.E. Moore, D.D. Bensard, B.O. Anderson, and A. Banerjee, Liver injury is a reversible neutrophil-mediated event following gut ischemia, Arch. Surg. 127:175 (1992).
8. K. Koike, E.E. Moore, F.A. Moore, V.S. Carl, J.M. Pitman, and A. Banerjee, Phospholipase A_2 inhibition decouples lung injury from gut ischemia-reperfusion, Surgery 112:173 (1992).
9. R.S. Poggetti, F.A. Moore, E.E. Moore, K.Koeike, and A. Banerjee, Simultaneous liver and lung injury following gut ischemia is mediated by xanthine oxidase, J. Trauma 32:723 (1992).
10. K. Koike, F.A. Moore, E.E. Moore, R.A. Read, V.S. Carl, and A. Banerjee, Gut ischemia mediates lung injury by a xanthine oxidase-dependent neutrophil mechanism, J. Surg. Res. 54:469 (1993).
11. M.G. Caty, K.S. Guice, K.T. Oldham, D.G. Remick, and S. Kunkel, Evidence for tumor necrosis factor-induced pulmonary microvascular injury after intestinal ischemia-reperfusion injury, Ann. Surgery 212:694 (1990).

12. J. Hill, T. Lindsay, C.R. Valeri, D. Shepro, and H.B. Hechtman, A CD18 antibody prevents lung injury but not hypotension after intestinal ischemia-reperfusion, J. Appl. Physiol. 74:659 (1993).
13. J. Hill, T. Lindsay, J. Rusche, C.R. Valeri, D. Shepro, and H.B. Hechtman, A Mac-1 antibody reduces liver and lung injury but not neutrophil sequestration after intestinal ischemia-reperfusion, Surgery 112:166 (1992).
14. D.L. Carden, J.A. Young, and D.N. Granger, Pulmonary microvascular injury after intestinal ischemia-reperfusion: role of P-selectin, J. Appl. Physiol. 75:2529 (1993).
15. J.M. Klausner, H. Anner, I.S. Paterson, L. Kobzik, C.R. Valeri, D. Shepro, and H.B. Hechtman, Lower torso ischemia-induced lung injury is leukocyte dependent, Ann. Surg. 208:761 (1988).
16. J.M. Klausner, I.S. Paterson, L. Kobzik, C.R. Valeri, D. Shepro, and H.B. Hechtman, Oxygen free radicals mediate ischemia-induced lung injury, Surgery 105:192 (1989).
17. J. Punch, R. Rees, B. Cashmer, K. Oldham, E. Wilkins, and D.J. Smith, Jr., Acute lung injury following reperfusion after ischemia in the hind limbs of rats, J. Trauma 31:760 (1991).
18. R. Welbourn, G. Goldman, L. Kobzik, I.S. Paterson, C.R. Valeri, D. Shepro, and H.B. Hechtman, Role of neutrophil adherence receptors (CD18) in lung permeability following lower torso ischemia, Circ. Res. 71:82 (1992).
19. H. Anner, R.P. Kaufman, Jr., L. Kobzik, C.R. Valeri, D. Shepro, and H.B. Hechtman, Pulmonary leukosequestration induced by hind limb ischemia, Ann. Surg. 206: 162 (1987).
20. H. Anner, R.P. Kaufman, Jr., L. Kobzik, C.R. Valeri, D. Shepro, and H.B. Hechtman, Pulmonary hypertension and leukosequestration after lower torso ischemia, Ann. Surg. 206:642 (1987).
21. J.M. Klausner, I.S. Paterson, C.R. Valeri, D. Shepro, and H.B. Hechtman, Limb ischemia-induced increase in permeability is mediated by leukocytes and leukotrienes, Ann. Surg. 208:755 (1988).
22. R. Welbourn, G. Goldman, M. O'riordain, T.F. Lindsay, I.S. Paterson, L. Kobzik, C.R. Valeri, D. Shepro, and H.B. Hechtman, Role for tumor necrosis factor as mediator of lung injury following lower torso ischemia, J. Appl. Physiol. 70:2645 (1991).
23. A. Seekamp, J.S. Warren, D.G. Remick, G.O. Till, and P.A. Ward, Requirements for tumor necrosis factor-α interleukin-1 in limb ischemia/reperfusion injury and associated lung injury, Am. J. Pathol. 143:453 (1993).
24. L.M. Colletti, D.G. Remick, G.D. Burtch, S.L. Kunkel, R.M. Strieter, and D.A. Campbell,Jr., Role of tumor necrosis factor-α in the pathophysiologic alterations after hepatic ischemia/reperfusion injury in the rat, J. Clin. Invest. 85:1936 (1990).
25. L.M. Colletti, G.D. Burtch, D.G. Remick, S.L. Kunkel, R.M. Strieter, K.S. Guice, K.T. Oldham, and D.A. Campbell, Jr., The production of tumor necrosis factor alpha and the development of a pulmonary capillary injury following hepatic ischemia/reperfusion, Transplantation 49:268 (1990).
26. N.B. Ratliff, J.W. Wilson, E. Mikat, D.B. Hackel, and T.C. Graham, The lung in hemorrhagic shock. IV. The role of neutrophilic polymorphonuclear leukocytes, Am. J. Pathol. 65:325 (1971).
27. H. Redl, G. Schlag, and D.E. Hammerschmidt, Quantitative assessment of leukostasis in experimental hypovolemic-tramatic shock, Acta Chir. Scand. 150:113 (1984).
28. J.W. Wilson, Leukocyte sequestration and morphologic augmentation in the pulmonary network following hemorrhagic shock and related forms of stress, Adv. Microcirc. 4:197 (1972).
29. R.S. Connell, R.L. Swank, and M.C. Webb, The development of pulmonary ultrastructural lesions during hemorrhagic shock, J. Trauma 15:116 (1975).
30. P.E. Pepe, R.T. Potkin, D.H. Reus, L.D. Hudson, and C.J. Carrico, Clinical predictors of the adult respiratory distress syndrome, Am. J. Surgery 144:124 (1982).
31. A.A. Fowler, R.F. Hamman, J.T. Good, K.N. Benson, M. Baird, D.J. Eberle, T.L. Petty, and T.M. Hyers, Adult respiratory distress syndrome: risk with common predispositions, Ann. Intern. Med. 98:593 (1983).
32. U. Haglund and B. Gerdin, Oxygen-free radicals (OFR) and circulatory shock, Circ. Shock 34:405 (1991).
33. G. Poli, F. Biasi, E. Chiarpotto, M.U. Dianzani, A.De Luca, and H. Esterbauer, Lipid peroxidation in human diseases: evidence of red cell oxidative stress after circulatory shock, Free Rad. Biol. Med. 6:167 (1989).
34. N.B. Vedder, B.W. Fouty, R.K. Winn, J.M. Harlan, and C.L. Rice, Role of neutrophils in generalized reperfusion injury associated with resuscitation from shock, Surgery 106:509 (1989).
35. B.O. Anderson, E.E. Moore, F.A. Moore, J.A. Leff, L.S. Terada, A.H. Harken, and J.E. Repine, Hypovolemic shock promotes neutrophil sequestration in lungs by a xanthine oxidase-related mechanism, J. Appl. Physiol. 71:1862 (1991).
36. M.K. Robinson, J.D. Rounds, R.W. Hong, D.O. Jacobs, and D.W. Wilmore, Glutathione deficiency increases organ dysfunction after hemorrhagic shock, Surgery 112:140 (1992).
37. K.J. Tracey, S.F. Lowry, T.J. Fahey III, J. D. Albert, Y. Fong, D. Hesse, B. Beutler, K.R. Manogue, S. Calvano, H. Wei, A. Cerami, and G.T. Shires, Cachectin/tumor necrosis factor induces lethal shock and stress hormone responses in the dog, Surg. Gyn. Obstet. 164:415 (1987).

38. K.J. Tracey, B. Beutler, S.F. Lowry, J. Merryweather, S. Wolpe, I.W. Milsark, R.J. Hariri, T.J. Fahey III, A. Zentella, J.D. Albert, G.T. Shires, and A. Cerami, Shock and tissue injury induced by recombinant human cachectin, Science 234:470 (1986).
39. A. Ferrante, M. Nandoskar, A. Walz, D.H.B. Goh, and I.C. Kowanko, Effects of tumour necrosis factor alpha and interleukin-1 alpha and beta on human neutrophil migration, respiratory burst, and degranulation, Int. Arch. Allergy Applied. Immunol. 86:82 (1988).
40. Z. Wankowicz, P.Megyeri, and A. Issekutz, Synergy between tumour necrosis factor α and interleukin-1 in the induction of polymorphonuclear leukocyte migration during inflammation, J. Leukocyte Biol. 43:349 (1988).
41. B.A. Pruitt, D.R. Erickson, and A. Morris, Progressive pulmonary insufficiency and other pulmonary complications of thermal injury, J. Trauma 15:369 (1975).
42. H.C. Polk, Consensus summary on infection, J. Trauma 19:894 (1979).
43. K.E. Fjellstrom and G. Arturson, Changes in the human complement system following burn trauma, Acta Path. 59:257 (1963).
44. A.B. Bjornson, W.A. Altemeier, H.S. Bjornson, Changes in humoral components of host defense following burn trauma, Ann. Surg. 186:88 (1977).
45. M. Heideman, The effect of thermal injury on hemodynamic, respiratory, and hematologic variables in relation to complement activation, J. Trauma 19:239 (1979).
46. G.O. Till, C. Beauchamp, D. Menapace, W. Tourtellotte, Jr., R. Kunkel, K.J. Johnson, and P.A. Ward, Oxygen radical dependent lung damage following thermal injury of rat skin, J. Trauma 23:269 (1983).
47. P.A. Ward, G.O. Till, R. Kunkel, and C. Beauchamp, Evidence for the role of hydroxyl radical in complement and neutrophil-dependent tissue injury, J. Clin. Invest. 72:789 (1983).
48. G.O. Till, W. Tourtellotte, Jr., M.J. Lutz, T. Annesley, and P.A. Ward, Acute lung injury secondary to skin burns: Evidence for role of hydroxyl radical and lipid peroxidation products, Circ. Shock 13:76 (1984).
49. L. Jin, C. Lalonde, and R.H. Demling, Lung dysfunction after thermal injury in relation to prostanoid and oxygen radical release, J. Appl. Physiol. 61:103 (1986).
50. J.A. Leff, L.K. Burton, E.M. Berger, B.O. Anderson, C.P. Wilke, and J.E. Repine, Increased serum catalase activity in rats subjected to thermal skin injury, Inflammation 17:199 (1993).
51. G.O. Till, and P.A. Ward, Systemic complement activation and acute lung injury, Federation Proc. 45:13 (1986).
52. D.C. Hohn, A.J. Meyers, S.T. Gherini, A. Beckmann, R.E. Markison, and A.M. Churg, Production of acute pulmonary injury by leukocytes and activated complement, Surgery 88:48 (1980).
53. S.Z. Perkowski, A.M. Havill, J.T. Flynn, and M.H. Gee, Role of intrapulmonary release of eicosanoids and superoxide anion as mediators of pulmonary dysfunction and endothelial injury in sheep with intermittent complement activation, Circ. Res. 53:574 (1983).
54. G.O. Till, K.J. Johnson, R. Kunkel, and P.A. Ward, Intravascular activation of complement and acute lung injury: Dependency on neutrophils and toxic oxygen metabolites, J. Clin. Invest. 69:1126 (1982).
55. C.W. Imrie, J.C. Ferguson, D. Murphy, and L.H. Blumgart, Arterial hypoxia in acute pancreatitis, Brit. J. Surg. 64:185 (1977).
56. J.H.C. Ranson, D.F. Roses, and S.D. Fink, Early respiratory insufficiency in acute pancreatitis, Ann. Surg. 178:75 (1973).
57. A.L. Warshaw, P.B. Lesser, M. Rie, D.J. Cullen, The pathogenesis of pulmonary edema in acute pancreatitis, Ann. Surg. 182:505 (1975).
58. P.G. Lankisch, G. Rahlf, and H. Koop, Pulmonary complications in fatal acute hemorrhagic pancreatitis, Digest. Dis. Sci. 28:111 (1983).
59. B. Interiano, I.D. Stuard, and R.W. Hyde, Acute respiratory distress syndrome in pancreatitis, Ann. Intern. Med. 77:923 (1972).
60. I.G. Renner, W.T. Savage, III, J.L. Pantoja, V.J. Renner, Death due to acute pancreatitis: A retrospective analysis of 405 autopsy cases, Digest. Dis. Sci. 30:1005 (1985).
61. K.S. Guice, K.T. Oldham, K.J. Johnson, R.G. Kunkel, M.L. Morganroth, and P.A. Ward, Pancreatitis-induced acute lung injury: An ARDS model, Ann. Surg. 208:71 (1988).
62. K.S. Guice, K.T. Oldham, M.G. Caty, K.J. Johnson, and P.A. Ward, Neutrophil-dependent, oxygen-radical mediated lung injury associated with acute pancreatitis, Ann. Surg. 210:740 (1989).
63. W. Marget, The last round against bacterial infections, Infection 18:197 (1990).
64. K.L. Brigham and B. Meyrick, Endotoxin and lung injury, Am. Rev. Respir. Dis. 133:913 (1986).
65. R.J. Ulevitch, Recognition of bacterial endotoxin in biologic systems, Lab. Invest. 65:121 (1991).
66. R.L. Danner, R.J. Elin, J.M. Hosseini, R.A. Wesley, J.M. Reilly, and J.E. Parrillo, Endotoxemia in human septic shock, Chest 99:169 (1991).
67. R.J. Stumacher, M.J. Kovnat, and W.R. McCabe, Limitations of the usefullness of the Limulus assay for endotoxin, N. Engl. J. Med. 288:1261 (1973).

68. S.J.H. vanDeventer,H.R. Buller, J. W. Cate, A. Sturk, and W. Pauw, Endotoxemia: an early predictor of septicaemia in febrile patients, Lancet 1:605 (1988).
69. H. U. Michie, K.R. Manogue, D.R. Spriggs, A. Revhaug, S. O'Dwyer, C.A. Dinarello, A. Cerami, S.M. Wolff, and D.W. Wilmore, Detection of circulating tumor necrosis factor after endotoxin administration, N. Engl. J. Med. 318:1481 (1988).
70. H.R. Michie, D.U. Spriggs, K.R. Manogue, M.L. Sherman, A. Revhaug, W.T. O'Dwyer, K. Arthur, C.A. Dinarello, A. Cerami, W.M. Wolff, D.W. Kufe, and D.W. Wilmore, Tumor necrosis factor and endotoxin induce similar metabolic responses in human beings, Surgery 104:280 (1988).
71. G.D. Martich, R.L. Danner, M. Ceska, and A.F. Wuffredini, Detection of interleukin-8 and tumor necrosis factor in normal humans after intravenous endotoxin: the effect of antiinflammatory agents, J. Exp. Med. 173:1021 (1991).
72. J.A. Leff, P.E. Parsons, C.E. Day, E.E. Moore, F.A. Moore, M.A. Oppegard, and J.E. Repine, Increased serum catalase activity in septic patients with the adult respiratory distress syndrome, Am. Rev Respir. Dis. 146:985 (1992).
73. J. A. Leff, P.E. Parsons, C.E. Day, N. Taniguchi, M. Jochum, H. Fritz, F.A. Moore, E.E. Moore, J.M. McCord, and J.E. Repine, Serum antioxidants as predictors of adult respiratory distress syndrome in patients with sepsis, Lancet 341:777 (1993).
74. P.G. Murphy, D.S. Myers, N.R. Webster, and J.G. Jones, Direct detection of free radical generation in an *in vivo* model of acute lung injury, Free Rad. Res. Comms., 15:167 (1991).
75. G. R. Bernard, W.D. Lucht, M.E. Niedermeyer, J.R. Snapper, M.L. Ogletree, and K.L. Brigham, Effect of N-acetylcysteine on the pulmonary response to endotoxin in the awake sheep and upon *in vitro* granulocyte function, J. Clin. Invest. 73:1772 (1984).
76. P.D. Carey, J.K. Jenkins, K. Byrne, C.J. Walsh, A.A. Fowler, and H.J. Sugerman, The neutrophil respiratory burst and tissue injury in septic acute lung injury: The effect of cyclooxygenase inhibition in swine, Surgery 112:45 (1992).
77. W. Leeper-Woodford, P.D. Carey, K. Byrne, B.J. Fisher, C. Blocher, H.J. Sugerman, and A.A. Fowler, Ibuprofen attenuates plasma tumor necrosis factor activity during sepsis-induced acute lung injury, J. Appl. Physiol. 71:915 (1991).
78. P.D. Carey, S.K. Leeper-Wooddford, C.J. Walsh, K. Byrne, A.A. Fowler, and H.J. Sugerman, Delayed cyclo-oxygenase blockade reduces the neutrophil respiratory burst and plasma tumor necrosis factor levels in sepsis-induced acute lung injury, J. Trauma 31:733 (1991).
79. Z. Xing, H. Kirpalani, D. Torry, M. Jordana, and J. Gauldie, Polymorphonuclear leukocytes as a significant source of tumor necrosis factor-α in endotoxin-challenged lung tissue, Am. J. Pathol. 143:1009 (1993).
80. K. J. Johnson, J.C. Fantone, J. Kaplan, and PA. Ward, *In vivo* damage of rat lungs by oxygen metabolites, J. Clin. Invest. 67:983 (1981).
81. R.M. Tate, K.M. Vanbenthuysen, C.M. Shasby, and I.F. McMurtry, Oxygen-radical-mediated permeability edema and vasoconstriction in isolated perfused rabbit lungs, Am. Rev. Respir. Dis. 126:80 (1982).
82. D.E. Gannon, X. He, P.A. Ward, J. Varani, and K.J. Johnson, Time-dependent inhibition of oxygen radical induced lung injury, Inflammation 14:509 (1990).
83. J. Varani, I. Ginsburg, D.F. Gibbs, P.S. Mukhopadhyay, C. Sulavik, K.J. Johnson, J.M. Weinberg, U.S. Ryan, and P.A. Ward, Hydrogen peroxide-induced cell and tisssue injury: Protective effects of Mn^{2+}, Inflammation 15:291 (1991).
84. H.E. May and P.B. McCay, Reduced triphosphopyridine nucleotide oxidase-catalyzed alterations of membrane phospholipids, J.Biol. Chem. 243:2296 (1968).
85. E.S. Stadtman, B.S. Berlett, and P.B. Chock, Manganese-dependent disproportionation of hydrogen peroxide in bicarbonate buffer, Proc. Natl. Acad. Sci. U.S.A. 87:384 (1990).
86. M.B. Yim, B.S. Berlett, P.B. Chock, and E.R. Stadtman, Manganese (II)-bicarbonate-mediated catalytic activity for hydrogen peroxide dismutation and amino acid oxidation: Detection of free radical intermediates, Proc. Natl. Acad. Sci. U.S.A. 87:394 (1990).
87. B.S. Berlett, P.B. Chock, M.B. Yim, and E.R. Stadtman, Manganese (II) catalyzes the bicarbonate dependent oxidation of amino acids by hydrogen peroxide and the amno acid-facilitated dismutation of hydrogen peroxide, Proc. Natl. Acad. Sci. U.S.A. 87:389 (1990).
88. P.B.L. Cheton and F.S. Archibald, Manganous complexes and the generation and scavenging of hydroxyl free radicals, Free Rad. Biol. Med. 5:325 (1988).
89. J.M.C. Gutteridge,and J.V. Bannister, Copper plus zinc and manganese superoxide dismutase inhibit deoxyribose degradation by the superoxide-driven fenton reaction at two different stages, Biochem. J. 234:225 (1986).
90. J.S. Bus, S.A. Aust andJ.F. Gibson, Superoxide and singlet oxygen catalyzed lipid peroxidation as a possible mechanism for paraquat (methyl viologen) toxicity, Biochem. Biophys. Res. Commun. 58:749 (1974).
91. L.L. Smith, The identification of an accumulation system for diamenes and polyamines into the rat lung and its relevance to paraquat toxicity, Arch. Toxicol. 5:1 (1982).

92. J.S. Bus, S.Z. Cagen, M. Olgaard, and J.E. Gibson, A mechanism of paraquat toxicity in mice and rats, Toxicol. Appl. Pharmacol. 35:5001 (1976).
93. T. J. Haley, Review of the toxicology of paraquat (1,1'-Dimethyl-4,4'-bipyridinium Chloride), Clin. Toxicol. 14:1 (1979).
94. E. Hoffer, I. Avidor, O. Benjaminov, L. Shenker, A. Tabak, A. Tamir, D, Merzbach, and U. Taitelman, N-acetylcysteine delays the infiltration of inflammatory cells into the lungs of paraquat-intoxicated rats, Toxicol. Appl. Pharmacol. 120:8 (1993).
95. J.R. Dawson, K. Norbeck, I. Anundi, and P. Moldeus, The effectiveness of N-acetylcysteine in isolated hepatocytes, against the toxicity of paracetamol, acrolein, and paraquat, Arch. Toxicol. 55:11 (1984).
96. E.LeB. Gray, Oxides of nitrogen: Their ocurrence, toxicity, hazard - a brief review, A.M.A. Arch. Ind. Health 19:479 (1959).
97. H.E. Stokinger, Ozone technology: A review of research and industrial experience: 1954-1964, Arch. Environ. Health 10:719 (1965).
98. H.E. Stokinger, Ozone toxicity: Review of literature through 1953, Arch. Indust. Hyg. 9:366 (1953).
99. J.N. Roehm, J.G. Hadley, and D.B. Menzel, Oxidation of unsaturated fatty acids by ozone and nitrogen dioxide, Arch. Environ. Health 23: 142 (1971).
100. B.D. Goldstein, C. Lodi, M.T. Collinson, and O.J. Balchum, Ozone and lipid peroxidation, Arch. Environ. Health 18: 631 (1969).
101. H.V. Thomas, P.K. Meuller, and R.L. Lyman, Lipoperoxidation of lung lipid in rats exposed to nitrogen dioxide, Science 159:532 (1968).
102. B.L. Fletcher and A.L. Tappel, Protective effects of dietary α-tocopherol in rats exposed to toxic levels of ozone and nitrogen dioxide, Environ. Res. 6:165 (1973).
103. L. Allegra, N.E. Moavero, and C. Rampoldi, Ozone-induced impairment of mucociliary transport and its prevention with N-acetylcysteine, Am. J. Med. 91:67 (1991).
104. C. von Ritter, R. Be, D.N. Granger, Neutrophilic proteases: mediators of formyl-methionyl-leucyl-phenylalanine-induced ileitis in rats, Gastrointerology 97:605 (1989).
105. J.A. Leff, M.A. Oppegard, E.C. McCarty, C.P. Wilke, P.F. Shanley, C.E. Day, N.K. Ahmed, L.M. Patton, and J.E. Repine, Dimethyl sulfoxide decreases lung neutrophil sequestration and lung leak, J. Lab. Clin. Med. 120:282 (1992).
106. J.E. Repine, J.W. Eaton, M.W. Anders J.R. Hoidal, and R.B. Fox, Generation of hydroxyl radical by enzymes, chemicals, and human phagocytes in vitro: detection with the anti-inflammatory agent, dimethyl sulfoxide, J. Clin. Invest. 64:1642 (1979).
107. J.C. Fantone and P.A. Ward, Role of oxygen-derived free radicals and metabolites in leukocyte-dependent inflammatory reactions, Am. J. Pathol. 107:397 (1982).
108. S.J. Weiss, Tissue destruction by neutrophils, N. Engl. J. Med. 320:365 (1989).
109. T. Church and W.A. Pryor Free-radical chemistry of cigarette smoke and its toxicological implications, Environ. Health Perspect. 64:111 (1985).
110. W. MacNee, M.M.E. Bridgeman, M. Marsden, E. Drost, S. Lannan, C. Selby, and K. Donaldson, The effects of N-acetylcysteine and glutathione on smoke-induced changes in lung phagocytes and epithelial cells, Am. J. Med. 91:60 (1991).
111. T. Schaberg, H. Haller, M. Rau. D. Kaiser, M. Fassbender, and H. Lode, Superoxide anion release induced by platelet-activating factor is increased in human alveolar macrophages from smokers, Eur. Respir. J. 5:387 (1992).
112. J.R. Hoidal, R.B. Le Marble, R. Pewrri, and I.E. Repine, Altered oxidative metabolic responses in vitro of alveolar macrophages from asymptomatic cigarette smokers, Am. Rev. Respir. Dis. 123:85 (1981).
113. K.J. Johnson, B.S. Wilson, G.O. Till, and P.A. Ward, Acute lung injury in rat caused by immunoglobulin A immune complexes, J. Clin. Invest. 74:358 (1984).
114. K.J. Johnson and P.A. Ward, Role of oxygen metabolites in immune complex injury of lung, J. Immunol. 126:2365 (1981).
115. J.E. Rinaldo and R.M. Rogers, Adult respiratory distress syndrome: changing concepts of lung injury and repair, N. Engl. J. Med. 306:900 (1982).
116. K.L. Brigham, Mechanisms of lung injury, Clin. Chest Med. 3:9 (1982).
117. M.B. Divertie, The adult respiratory distress syndrome, Mayo Clin. Proc. 57:371 (1982).
118. S. Chollet-Martin, P. Montravers, C. Gibert, C. Elbim, J.M. Desmonts, J.Y. Fagon, and M.A. Gougerot-Pocidalo, Subpopulation of hyperresponsive polymorphonuclear neutrophils in patients with adult respiratory distress syndrome: role of cytokine production, Am. Rev. Respir. Dis. 146:990 (1992).
119. M. C. Tagan, M. Markert, M.D. Schaller, F. Feihl, R. Chiolero, and C.H. Perret, Oxidative metabolism of circulating granulocytes in adult respiratory distress syndrome, Am. J. Med. 91:72 (1991).
120. G.R. Bernard, N-acetylcysteine in experimental and clinical acute lung injury, Am. J. Med. 91:54 (1991).
121. C.M. Grum, R.A. Ragsdale, L.H. Ketai, and R.H. Simon, Plasma xanthine oxidase activity in patients with adult respiratory distress syndrome, J. Crit. Care 2:22 (1987).

122. O.M. Pitkanen, M. Hallman, and S.M. Andersson, Correlation of free oxygen radical-induced lipid peroxidation with outcome in very low birth weight infants, J. Pediatr, 116:760 (1990).
123. R.W.I. Cooke, Factors associated with chronic lung disease in preterm infants, Arch. Dis. Child. 66:776 (1991).
124. J.R. Wispe, E.F. Bell, and R.J. Roberts, Assessment of lipid peroxidation in newborn infants and rabbits by measurements of expired ethane and pentane: influence of parenteral lipid infusion, Ped. Res. 19:374 (1985).
125. R.G. Crystal, J.D. Fulmer, W.C. Roberts, M.L. Moss, B.E. Line, and H.Y. Reynolds, Idiopathic pulmonary fibrosis: clinical, histologic, radiologic, physiologic, scintigraphic, cytological, and biochemical aspects, Ann. Intern. Med. 85:769 (1976).
126. L. Hamman and A.R. Rich, Acute diffuse interstitial fibrosis of the lungs, Bull. Johns Hopkins Hosp. 74:177 (1944).
127. J.D. Fulmer, W.C. Roberts, E.R. von Gal, and R.G. Crystal, Morphologic-physiologic correlates of the severity of fibrosis and degree of cellularity in idiopathic pulmonary fibrosis, J. Clin. Invest. 63:665 (1979).
128. J.D. Fuller, W.C. Roberts, E.R. von Gal, and R.G. Crystal, Small airways in idiopathic pulmonary fibrosis: comparison of morphologic and physiologic observations, J. Clin. Invest. 60;595 (1977).
129. A.M. Cantin, S.L. North, G.A. Fells, R.C. Hubbard, and R.G. Crystal, Oxidant-mediated epithelial cell injury in idiopathic pulmonary fibrosis, J. Clin. Invest. 79:1665 (1987).
130. A.M. Cantin, R.C. Hubbard, and R.G. Crystal, Glutathione deficiency in the epithelial lining fluid of the lower respiratory tract in idiopathic pulmonary fibrosis, Am. Rev Respir. Dis. 139:370 (1989).
131. Z. Borok, R. Buhl, G.J. Grimes, A.D. Bokser, R. C. Hubbard, K.J. Holroyd, J.H. Roum, D.B. Czerski, A.M. Cantin, and R.G. Crystal, Effect of glutathione aerosol on oxidant-antioxidant imbalance in idiopathic pulmonary fibrosis, Lancet 338:215 (1991).
132. D.A. Campbell, L.W. Poulter, and R.M. DuBois, Immunocompetent cells in bronchoalveolar lavage reflect the cell populations in transbronchial biopsies in pulmonary sarcoidosis, Am. Rev. Respir. Dis. 132:1300 (1985).
133. H. Nielsen, J. Frederiksen, D. Sherson, and N. Milman, Comparison of alveolar macrophage and blood monocyte oxidative burst response in pulmonary sarcoidosis, APMIS 98:401 (1990).
134. T. Schaberg, M. Rau, H. Stephan, and H. Lode, Increased number of alveolar macrophages expressing surface molecules of the CD11/CD18 family in sarcoidosis and idiopathic pulmonary fibrosis is related to the production of superoxide anions by these cells, Am. Rev. Respir. Dis. 147:1507 (1993).
135. J.A. Simpson, S.E. Smith, and R.T. Dean, Alginate may accumulate in cystic fibrosis lung because the enzymatic and free radical capacities of phagocytic cells are inadequate for its degradation, Biochem. Molec. Biology Int. 30:1021 (1993).
136. S.C. Langley, R.K. Brown, and F.J. Kelly, Reduced free-radical-trapping capacity and altered plasma antioxidant status in cystic fibrosis, Pediatr. Res. 33:247 (1993).
137. J.H. Roum, R. Buhl, N.G. McElvaney, Z. Borok, R.C. Hubbard, M. Chernick, A.M. Catin, and R.G. Crystal, Cystic fibrosis is characterized by a marked reduction in glutathione levels in pulmonary epithelial lining fluid, Am. Rev. Respir. Dis. 141:87 (1990).
138. B. Salh, K. Webb, P.M. Guyan, J.P. Day, C. Wickens, J. Griffin, J.M. Braganza, and T.L. Dormandy, Aberrant free radical activity in cystic fibrosis, Clin. Chim. Acta 181:65 (1989).
139. D.J. Riley and J.S. Kerr, Oxidant injury of the extracellular matrix: potential role in the pathogenesis of pulmonary emphysema, Lung 163:1 (1985).
140. A. Janoff, Elastase and emphysema, Am. Rev. Respir. Dis. 132:417 (1985).
141. V.V. Damiano, A. Tsang, U. Kucich, W.R. Abrams, J. Rosenbloom, P. Kimbel, M. Fallahnejad, and G. Weinbaum, Immunolocalization of elastase in human emphysematous lungs, J. Clin. Invest. 78:482 (1986).
142. R.M. Senior , H. Tegner, C. Kuhn, K. Ohlsson, B.C. Starcher, and J.A. Pierce, The induction of pulmonary emphysema with human leukocyte elastase, Am. Rev. Respir. Dis. 116:469 (1977).
143. K. Ohlsson, Neutral leukocyte proteases and elastase are inhibited by plasma α_1-antitrypsin, Scand. J. Clin. Lab. Invest. 28:251 (1970).
144. H. Carp, F. Miller, R. Hoidal, and A. Janoff, Potential mechanism of emphysema: alpha 1-proteinase inhibitor recovered from lungs of cigarette smokers contains oxidized methionine and has decreased elastase inhibitory capacity, Proc. Natl. Acad. Sci. U.S.A. 779:2041 (1982).
145. H. Kanazawa, N. Kurihara, K. Hirata, S. Fujimoto, and T. Takeda, The role of free radicals and neutrophil elastase in development of pulmonary emphysema, Jap J. Med. 31:857 (1992).
146. J.M. Petruska, S.H.Y. Wong, F.W. Sunderman, Jr., and B.T. Mossman, Detection of lipid peroxidation in lung and in bronchoalveolar lavage cells and fluid, Free Rad. Biol. Med. 9:51 (1990).

LIPID PEROXIDES IN HEPATIC, GASTROINTESTINAL, AND PANCREATIC DISEASES

Kunio Yagi

Institute of Applied Biochemistry
Yagi Memorial Park
Mitake, Gifu 505-01
Japan

INTRODUCTION

Although the liver, stomach, intestine, and pancreas are classified in the same group, i.e., digestive organs, their functions are largely different from each other. In spite of such differences, it seems that injury to these organs could be commonly provoked by lipid peroxidation with a burst of free radical production in situ. Therefore, administration of antioxidants would be useful treatment and even prophylactic for such injury. Some topics related to these problems are described.

LIPID PEROXIDES IN HEPATIC DISEASES

Since the liver is the most important center of metabolism in higher animals, lipid peroxides produced in any site of the body are mostly transferred to this organ and decomposed. However, if their amounts are excessive, they have a deleterious effect on it as evidenced by experimental burn injury (see the chapter "Lipid peroxides and related radicals in clinical medicine" in this volume). On the contrary, if the liver is injured by any cause and its lipid peroxide level is increased, these peroxides often leak from the liver into the bloodstream and thereby injure other intact organs or tissues.

Table 1 summarizes the data on the serum lipid peroxide level of patients suffering from hepatic diseases[1]. The values of these patients are markedly higher than those of normal subjects except for the case of compensated liver cirrhosis in which the level was only slightly increased. Among them, an extraordinary increase was found in fluminant hepatitis, which explains the situation of these patients who were gravely ill.

MEASUREMENT OF LIPID PEROXIDE LEVEL IN BIOPSIED SAMPLE

For measurement of the lipid peroxide level in a biopsied sample of the liver (or other

Free Radicals in Diagnostic Medicine, Edited by
D. Armstrong, Plenum Press, New York, 1994

Table 1. Serum lipid peroxide levels of patients suffering from hepatic diseases.

Subjects	n	Serum lipid peroxide level
Normal subjects	121	3.54 ± 0.72
Hepatitis		
Acute	14	6.40 ± 1.76**
Fluminant	8	13.10 ± 1.76**
Chronic active	16	7.46 ± 3.66**
Liver cirrhosis		
Compensated	39	3.92 ± 1.30*
Decompensated	4	8.22 ± 1.06**
Fatty liver	11	7.60 ± 1.58**

Normal subjects were 21-71 years of age of both sexes. Lipid peroxide levels were measured according to Yagi[2] and expressed in terms of malondialdehyde (nmol /ml serum). Mean ± SD is given. Significant difference vs. normal subjects: *$p<0.025$; **$p<0.001$.

organs or tissues), the assay method using the thiobarbituric acid (TBA) reaction under the conditions specified by us[3] is recommendable. As is the case for lipid peroxides in the serum (see the chapter "Lipid peroxides and related radicals in clinical medicine" in this volume), the TBA reaction should be performed in acetic acid solution at pH 3.5 and at 95°C for 60 min. As for interfering substances, we found that glucose less than 0.4 mg (2.2 μmol), sucrose less than 8.56 mg (25.0 μmol), and N-acetylneuraminic acid (sialic acid) less than 2.0 mg (6.47 μmol) in the reaction mixture of 4.0 ml did not affect the estimation of lipid peroxide level with this assay procedure. At the concentration of 1.0 mg (1.71 μmol) in the reaction mixture, bilirubin also reacts with TBA to interfere with the absorbance of 532 nm. In this case, fluorometry should be adopted, since the interference by bilirubin is negligible when fluorometric measurement (excitation: 515 nm; emission: 553 nm) is adopted. If trichloroacetic acid is used for the reaction of TBA with lipid peroxides, significant interference by sialic acid is observed, when the reaction product is determined by the absorption at 532 nm. This interference by sialic acid is negligible, if fluorometry is used. However, the author does not recommend to use trichloroacetic acid, because of the low yield of the reaction product of lipid peroxides with TBA.

Scheme 1 presents the recommended standard procedure for the assay of lipid peroxide levels in organs or tissues by the TBA reaction.

SUPPESSIVE EFFECT OF ANTIOXIDANTS ON LIPID PEROXIDATION IN THE LIVER

As demonstrated by the work of Maellaro et al.[4] using glutathione depleting agents, lipid peroxide-mediated liver injury accompanies a decrease in antioxidants such as vitamin E and ascorbic acid.

Therefore, such antioxidants are expected to be useful for the treatment of hepatic diseases and even as a prophylactic for these diseases. Among many data, only our recent findings are presented in Table 2, which shows that curcumin, ellagic acid, and α-tocopherol are all effective for suppression of lipid peroxidation provoked in the liver by carbon tetrachloride (CCl_4) administration[5]. The suppressive effects are in accord with the lowering of serum levels of glutamate oxaloacetate transaminase (GOT) and glutamate pyruvate transaminase (GPT), which are known to leak into the bloodstream from a damaged liver.

From the fact that plural antioxidants have such an effect, the author proposes that more than one antioxidant should be used simultaneously for the protection of the liver from

Scheme 1. Standard procedure for the assay of lipid peroxide levels in organs or tissues.

1. Biopsied sample (x g) is weighed and chilled in ice-cold 0.9% NaCl.
2. With ice-cold 1.15% KCl, 10% (w/v) homogenate is prepared using a glass homogenizer (total volume, X ml).
3. Homogenate (0.1 ml) is placed in a glass tube.
4. To this tube, 0.2 ml of 8.1% sodium dodecyl sulfate, 1.5 ml of 20% acetic acid solution adjusted to pH 3.5 with NaOH, 1.5 ml of 0.8% aqueous solution of TBA, and 0.7 ml of distilled water are added and mixed.
5. The mixture is heated at 95°C for 60 min in an oil bath.
6. After cooling with tap water, 1.0 ml of distilled water and 5.0 ml of the mixture of n-butanol and pyridine (15:1, v/v) are added and shaken vigorously.
7. After centrifugation at 4,000 rpm for 10 min, the n-butanol layer is taken for absorbance measurement at 532 nm.
8. Taking the absorbance of the standard solution, which is obtained by 5.0 nmol of tetramethoxypropane with TBA by steps 4-7, as A and that of the sample as a, the lipid peroxide level can be calculated and expressed in terms of malondialdehyde:

Lipid peroxide level in organ or tissue

$$= 5.0 \times \frac{a}{A} \times \frac{X}{0.1} \times \frac{1.0}{x} = \frac{a}{A} \times \frac{X}{x} \times 50 \text{ (nmol/g organ or tissue)}$$

Table 2. Effects of curcumin, ellagic acid, and α-tocopherol on liver lipid peroxide levels and serum GOT and GPT of rats administered CCl_4.

CCl_4	–	+	+	+	+
Antioxidant	–	–	Curcumin		
(μmol/kg body wieght/day)			20	100	200
Lipid peroxide level of liver (nmol/mg protein)	1.67 ±0.13	6.43 ±0.61	6.04 ±1.43	4.52 ±1.21*	2.27 ±0.08**
Serum GOT (I. U.)	76 ±11	1351 ±56	1201 ±109	586 ±59**	327 ±20**
Serum GPT (I. U.)	48 ±7	508 ±57	482 ±34	162 ±12**	86 ±9**

CCl_4	+	+	+	+
Antioxidant	Ellagic acid		α-Tocopherol	
(μmol/kg body wieght/day)	20	100	20	100
Lipid peroxide level of liver (nmol/mg protein)	2.40 0.08**	1.91 ±0.45**	4.48 ±1.21*	2.75 ±0.15**
Serum GOT (I. U.)	364 ±41**	288 ±42**	561 ±56**	301 ±48**
Serum GPT (I. U.)	87 ±10**	69 ±7**	151 ±13**	88 ±6**

Lipid peroxide levels were measured according to Ohkawa et al.[3] and expressed in terms of malondialdehyde (nmol/mg protein). Mean ± SD is given. n=4. Significant difference vs. the control group administered CCl_4: *p<0.05; **p<0.001.

radical-mediated injury with the expectation of their cooperative antioxidant action and the elimination of side effects due to their different chemical properties.

GASTROINTESTINAL DISEASES

Yoshikawa et al.[6-11] conducted extensive research on radical-mediated injury of the gastric mucosa. Using an experimental model of gastric mucosal injury caused by burn shock[6], they observed hemorrhagic erosions in the mucosa and a significant increase in the lipid peroxide level of the mucosa as measured by the method of Ohkawa et al.[3]. They also carried out experiments using a model of gastric mucosal injury caused by ischemia-reperfusion of the stomach, and found an increase in the lipid peroxide level in the mucosa[7,8]. Stress caused in rats by water immersion also led to an increase in the lipid peroxide level in the gastric mucosa[9]. Involvement of lipid peroxidation in the pathogenesis of acute gastric mucosal erosion was also suggested in rat models by treatment with regional hyperthermia, platelet-activating factor, compound 46/80, and indomethacin[10,11].

All these results indicate that the increase in lipid peroxide level in situ provokes gastric mucosal injury and suggest that the injury would be suppressed by administration of suitable antioxidants.

Similar data have already been reported by Granger et al.[12] on intestinal disorders. They found that regional ischemia increased the permeability of intestinal capillaries and that pretreatment with superoxide dismutase (SOD) significantly attenuated the capillary permeability. The experiments conducted by Keshavarzian et al.[13] demonstrated the role of reactive oxygen species in experimental colitis produced by introducing 5% acetic acid into the cavity of the colon. They found that either intraperitoneal injection of methoxypolyethylene glycol-SOD or oral administration of a reactive oxygen metabolite scavenger, sulfasalazide, decreased the severity of inflammation.

With respect to carcinogenesis, Babbs[14] proposed the involvement of free radicals generated in the cavity of the colon in colon cancer. In such cases, introduction of antioxidants into the colonic cavity might be useful for treatment of this form of cancer.

PANCREATIC DISEASES

Sanfey et al.[15] studied the pathogenesis of acute pancreatitis by using an animal model of this disease and observed the suppression of the symptoms by allopurinol; they suggested that active oxygen species are involved in the occurrence of acute pancreatitis. Schoenberg et al.[16,17] studied cerulein-induced pancreatitis in rats and found the lipid peroxide level to be increased in the pancreas. With the same pancreatitis model, Morita[18] found that the α-tocopherol content in the pancreas was significantly decreased. These results show that pancreatitis can be provoked by lipid peroxidation in situ and it can be cured or even prevented by suitable antioxidants.

REFERENCES

1. T. Suematsu, T. Kamada, H. Abe, S. Kikuchi, and K. Yagi, Serum lipoperoxide level in patients suffering from liver diseases, *Clin. Chim. Acta* 79: 267 (1977).
2. K. Yagi, A simple fluorometric assay for lipoperoxide in blood plasma, *Biochem. Med.* 15: 212 (1976).
3. H. Ohkawa, N. Ohishi, and K. Yagi, Assay for lipid peroxides in animal tissues by thiobarbituric acid reaction, *Anal. Biochem.* 95: 351 (1979).

4. E. Maellaro, A. F. Casini, B. D. Bello, and M. Comporti, Lipid peroxidation and antioxidant systems in the liver injury produced by glutathione depleting agents, *Biochem. Pharmacol.* 39: 1513 (1990).
5. I. Nishigaki, R. Kuttan, H. Oku, F. Ashoori, H. Abe, and K. Yagi, Suppressive effect of curcumin on lipid peroxidation induced in rats by carbon tetrachloride or ^{60}Co-irradiation, *J. Clin. Biochem. Nutr.* 13: 23 (1992).
6. T. Yoshikawa, N. Yoshida, H. Miyagawa, T. Takemura, T. Tanigawa, S. Sugino, and M. Kondo, Role of lipid peroxidation in gastric mucosal lesions induced by burn schock in rats, *J. Clin. Biochem. Nutr.* 2: 163 (1987).
7. T. Yoshikawa, M. Yasuda, S. Ueda, Y. Naito, T. Tanigawa, H. Oyamada, and M. Kondo, Vitamin E in gastric mucosal injury induced by ischemia-reperfusion, *Am. J. Clin. Nutr.* 53: 210S (1991).
8. T. Yoshikawa, Y. Naito, S. Ueda, H. Ichikawa, S. Takahashi, M. Yasuda, and M. Kondo, Ischemia-reperfusion injury and free radical involvement in gastric mucosal disorders, *in*: "Oxygen Transport to Tissue XIII", T. K. Goldstick, M. McCabe, and D. J. Maguire, ed., Plenum Press, New York (1992).
9. T. Yoshikawa, H. Miyagawa, N. Yoshida, S. Sugino, and M. Kondo, Increase in lipid peroxidation in rat gastric mucosal lesions induced by water-immersion restraint stress, *J. Clin. Biochem. Nutr.* 1: 271 (1986).
10. T. Yoshikawa, Y. Naito, S. Ueda, H. Oyamada, T. Takemura, N. Yoshida, S. Sugino, and M. Kondo, Role of oxygen-derived free radicals in the pathogenesis of gastric mucosal lesions in rats, *J. Clin. Gastroenterol.* 12 (Suppl. 1): S65 (1990).
11. T. Yoshikawa, Y. Naito, A. Kishi, T. Tomii, T. Kaneko, S. Iinuma, H. Ichikawa, M. Yasuda, S. Takahashi, and M. Kondo, Role of active oxygen, lipid peroxidation, and antioxidants in the pathogenesis of gastric mucosal injury induced by indomethacin in rats, *Gut* 34: 732 (1992).
12. D. N. Granger, G. Rutili, and J. M. McCord, Superoxide radicals in feline intestinal ischemia, *Gastroenterol.* 81: 22 (1981).
13. A. Keshavarzian, G. Morgan, S. Sedghi, J. H. Gordon, and M. Doria, Role of reactive oxygen metabolites in experimental colitis, *Gut* 31: 786 (1990).
14. C. F. Babbs, Free radicals and the etiology of colon cancer, *Free Radical Biol. Med.* 8: 191 (1990).
15. H. Sanfey, G. B. Bulkley, and J. L. Cameron, The pathogenesis of acute pancreatitis. The source and role of oxygen-derived free radicals in three different experimental models, *Ann. Surg.* 201: 633 (1985).
16. M. H. Schoenberg, M. Büchler, M. Gaspar, A. Stinner, M. Younes, I. Melzner, B. Bültman, and H. G. Beger, Oxygen free radicals in acute pancreatitis of the rat, *Gut* 31: 1138 (1990).
17. M. H. Schoenberg, M. Büchler, M. Helfen, and H. G. Beger, Role of oxygen radicals in experimental acute pancreatitis, *Eur. Surg. Res.* 24: 74 (1992).
18. Y. Morita, The role of oxygen-derived free radicals in the pathogenesis of cerulein-induced pancreatitis in rats, *J. Kyoto Pref. Univ. Med.* 101: 81 (1992).

REACTIVE OXYGEN MOLECULES IN THE KIDNEY

Wayne R. Waz and Leonard G. Feld

Division of Pediatric Nephrology
Children's Kidney Center
Children's Hospital of Buffalo
219 Bryant Street
Buffalo, New York 14222

INTRODUCTION

Reactive oxygen species (ROS) are produced as a normal consequence of aerobic respiration, as a response to immunologic stimulation, and as a by-product of many oxidation-reduction reactions in living organisms. The kidney is a site of significant aerobic metabolism. In its role of maintaining fluid and electrolyte homeostasis, the kidney accounts for 10% of whole body oxygen consumption while making up less than 1% of total body mass[1]. Circulating immune elements (including neutrophils, monocytes, immune complexes) and activated intrinsic glomerular cells (macrophages, mesangial cells) are involved in a majority of glomerulonephritides[2]. ROS can mediate some of the glomerular damage. The arachidonic acid cascade, responsible for both vasodilator and vasoconstrictor substances essential to normal renal function, is both initiated by and a source of ROS[3]. However, the production of ROS is, under most circumstances, balanced by intrinsic antioxidant defenses[4]. Understanding the relationship between ROS production and antioxidant defenses will clarify the role of ROS in renal disease.

This chapter will review evidence of the following:

1) ROS are produced in the kidney.
2) Antioxidant defenses are present in the kidney.
3) ROS cause renal injury in *in vitro* models.
4) ROS cause renal injury in *in vivo* models.
5) ROS mediate human renal disease.
6) Treatments for ROS-induced renal injury

ROS ARE PRODUCED IN THE KIDNEY

As in other organs, the kidney produces ROS as a consequence of normal cellular metabolism. Electron transport (in endoplasmic reticulum, nuclear membranes, and mitochondria), auto-oxidation of molecules (such as thiols and catecholamines), the actions of

Free Radicals in Diagnostic Medicine, Edited by
D. Armstrong, Plenum Press, New York, 1994

enzymes such as xanthine oxidase and amino acid oxidases, and arachidonic acid metabolism all generate reactive oxygen molecules[5]. Further ROS generation is contributed by resident glomerular cells such as macrophages and mesangial cells when exposed to opsonized bacteria, immunoglobulins, immune complexes, anti-neutrophil cytoplasmic antibodies, and Tamm-Horsfall glycoproteins, activating respiratory burst defense mechanisms[6-9]. As a filtering organ which is exposed to 20% of the cardiac output, additional oxidant stress is produced by circulating polymorphonuclear leukocytes and monocytes which infiltrate glomeruli and interstitial tissues. Its relatively high oxygen consumption and metabolic activity, combined with immune-mediated ROS production by both resident and infiltrating cells, may overwhelm antioxidant defenses and predispose the kidney to ROS injury.

Evidence of ROS production in glomerular cells was first demonstrated by Shah[10]. When stimulated by phorbol myristate acetate, isolated rat glomeruli showed increased chemiluminescence. However, this experiment did not isolate resident cells from infiltrating cells. Subsequent experiments have shown that the resident cells of the glomerulus, including mesangial cells , macrophages, endothelial and epithelial cells can produce ROS[6,11,12] in response to various stimuli including phorbol myristate acetate, zymosan, trypsin, chymotrypsin, platelet activating factor, immune complexes, membrane attack complex, adriamycin, puromycin aminonucleoside[8,12,13]. Likewise, renal tubular epithelial cells from proximal, distal, and collecting segments, as well as interstitial, cells produce ROS[12,14-16].

While experimental manipulations show that various kidney cells produce ROS, understanding of their role in normal renal physiology continues to evolve. For example, arachidonic acid metabolites are intimately involved in the regulation of renal hemodynamics, and macrophages and mesangial cells modify extracellular matrix and basement membranes. Any experimental manipulations which are intended to block ROS production or enhance antioxidant defenses must account for the physiologic, as well as the pathologic, functions of ROS.

ANTIOXIDANT DEFENSES ARE PRESENT IN THE KIDNEY

The function of intrinsic antioxidant enzymes (AOE) in the kidney continues to be defined. Superoxide dismutase, both manganese and copper-zinc dependent forms, catalase, glutathione peroxidase, and heme oxygenase have all been identified within the kidney. The regulation and relative importance of each remains unclear. Histidine, tryptophan, ascorbate, and alpha-tocopherol are ubiquitous free radical scavengers. However, as noted by Ichikawa[4], changes in the production of a single enzyme can alter the oxidant/antioxidant balance and lead to disease. As an example, he suggests that underproduction of superoxide dismutase leads to excess production of superoxide and reduced iron favoring hydroxyl radical formation, while overproduction of superoxide dismutase can produce amounts of hydrogen peroxide which overwhelm tissue hydroperoxidases such as catalase or glutathione peroxidase.

The ability of non-renal tissues to regulate antioxidant enzyme production in response to endogenous and exogenous stimuli including glucocorticoids[17-19], endotoxin[20], interleukin 1[21], lipopolysaccharide, and tumor necrosis factor[22] is well-demonstrated.

The importance of AOE regulation in renal physiology and pathophysiology is currently defined by 2 types of experiments: those which attempt to block production of endogenous AOE and assess the effect on renal function and histology, and those which attempt to induce endogenous AOE in experimental models of renal disease. Experiments which block production of AOE have used selenium-deficient and vitamin-E deficient diets[23] and diethyldithiocarbamate[24] to block superoxide dismutase production, and selenium deficient diet[25] and diethyl maleate[26] to prevent glutathione peroxidase production. These interventions resulted in development of proteinuria, reduced renal function, and acceleration of pre-existing renal disease. Experimental induction of AOE occurs in both renal cell culture and whole animals, and stimuli include glucocorticoids[27], and ischemia-reperfusion injury[28], (but not ischemia alone[29]). The importance of the balance among AOE is suggested by a study in which frogs were given aminotriazole to inhibit catalase production resulting in increased production of superoxide dismutase, glutathione reductase, GSH, and ascorbate[17]. Previous renal oxidant injury may also induce production of AOE[30,31]. Developmental regulation of AOE production also occurs. Neonatal rats show increased production of both Mn and Cu/Zn superoxide dismutase when compared with fetal rats[32], and both catalase and superoxide dismutase increase with development in hamsters[33].

Investigators have identified antioxidant defenses in the kidney, induced their production with both endogenous and exogenous stimuli, and demonstrated pathology in their absence. Future research will further characterize the relative importance of different enzymes and identify the specific mechanisms of control at the molecular level.

ROS CAUSE RENAL INJURY IN *IN VITRO* MODELS

As in other tissues, the kidneys are susceptible to oxidant damage when antioxidant defenses are overwhelmed. Damage may occur to lipid membranes, extracellular matrix, DNA, proteins, and amino acids[5]. *In vitro* studies of cellular and extracellular renal structures contribute to our understanding of the mechanisms through which ROS exert their effects. Damage to both glomerular and tubular structures occurs through ROS mediated mechanisms, and *in vitro* experiments help define the mechanisms.

Destruction of glomerular basement membranes may lead to proteinuria and progressive renal insufficiency. ROS contribute to the degradation of glomerular basement membranes, particularly in the presence of activated neutrophils. Glomerular basement membranes exposed to anti-glomerular basement membrane antibodies[34,35] (a model of autoimmune renal diseases), and PMA-stimulated neutrophils[36] demonstrate proteolysis. The presence of ROS in glomeruli, whether as a result of normal metabolism or in response to immunologic injury, can modify the activity of proteolytic enzymes[37]. A proposed mechanism for glomerular basement membrane proteolysis involves the inactivation of alpha-1 protease inhibitor, a mediator of elastase activity[38], allowing unopposed activity of elastase. However, in addition to decreasing the activity of protease inhibitors, ROS may also inactivate the proteases themselves[36]. ROS may play a role in the normal physiology of glomerular basement membrane turnover as well as in the initiation of pathologic injury.

The ability of ROS to alter the structure and function of renal tubular cells is well-established. LLC-PK1 cells, a cell line derived from proximal tubular cells, show impaired transport of glucose and phosphate, depletion of ATP, and decreased Na^+-K^+ adenosine triphosphatase (Na-K ATPase) when exposed to H_2O_2, and this impairment is prevented in the presence of catalase[39]. When comparing four cell lines representing various tubular segments, Andreoli and colleagues[40] found that cell lines representing proximal tubular segments exhibited greater degrees of cell lysis and detachment when exposed to xanthine-oxidase generated ROS. Studies of hypoxia/reoxygenation injury in rat proximal tubular cell primary cultures showed significant cellular injury which was attenuated by antioxidant substances including superoxide dismutase, catalase, DMTU, deferoxamine, allopurinol, glutathione, and pyruvate[15,16,41]. Re-oxygenation of hypoxic renal cortical slices caused increased lipid peroxidation and reduction of ATP[42]. However, isolated proximal tubule segments from both rats and rabbits, when exposed to hypoxia followed by reoxygenation, demonstrated cellular damage which was not significantly improved with superoxide dismutase, catalase, DMTU, deferoxamine, or allopurinol[43-47]. The conflicting results raise questions about the ability of ROS to damage renal tubules, particularly in models of hypoxia/reoxygenation injury. Culture conditions may account for part of the discrepancy, yet changes in intrinsic antioxidant defenses with experimental manipulation should be studied further[48].

An additional area of ROS-induced renal damage involves alterations in the balance of vasodilator and vasoconstrictor substances. While this area will be discussed further in the section on whole animal models, evidence of ROS-mediated alterations of arachidonic acid metabolites and other vaso-modulator and inflammatory mediators is also seen in *in vitro* systems. Rat glomeruli incubated with xanthine-oxidase generated O_2^- demonstrated increased synthesis of prostaglandins E_2, $F_{2\alpha}$, and thromboxane[49]. Rat mesangial cells in particular can be stimulated to increase PGE_2 synthesis. ROS increase intracellular cyclic AMP content by stimulating production of prostaglandin synthesis[7]. Other modification of vasodilation through ROS involves inactivation of endothelium-generated nitric oxide[50-52]. Through a series of as yet undefined alterations in vasodilator/vasoconstrictor balance, ROS can significantly alter renal blood flow and lead to structural and functional renal damage.

In vitro models of oxidant injury in kidney-based systems suggest that both glomerular and tubular structure and function may be significantly altered through ROS-mediated mechanisms. The next section will address whether these mechanisms play a role in animal models of human renal disease.

ROS CAUSE RENAL INJURY IN *IN VIVO* MODELS

The role of ROS in *in vivo* animal models of human renal disease continues to evolve. Evidence of ROS-induced damage to kidneys includes models of glomerulonephritis, nephrotic syndrome, acute renal failure, transplantation, toxic injury, infectious and obstructive nephropathies, and chronic renal failure. Several excellent reviews have been published[5,7,37,53].

Glomerulonephritis/Glomerular Injury

ROS are implicated in several models of glomerulonephritis or glomerular injury. The contribution of infiltrating neutrophils and the complement system varies among different models, and both intrinsic and extrinsic ROS-generating cells may be involved in pathogenesis. Glomerular injury may be induced by direct immunologic mechanisms (anti-glomerular basement membrane antibodies), indirect stimulation of the animal's immune system to create autoimmune renal damage (passive Heymann nephritis), treatment with nephrotoxic substances, and direct infusion of ROS into the kidney.

Anti-glomerular basement membrane antibody treatment of animals, a model for human Goodpasture's disease, causes proteinuria and decreased renal function both acutely and chronically. The acute (heterologous) phase of injury occurs hours after injection of antibody and is characterized by complement and B-cell activation, while the chronic (autologous) phase, occurring 2-5 days after antibody challenge, is mediated primarily by T-cells[2]. Injury is attenuated by neutrophil depletion and complement inactivation. Evidence of the role of ROS is demonstrated by the protective effect of antioxidant therapy including catalase, superoxide dismutase, and deferoxamine in the acute phase[54,55] and superoxide dismutase in the chronic phase[56]. Anti-glomerular basement membrane glomerulonephritis is mediated, in part, by hydrogen peroxide produced by activated neutrophils[57]. In other studies of the effect of neutrophil-derived ROS on renal injury, phorbol myristate acetate was directly infused into renal arteries of rats to stimulate neutrophil hydrogen peroxide production[58]. The studies showed significant polymorphonuclear leukocyte accumulation in the kidney, and micropuncture revealed decreased whole kidney and single nephron glomerular filtration rate, increased arteriolar resistances (efferent greater than afferent) and a decreased ultrafiltration coefficient. These functional deficits were reversed by catalase and neutrophil depletion. The relative contributions of complement activation and neutrophil infiltration were examined by renal arterial infusion of phorbol myristate acetate and cobra venom factor (to stimulate complement activation). These studies showed a primary role of neutrophil-generated hydrogen peroxide, enhanced by (but not dependent upon) complement activation[59,60]. Direct renal artery infusion of hydrogen peroxide induced a dose-dependent increase in urinary protein excretion which was not accompanied by changes in glomerular filtration rate, renal blood flow, or glomerular morphology. The proteinuric effect of hydrogen peroxide infusion was blocked by catalase and deferoxamine[61]. Taken together, studies of neutrophil-dependent glomerulonephritis demonstrate a role for ROS which includes, but is not limited to, hydrogen peroxide and its metabolites. A possible mechanism for ROS-induced injury in neutrophil-dependent models of glomerulonephritis was proposed by Johnson et.al.[62]: antigen-antibody complexes activate neutrophils in glomeruli. Released neutrophil enzymes, particularly myeloperoxidase, bind to the glomerular basement membrane and react with locally-produced hydrogen peroxide to damage the glomerular basement membrane and nearby cells. Additionally, alterations in the balance of local vasoconstrictor and vasodilator prostaglandins may shift to favor vasoconstriction (primarily of efferent arterioles) and further damage the glomerulus[37]. Neutrophil-independent models of glomerular injury also demonstrate a role for ROS. Passive Heymann nephritis is a model of human membranous nephropathy. In this model, rats are injected with antigen derived from rat proximal tubular brush borders. They develop subepithelial glomerular immune deposits, accompanied by proteinuria and renal insufficiency[2]. Immune complex deposition in the subepithelial spaces stimulates complement activation but not infiltration of circulating cells. The injury is complement-dependent but neutrophil-independent. Damage in this model is attenuated by deferoxamine and dimethylthiourea (DMTU)[63] and dimethyl sulfoxide (DMSO)[64], but not by SOD or catalase[63], suggesting that hydrogen peroxide and its metabolites may be responsible for injury. A similar pathologic lesion can be induced by injection of cationized bovine gamma globulin into previously-exposed rats. Activation of glomerular macrophages and mesangial cells may be responsible for pathogenesis. DMSO and DMTU, but not catalase or superoxide dismutase, attenuate proteinuria and renal insufficiency[65].

A single injection of puromycin aminonucleoside (PAN) into rats causes nephrotic syndrome and renal lesions similar to human minimal change disease. In a series of studies, Diamond and colleagues[53] demonstrated that PAN-induced nephrotic syndrome is mediated by xanthine-oxidase generated reactive oxygen metabolites including superoxide, hydrogen peroxide, and hydroxyl radical.

As discussed above, ROS are implicated in a variety of *in vivo* models of glomerulonephritis and primary glomerular injury which approximate human disease. The mechanisms include neutrophil-dependent, complement-dependent, and as yet unidentified mechanisms (Table 1). Antioxidant therapy of human glomerulonephritides may prove to be of some benefit, but as yet no definitive clinical trials have been published.

Ischemia-Reperfusion Injury

Ischemia-reperfusion injury, while not unique to renal tissue, is important in the pathogenesis of two major areas of human kidney disease: acute renal failure and transplantation. Acute renal failure frequently follows a compromise of renal blood flow or oxygenation, followed by a period of restored perfusion and subsequent reoxygenation. Transplantation of kidneys, particularly cadaveric kidneys, necessitates a period of complete ischemia, followed by re-establishment of blood flow in a new host. As in other tissues, renal ischemia-reperfusion injury involves a variety of mechanisms including conversion of xanthine dehydrogenase to xanthine oxidase, polymorphonuclear leukocyte recruitment and activation, disruption of mitochondrial membranes, adenosine triphosphate depletion, and altered production of arachidonic acid metabolites[5,48].

Table 1. Models of glomerulonephritis/glomerular injury (modified from [37])

Model	Complement-dependent?	Neutrophil-dependent?
Anti-glomerular basement membrane	yes	yes
Passive Heymann Nephritis	yes	no
Cationized gamma globulin	no	no
Phorbol myristate acetate	no	yes
Cobra venom factor	yes	yes
Puromycin aminonucleoside	no	no

Ischemia-reperfusion injury due to ROS is seen in animal models, and use of antioxidant molecules in reversing the injury raises hope for therapy of acute renal failure and post-transplant acute tubular necrosis in humans. Ischemia-reperfusion models demonstrate increased lipid peroxidation[66-68], and this lipid peroxidation can be decreased by superoxide dismutase, allopurinol[67], and deferoxamine[68]. Superoxide dismutase and allopurinol also improved renal function and enhanced regeneration of ATP[69]. Several investigators have shown that a significant component of the ROS-induced injury in ischemia-reperfusion models involves conversion of xanthine dehydrogenase to xanthine oxidase[70-74]. In one study closely approximating human disease, acute renal failure was induced in rats by hemorrhagic shock, with animals bled to maintain mean arterial blood pressure 40-80 mmHg for 180 minutes followed by re-infusion of blood. Allopurinol exerted a significant protective effect[73]. Using a novel electron spin resonance technique, Haraldsson[74] and co-workers showed that oxypurinol significantly reduced free radical production following ischemia-reperfusion. In addition to xanthine oxidase-induced ROS production, ischemia reperfusion injury may also be mediated through other sources of ROS production or destruction of intrinsic antioxidant defenses. Depletion of dietary vitamin E and glutathione peroxidase exacerbates ischemia-reperfusion injury[23]. Ischemia-reperfusion injury induces formation of 8-epi-prostaglandin $F_{2\alpha}$ through a non-cyclooxygenase pathway involving lipid peroxidation[75]. Infusion of this substance demonstrates thromboxane-like characteristics including decreased whole kidney and single nephron glomerular filtration and blood flow and increased afferent arteriolar resistance[75]. The role of complement activation in ROS-induced injury was demonstrated in studies which used zymosan and cuprophan membranes (a commonly used hemodialysis membrane) to activate the complement system, followed by ischemia-reperfusion injury. Complement activation increased lipid peroxidation and reduced renal function, and the effect was reversed by deferoxamine[76].

Several free-radical scavengers have been used effectively in organ-preservation solutions to minimize free-radical induced injury, including superoxide dismutase, mannitol, albumin, transferrin, ceruloplasmin, vitamin E, ascorbic acid, dimethyl sulfoxide, dimethylthioruea, deferoxamine. Xanthine oxidase inhibitors including allopurinol and oxypurinol have also been protective in animal models, as have glutathione, histidine, selenium, and adenosine[3,77-80]. An effective organ preservation solution should contain free radical scavengers and/or other antioxidants, and their use in human kidney transplantation will be discussed below. While animal models of ischemia-reperfusion injury demonstrate ROS-induced changes, their exact function in the initiation and propagation of cellular and subcellular injury is unclear. Not all investigators have found ROS-induced injury to be responsible for the pathology seen in these models[43,44,81-86], and the effects of lipid peroxidation, free radical generation, and arachidonic acid metabolite production may be only incidental, with renal injury mediated through other mechanisms. Despite conflicting studies, the weight of evidence favors some role for ROS in ischemia-reperfusion injury.

Progressive Renal Failure

With reduction of renal mass, be it from glomerular sclerosis, traumatic injury, chronic infection, or other mechanisms, oxygen consumption increases in the remaining kidney. The increased oxygen consumption may lead to increased ROS generation which overwhelms intrinsic antioxidant defenses and acts as a common mechanism for progressive renal insufficiency leading to end stage renal disease[87,88]. In 5/6 nephrectomized animals, individual nephron malondialdehyde content increased, as did urinary MDA excretion. When the additional stress of protein feeding was added, animals developed changes consistent with progression of chronic renal failure. Nath proposed that although decreasing renal mass demonstrated ROS-induced changes, the kidney compensates for increases in ROS by increasing intrinsic defenses. However, these defenses are more easily overwhelmed by additional stresses (such as high protein diets), and contribute significantly to the chronic progression of renal disease.

Other Models of ROS-Induced Renal Injury

Several nephrotoxic substances may induce renal injury through generation of free radicals or other reactive oxygen molecules, or by suppressing intrinsic antioxidant defenses[89]. Gentamicin can increase renal malondialdehyde content, particularly in mitochondria, with the effects reversed by hydroxyl radical scavengers and deferoxamine[90,91]. Adriamycin metabolism results in formation of semiquinone radical and superoxide anion[92]. Mercuric chloride, cephalosporins, and radiocontrast agents all contribute to lipid peroxidation[89]. Models of pigment-induced acute renal failure (glycerol injection or crush injury) release iron from hemoglobin and/or myoglobin, promoting the generation of ROS[90,93]. Not all toxins mediated damage through ROS. In uranyl acetate-induced renal failure, animals did not demonstrate increased hydrogen peroxide production, nor were they responsive to superoxide dismutase, glutathione peroxidase, or catalase[94].

Ample evidence shows that ROS can induce renal injury in animal models of human renal disease, and that antioxidants reverse these injuries. However, the role of ROS in the pathophysiology of human diseases, and thus the role of antioxidant therapies, continues to evolve.

ROS MEDIATE HUMAN RENAL DISEASE

Rather than emerging as etiologic agents for specific renal diseases, ROS seem ubiquitous in *in vitro* and *in vivo* models of nephropathy. Likewise with human nephropathy, theories of pathogenesis in a variety of kidney lesions may include free radical generation by both infiltrating immune cells and intrinsic kidney cells. Table 2 summarizes human renal diseases in which ROS have been implicated and, when available, associated experimental models.

Most of the data regarding human nephropathy involves protection of renal allografts from ischemia-reperfusion injury, and will be discussed in the section on antioxidant treatment of renal diseases[3,77,95]. Evidence continues to accumulate for the role of ROS in other renal diseases. In IgA nephropathy, circulating neutrophils have an amplified capacity for hydrogen peroxide production and accumulate in glomeruli[96]. In patients with insulin dependent diabetes mellitus, increased serum lipid peroxide concentration correlates with the degree of mesangiolysis seen on

Table 2. Experimental models of human renal diseases with possible ROS mechanisms.

Human disease	Experimental model
Glomerulonephritis	Anti-GBM antibody Passive Heymann Nephritis Cationized gamma globulin Phorbol myristate acetate Cobra venom factor
Nephrotic syndrome	Puromycin aminonucleoside Hydrogen peroxide
Ischemia-reperfusion/acute renal failure	Autotransplantation Renal artery clamping Hemorrhagic shock
Chronic renal failure	5/6 nephrectomy
Toxic nephropathy	Gentamicin, adriamycin, mercuric chloride, cephalosporin, radiocontrast dye, glycerol, crush injury
Obstructive nephropathy	Bilateral ureteral obstruction/probucol
Diabetic nephropathy	*
Hemolytic uremic syndrome	*
Radiation nephritis	*
IgA nephropathy	*
Pyelonephritis	*

***no experimental model currently implicates ROS in these human renal diseases.**

biopsy and with glomerular filtration rate. Furthermore, patients with proteinuria showed increased plasma levels of catalase[97]. Children with hemolytic uremic syndrome show decreased plasma vitamin E and red blood cell phosphatidyl ethanolamine[98], activated neutrophils[99], increased red blood cell malondialdehyde and oxidized glutathione concentration, and decreased activities of superoxide dismutase, catalase, and glutathione peroxidase[100]. Mechanisms of disease which include neutrophil infiltration or plasma membrane disruption may involve reactive oxygen metabolites, for example: radiation nephritis, pyelonephritis, obstructive uropathy, and lithotripsy-induced nephropathy[5].

TREATMENTS FOR ROS-INDUCED RENAL INJURY

The role of antioxidants in protecting transplant kidneys from ischemia-reperfusion injury continues to evolve. By cooling kidneys to 6 to 10° C. oxygen consumption is decreased by 90-95%, favoring decreased ROS production[77]. While human studies have not specifically addressed the effect of hypothermia on free radical production, its protective effect must be in part attributable to decreased oxidative metabolism. Early studies of the use of allopurinol as an additive to cadaveric kidney perfusate did not demonstrate a significant benefit[101], but more recently allopurinol has shown to improve graft function[3], particularly when used as part of a perfusate which also contains adenosine and glutathione[102-105]. This solution, developed at the University of Wisconsin (UW solution) has proven superior to similar solutions which do not contain allopurinol, glutathione, or adenosine. The relative importance of each of the three components is unknown. Superoxide dismutase has also been used to prevent oxidant damage. In a placebo-controlled trial of bovine superoxide dismutase (infused immediately after revascularization), treated kidneys showed improved long term function when compared to same-donor kidneys in other patients. Treated patients also showed a trend (although not statistically significant) toward improved creatinine clearance[106,107]. Because of its short half-life, superoxide dismutase has not been used in extensive clinical trials. Conjugation of SOD to polyethylene glycol or other methods of prolonging its half-life in the circulation may increase its efficacy[108]. A recent study infused a multivitamin cocktail containing vitamins E, C, A, and B complex into patients immediately prior to revascularization of grafts. Treated patients showed decreased plasma malondialdehyde and improved creatinine clearance during the first week post-transplant[109]. Other therapies often used post-transplantation that may, in addition to their intended use, reduce free radical-induced injury include mannitol, albumin, diltiazem, captopril, and probucol.

Along with their role in post-transplant injury, ROS may be involved in many renal diseases including acute and chronic glomerulonephritis, acute renal failure, IgA nephropathy, diabetic nephropathy, hemolytic uremic syndrome, radiation nephritis, and obstructive uropathy. However, the mere presence of reactive oxygen metabolites or a change in function induced by antioxidants observed in *in vitro* and *in vivo* models, or evidence of immune system activation in human diseases, does not sufficiently incriminate these molecules as primary mediators of human renal disease. With the exception of their use in transplantation, antioxidants do not currently play a well-defined role in the therapy of renal diseases.

REFERENCES

1. S.R. Gullans and S.C. Hebert, Metabolic basis of ion transport, *in:*"The Kidney, " B.M. Brenner and F.C. Rector eds., W. B. Saunders Company, Philadelphia (1991).
2. C.B. Wilson, The renal response to immunologic injury, *in:*"The Kidney, " B.M. Brenner and F.C. Rector eds., W. B. Saunders Company, Philadelphia (1991).
3. H.J. Schiller, K.A. Andreoni, and G.B. Bulkley, Free radical ablation for the prevention of post-ischemic renal failure following renal transplantation, *Klinische Wochenschrift.*69:1083-1094 (1991).
4. I. Ichikawa, S. Kiyama, and T. Yoshioka, Renal atioxidant enzymes: their regulation and function, *Kidney Int.*45:1-9 (1994).
5. S.P. Andreoli, Reactive oxygen molecules, oxidant injury, and renal disease, *Pediatr. Nephrol.* 5:733-742 (1991).
6. L. Baud, J. Hagege, J. Sraer, E. Rondeau, J. Perez, and R. Ardaillou, Reactive oxygen production by cultured rat glomerular mesangial cells during phagocytosis is associated with stimulation of lipoxygenase activity, *J. Exp. Med.*158:1836-1852 (1983).
7. L. Baud and R. Ardaillou, Reactive oxygen spcies: production and role in the kidney, *Am. J. Physiol.*251:F765-F776 (1986).
8. S.V. Shah, Oxidant mechanisms in glomerulonephritis, *Semin. Nephrol.*11:320-326 (1991).
9. J.K. Horton, M. Davies, N. Topley, D. Thomas, and J.D. Williams, Activation of inflammatory response of neutrophils by Tamm-Horsfall glycoprotein, *Kidney Int.*37:717-726 (1990).
10. S.V. Shah, Light emission by isolated rat glomeruli in response to phorbol myristate acetate, *J. Lab. Clin. Med.*98:46-57 (1981).
11. H.H. Radeke, B. Meier, N. Topley, J. Floge, G.G. Habermehl, and K. Resch, Interleukin 1-α and tumor necrosis factor-α induce oxygen radical production in mesangial cells, *Kidney Int.*37:767-775 (1990).

12. P. Stratta, C. Canavese, M. Dogliani, G. Mazzucco, G. Monga, and A. Vercellone, Role of free radicals in the progression of renal disease, *Am. J. Kidney Dis.*XVII (Suppl 1):33-37 (1991).
13. S. Adler, P.J. Baker, and R.J. Johnson, Complement membrane attack complex stimulates production of reactive oxygen metabolites by cultured rat masangial cells, *J. Clin. Invest.*77:762-767 (1986).
14. B.H. Rovin, E. Wurst, and D.E. Kohan, Production of reactive oxygen species by tubular epithelial cells in culture, *Kidney Int.*37:1509-1514 (1990).
15. M.S. Paller and T.V. Neumann, Reactive oxygen species and rat renal epithelial cells during hypoxia and reoxygenation, *Kidney Int.*40:1401-1049 (1992).
16. E.L. Greene and M.S. Paller, Xanthine oxidase produces O_2- in posthypoxic injury of renal epithelial cells, *Am. J. Physiol.*263:F251-F255 (1992).
17. L. Frank, P.L. Lewis, and I.R.S. Sosenco, Dexamethasone stimulation of fetal rat lung antioxidant enzyme activity in parallel with surfactant stimulation, *Pediatrics.*75:569-574 (1985).
18. P. Randhawa, M. Hass, L. Frank, and D. Massaro, Dexamethasone increases superoxide dismutase activity in serum-free rat fetal organ cultures, *Pediatr. Res.*20:895-898 (1986).
19. P. Randhawa, M. Hass, L. Frank, and D. Massaro, P_{02}-dexamethasone interactions in fibroblast growth and antioxidant enzyme activity, *Am. J. Physiol.*252:C396-C400 (1987).
20. Y. Shiki, B.O. Meyrick, K.L. Brigham, and I.M. Burr, Endotoxin increases superoxide dismutase in cultured bovine pulmonary endothelial cells, *Am. J. Physiol.*252:C436-C440 (1987).
21. A. Masuda, D.L. Longo, Y. Kobayashi, E. Appella, J.J. Oppenheim, and K. Matsushima, Induction of mitochondrial manganese superoxide dismutase by interleukin 1, *FASEB J.*2:3087-3091(1988).
22. G.A. Visner, W.C. Dougall, J.M. Wilson, I.M. Burr, and H.S. Nick, Regulation of manganese superoxide dismutase by lipopolysaccharide, interleukin 1, and tumor necrosis factor, *J. Biol. Chem.*265:2856-2864 (1990).
23. K.A. Nath and M.S. Paller, Dietary deficiency of antioxidants exacerbates ischemic injury in the rat, *Kidney Int.*38:1109-1117 (1990).
24. T. Hara, H. Miyai, T. Iida, A. Futenma, S. Nakamura, and K. Kato, Aggravation of puromycin aminonucleoside nephrosis by the inhibition of endogenous superoxide dismutase, *Proc. XIth Int. Congr. Nephrol.*442 (1990).
25. R. Baliga, M. Baliga, and S.V. Shah, Effect of selenium deficient diet in experimental glomerular disease, *Am. J. Physiol.*263:F56-F61 (1992).
26. H. Miyai, T. Hara, K. Yajada, S. Nakamura, A. Futenma, and K. Kato, Aggravation of puromycinaminonucleoside nephrosis by glutathione depleting agent, *Proc. XIth Int. Congr. Nephrol.*442 (1990).
27. T. Kawamura, T. Yoshioka, T. Bills, A. Fogo, and I. Ichikawa, Glucocorticoid activates glomerular antioxidant enzymes and protects glomeruli from oxidant injuries, *Kidney Int.*40:291-301 (1991).
28. T. Yoshioka, T. Bills, T. Moore-Jarrett, H.L. Greene, I.M. Burr, and I. Ichikawa, Role of intrinsic antioxidant enzymes in renal oxidant injury, *Kidney Int.*38:282-288 (1990).
29. M.L. Barnard, S. Snyder, T.D. Engerson, and J.F. Turrens, Antioxidant enzyme status of ischemic and postischemic liver and ischemic kidney in rats, *Free Radical Biol. Med.*15:227-232 (1993).
30. K.A. Nath, J. Zou, and M.E. Rosenberg, Prior exposure to hydrogen peroxide confers resistance to oxidative injury in LLC-PK1 cells and is associated with induction of heme oxygenase mRNA, *J. Am. Soc. Nephrol.*3:711 (1992).
31. N. Honda, A. Hishida, K. Ikuma, and K. Yonemura, Acquired resistance to acute renal failure, *Kidney Int.*31:1233-1238 (1988).
32. K. Asayama, H. Hayashibe, K. Dobashi, N. Uchida, M. Kobayashi, A. Kawaoi, and K. Kato, Immunohistochemical study on perinatal development of rat superoxide dismutases in lungs and kidneys, *Pediatr. Res.*29:487-491 (1991).
33. T.D. Oberly, L.W. Oberly, A.F. Slattery, L.J. Lauchner, and J.H. Elwell, Immunohistochemical localization of antioxidant enzymes in adult syrian hamster tissues and during kidney development, *Am. J. Pathol.*137:199-214 (1990).
34. M.G. Davies, G.A. Coles, and M.H. Harber, Effect of glomerular basement membrane on the initiation of chemiluminescence and lysosomal enzyme release in human polymorphonuclear leukocytes: an in vitro model of glomerular disease, *Immunology.*52:151-159.
35. H. Mossmann, B. Hoyer, W. Walz, K. Himmelspach, and D.K. Hammer, Antibody -dependent cellular cytotoxicity and chemiluminescence as a tool for studying the mechanism of anti-glomerular asement membrane nephritis. The role of the cytotoxic potential of polymorphonuclear granuylocytes and monocytes, *Immunlolgy.*53:545-552 (1984).

36. M.C.M. Vissers, R. Wiggins, and J.C. Fantone, Comparative ability of human monocytes and neutrophils to degrade glomerular basement membrane, *Lab. Invest.*60:831-383 (1989).
37. S. Shah, Role of reactive oxygen metabolites in experimental glomerular disease, *Kidney Int.* 35:1093-1106 (1989).
38. S.J. Weiss and S. Regiani, Neutrophils degrade subendothelial matrices in the presence of alpha$_1$-proteinase inhibitor, *J. Clin. Invest.*73:1297-1303 (1984).
39. S.P. Andreoli, J.A. McAteer, S.A. Seifert, and S.A. Kempson, Oxidant-induced alterations in glucose and phosphate transport in LLC-PK$_1$ cells: mechanisms of injury, *Am. J. Physiol.*265:F377-F384 (1993).
40. S.P. Andreoli and J.A. McAteer, Reactive oxygen molecule-mediated injury in endothelial and renal tubular epithelial cells in vitro, *Kidney Int.*38:785-794 (1990).
41. A.K. Salahudeen, E.C. Clark, and K.A. Nath, Hydrogen peroxide-induced renal injury: a protective role for pyruvate in vitro and in vivo, *J. Clin. Invest.*88:1886-1893 (1991).
42. A. Asakura, H. Ikeda, and G. Munekazu, In vitro oxygenation injury to slices prepared from ischemic kidney in rats, *Japanese Journal of Pharmacology.*60:149-151 (1992).
43. S.C. Borkan and J.H. Schwartz, Role of oxygen free radical species in in vitro models of proximal tubular ischemia, *Am. J. Physiol.*257:F114-F125 (1989).
44. R.B. Doctor and L.J. Mandel, Miminmal role of xanthine oxidase and oxygen free radicals in rat renal tubular reoxygenation injury, *J. Am. Soc. Nephrol.*1:959-969 (1991).
45. R. Zager, D.J. Gmur, B.A. Schimpf, C.R. Bredl, and C.A. Foerder, Evidence against increased hydroxyl radical production during oxygen deprivation-reoxygenation proximal tubular injury, *J. Am. Soc. Nephrol.*2:1627-1633 (1992).
46. R. Zager, B.A. Schimpf, C.R. Redl, and C.A. Foerder, Increased proximal tubular cell catalytic iron content: a result, not a mediator of, hypoxia-reoxygenation injury, *J. Am. Soc. Nephrol.*3:116-118 (1992).
47. J.M. Weinberg and H.D. Humes, Increases of cell ATP produced by adenine nucleotides in isolated rabbit kidney tubules, *Am. J. Physiol.*250:F720-F733 (1986).
48. K.J. Johnson and J.M. Weinberg, Postischemic renal injury due to oxygen radicals, *Current Opinion in Nephrology and Hypertension.*2:625-635 (1993).
49. L. Baud, M.P. Nivez, D. Chansel, and R. Ardaillou, Stimulation by oxygen radicals of prostaglandin production by rat renal glomeruli, *Kidney Int.*20:332-339 (1981).
50. L. Baud, B. Fouqueray, C. Philippe, and R. Ardaillou, Reactive oxygen species as glomerular autacoids, *J. Am. Soc. Nephrol.*2:S132-S138 (1992).
51. G.M. Rubanyi, Vascular effects of oxygen-derived free radicals, *Free Radical Biol. Med.*4:107-120 (1988).
52. R.J. Gryglewski, R. Palmer, M.J., and S. Moncada, Superoxide anion is involved in the breakdown of endothelium-derived vascular relaxing factor, *Nature.*320:454-456 (1986).
53. J.R. Diamond, The role of reactive oxygen species in animal models of glomerular disease, *Am. J. Kidney Dis.*XIX:292-300 (1992).
54. A. Rehan, K.J. Johnson, R.C. Wiggins, R.G. Kunkel, and P.A. Ward, Evidence for the role of oxygen radicals in acute nephrotoxic nephritis, *Laboratory Investigation.*51:396-403 (1984).
55. N.W. Boyce and S.R. Holdsworth, Hydroxyl radical mediation of immune renal injury by desferrioxamine, *Kidney Int.*30:813-817 (1986).
56. T. Adachi, M. Fukuta, Y. Ito, K. Hirano, M. Sugiura, and K. Sugiura, Effect of superoxide dismutase on glomerular nephritis, *Biochem. Pharmacol.*35:341-345 (1986).
57. R.J. Johnson, S.J. Klebanoff, R.F. Ochi, S. Adler, P. Baker, L. Sparks, and W.G. Couser, Participation of the myeloperoxidase-H_2O_2-halide system in immune complex nephritis, *Kidney Int.*32:342-349 (1987).
58. T. Yoshioka and I. Ichikawa, Glomerular dysfunction induced by polymorphonuclear leukocyte-derived reactive oxygen species, *Am. J. Physiol.*257:F53-F59 (1989).
59. A. Rehan, K.J. Johnson, R.G. Kunkel, and R.C. Wiggins, Role of oxygen radicals in phorbol myristate acetate-induced glomerular injury, *Kidney Int.*27:503-511 (1985).
60. A. Rehan, R.C. Wiggins, R.G. Kunkel, G.O. Till, and K.J. Johnson, Glomerular injury and proteinuria in rats after intrarenal injuction of cobra venom factor: evidence for the role of neutrophil-derived oxygen free radicals, *Am. J. Pathol.*123:57-66 (1986).
61. T. Yoshioka, I. Ichikawa, and A. Fogo, Reactive oxygen metabolites cause massive, reversible proteinuria and glomerular sieving defect without apparent ultrastructural abnormality, *J. Am.Soc. Nephrol.*2:902-912 (1991).

62. K.J. Johnson, A. Rehan, and P.A. Ward, The role of oxygen radicals in kidney disease, *Upjohn Symposium/Oxygen Radicals.*115-121 (1987).
63. S. Shah, Evidence suggesting a role for hydroxyl radical in passive Heymann nephritis in rats, *Am. J. Physiol.*254:F337-F344 (1988).
64. D. Lotan, B.S. Kaplan, J.S.C. Fong, P.R. Goodyer, and J.P. de Chadarevian, Reduction of protein excretion by dimethyl sulfoxide in rats with passive Heymann nephritis, *Kidney Int.*25:778-788 (1984).
65. M.A. Rahman, S.S. Emancipator, and J.R. Sedor, Hydroxyl radical scavengers ameliorate proteinuria in rat immune complex glomerulonephritis, *J. Lab. Clin. Med.*112:619-626 (1988).
66. J.M. McCord, Oxygen-derived free radicals in postischemic tissue injury, *N. Engl. J. Med.*312:159-163 (1985). 67. M.S. Paller, J.R. Hoidal, and T.F. Ferris, Oxygen free radicals in ischemic acute renal failure in the rat, *J. Clin. Invest.*74:1156-1164 (1984).
68. M.S. Paller and B.E. Hedlund, Role of iron in postischemic renal injury in the rat, *Kidney.Int.*34:474-480 (1988).
69. P.H. Lee, Y.C. Chung, M.T. Huang, R.H. Hu, and C.S. Lee, Protective effect of superoxide dismutase and allopurinol on oxygen free radical-induced damage to the kidney, *Transplantation Proceedings.*24:1353-1354 (1992).
70. R.N. McCoy, K.E. Hill, M.A. Ayon, J.H. Stein, and R.F. Burk, Oxidant stress following renal ischemia: changes in the glutathione redox ratio, *Kidney Int.*33:812-817 (1988).
71. T.G. McKelvey, M.E. Hollwarth, D.N. Granger, T.D. Engerson, U. Landler, and H.P. Jones, Mechanisms of conversion of xanthine dehydrogenase to xanthine oxidase in ischemic rat liver and kidney, *Am. J. Physiol.*254:G753-G760 (1988).
72. S.L. Linas, D. Whittenburg, and J.E. Repine, Role of xanthine oxidase in ischemia/reperfusion injury, *Am. J. Physiol.*258:F711-F716 (1990).
73. L. Yu, A.C. Seguro, and A.S. Rocha, Acute renal failure following hemorrhagic shock: protective and aggravating factors, *Renal Failure.*14:49-55 (1992).
74. G. Haraldsson, U. Nilsson, S. Bratell, S. Pettersson, T. Schersten, S. Akerlund, and O. Jonsson, ESR-measurement of production of oxygen radicals in vivo before and after renal ischemia in the rabbit, *Acta. Physiol. Scand.*146:99-105 (1992).
75. K. Takahashi, T.M. Nammour, M. Fukunaga, J. Ebert, J.D. Morrow, L.J. Roberts, R.L. Hoover, and K.F. Badr, Glomerular actions of a free radical-generated novel prostaglandin, 8_epi-prostaglandin $F_{2\alpha}$, in the rat: evidence for interaction with thromboxane A_2 receptors, *J. Clin. Invest.*90:136-141 (1992).
76. G. Schulman, A. Fogo, A. Gung, K. Badr, and R. Hakim, Complement activation retards resolution of acute ischemic renal failure in the rat, *Kidney Int.*40:1069-1074 (1991).
77. W.F. Finn, Prevention of ischemic injury in renal transplantation, *Kidney Int.*37:171-182 (1990).
78. A. Demirbas, S. Bozoklu, A. Ozdemir, N. Bilgin, and M. Haberal, Effect of α-tocopherol on the prevention of reperfusion injury caused by free oxygen radicals in the canine kidney autotransplantation model, *Transplantation Proceedings.*25:2274 (1993).
79. I. Koyama, G.B. Bulkley, G.M. Williams, and M.J. Im, The role of oxygen free radicals in mediating the reperfusion of cold-preserved ischemic kidneys, *Transplantation.*40:590-595 (1985).
80. W.I. Bry, G.M. Collins, N.A. Halasz, and M. Jellinek, Improved function of perfused rabbit kidneys by prevention of oxidative injury, *Transplantation.*38:579-582 (1984).
81. L.M. GAmelin and R.A. Zager, Evidence against oxidative injury as a critical mediator of postischemic acute renal failure, *Am. J. Physiol.*255:F450-F460 (1988).
82. M. Joannidis, G. Gstraunthaler, and W. Pfaller, Xanthine oxidase: evidence against a causative role in renal reperfusion injury, *Am. J. Physiol.*258:F232-F236 (1990).
83. M.A. Thornton, R. Winn, C.E. Alpers, and R.A. Zager, An evaluation of the neutrophil as a mediator of in vivo renal ischmeic-reperfusion injury, *Am. J. Pathol.*135:509-515 (1989).
84. M.S. Paller, Effect of neutrophil depletion on ischemic renal injury in the rat, *J. Lab. Clin. Med.*113:379-386 (1989).
85. S.L. Linas, D. Whittenburg, and J.E. Repine, O_2 metabolites cause reperfusion injury after short but not prolonged renal ischemia, *Am. J. Physiol.*253:F685-F691 (1987).
86. J.Z. Li, H.Y. Wang, J. Tang, W.Z. Zou, D.H. Lu, and D.W. Chen, The effect of calcitonin gene-related peptide on accute ischemia-reperfusion injury: ultrastructural and lipid peroxidation studies, *Renal Failure.*14:11-16 (1992).
87. K.A. Nath, A.J. Croatt, and T.H. Hostetter, Oxygen consumption and oxidant stress in surviving nephrons, *Am. J. Physiol.*258:F1354-F1362 (1990).

88. K.A. Nath and A.K. Salahudeen, Induction of renal growth and injury in the intact rat kidney by dietary deficiency of antioxidants, *J. Clin. Invest.*86:1179-1192 (1990).
89. A.R. Morrison and D. Portilla, Lipid peroxidation and the kidney, *in:*"Cellular Antioxidant Defense Mechanisms, " C.K. Chow ed., CRC press, Boca Raton (1988)90. P.D. Walker and S.V. Shah, Evidence suggesting a role for hydroxyl radical in gentamicin-induced acute renal failure in rats, *J. Clin. Invest.*81:334-341 (1988).
91. P.D. Walker, C. Das, and S.V. Shah, Cyclosporin A induced lipid peroxidation in renal cortical mitochondria, *Kidney Int.*29:311 (1986).
92. C.E. Myers, W.P. McGurie, R.H. Liss, I. Iprim, K. Grotzinger, and R.C. Young, Adriamycin: the role of lipid peroxidation in cardiac toxicity and tumor response, *Science.*197:165-167 (1977).
93. M. Odeh, The role of reperfusion-induced injury in the pathogenesis of the crush syndrome, *N. Engl.J. Med.*324:1417-1422 (1991).
94. N. Honda, A. Hishida, and A. Kato, Factors affecting severity of renal injury and recovery of function in acute renal failure, *Renal Failure.*14:337-340 (1992).
95. H. Rabl, G. Khoschsorur, T. Colombo, F. Tatzber, and H. Esterbauer, Human plasma lipid peroxide levels show a strong transient increase after successful revascularisation operations, *Free Radical Biol. Med.*13:281-288 (1992).
96. H.C. Chen, Y. Tomino, Y. Yaguchi, M. Fukui, K. Yokoyama, A. Watanabe, and H. Koide, Oxidative metabolism of polymorphonuclear leukocytes (PMN) in patients with IgA nephropathy, *Journal of Clinical Laboratory Analysis.*6:35-39 (1992).
97. T. Naito, H. Kida, and M. Takaeda, Role of reactive oxygen species system in the progression of diabetic nephropathy, *Abstract Book, XIth International Congress of Nephrology.*July 15-20:94A (1990).
98. S. O'Regan, R.W. Chesney, B.S. Kaplan, and K.N. Drummond, Red cell membrane phospholipid abnormalities in the hemolytic uremic syndrome, *Clin. Nephrol.*15:14-17 (1980).
99. K.D. Forsyth, A.C. Simpson, M.M. Fitzpatrick, T.M. Barratt, and R.J. Levinsky, Neutrophil-mediated endothelial injury in haemolytic uraemic syndrome, *Lancet* .II:411-414 (1989).
100. S. Turi, I. Nemeth, I. Vargha, and B. Matkovics, Oxidative damage of red blood cells in haemolytic uraemic syndrome, *Pediatr. Nephrol.*8:26-29 (1994).
101. L.H. Toledo-Pereyra, R.L. Simmons, L.C. Olson, and J.S. Najarian, Clinical effect of allopurinol on preserved kidneys, *Annals of Surgery.*185:128-131 (1977).
102. V.C. Marshall, M. Biguzas, P. Jabionski, D.F. Scott, B.O. Howden, A.C. Thomas, C.W. Cham, and K. Walls, UW solution for kidney preservation, *Transplant. Proc.*22:466-468 (1990).
103. R.J. Ploeg, Kidney preservation with the UW and Euro Collins solutions, *Transplantation.* 49:281-284 (1990).
104. M. Moukarzel, G. Benoit, H. Bensadoun, and et. al., Non-randomized comparative study between University of Wisconsin Cold Storage and Euro-Collins solution in kidney transplantation, *Transplant. Proc.* 22:2289-2290 (1990).
105. J.H. Southard, T.M. Van Gulik, M.S. Ametani, and et. al., Important components of the UW solution, *Transplantation.*49:251-257 (1990).
106. H. Schneeberger, W.D. Illner, D. Abendroth, G. Bulkley, F. Rutilli, M. Williams, M. Thiel, and W. Land, First clinical experience with superoxide dismutase in kidney transplantation-results of a double blind randomized study, *Transplant. Proc.*21:1245-1246 (1989).
107. H. Schneeberger, S. Schleibner, M. Schilling, W.D. Illner, D. Abendroth, E. Hancke, U. Jnicke, and Land, Prevention of acurte renal failure after kidney transplantation with rh_SOD; interim analysis of a double -blind placebo-controlled trial, *Transplant. Proc.*22:2224-2225 (1990).
108. R.A. Greenwald, Superoxide dismutase and catalase as therapeutic agents for human diseases. A critical review, *Free Radical Biol. Med.*8:201-209 (1990).
109. H. Rabl, G. Khoschsorur, T. Colombo, P. Petritsch, M. Rauchenwald, P. Koltringer, F. Tatzber, and H. Esterbauer, A multivitamin infusion prevents lipid peroxidation and improves transplantation performance, *Kidney Int.*43:912-917 (1993).

REACTIVE OXYGEN SPECIES (ROS) AND REPRODUCTION

Eve de Lamirande and Claude Gagnon

Urology Research Laboratory, Royal Victoria Hospital, and
Faculty of Medicine, McGill University,
Montréal, Qué, Canada, H3A 1A1

INTRODUCTION

The role of reactive oxygen species (ROS) in reproduction has long been the subject of investigation. As for other systems described in this book, there is compelling evidence for the involvement of ROS in physiology and pathology of both male and female reproductive systems. In this chapter, we will first briefly summarize informations linking ROS and the female reproductive functions, after which we will present in greater details the data in support for both the beneficial and detrimental effects of ROS on spermatozoa.

FEMALE REPRODUCTION

Uterus

The production of ROS in uterus appears closely associated with the presence of steroid hormones. In this organ, estradiol causes a 200-fold increase in peroxidase activity[1], most of which results from an estradiol-induced infiltration by eosinophils[2]. These peroxidases are thought to provide an effective bactericidal environment[3] and to exert a negative regulatory feedback to control estrogen levels[4]. The production of hydrogen peroxide (H_2O_2), the substrate for these peroxidases, is increased in vitro by estrogens and in vivo during estrus[5]. Histochemical studies demonstrate that H_2O_2 production originates from dismutation of the superoxide anion ($O_2^{-}\bullet$) generated by an NADPH oxidase located in the apical plasma membrane of the endometrial epithelium[6]. This H_2O_2 may be involved in both enzymatic (cyclooxygenase-catalyzed)[7] and non-enzymatic[8] generation of bioreactive prostaglandins, during parturition[9] and premature delivery[10].

Ovulation

Ovulation is often compared to acute inflammation because of the vascular swelling, the accumulation of immune cells, and the

Free Radicals in Diagnostic Medicine, Edited by
D. Armstrong, Plenum Press, New York, 1994

production of prostaglandins, all of which are associated with an inflammatory response[11]. In mammals, ROS may be involved in ovulation since both the number of mature follicules and the time needed for maturation are markedly reduced when rabbit ovary preparations are incubated in the presence of superoxide dismutase (SOD) or SOD + catalase[12]. It is hypothesized that ROS activate phospholipases causing the release of arachidonic acid which is converted, via the cyclooxygenase system, to prostaglandins, one of the factors required for ovulation[13].

ROS may also be involved in follicular atresia, a phenomenon by which follicules recruited for development and maturation, do not proceed to ovulation, but rather become atretic and regress. During this process, follicules lose their response to gonadotropins and granulosa cells, as well as oocytes, degenerate, possibly because of their extreme sensitivity to H_2O_2[14].

Oocyte

In sea urchin, H_2O_2 is of primordial importance for the formation of the fertilization envelope which prevents polyspermy and protects the embryo from the environment[15]. Following sperm penetration, a unique ovoperoxidase is secreted by the egg concomitantly with its substrate, H_2O_2 (from an NADPH oxidase in the egg plasma membrane) to cross-link tyrosyl moieties of proteins derived from cortical granules and produce a hard protective network of polymerized proteins[15].

In some mouse strains, the generation of ROS observed in vitro after fertilization[16,17] appears to be responsible for the block of embryo development that occurs at the two-cell stage. Modifications of the standard culture environment (20 % oxygen) by reducing oxygen tension to 5 % [18], adding ROS scavengers (SOD, L-cysteine)[18], or thioredoxin, a powerful protein disulfide reductase[19], protecting from light[19] and removing transition metals from the culture medium[19] attenuate the two-cell block and favour the development to the blastocyst level. Recently, Natsuyama et al.[20] suggested that one of the main cause for the two-cell block in vitro is the impairment of $p34^{cdc2}$ (a key regulator of the cell cycle) dephosphorylation due to the action of ROS.

Whereas early embryos (two-cell to blastocyst) are very sensitive to the toxic effects of ROS, more mature embryos (especially after the implantation stage) are more resistant to ROS. Furthermore, the superoxide anion is probably needed for their development since addition of SOD (> 10 μg/ml) to the incubation medium caused an important decrease in the formation rates of two germ cell layers and egg cylinder [21].

Corpus Luteum

After ovulation, granulosa and theca cells surrounding the oocyte differentiate to form the corpus luteum, a temporary, LH-stimulated, endocrine gland secreting progesterone and, in primates, estradiol as well. If no pregnancy occurs, the corpus luteum regresses in a way that shares resemblances with cell senescence. There is direct evidence for the production of $O_2^{-\bullet}$ [22,23] and H_2O_2 [24] at the beginning of the luteolytic process, both of which originating from luteal membranes. Interestingly, activated neutrophils, through the ROS released during their oxidative burst [25], ovarian ischemia-reperfusion [26] as well as treatment with xanthine oxidase [27] also induce the changes observed during luteolysis [25].

The superoxide anion is not directly toxic to cells and, in this system, is mostly known for its ability to activate phospholipase A_2 [22,23]. On the other hand, its dismutation product, H_2O_2, is directly luteolytic and, even at low concentrations, inhibits cAMP and progesterone production and causes an intracellular ATP depletion[28]. Hydrogen peroxide appears to inhibit membrane translocation and/or transport of cholesterol to the mitochondria where it is normally metabolized into progesterone [29]. The ATP depletion is mostly a result of peroxide-induced DNA damage that was not related to luteolysis itself [28]. Hydrogen peroxide is not very reactive but can oxidize essential cellular sulphydryl groups and antioxidant vitamins, and/or be involved in the formation of the very toxic hydroxyl radical (•OH) through the Fenton reaction in the presence of iron ions. The detection of important levels of lipid peroxidation in the regressing corpus luteum and the coincident depletion of luteal ascorbic acid are consistent with these reactions [22,23,27]. The increased lipid peroxidation may also serve a functional role since the products of lipid peroxidation can stimulate activities of phospholipase A_2 and cyclooxygenase and therefore the local production of prostaglandins [8,30]. By this mechanism, H_2O_2 may cause a positive feedback induction of ROS and prostaglandin PGF_{2a} synthesis. These data suggest that H_2O_2 may induce luteal arrest directly and/or may act with extraluteal cellular components that secondarily initiate luteolysis.

Endometriosis

Endometriosis involves implantation and cyclic changes of endometrial tissue at ectopic sites in the pelvic cavity. Recent experimental evidence associate endometriosis with an inflammatory response. Halme et al.[31] observed that, during endometriosis, there is an increase influx of peritoneal macrophages which undergo further maturation-activation stages; they hypothesized that the resultant important population of large macrophages may contribute to the maintenance of the disease or of the related infertility. In a rabbit model, the combined instillation of SOD and catalase significantly reduce the formation of intraperitoneal adhesions at endometrial sites, emphasizing the important role ROS may play in this disease [32]. Furthermore, endometrial cysts produced 2.5- to 10-fold more eicosanoids (PGE_2, PGF_{2a}, 6-keto-PGF_{1a} and TXB_2) than non-endometrial cysts and normal ovaries [33]. Whether increased ROS production is a cause or a consequence of endometriosis remains to be determined. Nevertheless, increase in ROS production may be related to the decreased fertility observed in the more advanced stages of the disease.

MALE REPRODUCTION

In contrast to the numerous studies on the effects of ROS on female reproductive system, there are much less reports on the actions and roles of ROS in the male reproductive organs. In cases of chronic alcohol administration to animals, testicular lipid peroxidation correlates well with the extent of gonadal injury. Alcohol intake causes a decrease in the testicular glutathione content and modifies the precarious antioxidant balance of testicular tissue [34].

By immunohistochemical techniques using a polyclonal antibody, SOD was found at various levels of the male reproductive system including, the ductus deferens, the prostate, seminal vesicles, and spermatogonia [35]. This broad distribution pattern of SOD could reflect a role in local defence mechanisms against $O_2^{-\bullet}$ attack along the reproductive tract.

Incidence of ROS in Semen and Infertility

The toxic effects of fatty acid peroxides on spermatozoa has been recognized fifteen years ago[36]. Due to their high content of polyunsaturated fatty acids and their relatively low levels of ROS scavengers, human spermatozoa are particularly sensitive to oxygen-induced damages[37,38]. Hydrogen peroxide [39-42], lipid peroxides and their degradation products (hydroxy alkenals, malonaldehyde, etc.) [36,38] are highly toxic to spermatozoa and cause an irreversible arrest of motility.

Several investigators have suggested that excessive production of ROS by spermatozoa is a cause for idiopathic infertility. ROS generation in human semen is negatively associated with both the outcome of the sperm-oocyte fusion assay and fertility in vivo [43]. As many as 25% of semen samples from an unselected population of men consulting for infertility produce significant levels of ROS whereas none of the semen from azoospermic patients or from fertile donors generate detectable levels of ROS [44]. Furthermore, there is an inverse correlation between the level of ROS and the percentage of motile spermatozoa in semen. Cellular elements, including round cells, immotile and/or abnormal spermatozoa, leukocytes and even morphologically normal spermatozoa, but not the seminal plasma or dead spermatozoa, are responsible for ROS production[44-46].

All human semen samples, contain an heterogenous population of normal and morphologically and/or biochemically abnormal spermatozoa. These abnormal spermatozoa, as well as spermatozoa from infertile patients with inflammation of the genital tract, generate more ROS than normal spermatozoa[44,45,47-49]. This production of ROS is associated with a loss of sperm motility and decreased capacity for sperm-oocyte fusion in vitro[39].

ROS Scavengers

To counteract the effects of ROS, spermatozoa and seminal plasma possess systems to scavenge ROS. Semen contains enzymes such as SOD [38,50], catalase[51] and the glutathione peroxidase/reductase system [40], but also other substances that can act as ROS scavengers including albumin [52], glutathione[53], pyruvate[54], taurine and hypotaurine[52], vitamins E [55], and C [56], etc.. The importance of these ROS scavengers is emphasized by the positive correlation between the level of SOD in spermatozoa and the duration of sperm motility in semen[38]. SOD also prevents the loss of motility and the malonaldehyde accumulation in washed spermatozoa (removal of seminal plasma and of other cell components by Percoll density centrifugation) incubated under aerobic conditions [57]. The high level of ROS measured in the semen of infertile patients does not appear to be due to a lowered scavenging capacity towards $O_2^{-\bullet}$ and H_2O_2 but rather to an increased production of ROS by cells present in semen[58].

Sources of ROS in Semen

The origin of the increased production of ROS in spermatozoa of infertile patients is presently unknown. The presence of an NADPH oxidase at the level of the sperm membrane is suggested only by the fact that phorbol myristate acetate (PMA), at micromolar concentrations, can cause an increase in sperm ROS generation and that this increase is more important in infertile than in fertile men[59,60]. The human sperm diaphorase (a mitochondrial NADH-dependent oxidoreductase) probably represents a more important source of ROS, and is 3-fold more active in spermatozoa from oligospermic semen than from normal semen[61].

White blood cells, and specially neutrophils, present in most semen samples can be a very important source of ROS depending on their activation state and concentration [45,59,62,63]. When isolated blood neutrophils are activated with PMA and mixed with washed spermatozoa, sperm motility decreases in a time- and concentration-dependent manner[63]. Addition of the combination catalase and dimethylsulfoxide to spermatozoa confers full protection against the effects of neutrophils [63]. Similarly, addition of seminal plasma to spermatozoa incubated with activated neutrophils decreases the effects of the latter on sperm motility. However, this protection conferred by seminal plasma varies enormously from one specimen to the other. Taken together, these data suggest that the presence of neutrophils along the reproductive tract (due to inflammation or infection) could be a cause for decreased fertility [63].

Semen also contains other cell types, such as epithelial and immature germ cells, but nothing is known on their ROS production and scavenging properties.

Mechanisms of ROS Toxicity on Spermatozoa

Lipid peroxidation of the sperm membrane is the first mechanism suggested to be responsible for the reduced sperm function after exposure to ROS[36,38]. Lipid peroxidation is observed after spermatozoa are incubated in conditions (vigorous shaking of a thin film of spermatozoa in a medium deficient in energy substrate or treatment with high concentrations of H_2O_2) leading to an irreversible loss of motility as well as loss of viability [36,38,40]. Treatment of spermatozoa with the combination iron (Fe^{2+})/ascorbic acid is also associated with an increased lipid peroxidation and malonaldehyde formation (as measured by the thiobarbituric acid assay)[39]. The malonaldehyde generated does not appear to involve first-chain initiation of a peroxidation sequence through the generation of •OH [42]. Iron rather stimulates the peroxidative chain reaction, resulting in the generation and release of further lipid peroxides from which malonaldehyde is subsequently derived[42].

We investigated the the mode of action of ROS on spermatozoa under conditions in which sperm viability is not compromised. Spermatozoa move as a result of ondulating waves generated by the sliding of microtubules within the axonemes, the core structure at the center of the flagellum which is responsible for motility. Axonemal function and integrity can easily be assessed using demembranated sperm models in which spermatozoa are first immobilized in a medium containing the detergent Triton X-100, and flagellar movement is immediately reactivated upon addition of Mg.ATP [64].

When washed motile human spermatozoa are incubated at room temperature with the combination xanthine (0.6 mM) + xanthine oxidase (0.05 U/ml) (X+XO), sperm flagellar beat frequency progressively decreases, leading to a complete arrest of motility 1 to 4 hours after the beginning of incubation, even though $O_2^{-\bullet}$ is generated only during the first 15-20 min of incubation (Fig. 1). Intracellular levels of ATP drop even faster than the beat frequency (Fig. 1). Under these conditions, sperm viability is not significantly affected. These observations, as well as others[41,54], suggest that the depletion of intracellular ATP is one of the early cellular modifications that triggers a cascade of events leading to sperm immobilization. The motility of intact spermatozoa stops when their ATP content drops by 90% $\pm$ 5% (mean $\pm$ SEM). Since ROS immobilized spermatozoa cannot be reactivated after demembranation in the presence of Mg.ATP (as described above) the data suggest that sperm axonemes are altered. However, when these same spermatozoa are demembranated in conditions allowing rephosphorylation of axonemal proteins (medium supplemented with cAMP, protein kinase, and sperm extract), spermatozoa reactivate motility although after a delay.

Surprisingly, in 50% of the cases, motility of intact spermatozoa spontaneously reinitiates, 6 to 24 hours after ROS immobilization (30-70% of control sperm motility); however, in these samples, sperm tails beat at low frequency and ATP levels are increased to a level compatible with motility [54]. Catalase, but not SOD nor DMSO, can totally prevent ROS induced loss of sperm motility [41].

These data suggest that: 1) H_2O_2 is the toxic ROS for human spermatozoa, 2) ATP depletion appears to play an important role in sperm immobilization, 3) ROS treatment indirectly causes a reduction in the phosphorylation of axonemal proteins which results in sperm immobilization, and 4) spermatozoa have limited endogenous mechanisms to reverse these damages.

Aitken et al.[42] using the same combination, but at much higher concentrations of X (2mM) and XO (0.08 U/ml), observed a major decrease in all sperm motility parameters (percentage, velocity, etc.) and production of significant levels of lipid peroxides in spermatozoa (as measured by the presence of thiobarbituric acid reactive substances) 15-30 min after the beginning of treatment. The Fenton rather than Harber-Weiss reaction is probably involved in initiating the peroxidative damage induced by the H_2O_2 generated with X+XO.

ROS and Sperm Capacitation

Mammalian spermatozoa must complete a series of membranous and metabolic changes (termed capacitation) in the female genital tract before fertilization is possible[65]. These processes are associated with hyperactivation (HA), a vigorous, poorly progressive type of motility [66]. Once capacitation is completed, spermatozoa can proceed to the acrosome reaction, which is an exocytotic process involving fusion of the outer acrosomal membrane with the sperm head plasma membrane, fenestration and vesiculation of the fused membranes, and dispersion of the acrosomal matrix[67]. The enzymes liberated (acrosin and hyaluronidase) help digest the cumulus oophorus and the zona pellucida, allowing spermatozoa to reach, bind and fertilize the oocyte.

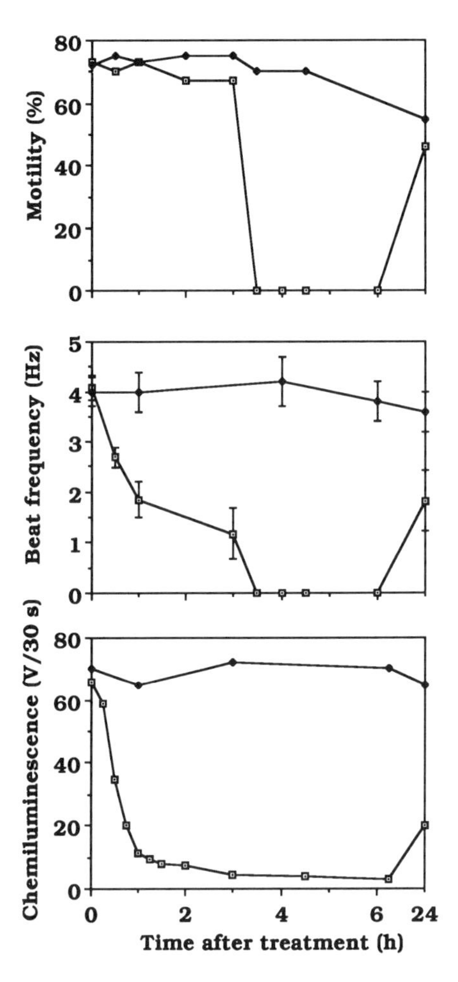

Figure 1. Percentage of motility, flagellar beat frequency and ATP levels in spermatozoa after treatment with the combination xanthine/xanthine oxidase. Spermatozoa were diluted in HEPES-buffered saline (control, -♦-) or treated with X (0.6 mM) + XO (0.05 U/ml)(-▣ -). ATP levels were measured by chemiluminescence using the luciferin/luciferase assay. Control spermatozoa had $0.90 \pm 0.05 \times 10^{-10}$ mole ATP/10^6 spermatozoa before the treatment. The ATP level dropped to $10 \pm 5\%$ of control sperm ATP levels at the time of immobilization and increased to $21 \pm 3\%$ when motility spontaneously reinitiated. From refs 54 and 64.

During our studies on the toxic effects of ROS (described above), we noticed that, for the first few minutes following the addition of X (0.5mM) + XO (0.05 U/ml), spermatozoa show high levels of vigorous motility before the beat frequency is affected. Addition of catalase (CAT) to the medium to scavenge the toxic H_2O_2, allows the observation of this vigorous motility that bears all the characteristics of hyperactivated motility[68]. The combination X+XO+CAT induces higher levels of sperm HA (16% ± 2%) than foetal cord serum (FCS, 8% ± 1%), a known biological inducer of this process[66] or Ham's F-10 medium alone (control, 5.4% ± 0.6%)[68]. If X+XO+CAT are incubated for 30 min (no ROS remains in solution after this period of time) prior to their addition to spermatozoa, no induction of sperm HA takes place. Furthermore, addition of SOD, but not inactivated SOD, prevents the rise in HA induced not only by X+XO+CAT, but also by FCS, and reverses the HA of already hyperactivated spermatozoa[68,69].

The HA observed after treatment with X+XO+CAT increases in parallel with sperm capacitation (X+XO+CAT: 23% ± 1%; FCS: 9% ± 1%; control: 5% ± 1%) as measured by the incidence of acrosome reaction obtained after a 30 min incubation with lysophosphatidylcholine (a compound known to induce the acrosome reaction only in capacitated spermatozoa)[68]. As it is observed for HA, addition of SOD prevents the rise of sperm capacitation observed after treatment with X+XO+CAT and with FCS. These data suggest that $O_2^{-\bullet}$ can trigger human sperm HA and capacitation and that a sustained $O_2^{-\bullet}$ generation is needed to maintain the progression of these events[68,69].

The 3 μM iron present in Ham's F-10 medium does not appear to play an important role in HA and capacitation induced by X+XO+CAT or by FCS since desferrioxamine (0.1 mM) has no effect on the percentages of HA and capacitation observed[69].

The requirement of $O_2^{-\bullet}$ for human sperm capacitation may be species-specific, since H_2O_2 appears to be the ROS responsible for increasing capacitation in golden hamster spermatozoa[70].

The important role of $O_2^{-\bullet}$ in human sperm capacitation is further emphasized by another series of experiments. The ability of FCS and two biological fluids encountered by spermatozoa, seminal plasma and follicular fluid (whole or fractionated into their high and low molecular weight components), to promote human sperm capacitation are determined and compared (Table 1)[71].

Table 1. Human sperm capacitation induced by biological fluids

Fluid	Capacitation (%) Whole	Dialyzed	Ultrafiltrate
Seminal plasma	21 ± 1	19 ± 1	32 ± 1
FCS	9 ± 1	10 ± 1	23 ± 1
Follicular fluid	30 ± 5	27 ± 4	48 ± 6

Washed human spermatozoa are incubated at 37°C in Ham's F-10 medium supplemented with 7.5% of biological fluid. Capacitation is measured 3.5 h later. Capacitation of control spermatozoa (in Ham's F-10 alone) is 5 ± 1%. Five samples of each fluids are tested and aliquotes of the same fluids are used to obtain the high molecular weight (dialysis, > 12 kD exclusion limit) and the low molecular weight (ultrafiltration, < 3kD exclusion limit) components of the fuids. Mean ± SEM.

Very different levels of capacitation are obtained with the various fluids and ultrafiltrates are consistently more powerful inducers than the corresponding whole or dialyzed fluid[71]. Addition of SOD to these fluids, always causes an important reduction in sperm capacitation (to control level). These data suggest that the SOD-like activity (total $O_2^{-\bullet}$ scavenging capacity) of these fluids may be important for their ability to induce capacitation[71]. To test this hypothesis, the SOD-like activity is determined, the inhibition of nitroblue tetrazolium reduction due to $O_2^{-\bullet}$ generated by X+XO in the presence of 7.5% of biological fluid being a measure of this activity. The capacitation triggered by the different fluids tested inversely correlates with the SOD-like activity of these fluids tested (Fig. 2)[71]. These results indicate that biological fluids contain inducers of sperm capacitation, and that these probably act by a common mechanism, possibly through direct or indirect induction of an NADPH oxidase at the level of sperm membrane. The data also strongly suggest that the SOD-like activity of a specific fluid is one of the most important factors that will determine its ability to induce capacitation [71].

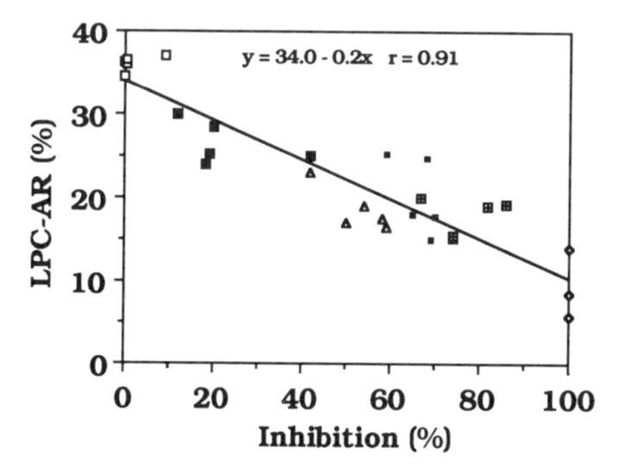

Figure 2. Correlation between human sperm capacitation obtained with various biological fluids and the SOD-like activity of these fluids.
The percentages of capacitation (measured by the lysophosphatidylcholine-induced acrosome reaction) were obtained after treatment of at least two sperm preparations with five individual samples of seminal plasma (⊞), dialyzed seminal plasma (▪), ultrafiltrates from seminal plasma (■), fetal cord serum (- Δ-) and follicular fluid (□), and pure SOD (0.01 mg/ml)(- ◊-). The SOD-like activity (total $O_2^{-\bullet}$ scavenging capacity) of the fluids (7.5%, v/v) is expressed as the percentage of inhibition of nitroblue tetrazolium reduction due to the $O_2^{-\bullet}$ generated by X+XO. From ref 71.

Some infertile patients (13% of the cases) whose spermograms are considered "normal" according to the criteria defined by the World Health Organization, have levels of spontaneous sperm HA occurring in whole semen (16% $\pm$ 3%) that are much higher than those observed in semen from fertile volunteers (2.6% $\pm$ 0.3%)[72]. Interestingly, the SOD-like activity found in the seminal plasma and spermatozoa of these patients are respectively 37% and 40% lower than that found in

HA observed in whole semen impairs the transport of spermatozoa through the female genital tract and/or leads to premature capacitation is not yet known, but it could be one cause for idiopathic infertility generally not detected by a standard spermogram.

Conclusions

In female reproduction, ROS play a physiological role in many phenomena, such as follicular atresia, corpus luteum regression and formation of the fertilization envelope, but also a pathological role in others such, as premature delivery, endometriosis, and block of early embryo development. Spermatozoa are very sensitive to the actions of ROS. The nature and the level of the ROS involved as well as the time and site of ROS exposure, will determine the effects observed. Even though seminal plasma has enormous scavenging capacity, it cannot always provide efficient protection to spermatozoa, especially if ROS are produced from within the cell. Low concentrations of hydrogen peroxide cause a reversible loss of motility mostly due to ATP depletion and insufficient axonemal phosphorylation, but higher concentrations induce lipid peroxidation, permanent loss of sperm functions, and cell death. On the other hand, $O2^{-\bullet}$ has a positive role towards human spermatozoa, being involved in induction of hyperactivation and capacitation, therefore playing a crucial role in the preparation of spermatozoa for fertilization.

REFERENCES

1. C. Lyttle and E. DeSombre, Uterine peroxidase as a marker for estrogen action, *Proc. Natl. Acad. Sci. USA* 74:3162 (1977).
2. Y. Lee, R. Howe S.-J. Sha, C. Teuscher, D. Sheehan and C. Lyttle, Estrogen regulation of an eosinophil chemotactic factor in the immature rat uterus, *Endocrinology* 125:3022 (1989).
3. S. Klebanoff and D. Smith, Peroxidase-mediated antimicrobial activity of rat uterine fluid, *Gynecol. Invest.* 1:21 (1970).
4. S. Klebanoff, Inactivation of estrogen by rat uterine preparations, *Endocrinology* 76:301 (1985).
5. R. Johri and P. Dasgupta, Hydrogen peroxide formation in the rat uterus under hormone-induced conditions, *J. Endocrinol.* 86:477 (1980).
6. Y. Ishikawa, K. Hirai and K. Ogawa, Cytochemical localization of hydrogen peroxide production in the rat uterus, *J. Histochem. Cytochem.* 32:674 (1984).
7. M.E. Hemler, H.W. Cook and W.E.M. Lands, Prostaglandin biosynthesis can be triggered by lipid peroxides, *Arch. Biochem. Biophys.* 193:340 (1979).
8. J.D. Morrow, K.E. Hill, R.F. Burk, T.M. Nammour, K.F. Badr and L.J. Roberts, II. A series of prostaglandin F2-like compounds are produced in vivo in humans by a non-cyclooxygenase, free radical-catalyzed mechanism, *Proc. Natl. Acad. Sci. USA* 87:9383 (1990).
9. P. Cherouny, R. Ghodgaonkar, J. Niebyl and N. Dubin, Effect of hydrogen peroxide on prostaglandin production and contractions of the pregnant rat uterus, *Am. J. Obstet. Gynecol.* 159:1390 (1988).
10. H. Minkoff, Prematurity: Infection as an etiologic factor, *Obstet. Gynecol.* 62:137 (1983).
11. L.L. Espey, Ovulation as an inflammatory reaction - A hypothesis, *Biol. Reprod.* 22:73 (1980).
12. T. Miyazaki, K. Sueoka, A.M. Dharmarajan, S.J. Atlas, G.B. Bulkley and E.E. Wallach, Effect of inhibition of oxygen free radical on ovulation and

progesterone production by the in-vitro perfused rabbit ovary. *J. Reprod. Fertil.* 91:207 (1991).
13. H.R. Behrman, Prostaglandins in hypothalamo-pituitary and ovarian function, *Annu. Rev. Physiol.* 41:685 (1979).
14. Y. Margolin, R. Aten and H. Behrman, Antigonadotropic and antisteroidogenic actions of peroxide in rat granulosa cells, *Endocrinology* 127:245 (1990).
15. B.M. Shapiro, The control of oxidant stress at fertilization, *Science* 252:533 (1991).
16. M.H. Nasr-Esfahani, J.R. Aitken and M.H. Johnson, Hydrogen peroxide levels in mouse oocytes and early cleavage stage embryos developed in vitro and in vivo, *Development* 109:501 (1990).
17. Y. Goto, Y. Noda, T. Mori and M. Nakano, Increased generation of reactive oxygen species in embryos cultured in vitro, *Free Radic. Biol. Med.* 15:69 (1993).
18. Y. Noda, H. Matsumoto, Y. Umaoka, K. Tatsumi, J. Kishi and T. Mori, Involvement of superoxide radicals in the mouse two-cell block, *Molec. Reprod. Develop.* 28:356 (1991).
19. Y. Goto, Y. Noda, K. Narimoto, Y. Umaoka and T. Mori, Oxidative stress on mouse embryo development in vitro, *Free Radic. Biol. Med.* 13:47 (1992).
20. S. Natsuyma, Y. Noda, M. Yamashita, Y. Nagahama and T. Mori, Superoxide dismutase and thioredoxin restore defective $p34^{cdc2}$ kinase activation in mouse two-cell block, *Biochem. Biophys. Acta* 176:90 (1993).
21. T. Tokura, Y. Noda, K. Narimoto, Y. Umaoka, T Mori and K Ogawa, Effect of superoxide dismutase on the developpement following the implantation stage in mice, *Acta Histochem. Cytochem.* 25: 491 (1992).
22. M. Sawada and J.C. Carlson, Rapid plasma membrane changes in superoxide radical formation, fluidity and phospholipase A_2 activity in the corpus luteum of the rat during induction of luteolysis, *Endocrinology* 128:2992 (1991).
23. J.C. Carlson, X.M. Wu and M. Sawada, Oxygen radicals and the control of ovarian corpus luteum function, *Free. Radic. Biol. Med.* 14:79 (1993).
24. J.C.M. Riley and H.R. Behrman, In vivo generation of hydrogen peroxide in the rat corpus luteum during luteolysis, *Endocrinology* 128:1749 (1991).
25. J.R. Pepperell, K. Wolcott and H. R. Berhman, Effects of neutrophils in the rat luteal cells, *Endocrinology* 130: 1001 (1992).
26. N. Sugino, Y. Nakamura, N. Okuno, M. Ishimatu, T. Teyama and H. Kato, Effects of ovarian ischemia-reperfusion on luteal function in pregnant rats, *Biol. Reprod.* 49: 345 (1993).
27. X. Wu, K. Yao and J.C. Carlson, Plasma membrane changes in the rat corpus luteum induced by oxygen radical generation, *Endocrinology* 133: 491 (1993).
28. H. Behrman and S. Preston, Luteolytic actions of peroxide in rat ovarian cells, *Endocrinology* 124:2895 (1989).
29. H. Behrman and R. Aten, Evidence that hydrogen peroxide blocks hormone-sensitive cholesterol transport into mitochondria of rat luteal cells, *Endocrinology* 128:2958 (1991).
30. A. Sevanian, M.L. Wratten, L.L. McLeod and E. Kim, Lipid peroxidation and phospholipase A_2 activity in liposomes composed of unsaturated phospholipids: A structural basis for enzyme activation, *Biochem. Biophys. Acta* 961:316 (1988).
31. J. Halme, S. Becker and S. Haskill, Altered maturation and function of peritoneal macrophages: Possible role in pathogenesis of endometriosis, *Am. J. Obstet. Gynecol.* 156:783 (1987).
32. D.M. Portz, T.E. Elkins, R. White, J. Warren, S. Adadevoh and J. Randolph, Oxygen free radicals and pelvic adhesion formation: I. Blocking oxygen free radical toxicity to prevent adhesion formation in an endometriosis model, *Int. J. Fertil.* 36:39 (1991).
33. H. Koike, H. Egawa, T. Ohtsuka, M. Yamaguchi, T. Ikenoue and N. Mori, Eicosanoids production in endometriosis, *Prostaglandins leukotrienes & essential fatty acids* 45:313 (1992).
34. E.R. Rosenblum, J.S. Gavaler and D.H. Van Thiel, Lipid peroxidation: A mechanism for alcohol-induced testicular injury, *Free Radic. Biol. Med.* 7:569 (1989).
35. T. Nonogaki, Y. Noda, K. Narimoto, M. Shiotani, T. Mori, T. Matsuda and O. Yoshida, Localization of CuZn-superoxide dismutase in the human male genital organs, *Human Reprod.* 7:81 (1992).

36. R. Jones, T. Mann and R. Sherins, Peroxidative breakdown of phospholipids in human spermatozoa, spermicidal properties of fatty acid peroxides, and protective action of seminal plasma, *Fertil. Steril.* 31:531 (1979).
37. R. Aitken and J. Clarkson, Cellular basis of defective sperm function and its association with the genesis of reactive oxygen species by human spermatozoa, *J. Reprod. Fertil.* 81: 459 (1987).
38. J.G. Alvarez, J.C. Touchstone, L. Blasco and B.T. Storey, Spontaneous lipid peroxidation and production of hydrogen peroxide and superoxide in human spermatozoa. Superoxide dismutase as major enzyme protectant against oxygen toxicity, *J. Androl.* 8:338 (1987).
39. J.R. Aitken, J.S. Clarkson and S. Fishel, Generation of reactive oxygen species, lipid peroxidation and human sperm function, *Biol. Reprod.* 40:183 (1989).
40. J.G. Alvarez and B.T. Storey, Role of glutathione peroxidase in protecting mammalian spermatozoa from loss of motility caused by spontaneous lipid peroxidation, *Gamete Res.* 23:77 (1989).
41. de Lamirande and C. Gagnon, Reactive oxygen species and human spermatozoa I. Effects on the motility of intact spermatozoa and on sperm axonemes, *J. Androl.* 13:368 (1992).
42. R.J. Aitken, D. Buckingham and D. Harkiss, Use of a xanthine oxidase free radical generating system to investigate the cytotoxic effects of reactive oxygen species on human spermatozoa, *J. Reprod. Fert.* 97:441 (1993).
43. R.J. Aitken, D.S. Irvine and F.C. Wu, Prospective analysis of sperm-oocyte fusion and reactive oxygen species generation as criteria for the diagnosis of infertility, Am. *J. Obstet. Gynecol.* 164:542 (1991).
44. A. Iwasaki and C. Gagnon, Formation of reactive oxygen species in spermatozoa of infertile patients, *Fertil. Steril.* 57:409 (1992).
45. R.J. Aitken and K.M. West, Analysis of the relationship between reactive oxygen species production and leukocyte infiltration in fractions of human semen separated on Percoll gradients, *Int. J. Androl.* 13:433 (1990).
46. R.J. Aitken, D. Buckingham, K. West, F.C. Wu, K. Zikopoulos and D.W. Richardson, Differential contribution of leucocytes and spermatozoa to the generation of reactive oxygen species in the ejaculates of oligozoospermic patients and fertile donors, *J. Reprod. Fert.* 94:451 (1992).
47. J.R. Aitken and J.S. Clarkson, Significance of reactive oxygen species and antioxidants in defining the efficacy of sperm preparation techniques, *J. Androl.* 9:367 (1988).
48. B. Rao, J.C. Soufir, M. Martin and G. David, Lipid peroxidation in human spermatozoa as related to midpiece abnormalities and motility, *Gamete Res.* 24: 127 (1989).
49. R. D'Agata, E. Vicari, M.L. Moncada, G. Sidoti, A.E. Calogero, M.C. Fornito, G. Minacapilli, A. Mongioi and P. Polosa, Generation of reactive oxygen species in subgroups of infertile men, *Int. J. Androl.* 13:344 (1990).
50 H.P. Nissen and H.W. Kreysel, Superoxide dismutase in human semen, *Klin. Wochenschr.* 61:63 (1983).
51. C. Jeulin, J.C. Soufir, P. Weber, D. Laval-Martin and R. Calvayrac, Catalase activity in human spermatozoa and seminal plasma, *Gamete Res.* 24:185 (1989).
52. J.G. Alvarez and B.T. Storey, Taurine, hypotaurine, epinephrine and albumin inhibit lipid peroxidation in rabbit spermatozoa and protect against loss of motility, *Biol. Reprod.* 29:548 (1983).
53. B. Halliwell and J.M.C. Gutteridge, "Free Radicals inBiology and Medicine", 2nd edition, Clarendon Press, Oxford (1989).
54. E. de Lamirande and C. Gagnon, Reactive oxygen species and human spermatozoa II. Depletion of adenosine triphosphate plays an important role in the inhibition of sperm motility, *J. Androl.* 13:379 (1992).
55. C.K. Chow, Vitamin E and oxidative stress, *Free Radic. Biol. Med.* 11:215 (1991).
56. E.B. Dawson, W.A. Harris, M.C. Teter and L.C. Powell, Effect of ascorbic acid supplementation on the sperm quality of smokers, *Fertil. Steril.* 58:1034 (1992).
57. T. Kobayashi, T. Miyazaki, M. Natori and S. Nozawa, Protective role of superoxide dismutase in human sperm motility: superoxide dismutase activity and lipid peroxide in human seminal plasma and spermatozoa, *Human Reprod.* 6:987 (1991).

58. A. Zini, E. de Lamirande and C. Gagnon, Reactive oxygen species in semen of infertile patients: levels of superoxide dismutase- and catalase-like activities in seminal plasma and spermatozoa, *Int. J. Androl.* 16:183 (1993).
59. R.J. Aitken, D.W. Buckingham and K.M. West, Reactive oxygen species and human spermatozoa: analysis of the cellular mechanisms involved in luminol- and lucigenin-dependent chemiluminescence, *J. Cell Physiol.* 151:466 (1992).
60. M. Gavella, V. Lipovac and T. Marotti, Effect of pentoxifylline on superoxide anion production by human sperm, *Int. J. Androl.* 14:320 (1991).
61. M. Gavella and V. Lipovac, NADH-dependent oxidoreductase (diaphorase) activity and isozyme pattern of sperm in infertile men, *Arch. Androl.* 28:135 (1992).
62. E. Kessopoulou, M.J. Tomlinson, C.L.R. Barratt, A.E. Bolton and I.D. Cooke, Origin of reactive oxygen species in human semen: spermatozoa or leucocytes? *J. Reprod. Fert.* 94:463 (1992).
63. N.N. Kovalski, E. de Lamirande and C. Gagnon, Reactive oxygen species generated by human neutrophils inhibit sperm motility: protective effect of seminal plasma and scavengers, *Fertil. Steril.* 58:809 (1992).
64. E. de Lamirande, C.W. Bardin and C. Gagnon, Aprotinin and a seminal plasma factor inhibit the motility of demembranated reactivated rabbit spermatozoa, *Biol. Reprod.* 28:788 (1983).
65. R. Yanagimachi, Mammalian fertilization, *in*: "The Physiology of Reproduction," E. Knobil, J.D. Neil, L.L. Ewing, C.L. Markert, G.S. Greenwald and D.W. Pfaff, eds., Raven Press, New York (1988).
66. L.J. Burkman, Hyperactivated motility of human spermatozoa during in vitro capacitation and implications for fertility, *in*: "Controls of Sperm Motility," C. Gagnon, ed., CRC Press, Boca Raton, FL (1990).
67. A.I. Yudin, W. Gottlieb and S. Meizel, Ultrastructural studies of the early events of the human sperm acrosome reaction as initiated by human follicular fluid, *Gamete Res.* 20:11 (1988).
68. E. de Lamirande and C. Gagnon, A positive role for the superoxide anion in triggering hyperactivation and capacitation of human spermatozoa, *Int. J. Androl.* 16:21 (1993).
69. E. de Lamirande and C. Gagnon, Human sperm hyperactivation and capacitation as parts of an oxidative process, *Free Radic. Biol. Med.* 14:157 (1993).
70. I. Bize, G. Santander, P. Cabello, D. Driscoll and C. Sharpe, Hydrogen peroxide is involved in hamster sperm capacitation in vitro, *Biol. Reprod.* 44:398 (1991).
71. E. de Lamirande, D. Eiley, and C. Gagnon, Inverse relationship between the induction of human sperm capacitation and spontaneous acrosome reaction by various biological fluids and the superoxide scavenging capacity of these fluids, *Int. J. Androl.* 16: 258 (1993).
72. E. de Lamirande and C. Gagnon, Human sperm hyperactivation in whole semen and its association with low superoxide scavenging capacity in seminal plasma, *Fertil. Steril.* 59:1291 (1993).

AUTOIMMUNE AND INFLAMMATORY DISEASES

Jonathan A. Leff
Assistant Professor of Medicine
Webb-Waring Institute for Biomedical Reearch
University of Colorado School of Medicine
4200 East Ninth Avenue - Box C322
Denver, Colorado 80262

INTRODUCTION

Toxic oxygen radicals (collectively called reactive oxygen species, ROS) are intricately related to a vast number of disease processes including a variety of autoimmune and inflammatory disorders[1]. It is difficult to directly measure ROS because of their extremely short half-life. However, the reactions in which they participate often leave measurable markers ("footprints") that reflect oxidant activity. Primarily four strategies have evolved to detect oxidant activity for diagnostic purposes: 1) measuring products of oxidation, for example lipid peroxides or oxidized glutathione (GSSG); 2) measuring the release of products from damaged cells, for example the release of hepatic enzymes during acute hepatitis; 3) measuring antioxidant enzyme activities which may reflect the systemic compensation to oxidant stress; 4) ROS can be directly measured although the technology is complex and imperfect, the methodology somewhat complicated, and the equipment cost often prohibitive. In this review, I will discuss sources of selected oxidant markers, methodologies involved in their measurement, and their relevance to the diagnosis and management of several autoimmune and inflammatory diseases.

SELECTED MARKERS OF OXIDANT INDUCED DISEASE

Extracellular Catalase

Methodology of Measurement. Catalase is an enzyme which catalyzes the destruction of hydrogen peroxide (H_2O_2) by one of two reactions. In the first, H_2O_2 is converted to oxygen and H_2O (catalatic reaction). In the second H_2O_2 oxidizes a variety of substrates including methanol, ethanol, formic acid, thiols, nitrites, quinones and phenols (peroxidatic reaction). Which of these two pathways predominates is currently unclear. There are many assays for the measurement of catalase activity, usually based on the rate of liberation of oxygen, the disappearance rate of H_2O_2, or the heat liberated by the reaction[2]. The rate of liberation of O_2 can be measured polarographically with an oxygen electrode, manometrically, volumetrically, or with an

Free Radicals in Diagnostic Medicine, Edited by
D. Armstrong, Plenum Press, New York, 1994

approximate paper disk method. Alternatively, the disappearance rate of H_2O_2 can be determined spectrophotometrically, polarographically, electrochemically, or by titration. Since H_2O_2 absorbs at 240 nm, its destruction can be followed spectrophotometrically[2]. The disadvantage of this approach is its theoretical non-specificity since any consumption of H_2O_2 by other peroxidases, or by non-specific means will appear as catalase activity. In addition, this assay is not particularly sensitive. The polarographic assay offers the advantage of being specific for catalase (no other peroxidase liberates oxygen) and being more sensitive than the spectrophotometric assay[3].

Source of extracellular catalase. The presence of catalase activity in serum or plasma may be easily overlooked due to the massive quantities of catalase present in erythrocytes and the frequent contamination of serum and plasma with lysed erythrocytes during sample preparation. However, elevated serum and plasma catalase activity has been associated with several disease processes (Table 1). The source of extracellular catalase is frequently unclear. In hemolytic diseases and acute pancreatitis the source appears to be lysed erythrocytes[4,5]. However, in other inflammatory diseases the source remains unknown[6]. In one case, a lymphocyte T-cell line has been shown to release catalase which protects the cell from programmed cell death[7].

Extracellular Manganese Superoxide Dismutase (MnSOD)

Methodology of measurement. Mammalian systems possess three forms of superoxide dismutases. All catalyze the same reaction; the dismutation of superoxide anion to hydrogen peroxide. The first, described in 1969 by McCord and Fridovich, is a copper and zinc containing enzyme present in the cytosol (Cu,Zn-SOD)[8]. The second is also a copper containing enzyme present in extracellular fluid, called extracellular SOD (ECSOD)[9]. Lastly, a manganese containing SOD (MnSOD) located mainly in the matrix of the mitochondria is present[10]. While all catalyze the same reaction they are products of separate genes and have differing sensitivities to inhibitors which can help distinguish their activity *in vitro*. For example, Cu,Zn-SOD is sensitive to hydrogen peroxide and cyanide while MnSOD is not. While assays based on enzymatic activities are widespread, they suffer from several disadvantages. First, all three enzymes have similar activity making it difficult to distinguish their separate activities. Second, many tissues contain compounds with SOD-like activity[11]. Finally, assays based on activity require careful tissue preparation and storage as enzyme activity declines with time and improper storage. Assays based on immunologic recognition with antibodies have the advantage of demonstrating no cross-reactivity since the SOD enzymes are structurally and antigenically unique. These assays are reproducible, quick and convenient for clinical and laboratory work. Since these assays are specific and detect immunogenicity and not activity, for the particular enzyme, they are unaffected by tissue inhibitors or activators. Several types of immunochemical assays are used including immunodiffusion, radioimmunoassay or a sandwich type enzyme-linked immunosorbent assay (ELISA)[11]. The accuracy of these assays is, of course, dependent on the selectivity of the antibodics. Both polyclonal and monoclonal antibodies may be used but given the cross species similarities these antibodies are difficult to raise. These antibodies have indeed been produced and hopefully will be commercially available. One ELISA utilizing a monoclonal antibody has a sensitivity of 2 ng/ml with a working range of 2-200 ng/ml[12]. The coefficient of variation is 3% and within run reproducibility 5%.

Source of extracellular MnSOD. MnSOD is a component of the mitochondrial matrix. Since erythrocytes contain no MnSOD serum measurements are unaffected by hemolysis. The source of serum MnSOD is presumed to be mitochondrial in all cases (Table 2). In acute myocardial infarction the specific source has been shown to be release from cardiac tissue[11]. In other disease states, presumably affected organs are responsible for the increase in serum values. The mechanism of release of MnSOD from mitochondria remains unknown. Serum MnSOD has a half-life of 6 hours[11].

Hydrogen Peroxide (H_2O_2)

Methodology of measurement. Although H_2O_2 is not a free radical, it is a potent oxidant and plays a central role in the metabolism of oxygen free radicals generating several toxic free radicals in the process. It is metabolized by glutathione peroxidase, catalase and participates in other reactions as well. It is the most stable of the primary reactive oxygen species and, as such, can sometimes be detected in human blood, urine, and since it is relatively volatile, even expired breath[13-20]. The detection of H_2O_2 in blood, however, is the subject of some controversy[15]. Many techniques have been adapted to measure small quantities of H_2O_2, often in the picomole range. These include high-performance liquid chromatography, isoluminol chemiluminescence, fluorescence spectroscopy among others[21].

Source of H_2O_2. Hydrogen peroxide can be formed by the dismutation of superoxide anion. Neutrophils and other inflammatory cells as well as xanthine oxidase are significant sources. The precise source of H_2O_2 in specific disease processes has not been elucidated.

SELECTED DISEASES

Sepsis and the adult respiratory distress syndrome (ARDS)

Serum catalase activity in sepsis and ARDS. Both sepsis and ARDS represent syndromes with an altered oxidant-antioxidant balance. Many investigators have revealed evidence of an increased oxidant burden in ARDS[22]. Less attention has been paid to the status of the enzymatic antioxidant defense systems. We studied septic patients with and without established ARDS[6]. We found that serum catalase activity was elevated in those septic patients with ARDS compared to those without (Figure 1). Serum catalase activity was also elevated in both septic groups compared to healthy control subjects (normal value 7.3 U/ml, s.d. 3.5). In contrast, serum activity of another antioxidant enzyme, glutathione peroxidase, was similar in septic patients with or without ARDS and control subjects.

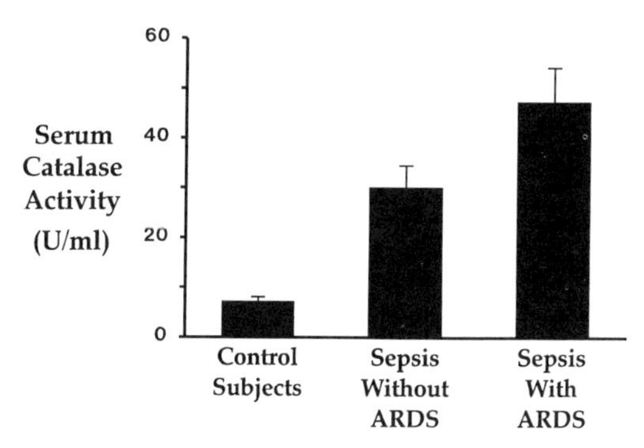

Figure 1. Serum catalase activity was increased ($p \leq 0.05$) in septic patients with ARDS (n=9) compared with septic patients without ARDS (n=20). Serum from healthy control subjects had less ($p \leq 0.05$) serum catalase activity compared to both septic groups. Values are mean ± SEM. Modified from reference 6.

We also studied several less acute pulmonary diseases and found normal values for serum catalase in pulmonary fibrosis, alpha-1-antitrypsin deficiency, sarcoidosis, and cystic fibrosis. To investigate erythrocyte (RBC) hemolysis as a potential source of serum catalase in these septic patients we evaluated serum haptoglobin levels and a RBC fragility index. Serum haptoglobin levels were similar in septic patients with or without ARDS and healthy control subjects. The RBC fragility index (the tonicity of saline required to lyse 50% of RBC) was also similar in all three groups. While not ruling out subtle hemolysis as a source of the elevated serum catalase activity, we could find no evidence to support this premise. Several alternative explanations are possible including tissue injury with release of catalase, increased cellular excretion of catalase, or upregulation of catalase activity in the face of an oxidant stress.

Since septic patients with ARDS have elevated serum catalase activity, we explored the possibility that this elevation might occur before the clinical diagnosis of ARDS, and thus provide a blood test that might predict the development of ARDS in septic patients. We studied patients admitted to an intensive care unit with the diagnosis of sepsis[23]. Roughly twenty percent of these patients subsequently developed ARDS. Serum was withdrawn at study entry and collected every six to twenty four hours thereafter. Since some of the septic patients at study entry were destined to develop ARDS but had not yet done so, we were able to ask the question whether serum antioxidant enzyme levels could predict who was destined to develop ARDS before they had done so clinically. We found that serum catalase activity measured with a sensitive and specific polarographic assay was elevated in septic patients compared to serum catalase activity in healthy control subjects. We also found that serum catalase activity was higher in septic patients who subsequently developed ARDS compared to septic patients who did not (Figure 2). Elevated serum catalase activity (>30 units/ml) in septic patients had a 83% sensitivity and 65% specificity for the subsequent development of ARDS. The positive predictive value was 42% while the negative predictive value was 93%. Despite significant overlap in the groups, blood markers predicting various processes will be an important area of future research. In this way a highly selected group of patients more likely to develop ARDS or other syndromes can be identified and subjected to experimental therapies. Treating all at-risk patients for ARDS would be too impractical since the vast majority (<10%) do

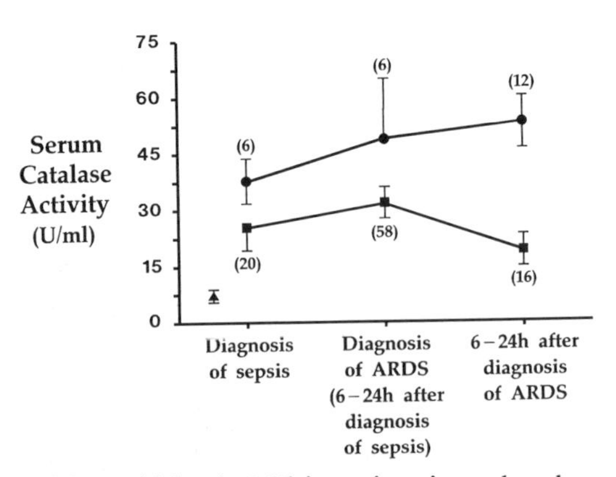

Figure 2. Serum catalase activity was higher ($p \leq 0.05$) in septic patients who subsequently developed ARDS (closed circles) compared to septic patients who did not develop ARDS (closed squares). Both septic groups had higher ($p \leq 0.05$) serum catalase activity compared to healthy control subjects (closed triangles). Serum catalase activity continued to rise in septic patients who developed ARDS and decreased over time in septic patients without ARDS. Values are mean ± SEM with numbers of patients in parentheses. Modified from reference 23.

not proceed to ARDS. The serum catalase activity in septic patients with ARDS continued to rise during the course of their illness while declining in the septic patients who did not develop ARDS (Figure 2). Serum catalase activity did not predict survival in this series of patients. Serum catalase activity also appears to be elevated in patients with ARDS associated with other risk factors (unpublished observations).

Serum manganese superoxide dismutase (MnSOD) in sepsis and ARDS. Serum levels of MnSOD measured by ELISA have recently emerged as another marker of sepsis and ARDS. In the study mentioned above MnSOD levels were also measured in the same septic patients some of whom subsequently developed ARDS[23]. Similar to serum catalase activity, we found that the serum MnSOD level measured by ELISA was elevated in septic patients compared to the serum MnSOD level in healthy control subjects. We also found that the serum MnSOD level was higher in septic patients who subsequently developed ARDS compared to septic patients who did not (Figure 3). If a cutoff of 450 ng/ml was used, an elevated serum MnSOD level in a septic patient without ARDS had a 67% sensitivity, 88% specificity, 67% positive predictive value and a 88% negative predictive value for the subsequent development of ARDS. In septic patients who did not develop ARDS, the serum MnSOD level rose over time and then decreased into the normal range. In septic patients after the development of ARDS, the serum MnSOD level remained elevated during the course of the study. Serum MnSOD levels also did not predict patient survival in this study.

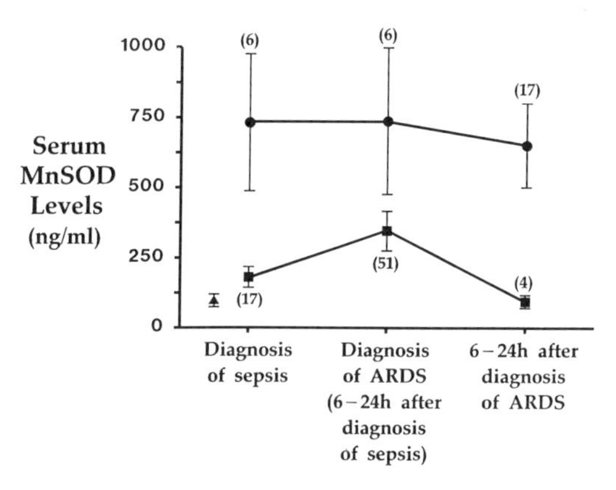

Figure 3. The serum manganese superoxide dismutase (MnSOD) level measured by ELISA was higher ($p \leq 0.05$) in septic patients who subsequently developed ARDS (closed circles) compared to septic patients who did not develop ARDS (closed squares). Both septic groups had higher ($p \leq 0.05$) serum MnSOD levels compared to healthy control subjects (closed triangles). Serum MnSOD levels remained elevated in septic patients who developed ARDS and decreased over time in septic patients without ARDS. Values are mean ± SEM with numbers of patients in parentheses. Modified from reference 23.

Exhaled breath H_2O_2 in ARDS. There have been three reports of exhaled breath H_2O_2 during the course of ARDS[17,19,24]. In the first Baldwin and associates studied 43 hospitalized patients, 16 with a diagnosis of ARDS and 27 without. All patients were supported on mechanical ventilation. The 16 patients with ARDS had elevated levels of breath H_2O_2 on the day of diagnosis of ARDS compared to the levels seen in patients without ARDS (Figure 4).

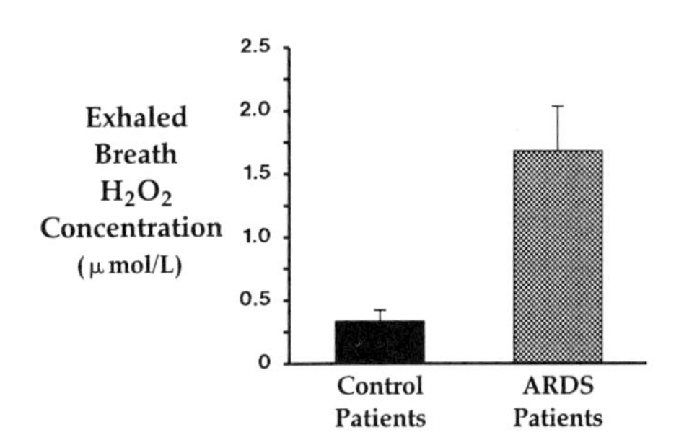

Figure 4. Exhaled breath H_2O_2 concentrations were increased ($p \leq 0.05$) in 16 patients on the day of diagnosis of ARDS compared to H_2O_2 concentrations in 27 patients without ARDS. Values are mean ± SEM. Replotted from reference 17.

The correlation of breath H_2O_2 concentrations with plasma lysozyme levels (a marker of neutrophil turnover) suggested activated neutrophils as the source of H_2O_2 in these patients. In another study, Sznajder and colleagues confirmed these findings[19]. Fifty-five ventilated patients with ARDS had a mean breath H_2O_2 concentration of 2.34 ± 1.15 (sd) μmol/L compared to 0.99 ± 0.72 (sd) μmol/L in patients without ARDS. These are difficult studies since the H_2O_2 concentrations are near the limits of detection and since bacteria and saliva both contain catalase thus making interpretation difficult.

Other markers. Various other markers of oxidative damage in ARDS have been proposed including oxidized glutathione (GSSG) in alveolar fluid, depressed plasma vitamin E levels, increased plasma lipid peroxidation products, as well as other oxidized proteins in lavage fluid (e.g. alpha-1-protease inhibitor)[25-29].

Human Immunodeficiency Virus (HIV) Infection

HIV infection is an insidiously progressive disorder whose pathogenesis in part involves oxygen radicals[30]. The immunological defense of these patients is further hindered by deficiencies in several antioxidants including glutathione, vitamin E, selenium, and manganese-containing superoxide dismutase[30-33]. Buhl and colleagues studied 14 individuals with asymptomatic HIV infection measuring glutathione (GSH)

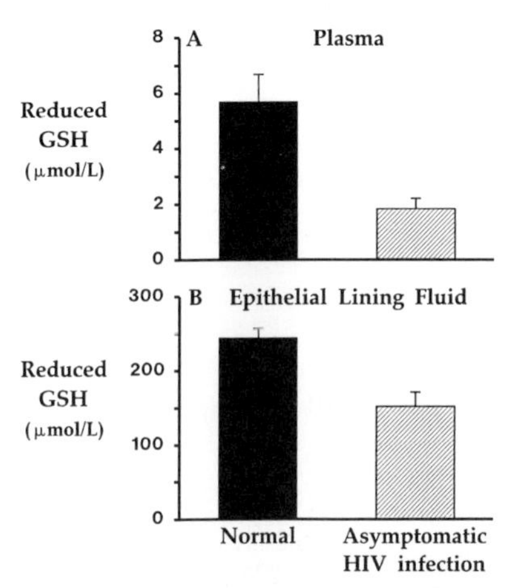

Figure 5. Reduced glutathione (GSH) levels were measured in healthy control subjects and patients with asymptomatic human immunodeficiency virus (HIV) infection. Reduced GSH levels were significantly ($p<0.05$) reduced in both plasma and epithelial lining fluid of individuals with asymptomatic HIV infection compared to healthy control subjects. Values are mean ± SEM. Replotted from reference 31.

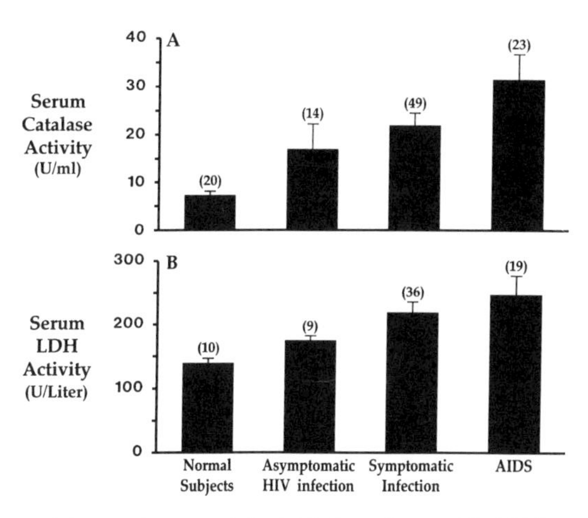

Figure 6. Serum catalase activity was increased ($p \leq 0.05$) in patients with AIDS compared to other HIV-infected individuals and healthy control subjects. Individuals with AIDS and symptomatic infection had greater ($p \leq 0.05$) serum catalase activity than healthy control subjects (panel A). Serum LDH activity was increased ($p \leq 0.05$) in all HIV-seropositive individuals compared to healthy control subjects (panel B). Values are mean ± SEM with numbers of patients in parentheses. Modified from reference 34.

levels in plasma and epithelial lining fluid (obtained by bronchoscopy)[31]. They found that individuals with asymptomatic HIV infection had significantly depressed levels of reduced GSH in both plasma and epithelial lining fluid (Figure 5).

This appears to be an early manifestation of HIV infection and may influence the progression of the process. Given this perturbed oxidant-antioxidant balance we examined the serum catalase activity in HIV infected patients[34]. Serum was collected during routine clinic visits in patients free of acute disease. Patients were classified as having asymptomatic infection, symptomatic infection, or AIDS by standard criteria. We found that serum catalase activity increased progressively as HIV infection progressed (Figure 6). We found a similar pattern with serum lactate dehydrogenase activity (LDH, Figure 6). In contrast, serum glutathione peroxidase activity was not elevated in HIV infected individuals and did not increase with progression of infection.

Serum haptoglobin levels and RBC fragility indices were similar in HIV infected individuals and healthy control suggesting that RBC hemolysis was not the source of serum catalase in these patients. The fact that serum catalase activity correlated ($r=0.67$, $p \leq 0.05$) with serum LDH activity suggested that progressive tissue injury with release of catalase may be a source of serum catalase activity in these patients[34].

Rheumatoid arthritis

Rheumatoid arthritis (RA) is a systemic autoimmune disease of unknown etiology. Although RA is often a multisystem disease, it is primarily associated with intermittent acutely inflamed joints. An actively inflamed rheumatoid joint is massively infiltrated with neutrophils and neutrophil-derived products such as lysozyme and stable prostaglandins. These activated neutrophils would be expected to release toxic oxygen metabolites and in fact there is a great deal of evidence that oxygen radicals play an important part in the inflammation of the rheumatoid joint. Interleukin-1 (IL-1) present in inflamed joints may prime neutrophils[35] for oxidant release and activate other local cells such as synoviocytes and chondrocytes to release proteolytic enzymes. Besides neutrophils, activated macrophages may produce oxygen radicals in the inflamed rheumatoid joint[36]. It has also been proposed that an ischemia-reperfusion injury may occur in RA joints providing substrate and conditions favorable for xanthine oxidase-derived free radical generation[37]. Many markers of this oxidant-induced damage have been measured both in serum and joint (synovial) fluid and may be important in understanding the pathogenesis of RA as well as following disease activity (Table 3).

Lunec and associates studied 58 patients with RA and compared their sera and synovial fluid to healthy control subjects and patients with degenerative (non-inflammatory) joint disease (mainly osteoarthritis)[38]. Using techniques to measure conjugated dienes, visible and ultraviolet fluorescence, they found that greater than 90% of synovial fluids from all patients had evidence of free radical oxidation (peroxidation) products. In particular, however, patients with RA had greater free radical oxidation products than patients with non-inflammatory joint disease and healthy controls. Free radical oxidation products were also elevated in serum of RA patients compared to healthy controls and tended to decrease when patients were treated with a non-steroidal anti-inflammatory agent (feprazone) or gold (Figure 7).

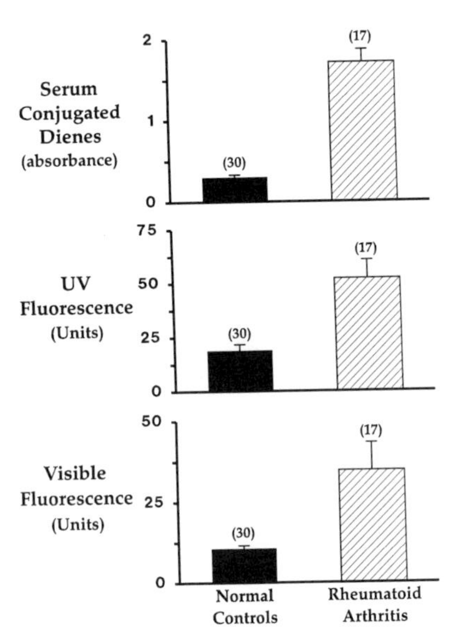

Figure 7. Serum was withdrawn from healthy control subjects and patients with rheumatoid arthritis. Conjugated dienes, UV fluorescence and visible fluorescence (as measures of free radical oxidation products) were all increased ($p \leq 0.05$) in RA patients compared to normal control subjects. Values are mean ± SEM with numbers of patients in parentheses. Replotted from reference 38.

There also was a correlation between serum free radical oxidation products and clinical activity of the disease. This provides a rationale, if corroborated, for using serum markers to follow disease activity and response to therapy. Measurement of lipid peroxidation products is, however, fraught with difficulties. The techniques of measurement are tedious, artifacts are frequent, the products measured are unstable and subject to degradation, and the exact products to measure are frequently unknown in specific conditions.

Rowley and co-workers extended these observations by measuring thiobarbituric acid-reactive (TBAR) material in serum and synovial fluid from RA patients[39]. TBAR material is thought to largely measure malondialdehyde (MDA), a product of lipid peroxidation. Although widely used as an index of lipid peroxidation, the measure is subject to artifact and is non-specific. Nonetheless, in many systems it can be a useful marker. They found detectable TBAR material in the serum and synovial fluid of RA patients. Of interest, they found correlations of TBAR material in joint fluid with clinical activity of disease.

Ambanelli and colleagues studied serum markers in 25 RA patients and compared the results to values from 15 subjects matched for sex and age[40]. They found elevations in "serum antioxidant activity" and decreases in serum SH groups (thiols) in RA patients compared to controls. Moreover, serum SH levels correlated inversely with disease activity (r=-0.57). In addition, RA patients who responded to therapy had increased SH levels compared to pre-therapy values (Figure 8).

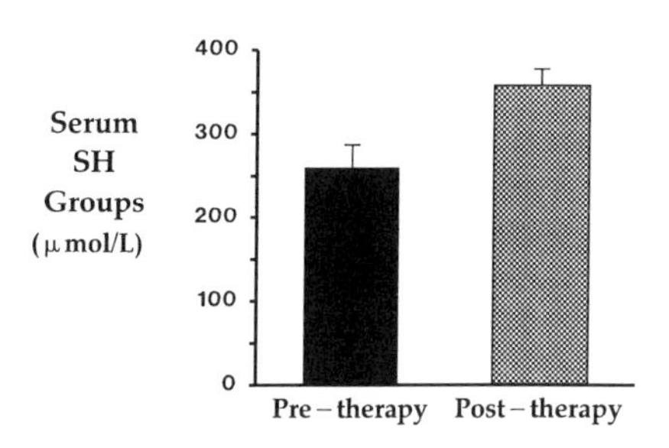

Figure 8. Serum SH groups (thiols) were measured before and after therapy (with tiopronin) in patients with rheumatoid arthritis. Non-responders did not increase serum SH levels (not shown), however, SH levels in responders increased ($p \leq 0.05$) after therapy compared to values before therapy. Values are mean ± SEM. Replotted from reference 40.

Imadaya and associates investigated antioxidant enzyme levels in erythrocytes from RA patients and found reduced levels of superoxide dismutase, catalase and glutathione peroxidase compared to patients with degenerative arthritis and healthy controls[41]. This conflicts with previous work[42] but has been confirmed by others [43].

Situnayake and co-workers evaluated the ability of serum from RA patients to resist attack in vitro by controlled peroxidation induced by peroxyl radicals (the "TRAP" assay)[44]. They found that serum from RA patients had reduced ability to resist attack by peroxyl radicals (TRAP) compared to normal controls. The major determinant of TRAP in RA patients was found to be uric acid while in control serum it was vitamin E. They also found reduced plasma ascorbic acid, serum vitamin E and serum sulfhydryl levels in RA patients compared to healthy controls. In these studies, no relation between these free radical markers and disease activity were noted, however.

Lunec and colleagues have studied serum and synovial fluid from RA patients[45]. They found that when human immunoglobulin G (IgG) was exposed *in vitro* to free radicals, auto-fluorescent monomeric and polymeric IgG was formed. These IgG aggregates can stimulate neutrophils to release toxic oxygen metabolites *in vitro*. Using high-performance liquid chromatography (HPLC) they were able to isolate these complexes from sera and synovial fluid from RA patients. The results suggest neutrophils invading the rheumatoid joint may release oxygen radicals causing aggregation of IgG which in turn activates other neutrophils to release more oxygen radicals creating a perpetuating cycle of inflammation and damage.

Grootveld and associates have suggested both synovial fluid formate levels and a hyaluronate-derived low molecular mass oligosaccharide species as novel markers of reactive oxygen radical activity in the inflamed rheumatoid joint during exercise detecting these species using proton-nuclear magnetic resonance imaging[46]. In separate studies Grootveld and Halliwell have shown that allantoin, the product of free radical attack on uric acid, is elevated in serum and synovial fluid of RA patients[47]. The practical value of these studies remains to be shown. In summary, numerous alterations in free radical markers are present in rheumatoid disease. Many of these markers correlate with clinical disease activity and some decrease after effective treatment. It is not yet clear whether these markers add anything to the routine clinical assessment of these patients. Further study will hopefully answer this question.

Primary Biliary Cirrhosis (PBC)

PBC is a chronic disorder characterized by clinical and laboratory evidence of impaired bile excretion and progressive liver destruction centered around intrahepatic bile ducts. PBC is associated with a number of specific and nonspecific immunologic abnormalities including anti-mitochondrial antibodies, elevated serum IgM levels, antibodies directed against bile canaliculi, and impaired lymphocyte transformation. This and other evidence suggests that PBC represents an autoimmune disorder. There is mounting evidence that oxygen free radicals may be important in its pathogenesis. Hepatic copper levels are greatly elevated (albeit late) in PBC and copper can reduce O_2 to superoxide anion as one potential mechanism. Ono and colleagues measured serum MnSOD levels in patients with PBC and found them to be greatly elevated (407 $\pm$ 35 ng/ml) compared to control patients (<150 ng/ml)[48]. This was a very early finding in the disease. This is interesting since antibodies directed against mitochondria are present in this disease and MnSOD is a mitochondrial product. Cytokines such as interleukin-1 and tumor necrosis factor induce MnSOD gene expression and also induce an intracellular oxidant stress in endothelial cells and neutrophils possibly relating these observations[49,50].

Table 1. Disorders Associated With Elevated Serum or Plasma Catalase Activity.

Disorder	Reference
Fatty liver	51
Acute alcoholic hepatitis	51
Toxic hepatitis	51
Acute pancreatitis	4,52
Acute cholecystitis	52
Congestive heart failure	51
Sepsis	6,23
Adult respiratory distress syndrome	6,23
Human immunodeficiency virus (HIV) infection	34
Skin burn	53
Hemolytic anemia	55

CONCLUSION

The involvement of oxidant species in human autoimmune and inflammatory diseases is now undeniable. With the advent of new techniques to measure either these reactive species themselves or their reaction products we are now in a position to assess the utility of these oxidant markers in the diagnosis and management of these challenging diseases.

Table 2. Disorders Associated With Elevated Serum MnSOD Levels.

Disorder	% Elevated*	Reference
Myocardial infarction	63	12
Liver disease		
Primary biliary cirrhosis	97	11
Hepatoma	60	11,48
Hepatitis	?	11,48
Cirrhosis	33	11,48
Gastric cancer	27	11
Lung cancer	33	11
Lymphoma	17	11
Sepsis	35	23
Adult respiratory distress syndrome	89	23
Acute myeloid leukemia	60	11,55
Acute lymphocytic leukemia	27	11,55
Ovarian cancer	62	11,55
Cervical cancer	19	11
Endometrial cancer	22	11
Benign ovarian tumors	13	11

Normal values: Females 99.8 ± 24.8 (s.d.) ng/ml ; Males 88.8 ± 20.8 (s.d.) ng/ml
* Abnormal value defined as >2 s.d. above mean.

Table 3. Free Radical Markers in Rheumatoid Arthritis.

Marker	Reference
Lipid peroxidation products in blood and synovial fluid	38,39
Reduced serum thiols	40,44
Increased serum ceruloplasmin levels	40
Decreased erythrocyte antioxidants	41-43
Decreased plasma and synovial fluid ascorbic acid levels	44
Decreased serum "antioxidant" activity	44
Decreased serum vitamin E levels	44
Increased neutrophil oxidant activity	56
Increased synovial fluid hyaluronate-derived degradation products	46
Increased synovial fluid formate levels	46
Increased serum and synovial fluid allantoin levels	47
Aggregated serum and synovial fluid IgG	45

REFERENCES

1. C.E. Cross, Oxygen radicals and human disease, Ann. Intern. Med. 107:526-545 (1987).
2. H. Aebi, Catalase in Vitro, in: "Methods in Enzymology: Oxygen radicals in biological systems. Volume 105", L. Packer., ed., Academic Press, Inc., Orlando, pp. 121-126 (1984).
3. J.A. Leff, M.A. Oppegard, L.S. Terada, E.C. McCarty, and J.E. Repine, Human serum catalase decreases endothelial cell injury from hydrogen peroxide, J. Appl. Physiol. 71(5):1903-1906 (1991).
4. L. Goth, Origin of serum catalase activity in acute pancreatitis, Clin. Chim. Acta 186:39-44 (1989).
5. L. Goth, H. Nemeth, and I. Meszaros, Serum catalase activity for detection of hemolytic diseases [letter], Clin. Chem. 29:741-743 (1983).
6. J.A. Leff, P.E. Parsons, C.E. Day, E.E. Moore, F.A. Moore, M.A. Oppegard, and J.E. Repine, Increased serum catalase activity in septic patients with the adult respiratory distress syndrome, Am. Rev. Respir. Dis. 146:985-989 (1992).
7. P.A. Sandstrom and T.M. Buttke, Autocrine production of extracellular catalase prevents apoptosis of the human CEM T-cell line in serum-free medium, Proc. Natl. Acad. Sci. U. S. A. 90:4708-4712 (1993).
8. J.M. McCord and I. Fridovich, Superoxide dismutase: an enzymic function for erythrocuprein (hemocuprein), J. Biol. Chem. 244:6049-6055 (1969).
9. S.L. Marklund, Human copper-containing superoxide dismutase of high molecular weight, Proc. Natl. Acad. Sci. USA 79:7634-7638 (1982).
10. B.B. Keele,Jr., J.M. McCord, and I. Fridovich, Superoxide dismutase from escherichia coli B: A new manganese-containing enzyme, J. Biol. Chem. 245:6176-6181 (1970).

11. N. Taniguchi, Clinical significances of superoxide dismutases: changes in aging, diabetes, ischemia, and cancer, Adv. Clin. Chem. 29:1-59 (1992).
12. T. Kawaguchi, K. Suzuki, Y. Matsuda, T. Nishiura, T. Uda, M. Ono, C. Sekiya, M. Ishikawa, S. Iino, Y. Endo, and N. Taniguchi, Serum Mn-superoxide dismutase: Normal values and increased levels in patients with acute myocardial infarction and several malignant diseases determined by enzyme-linked immunosorbent assay using a monoclonal antibody, J. Immunol. Meth. 127:249-254 (1990).
13. B. Frei, Y. Yamamoto, D. Niclas, and B.N. Ames, Evaluation of an isoluminol chemiluminescence assay for the detection of hydroperoxides in human blood plasma, Anal. Biochem. 175:120-130 (1988).
14. Y. Yamamoto, B. Frei, and B.N. Ames, Assay of lipid hydroperoxides using high-performance liquid chromatography with isoluminol chemiluminescence, Methods Enzymol. 186:371-380 (1990).
15. A. Nahum, L.D.H. Wood, and J.I. Sznajder, Measurement of hydrogen peroxide in plasma and blood, Free Radic. Biol. Med. 6:479-484 (1989).
16. S.D. Varma and P.S. Devamanoharan, Excretion of hydrogen peroxide in human urine, Free Rad. Res. Comms. 8:73-78 (1990).
17. S.R. Baldwin, R.H. Simon, C.H. Grum, L.H. Ketai, L.A. Boxer, and L.J. Devall, Oxidant activity in expired breath of patients with adult respiratory distress syndrome, Lancet 1:11-14 (1986).
18. M.D. Williams and B. Chance, Spontaneous chemiluminescence of human breath: spectrum, lifetime, temporal distribution and correlation with peroxide, J. Biol. Chem. 258:3628-3631 (1983).
19. J.I. Sznajder, A. Fraiman, J.B. Hall, W. Sanders, G. Schmidt, G. Crawford, A. Nahum, P. Factor, and L.D.H. Wood, Increased hydrogen peroxide in the expired breath of patients with acute hypoxemic respiratory failure, Chest 96:606-612 (1989).
20. J.A. Leff, C.P. Wilke, B.M. Hybertson, P.F. Shanley, C.J. Beehler, and J.E. Repine, Post-insult treatment with N-acetylcysteine decreases interleukin-1-induced lung neutrophil sequestration and oxidative lung leak in rats, Am. J. Physiol. 265:L501-L506 (1993).
21. J.J. Hageman, A. Bast, and N.P.E. Vermeulen, Monitoring of oxidative free radical damage in vivo: Analytical aspects, Chem. Biol. Interactions 82:243-293 (1992).
22. J.E. Repine, Scientific perspectives on adult respiratory distress syndrome, Lancet 339:466-469 (1992).
23. J.A. Leff, P.E. Parsons, C.E. Day, N. Taniguchi, M. Jochum, H. Fritz, F.A. Moore, E.E. Moore, J.M. McCord, and J.E. Repine, Serum antioxidants as predictors of adult respiratory distress syndrome in patients with sepsis, Lancet 341:777-780 (1993).
24. W.C. Wilson, J.F. Swetland, J.L. Benumof, P. Laborde, and R. Taylor, General anesthesia and exhaled breath hydrogen peroxide, Anesthesiology 76:703-710 (1992).
25. G.R. Bernard, B.B. Swindell, M.J. Meredith, F.E. Carroll, and S.B. Higgins, Glutathione (GSH) repletion by N-acetylcysteine in patients with the Adult Respiratory Distress Syndrome, Am. Rev. Resp. Dis. 139:A221 (1989) (Abstract).
26. E.R. Pacht, A.P. Timerman, M.G. Lykens, and A.J. Merola, Deficiency of alveolar fluid glutathione in patients with sepsis and the adult respiratory distress syndrome, Chest 100:1397-1403 (1991).
27. E. Bunnell and E.R. Pacht, Oxidized glutathione is increased in the alveolar

fluid of patients with the adult respiratory distress syndrome, Am. Rev. Resp. Dis. 148:1174-1178 (1993).

28. C. Richard, F. Lemonnier, M. Thibault, M. Couturier, and P. Auzepy, Vitamin E deficiency and lipoperoxidation during adult respiratory distress syndrome, Crit. Care. Med. 18:4-9 (1990).
29. C.G. Cochrane, R.G. Spragg, and S.D. Revak, Pathogenesis of the adult respiratory distress syndrome: evidence of oxidant activity in bronchoalveolar lavage fluid, J. Clin. Invest. 71:754-758 (1983).
30. B. Halliwell and C.E. Cross, Reactive oxygen species, antioxidants, and acquired immunodeficiency syndrome. Sense or speculation? Arch. Intern. Med. 151:29-31 (1991).
31. R. Buhl, K.J. Holroyd, A. Mastrangeli, A.M. Cantin, H.A. Jaffe, F.B. Wells, C. Santini, and R.G. Crystal, Systemic glutathione deficiency in symptom-free HIV-seropositive individuals, Lancet 2:1294-1298 (1989).
32. J.J. Javier, M.K. Fodyce-Baum, R.S. Beach, M. Gavancho, C. Cabrejos, and E. Mantero-Atienza, Antioxidant micronutrients and immune function in HIV-1 infection, FASEB Proc. 4:A940 (1990).
33. B.M. Dworkin, W.S. Rosenthal, G.P. Wormser, and L. Weiss, Selenium deficiency in the acquired immune deficiency syndrome, J. Parenter. Ent. Nutr. 10:405-407 (1986).
34. J.A. Leff, M.A. Oppegard, T.J. Curiel, K.S. Brown, R.T. Schooley, and J.E. Repine, Progressive increases in serum catalase activity in advancing human immunodeficiency virus infection, Free Radical Biol. Med. 13:143-149 (1992).
35. Y. Ozaki, T. Ohashi, and S. Kume, Potentiation of neutrophil function by recombinant DNA-produced interleukin-1a, J. Leukocyte Biol. 42(6):621-627 (1987).
36. B. Halliwell, J.R. Hoult, and D.R. Blake, Oxidants, inflammation, and anti-inflammatory drugs, FASEB J. 2:2867-2873 (1988).
37. J. Unsworth, J. Outhwaite, D.R. Blake, C.J. Morris, J. Freeman, and J. Lunec, Dynamic studies of the relationship between intraarticular pressure, synovial fluid oxygen tension and lipid peroxidation in the inflamed knee: an example of reperfusion injury, Annu. Clin. Biochem. 25:8S-11S (1988).
38. J. Lunec, S.P. Halloran, A.G. White, and T.L. Dormandy, Free-radical oxidation (peroxidation) products in serum and synovial fluid in rheumatoid arthritis, J. Rheumatol. 8:233-245 (1981).
39. D. Rowley, J.M.C. Gutteridge, D. Blake, M. Farr, and B. Halliwell, Lipid peroxidation in rheumatoid arthritis: thiobarbituric acid-reactive material and catalytic iron salts in synovial fluid from rheumatoid patients, Clin. Sci. 66:691-695 (1984).
40. U. Ambanelli, A. Spisni, and G.F. Ferraccioli, Serum antioxidant activity and related variables in rheumatoid arthritis. Behaviour during sulphydrylant treatment, Scand. J. Rheumatol. 11:203-207 (1982).
41. A. Imadaya, K. Terasawa, H. Tosa, M. Okamoto, and K. Toriizuka, Erythrocyte antioxidant enzymes are reduced in patients with rheumatoid arthritis, J. Rheumatol. 15:1628-1631 (1988).
42. P. Scudder, J. Stocks, and T.L. Dormandy, The relationship between erthrocyte superoxide dismutase activity and erythrocyte copper levels in normal subjects and in patients with rheumatoid arthritis, Clin. Chem. Acta 69:397-403 (1976).
43. J.C. Banford, D.H. Brown, R.A. Hazelton, C.J. McNeil, R.D. Sturrock, and W.E. Smith, Serum copper and erythrocyte superoxide dismutase in rheumatoid disease, Ann. Rheum Dis. 41:458-462 (1982).
44. R.D. Situnayake, D.I. Thurnham, S. Kootathep, S. Chirico, J. Lunec, M. Davis, and B. McConkey, Chain breaking antioxidant status in rheumatoid arthritis:

clinical and laboratory correlates, Ann. Rheum Dis. 50:81-86 (1991).
45. J. Lunec, D.R. Blake, S.J. McCleary, S. Brailsford, and P.A. Bacon, Self-perpetuating mechanisms of immunoglobulin G aggregation in rheumatoid arthritis, J. Clin. Invest. 76:2084-2090 (1985).
46. M. Grootveld, E.B. Henderson, A. Farrell, D.R. Blake, H.G. Parkes, and P. Haycock, Oxidative damage to hyaluronate and glucose in synovial fluid during exercise of the inflamed rheumatoid joint. Detection of abnormal low-molecular-mass metabolites by proton-n.m.r. spectroscopy, Biochem. J. 273:459-467 (1991).
47. M. Grootveld and B. Halliwell, Measurement of allantoin and uric acid in human body fluids- A potential index of free radical reactions in vivo? Biochem. J. 243:803-808 (1987).
48. M. Ono, C. Sekiya, M. Ohhira, M. Namiki, Y. Endo, K. Suzuki, Y. Matsuda, and N. Taniguchi, Elevated level of serum Mn-superoxide dismutase in patients with primary biliary cirrhosis: possible involvement of free radicals in the pathogenesis in primary biliary cirrhosis, J. Lab. Clin. Med. 118:476-483 (1991).
49. T. Matsubara and M. Ziff, Increased superoxide anion release from human endothelial cells in response to cytokines, J. Immunol. 137:3295-3298 (1986).
50. S.J. Klebanoff, M.A. Vadas, J.M. Harlan, L.H. Sparks, J.R. Gamble, J.M. Agosti, and A.M. Waltersdorph, Stimulation of neutrophils by tumor necrosis factor, J. Immunol. 136:4220-4225 (1986).
51. L. Goth, I. Meszaros, and H. Nemeth, Serum catalase enzyme activity in liver diseases, Acta. Biol. Hung. 38:287-290 (1987).
52. I. Meszaros, L. Goth, and G. Vattay, The value of serum catalase activity determinations in acute pancreatitis, Digestive Diseases 18:1035-1041 (1973).
53. J.A. Leff, L.K. Burton, E.M. Berger, B.O. Anderson, C.P. Wilke, and J.E. Repine, Increased serum catalase activity in rats subjected to thermal skin injury, Inflammation 17:199-204 (1993).
54. T. Nishiura, K. Suzuki, T. Kawaguchi, H. Nakao, N. Kawamura, M. Taniguchi, Y. Kanayama, T. Yonezawa, S. Iizuka, and N. Taniguchi, Elevated serum manganese superoxide dismutase in acute leukemias, Cancer Lett. 62:211-215 (1992).
55. M. Ishikawa, Y. Yaginuma, H. Hayashi, T. Shimizu, Y. Endo, and N. Taniguchi, Reactivity of a monoclonal antibody to manganese superoxide dismutase with human ovarian carcinoma, Cancer Res. 50:2538-2542 (1990).
56. P. Suryaprabha, U.N. Das, G. Ramesh, K.V. Kumar, and G.S. Kumar, Reactive oxygen species, lipid peroxides and essential fatty acids in patients with rheumatoid arthritis and systemic lupus erythematosus, Prostaglandins Leukot. Essent. Fatty. Acids 43:251-255 (1991).

ANTIOXIDANTS AND CANCER: MOLECULAR MECHANISMS

Peter J. O'Brien

Faculty of Pharmacy
University of Toronto
Toronto, Ontario M5S 2S2
Canada

INTRODUCTION

Because chemotherapy is often ineffective for the treatment of cancer, considerable efforts are being made to develop agents for "chemoprevention" that would prevent cancers from developing in the first place. Furthermore understanding the antitumorigenic mechanisms should further our comprehension of carcinogenesis mechanisms. In the following, possible molecular mechanisms by which antioxidants inhibit carcinogenesis are reviewed.

Synthetic antioxidants are widely used in industry and food processing as substances which slow the oxidation rate of autoxidisable substrates. Antioxidants are also found in all biological species and protect against the potentially harmful effects of processes or reactions that cause excessive oxidations. Because of this, biological antioxidants form an important part of our diet and together with intracellular antioxidants and antioxidant enzyme systems play an important role in preventing various pathological diseases and cancer. For the purpose of this review and because carcinogenesis may often be a consequence of oxidative stress, an antioxidant is also defined as a substance which prevents the oxidation of a cell and/or normalises a cell's redox potential.

Antioxidants or agents which are metabolised to antioxidants may therefore function by at least one or more of the following mechanisms:

1. scavenging or reducing lipid free radicals (chain-breaking primary antioxidants) eg. α-tocopherol.
2. scavenging reactive oxygen species eg. polyphenolics
3. quenching the formation of singlet oxygen eg. ß-carotene, retinol
4. scavenging prooxidant metals eg. polyphenolics, flavonoids
5. oxidising ferrous iron activity eg. ceruloplasmin, apoferritin
6. inhibiting prooxidant enzymes eg. allopurinol
7. reducing oxidatively stressed cells eg. ethanol, sorbitol, xylitol (NADH generators)
8. sparing or regenerating intracellular antioxidants eg. ascorbate, Nacetylcysteine
9. inducing or enhancing the cell's protective enzymic defense against oxygen or oxidants eg. butylated hydroxyanisole
10. stabilizing membranes against lipid peroxidation by decreasing membrane fluidity eg. cholesterol, 17-ß-estradiol, tamoxifen

Free Radicals in Diagnostic Medicine, Edited by
D. Armstrong, Plenum Press, New York, 1994

11. inhibiting enzymes that mediate gene expression as a result of oxidative stress eg. eg. tamoxifen, methoxybenzamide.

It is likely that other anticarcinogenesis mechanisms exist for antioxidants. Furthermore anticarcinogenic agents which are not antioxidants may act by one of the above mechanisms eg. by metabolism to an antioxidant.

The antitumorigenic effect of antioxidants were probably first described in 1934 when wheat germ oil was found to prevent tar carcinoma in mice and was tentatively attributed to vitamin E in the oil[1]. Subsequently it was shown that adding wheat germ oil to the diet reduced the number of a mixed group of tumours resulting from the intraperitoneal injection of mice with 3-methylcholanthrene[2]. Dietary vitamin E also prevented subcutaneous sarcomas induced by 3-methylcholanthrene[3]. The antioxidants vitamin E[4], selenium[5] and vitamin C[5] applied to mouse skin also reduced tumour formation induced by 7,12-dimethylbenz(a)anthracene initiation in a 2 stage system of tumorigenesis. The synthetic antioxidants butylated hydroxyanisole, butylated hydroxytoluene and ethoxyquin also prevented the carcinogenicity of 7,12-dimethylbenzanthracene on the forestomach of the mouse[6]. The latter antioxidants are commonly added to foods to maintain freshness and prevent spoilage by oxidation. They are frequently used in dried cooking oils, cereals, canned goods and animal feed. Their metabolism is slow and they accumulate in adipose tissue.

Several epidemiological studies now indicate that a human diet rich in vegetables and fruits lowers the incidence of cancer[7]. In a review of 172 published epidemiological studies it was apparent that the quarter of the population with low dietary intake of fruits and vegetables had double the cancer rate for those with high intake for most types of cancer (lung, larynx, oral cavity, oesophagus, stomach, colon and rectum, bladder, pancreas, cervix and ovary)[9]. Clinical trials using various antioxidant supplements have been promising but larger studies may be needed. Only 9% of Americans are eating 2 fruits and 3 vegetables a day as recommended by the National Cancer Institute and the National Research Council/National Academy of Sciences[8]. Those individuals exposed to oxidants as a result of smoking, UV exposure, chronic infections, dietary iron etc. would be expected to have a deficiency of endogenous antioxidants[9] and would benefit from antioxidant supplements. Indeed much of the cancer resulting from liver, colon, bladder, stomach and lung has been attributed to oxidative stress resulting from infection or inflammation[10]. In addition to plant antioxidants the chemopreventive action of many edible plants and vegetables has been attributed to increased fibre content as well as decreased meat, fat and total energy intake. However, this could be offset by various mutagenic compounds present in some plants[10]. On balance though, the plant antioxidants probably play a major role in the anticarcinogenic effects of dietary fruits and vegetables.

Various antioxidant compounds in fruits and vegetables have now been shown to inhibit several types of cancer and include phenols, indoles, aromatic isothiocyanates and dithiolethiones (reviewed[11]). The various vegetables and fruits that are sources of these compounds are given in table 1 and could form the basis of various "anticarcinogenic diets or recipes". Beverages particularly green tea which are richer in polyphenols such as epigallocatechin-3-gallate than black tea also prevent human gastric cancer and animal chemical carcinogenesis[12]. Although animal research and human population studies make it clear that increasing levels of dietary antioxidants lead to increasing levels of protection against cancers, the molecular mechanisms responsible are not understood. Several reviews classifying antioxidants by anticarcinogenesis mechanisms have been published previously[11-15].

Table. 1. **Plant Anticarcinogens**

1. All plants: eg. *flavonoids* (1g per day human diet), *catechols, α-tocopherol, ß-carotene, ascorbic acid, tannins*

2. Cruciferous vegetables: eg. Cruciferae--mustards, Brassica -- broccoli, cauliflower, brussel sprouts; processing releases eg. *allyl isothiocyanate, sulforaphane, indole 3-carbinol*

3. Allium species, including garlic, onions, leeks, shallots eg. *diallyl sulphide*

4. Citrus fruit oils, caraway seed oil, eg. *limonene and carvone.* (monoterpenes)

5. Strawberries, black currants, raspberries, grapes, black walnuts eg. *ellagic acid*

6. Cabbage and other cruciferous vegetables eg. *dithiolethiones (oltipraz)*

7. Soybean -- eg. *isoflavones*

8. Green tea -- eg. *epigallocatechin* and *hydrolysable tannins* (polyphenolic)
   Coffee -- eg. *chlorogenic acid*
   Wine eg. *flavonoids*

9. Spices eg.
   turmeric -- *curcumin* (a phenolic), yellow colour of curry
   rosemary, sage eg. *carnosol* (diphenolic)

10. Medicinal plants
    milk thistle seeds eg. *silymarin* (a flavonoid)
    creosote bush eg. *nordihydroguaiaretic acid* (diphenolic)
    boldo leaves eg. *boldine* (an alkaloid)
    Kampo medicines eg. *tannins, flavonoids* etc.
    Ayurvedic food supplements eg. *tannins, flavonoids, curcumin* etc.

Two major theories for chemical carcinogenesis have evolved. In the genetic or somatic mutation theory it is supposed that procarcinogens are metabolically activated by a family of inducible cytochrome P-450s (designated phase 1 enzymes) to electrophilic species which damage nucleophilic centres on the bases of DNA[16]. These mutations are then inherited by all the progeny of the altered cells. The epigenetic theory on the otherhand proposes that there is an aberrant differentiation of the cell, in which the inducer or repressor regions of the genome are modified but the basic gene products are unchanged. In this theory oxidants trigger reactions which resemble those induced by growth and differentiation factors.

In several model systems, the induction of cancer by chemicals seems to proceed in multistages or multisteps and which can involve both theories. Recent molecular genetic studies of human tumors have also established that alterations in protooncogenes and tumour suppressor genes and successive clonal expansion contribute to each step in multi-step carcinogenesis. Mutations in the tumour suppressor gene p53 can be found in 50% of all human cancer[17] and serum p53 antibodies may be associated with a relatively poor prognosis and metastases. Three stages have been described. Initiation typically requires only a single exposure to a subthreshold dose of carcinogen and results from interaction of a metabolite of the procarcinogen with DNA. The chemoprevention of chemical carcinogenesis by 3-methylcholanthrene or flavones was first attributed to the induction of phase 1 enzymes which detoxified the procarcinogen[17]. Promotion, which follows initiation, is a phenomenon of gene activation eg. by DNA oxidation in which the latent phenotype of the initiated cell becomes expressed through selection and clonal expansion. Progression, involves the conversion of benign tumors into malignant neoplasms and could involve mutation of tumour suppressor genes eg. by DNA oxidation.

The following reviews the current state of knowledge on how antioxidants prevent mutagenicity and whether there are any potential hazards for their use. The topics discussed are subdivided as follows:

I. Mechanisms of chemoprevention by antioxidants
   i. antiinitiator effects of antioxidants: do they prevent the metabolic activation of carcinogens to free radicals or electrophiles which interact with DNA resulting in gene mutations?
   ii. antipromoter effects of antioxidants: do they decrease oxidative DNA damage or cell proliferation responsible for tumour promotion?
   iii. induction of carcinogenesis resistance by antioxidant metabolites: what molecular mechanisms are suggested from cellular mechanistic studies?
   iv. antinitrosating effects of antioxidants: do antioxidants also prevent carcinogenicity associated with inflammation?

II. antioxidant increased carcinogenesis and prooxidant activity: are there potential dangers in antioxidant therapy?

III. antioxidants and chemotherapy: can antioxidants protect normal tissue from chemotherapy?

I. MECHANISMS OF CHEMOPREVENTION BY ANTIOXIDANTS

i. Mechanisms Of Antiinitiator Effects

1) Polycyclic aromatic hydrocarbon initiators and phenolic or flavonoid antioxidants. Synthetic phenolic antioxidants such as *butylated hydroxyanisole* (BHA) inhibit the microsomal mixed-function oxidase (phase 1 enzymes) catalysed hydroxylation of benzo(a)pyrene to mutagenic products[18]. This was attributed to the ability of these antioxidants at discharging the activated oxygen complex of cytochrome P450s[19]. This mechanism could explain the prevention of carcinogenesis if the initiator is administered within a few hours of the antioxidant. However feeding these antioxidants for several days, also induced the phase 2 detoxication enzymes epoxide hydrolase and GSH S transferase. Benzo(a)pyrene hydroxylase was still partly inhibited at this time but was induced by 2,6-di-butylphenol[19,20]. The ability of antioxidants to prevent benzo(a)pyrene carcinogenesis, if given several days before benzo(a)pyrene correlates best with the induction of phase 2 detoxication enzymes which could be enhanced or offset by effects on cytochrome P450 levels.

The phase 2 enzymes induced in vivo by most antioxidants include the following:

1) Glutathione transferases which conjugate mostly hydrophobic electrophiles with GSH.

2) NAD(P)H: quinone reductase (DT-diaphorase) which promote obligatory two-electron reduction of quinones to hydroquinones. The latter can then be detoxified by glucuronidation. Quinone reductase therefore competes with reductases which catalyse the one-electron reduction of quinones and thereby cause oxygen activation as a result of futile redox cycling.

3) UDP glucuronyltransferases which conjugate xenobiotics with glucuronic acid, thus facilitating their excretion.

4) Epoxide hydrolase which inactivates epoxides and arene oxides by hydration to diols.

Because BHA, t butylhydroquinone or catechol induce phase 2 enzymes but not the phase 1 enzyme aryl hydrocarbon hydroxylase (AHH), antioxidants were originally classified as monofunctional inducers. This distinguished them from bifunctional inducers (eg. polycyclic aromatic hydrocarbons, ß-naphthoflavone and azo dyes) which induced both phase 1 and phase 2 enzymes[14-16,21]. The antioxidants thiocarbamates, 1,2-dithiole-3-thione, benzyl isothiocyanate and sulforaphane, (a newly discovered isothiocyanate) were also found to be monofunctional inducers[22,23]. However the antioxidants diallyl sulfide[24], indole-3-carbinol[25] or butylated hydroxytoluene[25] were found to be bifunctional inducers. The induction of AHH probably partly reflects binding to the Ah receptor of large planar aromatics. The ligand Ah receptor complex is then transported to the nucleus where it binds and activates "xenobiotic responsive elements" in the upstream regions of the cytochrome P450 1A1/2 genes. The reaction is specific and the transcription of genes for other phase 1 enzymes is not induced. By contrast the antioxidants phenethyl isothiocyanate have recently been shown to markedly induce P450 2B1 whereas 1A1/2 was down regulated[27] Other bifunctional inducers which induce P450 2B1 and phase 2 enzymes are phenobarbital and the antioxidants ethoxyquin and oltipraz[28].

Could the use of antioxidants (eg. flavonoids in cosmetic products, toiletries or ointments) diminish the risk of melanoma or carcinogenesis at other sites? The *flavonoids quercetin* and *myricetin* inhibit skin tumor-initiation activities of various polycyclic aromatic hydrocarbons in mice[29,30] and the formation of mammary tumors in rats[31]. Flavonoid antioxidants also inhibit microsomal mixed function oxidase activity in vitro[32-34]. *Catechin* forms a modified type II spectra, whereas quercetin forms a type I spectra with cytochrome P450 in vitro[35]. Catechin may therefore act as a Fe(III) ligand by stabilising the low-spin form of P450 and preventing its reduction[35]. However other investigators found that catechin formed a reversed type I spectra suggesting displacement of an endogenous substrate[38] but the inhibition of P450 reductase by catechin may be a better explanation of the inhibition of hepatocyte catalysed aromatic amine mutagenicity[36] and the metabolic activation of benzo(a)pyrene in hamster embryo cell cultures[29]. However there is no evidence yet that flavonoids inhibit P450 or P450 reductase in vivo. Feeding quercetin induces P450 1A 1/2 in intestine but not liver[22]. It is not known if phase 2 enzymes are induced by quercetin or other flavonoids.

Plant phenolics eg. *ellagic acid, caffeic acid, tannic acid* or green tea *polyphenols* when applied topically to mice also caused a substantial inhibition of epidermal metabolism and DNA binding by polycyclic aromatic hydrocarbons[13,37,38,29]. Feeding tannic acid for 45 days markedly inhibits P450 IA1/2 (lung, forestomach). Dietary ellagic acid for 3 weeks also decreases total P450 levels (liver, esophagus). However ellagic acid and tannic acid also rapidly destroy the mutagenicity of benzo(a)pyrene diol epoxide, by forming an inactive covalent adduct[39]. The diol epoxide is believed to be the ultimate carcinogenic metabolite formed from benzo(a)pyrene by the action of P450 1A1 and epoxide hydrase. *Ellagic acid* also inhibits most constitutive P450 isoenzymes in vitro[40]. Feeding ellagic acid in the drinking water for 16 weeks inhibits P450 IA1/2 and induces GSH S transferase in the liver[41]. Similar results were obtained following the oral administration of green tea phenols in the drinking water for 4 weekss[13]. It is not known if other plant phenolics induce phase II enzymes or inhibit or downregulate cytochrome P450. *Nordihydroguaiaretic acid* and *propyl gallate* form a type 1 binding difference spectrum with P450 in vitro[42,43], whereas *epicatechin* and *epigallocatechin* form a modified type II binding spectra indicating the formation of a ligand complex with P450 heme iron[12].

2) Nitrosamine Initiators And Antioxidants Of The Allium And Brassica Family. Tobacco specific nitrosamines are likely etiological factors in tobacco related human cancers. *Diallyl sulfide* (a component of garlic oil)[44] or *catechin* (a flavonoid)[45] inhibit lung tumorigenesis and the metabolic activation and DNA damage induced by the potent tobacco specific carcinogen 4-(methylnitrosamino)-1-(3 pyridyl)-1-butanone (NNK). *Ellagic acid* however was much less effective[46] but did inhibit the oesophageal microsomal metabolism of the oesophageal carcinogen methylbenzylnitrosamine[47]. Several P450 isozymes, including 1A2, may be involved in the activation of NNK whereas 2E1 is involved in the activation of methylbenzylnitrosamine.

The consumption of allium vegetables, including garlic and onion seems to be protective against stomach cancer in a region of China[48]. *Diallyl sulfide*, an antioxidant derived from garlic, inhibits nuclear aberration and development of tumors induced by 1,2 dimethylhydrazine in the colon and by N-nitrosomethylbenzylamine in the esophagus[49]. The activating P450 is believed to be 2El which is inactivated by diallyl sulfide via suicide inhibition by the product diallyl sulphone. The latter may mediate the transcriptional activation of 2B1/2 genes which occurs within a few hours of diallyl sulfide administration[24,49]. Other P450's induced include 3A1/2 and 1A2. Phase 2 enzymes such as glutathione S transferase, UDP glucuronyltransferase, epoxide hydrolase were also induced[24].

Anticancer agents found or formed from cruciferous vegetables, such as broccoli, brussel sprouts, cabbage and cauliflower that have been studied include indoles, 1,2-dithiole-3-thiones and phenethyl isothiocyanate. Consumption of cruciferous vegetables induced the activity of both phase 1 and phase 2 drug detoxifying enzymes in animals and humans. This is associated with a decreased risk of cancer of the gastro-intestinal tract in humans. Rat P450 1A1/2, 2B1/2 and 2E1 were induced in vivo by broccoli[50] and 1A2, 2B1, GS transferases and DT diaphorase were induced by cooked brussel sprouts[51]. The above agents also decreased carcinogenesis in rats induced by aflatoxin, dimethylbenz(a)anthracene or NNK.

The decreased N-nitrosodimethylamine induced methylation of hepatic DNA by *phenethyl isothiocyanate* was attributed to suppression of P450 2E1 and the induction of 2B1 and the phase 2 enzymes glutathione S transferase[52,53].

Decreased aflatoxin induced carcinogenesis by indoles such as *indolecarbinol* has been attributed to induction of P450 1A1[53]. Indolecarbinol also inhibits diethylnitrosamine induced hepatocarcinogenesis in male rats[54] but it is not known if indole carbinol inactivates 2E1. Cytochrome P450 1A1/2, 2B1, 3A and phase II enzymes were however induced[55].

Dithiolethiones have also been shown to prevent benzo(a)pyrene (1A1) or dimethylnitrosamine (2E1) or aflatoxin B_1 (3A4 in humans) induced pulmonary, forestomach or hepatocarcinogenesis respectively[56,28]. The terms in parenthesis refer to the P450 isoenzymes responsible for activation. The cytoprotective mechanism of dithiolethiones probably involves the induction of phase 2 enzymes such as DT diaphorase, epoxide hydrase, GS transferases as well as an elevation of GSH[57]. Cytochrome P450 levels were little affected. P450 isoenzyme composition was not studied but aflatoxin B_1 activation was inhibited.[28] This antioxidant therefore seems to be promising and is in phase 1 clinical trials.

In order to predict from rodent chemoprotective studies which antioxidants to use in clinical trials as potential anticarcinogenic protective agents it may be wiser to choose those that induce phase II enzymes and increase GSH levels rather than rely on phase I modulators. This is because different isoforms of cytochrome P450 can simultaneously serve to metabolically inactivate a procarcinogen while also participating in its metabolic activation.

3) Radicals As Mediators Of Tumor Initiation. Carcinogens may also undergo metabolic activation or mixed function oxidase independent pathways particularly in extrahepatic tissues. These pathways involve enzymes that carry out a one electron oxidation of arylamine, heterocyclic amines and polycyclic hydrocarbons to free radicals that form mutagenic products[58,59]. These enzymes include peroxidases, prostaglandin H synthase, lipoxygenase. Lipid peroxidation can also activate procarcinogens[60]. Tumors are also frequently observed in tissues that have relatively high prostaglandin synthetase activity. Non-steroidal anti-inflammatory drugs such as indomethacin or aspirin inhibit the cyclooxygenase activity of prostaglandin synthetase and may inhibit arylamine induced bladder carcinogenesis[61]. Peroxidase-catalyzed DNA adduct formation with benzidine or 2 naphthylamine or aminofluorene were formed in the bladder urothelium after treatment of dogs with radiolabeled carcinogens[62,63]. Reactive metabolites formed in the liver and transported to the bladder may also be activated. The mutagenic activation of benzidine by prostaglandin synthetase may require prior N-acetylation[64]. Many studies have shown that antioxidants inhibit lipid peroxidation <u>in vivo</u>[65], as well as lipoxygenase or prostaglandin synthase <u>in vitro</u>[66]. Antioxidants could also react directly with the mutagenic polycyclic aromatic hydrocarbon cation radical[12]. Carcinogenesis initiated by lipid peroxidation, prostaglandin synthase or lipoxygenase should readily be prevented by antioxidants.

The enzyme myeloperoxidase present in neutrophils and monocytes becomes activated during inflammation when these cells invade various tissues eg. the lungs of smokers. This peroxidase in activated leukocytes activates arylamine toxins and procarcinogens[67,68] and have been suggested to cause bone marrow toxicity during hypersensitivity adverse drug reactions resulting from arylamine drug therapy of susceptible individuals[69]. Antioxidants can prevent this activation by acting as peroxidase donors, trapping the reactive metabolites formed or by inhibiting the superoxide generating oxidase eg. by scavenging superoxide or inhibiting the oxidase[67,68,70].

ii. Antipromoter Effects Of Antioxidants

The ability of antioxidants to increase metabolic detoxication and decrease the activation of carcinogens does not explain why antioxidants are often still protective when administered after the carcinogen has been metabolised. Because these antioxidants act on the promotion stage of carcinogenesis they should be regarded as antipromoters. Evidence that reactive oxygen species are involved in the tumor promotion stage of carcinogenesis largely originated from studies of mouse skin carcinogenesis. Thus tetradecanoylphorbol acetate (TPA), organic hydroperoxides and H_2O_2 are known to be promoters and progressors of mouse skin tumours but not initiators of carcinogenesis[71]. Oxidative DNA damage was found in mouse skin treated with tumour promoters and the superoxide dismutase mimic Cu(II) diisopropyl salicylate prevented carcinogenesis[72]. Reactive oxygen species or lipid peroxidation systems directly cause oxidative DNA damage. Some so called nongenotoxic carcinogens that seem to act by the epigenetic theory of carcinogenesis have now been shown to cause oxidative DNA damage indirectly. Oxidative DNA damage preceded rat hepatocarcinogenesis induced by peroxisome proliferators[73] or a choline deficient diet[74]. The oxidants $KBrO_3$ or ferric nitrilo-triacetate also induced oxidative DNA damage before kidney carcinogenesis ensued in rats[75,76].

Oxidative DNA damage includes strand breaks and oxidative modification of DNA bases eg. 8OH deoxyguanosine formation. The latter can serve as a mispairing lesion (G-T and A-T substitution) during cellular DNA replication but are usually repaired by DNA glycosylases and/or endonucleases[77]. Oxidative DNA damage by oxygen radicals or nitrogen oxides generated by chronic inflammation also results in deamination of cytosine and guanine and could cause mutagenesis of the key tumour suppressor gene p53[17]. One consequence would be the removal of the G_1, checkpoint thereby shortening the time for DNA repair, preventing apoptosis and causing radiation resistance in some cell types. The resulting increased frequency of mutations and cell proliferation could increase the clonal expansion of preneoplastic and neoplastic cells. This mechanism would explain why rats that inhale particulate materials (that do not act directly on DNA) and cause inflammation have a high incidence of lung cancer[17].

Antioxidants could prevent tumor promotion by preventing oxidative DNA damage or by preventing the various steps that lead to increased cell replication. Butylated hydroxytoluene, butylated hydroxyanisole, nordihydroguaiaretic acid and diallyl sulfide inhibited the transformation of mouse skin cells promoted by phorbol myristate acetate[78] or benzoyl peroxide[79]. Topically applied ascorbate, curcumin, chlorogenic acid, caffeic acid and ferulic acid also prevented the promotion of skin tumors and the induction of epidermal ornithine decarboxylase by tetradecanoylphorbol acetate[80]. The antioxidants ethoxyquin and BHA prevented hepatic tumorigenesis in rats fed the peroxisome proliferator ciprofibrate[81]. However the antioxidants BHA, diphenylphenylenediamine and trolox C did not affect liver necrosis and nuclear lipid peroxidation induced by choline deficiency although the antioxidant N methoxy phenylacetyl-dehydroalanine and the spin trap PBN prevented nuclear lipid peroxidation[82]. Dietary vitamin E prevented ferric nitrilotriacetate induced nephrotoxicity in rats[83]. The dietary administration of plant phenolics ie. caffeic, ellagic, chlorogenic and ferulic acids also prevented tongue carcinogenesis induced by 4 nitroquinoline-1-oxide[84] which also causes oxidative DNA damage[85]. Polyphenolic fractions isolated from green tea also prevented tetradecanoyl-phorbol acetate tumor promotion in 7,12 dimethylbenzanthracene initiated mouse skin[13]. However allyl disulfide was ineffective as an antipromoter and indole 3 carbinol and retinoids acted as promoters[86]. It is not known if dithiolethiones are antipromoters.

It is generally believed that carcinogenic oxidants such as reactive oxygen species cause tumor promotion by acting as a stimulus for increased cell replication and DNA synthesis[87]. Redox mechanisms suggested for the regulation of growth promotion (Fig. 1) are as follows:

(1) oxidants trigger the Ca^{2+} dependent epigenetic pathways used by polypeptide growth factors and hormones to trigger cell replication. Protooncogene activation may occur as a result of protein phosphorylation by Ca^{2+} dependent protein tyrosine kinase. Protooncogene activation may also occur as a result of the repair of endonucleolytic DNA breakage (caused by Ca^{2+} dependent endonucleases) by histone poly ADP ribosylation. Peroxisome proliferators may also increase cytosolic Ca^{2+} by accumulating in mitochondrial membranes and uncoupling oxidative phosphorylation[88].

(2) oxidation of DNA activates protooncogenes or inactivates tumour suppressor genes. Oxidant induced DNA strand breaks are believed to be repaired by endonucleolytic endonucleases. Repair of the endonucleolytic DNA breakage involves the poly ADP ribosylation of chromosomal proteins which could activate protoonocogenes[87]. The oxidative deamination of cytosine and guanine may contribute to the mutagenesis of the p53 tumour suppressor gene[17].

(3) Oxidants can inactivate protein tyrosine phosphatase, whose catalytic activity is dependent on a specific cysteine residue. This could result in protooncogene activation by increasing the phosphorylation of proteins involved in signal transduction pathways[89].

(4) Oxidants (eg. H_2O_2) may oxidatively activate directly the transcription factors (eg. NF-kB or OxyR proteins) that activate protooncogenes. This could occur if transcription factor sulfhydryls prevent DNA binding and represents a post-translational control mechanism.

At the molecular level the above mechanisms of carcinogenesis are proposed to arise by three different routes. a) Genotoxic carcinogenesis is initiated by DNA base mutations of the gene which results in the formation of different gene products. b) In epigenetic carcinogenesis the gene regulatory regions are modified by reactive oxygen species so that transcription is prevented or promoted. c) In paragenetic carcinogenesis the regulatory or transcription factors are modified so that transcription is prevented or promoted. The carcinogenicity of oxidants therefore seems to result from their ability to increase cell replication and DNA synthesis. Indeed cell growth could be promoted by cellular oxidative stress which results from an increase in prooxidant aerobic metabolism resulting in the production of reactive oxygen species[87]. The promotion of tumorigenesis by oxidative stress would be offset by the antioxidant capacity of the cell.

The prevention of nitrosoguanidine-induced cellular transformation of human fibroblasts by quercetin has also been attributed to inhibition of poly (ADP-ribose) polymerase[90] and to a lesser extent the tyrosine-specific protein kinase[91]. The serine/threonine specific protein kinase C is directly activated by tetradecanoylphorbol acetate due to its diacylglycerol structure which in turn phosphorylates regulatory proteins that affect cell proliferation. Quercetin and the chemopreventive agent tamoxifen inhibit protein kinase C[11].

The arachidonic acid cascade with lipoxygenase and cyclooxygenase also generates active oxygen species and plays a role in tumor promotion. Epidermal ornithine decarboxylase (ODC) induction caused by tetradecanoylphorbol acetate seems to be closely related to its tumor-promoting action in mouse skin carcinogenesis. Catecholic and flavonoid antioxidants but not catechin or butylated hydroxytoluene prevent ODC

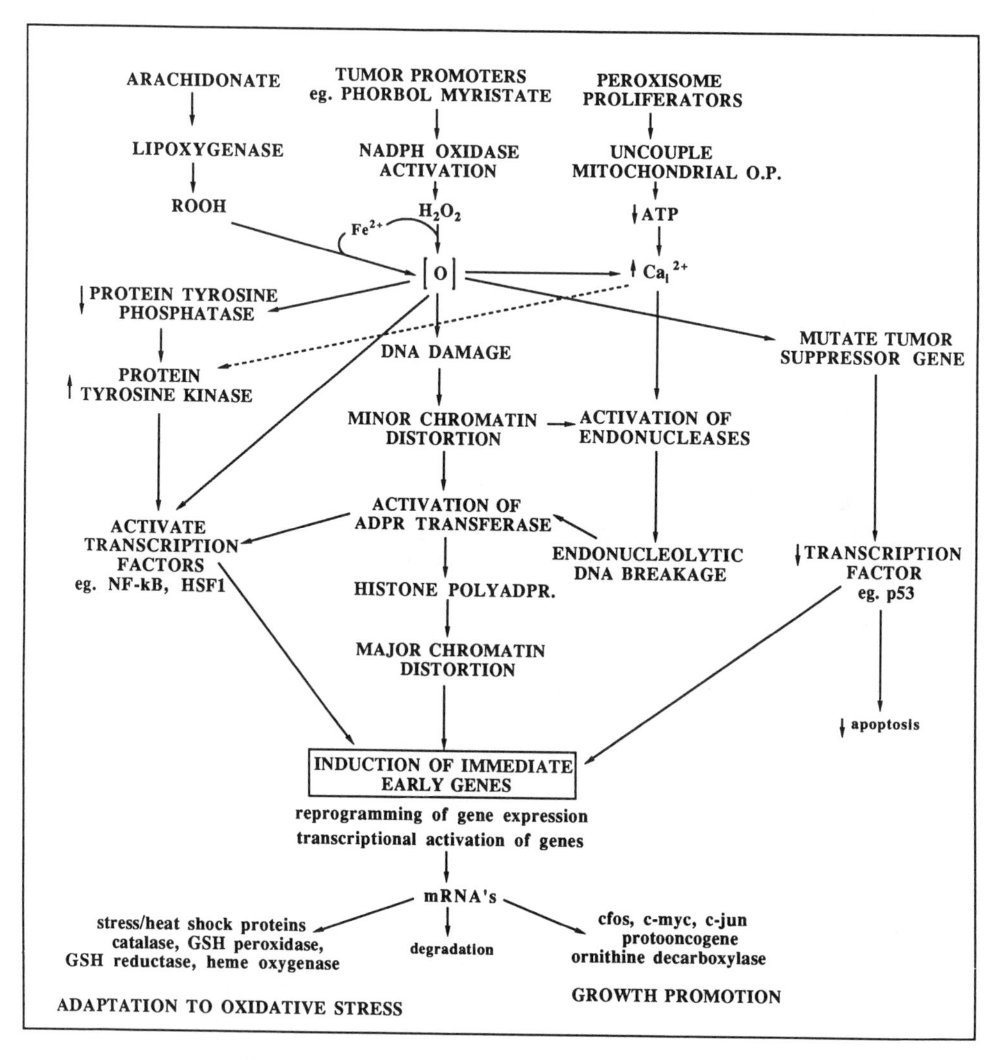

Fig. 1. Suggested Molecular Tumor Promotion Mechanisms.

induction and their antipromoter activity has been attributed to their action as lipoxygenase inhibitors[76].

Prostaglandin synthetase inhibitors such as indomethacin, sulindac and aspirin reduces colon carcinogenesis in rats or humans[92,93] whereas ibuprofen reduces bladder carcinogenesis[13]. This suggests that prostaglandins can play an oxidant role in tumor promotion and progression.

iii. Resistance To Carcinogenesis Induced By Antioxidant Metabolites

Cellular studies investigating the anticarcinogenic mechanism of many antioxidants seems to indicate that antioxidants induce phase 2 detoxification enzymes by enhanced transcription. To study the molecular events that lead to glutathione S-transferase gene activation in response to xenobiotics, Pickett et al. isolated and characterised a gene encoding a rat glutathione S-transferase Ya subunit[94]. Analysis of the 5'-flanking region of this gene by transfection experiments demonstrated at least two cis-acting regulatory elements in the upstream region of the Ya gene. The "xenobiotic response element XRE" contains a core sequence (5'-TNGCGTG-3') that is also found in multiple copies in the 5'-flanking region of the cytochrome P-450 1A1 gene. Planar aromatic compounds (eg. ß-naphthoflavone and 3-methylcholanthrene) activate transcription via XRE. In addition, H_2O_2 or phenolic antioxidants such as tert-butylhydroquinone (possibly by generating H_2O_2 or reactive oxygen species formed by redox cycling) activate transcription through the "antioxidant response element ARE" which contains a core sequence (5'-GTGACAAAGC-3')[94]. The latter could therefore mediate an inducible response to metabolites of the planar aromatic compounds formed by the P-450 1A1 gene. Talalay et al. however believe that a similar but not identical regulatory element identified in the 5'-flanking regions of the mouse gene responds to Michael reaction acceptors that are G S transferase substrates. They described the regulatory element as the "electrophile-responsive element EpRE although it was later found to respond to H_2O_2[16,95]. Indeed H_2O_2 or organic hydroperoxides readily induced phase 2 detoxification enzymes[94,95] and functional ARE sequences have also been demonstrated in the human NAD(P)H quinone reductase genes[96]. The phorbol ester tetradecanoylphorbol acetate which induces the AP-1 family of transcription factors could also interact with the ARE as a result of reactive oxygen species formation. Molecular mechanisms suggested for the transcriptional regulation of AHH and phase 2 enzymes are shown in Fig. 2.

a) Adaptive Response To Electrophiles. The mechanism proposed involves oxidative metabolism of the antioxidant to electrophiles which activate the "electrophile-responsive element" of the regulatory regions of phase 2 enzyme genes and results in enhanced transcription of these enzymes. Induction of these enzymes has been suggested to be useful for screening for potential natural anticarcinogenic agents and recently this method led to the discovery of a new highly potent isothiocyanate, sulforaphane, that was isolated from broccoli[23]. Phase 1 enzymes are also induced by many antioxidants and are therefore regarded as bifunctional in contrast to flavones or polycyclic aromatic hydrocarbons which are monofunctional and only induce phase 1 enzymes[21]. Electrophiles formed by phase 1 enzymes responsible for the induction of phase 2 enzymes include quinones, quinoneimines, diimines which are oxidation metabolites of the antioxidants butylated hydroxyanisole or various catechols, hydroquinones and 1,4 phenylenediamines. Evidence that electrophilic metabolites are responsible is that 1,3 diphenols and 1,3 phenylenediamines do not induce phase 2 enzymes and they cannot undergo a two electron oxidation to an electrophile. Furthermore induction also readily occurs with olefins conjugated to electron-

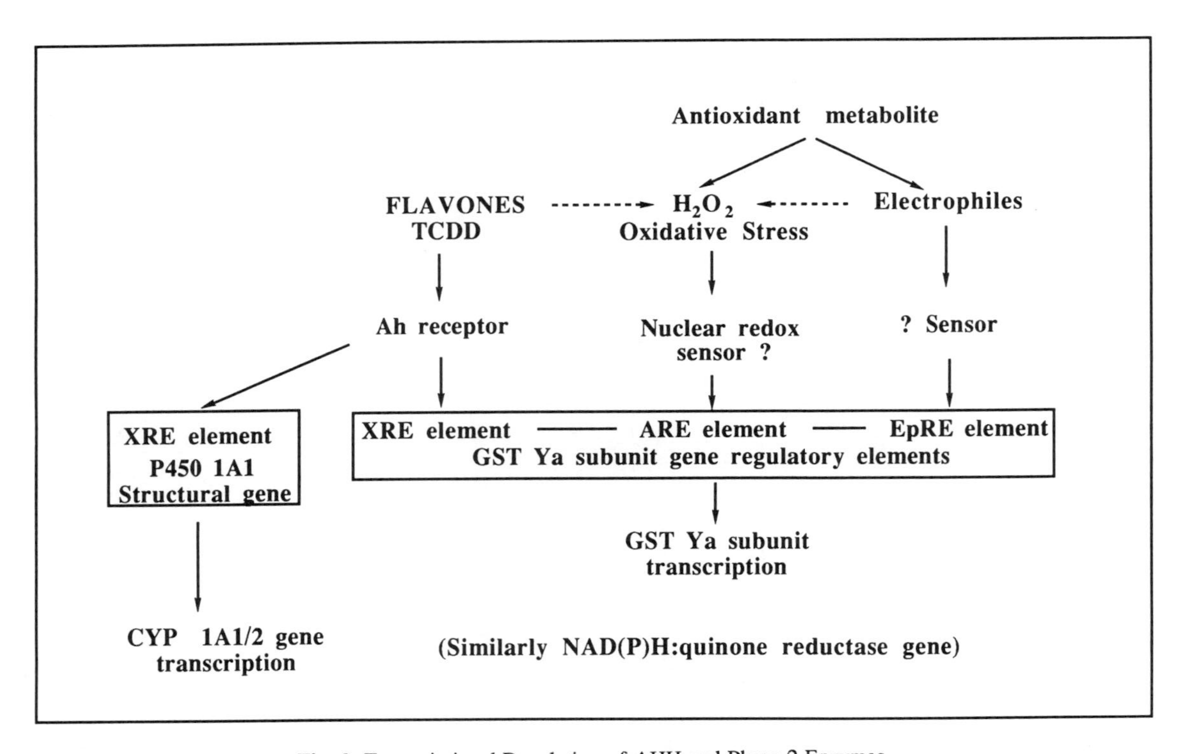

Fig. 2. Transcriptional Regulation of AHH and Phase 2 Enzymes.

withdrawing functions i.e. Michael reaction acceptors[97]. Dithiolethione analogs also induce NAD(P)H quinone reductase even though it is unlikely that this reductase is involved in the metabolic detoxication of the dithiolethione[98]. Dithiolethiones were the most effective[99] agents at inducing GS transferases particularly class α-transcripts in cultured human hepatocytes. These isozymes can detoxify aflatoxin B_1 epoxide. Presumably dithiolthiones oxidise a sulfhydryl group of the redox sensor protein involved in the transcriptional activation of phase 2 enzymes through the ARE or EpRE elements.

b) Adaptive Response To Oxidative Stress. The anticarcinogenic mechanism of other antioxidants may result from their ability to induce "antioxidant enzymes". The mechanism proposed involves oxidative metabolism of the antioxidant to radicals which reductively activate oxygen. These radicals could be formed from electrophiles by reductases and thus involve futile redox cycling resulting in oxidative stress.

Adaptive responses to oxidative stress, hyperthermia, heavy metals and sodium arsenite have been identified in various cells which could prevent an increased expression of proto-oncogenes resulting from redox signalling induced by oxidative stress (Fig. 1). Normally superoxide oxygen radicals are rapidly removed by superoxide dismutase working in conjunction with catalase or glutathione peroxidase[100]. Exposure of cultured mammalian cells to hydrogen peroxide results in the selective and reversible expression of specific stress polypeptides while lowering the production of most other cellular proteins[101]. The resulting resistance of the cloned cell lines to hydrogen peroxide has been attributed to large increases in catalase[102,103], glutathione peroxidase[104-106] and glutathione[107]. The latter increase has been attributed to induction of the cystine transporter[108]. The 32KDa stress protein induced may be heme oxygenase[109] which catalyses the formation of the physiological antioxidant bilirubin. The function of the stress proteins are unknown but may play a "chaperone" role in stabilising the tertiary structure of newly synthesised proteins so that they are more resistant to oxidative stress. Resistance may therefore occur via induction of an increased detoxification system. However acquired tumor resistance to reactive oxygen species then becomes a significant problem in the treatment of cancer patients with oxidant antitumour drugs[110,111].

The chronic oral administration of green tea polyphenols to mice for 4 weeks resulted in moderate to significant enhancement in glutathione peroxidase, catalase, NADPH-quinone oxidoreductase and glutathione S transferase activities in small bowel, lung and liver[13].

iv. Antinitrosating Effects Of Antioxidants

An attractive hypothesis for cancer occurrence involves the endogeneous formation of activated N-nitroso compounds in the stomach of man. Prevention of nitroso carcinogen formation by dietary antioxidants could therefore occur in the alimentary tract. Thus the antioxidants vitamin C, vitamin E, butylated hydroxyanisole, propyl gallate can prevent the methylation of liver DNA or hepatotoxicity following the oral administration of diethylamine plus nitrite that is believed to result in diethylnitrosamine formation[112,113]. Vitamin C, vitamin E, chlorogenic acid, ferulic acid or caffeic acid also prevented normal urinary excretion of nitrosoproline after its formation in the stomach from proline and nitrite[114-116].

The antioxidant mechanisms probably involve reduction or trapping of the nitrosating agent as follows:

N_2O_3 + catechol   NO + o-quinone

N_2O_3 + phenol   nitrosophenol.

It has been suggested that the types of p53 mutations found in human cancers could be explained by the nitrosative deamination of bases by superoxide and/or nitrogen oxide radicals[117]. Furthermore carcinogenicity has also been associated with inflammatory processes. Inflammation attracts many types of cells to the affected tissue or organ. These cells form oxygen and nitrogen oxide radicals to help destroy tumor cells or invading microbes. In macrophages and certain other leukocytes, nitric oxide is formed by an inducible nitric oxide synthase activated by endotoxins or cytokines produced as a result of inflammation. Peroxynitrite, nitric oxide and hydrogen peroxide are also released during the respiratory burst of neutrophils activated by tumor promoters.

II. ANTIOXIDANT INCREASED CARCINOGENESIS AND PROOXIDANT ACTIVITY

a) Increased Initiation By Antioxidants

Some flavonoids or phenolic antioxidants[20] are potent inducers of cytochrome P450 dependent monooxygenases (P450IA1/2) which are responsible for the activation of many chemical carcinogens. Thus indole 3 carbinol, diallyl sulfide and phenethyl isothiocyanate whilst inducing phase II drug metabolising enzymes also induces P450IA1/2[55]. Indole 3 carbinol, an autolysis product of dietary glucobrassicin, is activated by the acid environment of the stomach to form condensation products which have planar aromatic structures. This planarity enables these products to bind to the Ah receptor albeit with a lower affinity than tetrachlorodibenzodioxin, a carcinogenesis promoter and antiinitiator[25]. This ligand-receptor complex activates the transcription of cytochrome-P450 IA1 genes[21]. Interestingly induction of IA1 by indole carbinol could be beneficial as the induction of estradiol hydroxylase in humans with indole carbinol[119] could prevent estrogen responsive breast cancer by increasing levels of anti-estrogenic catechol estrogens such as 2 hydroxyestrone. Indolecarbinol in mice also induces estadiol hydroxylase[120] and prevents the induction of spontaneous mammary tumors[121].

Furthermore using antioxidants as cytochrome P450 inhibitors to prevent carcinogen metabolic activation in the liver could result in a decreased first-pass clearance of many xenobiotics by the liver. This in turn could result in greater delivery of the carcinogen to extrahepatic tissues and enhance extrahepatic carcinogenesis.

b) Increased Promotion By Antioxidants

Recently diallyl sulfide has been found to promote N-diethylnitrosamine induced hepatocarcinogenesis[122] and indole carbinol has been found to promote aflatoxin B_1 hepatocarcinogenesis[123].

Butylated hydroxytoluene, butylated hydroxyanisole and catecholic antioxidants can also promote carcinogenesis of the stomach, esophagus, bladder, thyroid even though they inhibit carcinogenesis of the skin, mammary gland and ear duct[124]. The molecular mechanisms involved are not known but polyphenolic antioxidants or metabolites of phenolic antioxidants have been shown to autooxidize or undergo redox cycling to generate reactive oxygen species. Because of this these antioxidants could act as prooxidants and enhance tumor promotion. Ellagic, gallic, ferulic and chlorogenic acids are antioxidants of potential interest because they did not promote carcinogenesis.

III. ANTIOXIDANTS AND CHEMOTHERAPY

Hepatocellular carcinoma arises from a rare hepatocyte transformed in a way that is characterized by a decrease in the expression and activity of phase 1 enzymes and an increase in the activity of phase 2 enzymes. The transformed hepatocyte is therefore more resistant to chemical carcinogens[125]. Whilst antioxidants are highly effective at preventing chemical carcinogenesis, it is not clear whether they would be effective in the treatment of cancer once the tumor has formed. Tumor cells generally undergo lipid peroxidation less readily than normal cells even in the presence of prooxidant agents, such as ascorbate/$FeSO_4$[126]. This is believed to result from their decreased polyunsaturated fatty acid composition and lower levels of the prooxidant enzymes P450, P450 reductase and xanthine oxidase[127].

By contrast the enzymic antioxidant enzyme system superoxide dismutase, catalase and GSH peroxidase of tumour cells is often severely impaired[128]. Tumor growth and metastases are therefore inhibited in vivo by w-3 fatty acid diets (eg. fish oil containing diets). Cis unsaturated fatty acids are readily taken up by tumor cells, oxidised and cause tumor cell toxicity or prevent tumor cell proliferation without affecting normal cells[129]. Hepatoma cells also have much less alcohol dehydrogenase and aldehyde reductase activity than that of normal cells so that they are much more susceptible to lipid peroxidation aldehydic decomposition products[129].

The α-tocopherol content of hepatoma cells is higher than hepatocytes but is more readily depleted during iron induced lipid peroxidation than occurs with normal hepatocytes[130]. The poorer maintenance of α-tocopherol levels in tumor cells could indicate slower redox cycling and/or lower ascorbate levels. In view of the above it would be difficult to predict the effect of dietary antioxidants once the tumor has formed. Dietary antioxidants could increase tumor cell resistance to lipoxygenase containing cells or oxygen activating cells of the immune system particularly if the tumor cell selectively concentrates the antioxidant. On the other hand flavonoids eg. quercetin have been found to prevent the growth of various malignant human cells[131] possibly as a result of ATP depletion and unrelated to its antioxidant activity.

Radiotherapy and/or chemotherapy targets tumor cell DNA and prevents DNA replication and tumor cell proliferation. The use of radiotherapy and chemotherapy however is dose limited. Alkylating or quinone anticancer agents above this dose can cause lipid peroxidation toxicity[132] or oxidative stress toxicity[133] respectively in normal cells which could be life threatening. Dietary antioxidants, prior to and at the time of chemotherapy could be useful in preventing chemotherapy toxicity as antioxidants prevent alkylating agent toxicity in normal cells[132].

CONCLUSIONS

Possible targets for blocking the initiation of carcinogenesis with antioxidants include inhibiting or inducing cytochrome P450 isoenzymes; induction of phase II enzymes; scavenging electrophiles, reactive oxygen or nitrosating species; and inhibiting arachidonic acid metabolism, lipoxygenase, prostaglandin synthetase, myeloperoxidase. Possible targets for suppressing the promotion or progression of carcinogenesis with antioxidants include inhibiting prooxidant enzymes; inhibiting arachidonic acid metabolism; inhibiting oxidative DNA damage; inhibiting protein kinases and poly(ADP-ribose) polymerase and inhibiting oncogene expression. Most of these antioxidants probably prevent oxidation of the cell and thereby help maintain the cell's redox potential. Although other cytoprotective mechanisms cannot be excluded, cancer protection by agents that are not antioxidants may subsequently be shown to act by averting redox stress eg. by forming antioxidant metabolites or by modulating DNA repair or gene expression that promotes carcinogenesis. A diet rich in antioxidants particularly for groups subject to high oxidant loads such as smokers, should increase the protection of our society against cancer. Chemoprevention trials are however needed to determine which combination of antioxidants or antioxidant diets are the most beneficial. The cost of chemopreventative programs in a given population particularly if it involves a nutritional approach is far less costly than the health care costs of treating cancer.

REFERENCES

1. Jaffe, W. The influence of wheat germ oil on the production of tumors in rats by methylcholanthrene. Proc. Soc. Exp. Biol. Med. 111, 714-715 (1946).

2. Haber, S.L. and Wissler, R.W. Effect of vitamin E on carcinogenicity of methylcholanthrene. Proc. Soc. Exp. Biol. Med. 111, 774-775 (1962).

3. Graham, S. Results of case-control studies of diet and cancer in Buffalo, N. Y., Cancer Res. 43, 2409s-2413s (1983).

4. Shamberger, R.J. Relationship of selenium to cancer I inhibitory effect of selenium on carcinogenesis. J. Natl. Cancer Inst. 44, 931-936 (1970).

5. Shamberger, R.J. Increase of peroxidation in carcinogenesis. J. Natl. Cancer Inst. 48, 1491-1497 (1972).

6. Wattenberg, L.W. Inhibition of carcinogenic and toxic effect of polycyclic hydrocarbons by phenolic antioxidants and ethoxyquin. J. Natl. Cancer Inst. 48, 1425-1430 (1972).

7. Hocman, G. Prevention of Cancer: Vegetables and Plants. Comp. Biochem. Physiol. 93B, 201-212 (1989).

8. Block, G., Patterson, B. and Subar, A. Fruit, vegetables, and cancer prevention: a review of the epidemiological evidence. Cancer 18, 1-29 (1992).

9. Ames, B.N., Shigenaga, M.K. and Hagen, T.M. Oxidants, antioxidants, and the degenerative diseases of aging. Proc. Natl. Acad. Sci. 90, 7915-7922 (1993).

10. Ames, B.N. Dietary carcinogens and anticarcinogens: oxygen radicals and degenerative diseases. Science 221, 1256-1264 (1983).

11. Morse, M.A. and Stone, G,D. Cancer chemoprevention: principles and prospects. Carcinogenesis 14, 1737-1746 (1993).

12. Mukhtar, H., Katiyar, S.K. and Agarwal, R. Green tea and skin-anticarcinogenic effects. J. Inv. Dermatol. 102, 3-7 (1994).

13. Wattenberg, L.W. Chemoprevention of cancer. Cancer Res. 45, 1-8 (1985).

14. Wattenberg, L.W. Inhibition of carcinogenesis by minor anutrient constituents of the diet. Proc. Nutr. Soc. 49, 173-183 (1990).

15. Prestera, T., Holtzclaw, W.D., Zhang, Y. and Talalay, P. Chemical and molecular regulation of enzymes that detoxify carcinogens. Proc. Natl. Acad. Sci. 90, 2965-2969 (1993).

16. Miller, E.C., Miller, J.A., Brown, R.R. and MacDonald, J.C. On protective action of certain polycyclic aromatic hydrocarbons against carcinogenesis by aminoazo dyes and acetylaminofluorene. Cancer Res. 18, 469-473 (1958).

17. Harris, C.C. p53: at the crossroads of molecular carcinogenesis and risk assessment. Science 262, 1980-1982 (1993).

18. Rahimtula, A.D., Zachariah, P.K. and O'Brien, P.J. The effects of antioxidants on the metabolism and mutagenicity of benzo(a)pyrene in vitro. Biochem. J. 164, 473-475 (1977).

19. Rahimtula, A.D., Zachariah, P.K. and O'Brien, P.J. Differential effects of antioxidants on benzo(a)pyrene-3 hydroxylase activity in various tissues of rat. Br. J. Cancer 40, 105-112 (1979).

20. Rahimtula, A.D., Jernstrom, B., Dock, L. and Moldeus, P. Effects of dietary and in vitro 2-t butylhydroxyanisole and other phenols on hepatic enzyme activities in mice. Br. J. Cancer. 45, 935-944 (1982).

21. Prochaska, H.J. and Talalay, P. Regulatory mechanisms of monofunctional and bifunctional anticarcinogenic enzyme inducers in murine liver. Cancer Res. 48, 4776-4782 (1988).

22. Egner, P.A., Kensler, T.W., Prestera, T., Talalay, P., Libby, A.H. and Curphay, T.J. Regulation of phase 2 enzyme induction by oltipraz and other dithiolethiones. Carcinog. 15, 177-181 (1994).

23. Zhang, Y, Talalay, P., Cho, C.G. and Posner, G.H. A major inducer of anticarcinogenic protective enzymes from broccoli: isolation and elucidation of structure. Proc. Natl. Acad. Sci. 89, 2399-2403 (1992).

24. Haber, D., Siess, M.H., Waziers, I., Beaune, P. and Suschetet, M. Modification of hepatic drug metabolizing enzymes in rats fed naturally occurring allyl sulfides. Xenobiotica 24, 169-182 (1994).

25. Bjeldanes, L.F., Kim, J., Grose, K.R., Bartholomew, J.C. and Bradfield, C.A. Aromatic hydrocarbon responsiveness receptor agonists generated from indole-3-carbinol in vitro and in vivo: comparisons with tetrachlorobenzodioxin. Proc. Natl. Acad. Sci. 88, 9543-9547 (1991).

26. Cha, Y. and Heine, H.S. Comparative effects of dietary administration of butyl-hydroxyanisole and butylhydroxytoluene on several hepatic enzyme activities in mice and rats. Cancer Res. 42, 2609-2615 (1982).

27. Guo, Z., Smith, T.J., Wang, E., Thomas, P.E. and Yang, C.S. Effects of phenethyl isothiocyanate, a carcinogenesis inhibitor, on xenobiotic metabolising enzymes and nitrosamine metabolism in rats. Carcinog. 13, 2205-2210 (1992).

28. Putt, D.A., Kensler, T. and Hollenberg, P. Effects of three chemoprotective antioxidants, ethoxyquin, oltipraz and 1,2 dithiole-3-thione on cytochrome P-450 levels and aflatoxin ß, metabolism. FASEB J. 5, A1517 (1991).

29. Chae, Y., Ho, D.K., Cassady, J.M., Cook, V.M., Craig, B.M. and Baird, W.M. Effects of synthetic and naturally occurring flavonoids on metabolic activation of benzo(a)pyrene in hamster embryo cell cultures. Chem.-Biol. Interacns. 82, 181-193 (1992).

30. Mukhtar, H. Das, M., Khan, W., Wang, Z.Y., Bik, D.P. and Bickers, D. Exceptional activity of tannic acids among naturally occurring plant phenols in protecting against dimethylbenzanthracene, benzopyrene, 3 methylcholanthrene induced skin tumorigenesis in mice. Cancer Res. 48, 2361-2365 (1988).

31. Verma, A.K., Johnson, J.A., Gould, M.N. and Tanner, M.A. Inhibition of dimethylbenzanthracene and N-nitrosomethylurea induced rat mammary cancer by dietary flavonol quercetin. Cancer Res. 48, 5754-5761 (1988).

32. Chae, Y.H., Marcus, C.B., Ho, D.K., Cassady, J.M. and Baird, W.M. Effects of synthetic and naturally occurring flavonoids on benzopyrene metabolism by hepatic microsomes. Cancer Lett. 60, 15-24 (1991).

33. Shah, G.M. and Bhattacharya, R.K. Modulation by plant flavonoids and related phenolics of microsome catalysed adduct formation between benzo(a)pyrene and DNA. Chem.-Biol. Interacns. 59, 1-15 (1986).

34. Siess, N.H. and Vernevaut, M.F. The influence of food flavonoids on the activity of some hepatic microsomal monooxygenases in rats. Fd. Chem. Toxic. 20, 883-886 (1982).

35. Beyeler, S., Testa, B. and Perrissoud, D. Flavonoids as inhibitors of rat liver monooxygenase activities. Biochem. Pharmacol. 37, 1971-1979 (1988).

36. Steele, C.M., Lalies, M. and Ioannides, C. Inhibition of the mutagenicity of aromatic amines by the plant flavonoid(+)-catechin. Cancer Res. 45, 3573-3577 (1985).

37. Das, M., Khan, W.A., Asokan, P., Bickers, D.R. and Mukhtar, H. Inhibition of epidermal xenobiotic metabolism in mice by naturally occurring plant phenols. Cancer Res. 47, 760-766 (1987).

38. Chang, R.L., Huang, M.T., Wood, A.W., Wong, C.Q., Newmark, H.L., Yagi, H., Sayer, J.M., Jerina, D.M. and Conney, A.H. Effect of ellagic acid and hydroxylated flavonoids on the tumorigenicity of benzo(a)pyrene on mouse skin. Carcinog. 6, 1127-1133 (1985).

39. Sayer, J.M., Whalen, D.L. and Jerina, D.M. Chemical strategies for the inactivation of bay-region diol epoxides. Drug. Metab. Rev. 20, 155-182 (1989).

40. Zhang, Z., Hamilton, S.M., Stewart, C., Strothers, A. and Teel, R.W. Inhibition of liver microsomal cytochrome P450 activity and metabolism of the tobacco specific nitrosamine NNK by ellagic acid. Anticancer Res. 13, 2341-2346 (1993).

41. Das, M., Bickers, D.R. and Mukhtar, H. Effect of ellagic acid on hepatic and pulmonary xenobiotic metabolism in mice: studies on the mechanism of its anticarcinogenic action. Carcinog. 6, 1409-1413 (1985).

42. Agarwal, R., Wang, Z.Y., Bik, D.P. and Mukhtar, H. Nordihydroguaiaretic acid, an inhibitor of lipoxygenase, also inhibits cytochrome P450 mediated monooxygenase activity. Drug Metab. Disp. 19, 620-624 (1991).

43. Yang, C.S. and Strickhart, F.S. Inhibition of hepatic mixed function oxidase activity by propyl gallate. Biochem. Pharmacol. 23, 3129-3138 (1974).

44. Hong, J., Wang, Z., Smith, T.J., Zhou, S., Shi, S., Pan, J. and Yang, C.S. Inhibitory effects of diallyl sulfide on the metabolism and tumorigenicity of the tobacco specific carcinogen 4-(methylnitrosamino)-1-(3 pyridyl)-1-butanone in A/J mouse lung. Carcinogenesis 13, 901-904 (1992).

45. Liu, L. and Castonguay, A. Inhibition of the metabolism and genotoxicity of 4 methyl-nitrosamino pyridyl butanone in rat hepatocytes by catechin. Carcinog. 12, 1203-1208 (1991).

46. Castonguay, A., Allaire, L., Charest, M., Rossignol, G., and Boutet, M. Metabolism of methylnitrosamino-1-(3 pyridyl)-1 butanone by hamster respiratory tissues cultured with ellagic acid. Cancer Lett., 46, 93 (1989); for errata, see 47, 161.

47. Barch, D.H. and Fox, C.C. Dietary ellagic acid reduces the esophageal microsomal metabolism of methylbenzylnitrosamine. Cancer Lett. 44, 39-44 (1989).

48. You, W., Blot, W., Chang, A.E., Yang, Z.T., Qi, A., Henderson, B.E., Fraumeni, J.F., Wang, T. Allium vegetables and reduced risk of stomach cancer. JNCI 81, 162-164 (1989).

49. Pan, J., Hong, J.Y., Ma, B.L., Ning, S.M., Paranawithana, S.R. and Yang, C.S. Transcriptional activation of cytochrome P450 2B1/2 genes in rat liver by diallyl sulfide, a compound derived from garlic. Arch. Biochem. Biophys. 302, 337-342 (1993).

50. Vang, O., Jensen, H. and Autrup, H. Induction of cytochrome P450 1A1, 1A2, 2B1, 2B2 and 2E1 by broccoli in rat liver and colon. Chem. Biol. Interacns. 78, 85-96 (1991).

51. Wortelboer, H.M., De Kruif, C.A., Iersel, A.A., Noordhoek, J., Blaauboer, B., Van Bladeren, P.J. and Falke, H.E. Effects of cooked brussel sprouts on cytochrome P450 profile and phase II enzymes in rat. Fd. Chem. Toxic. 30, 17-27 (1992).

52. Ishizaki, H., Brady, J.F., Ning, S.M. and Yang, C.S. Effect of phenethyl isothiocyanate on microsomal N-nitrosodimethylamine metabolism and other monooxygenase activities. Xenobiotica 20, 255-264 (1990).

53. Salbe, A.D., and Bjeldanes, L.F. Effect of diet on the DNA binding of aflatoxin B_1 in the rat. Carcinogenesis 10, 629-634 (1989).

54. Tanaka, T., Mori, Y., Morishita, Y., Hara, A., Ohno, T., Kojima, T. and Mori, H. Inhibitory effect of sinigrin and indolecarbinol on diethylnitrosamine-induced hepatocar-cinogenesis in male rats. Carcinogenesis 11, 1403-1406 (1990).

55. Wortelboer, H.M., Kruif, C.A., Falke, H.E., Noordhoek, J. and Blaauboer, B.J. Acid reacton products of indole carbinol and their effects on cytochrome P450 and phase II enzymes in hepatocytes. Biochem. Pharmacol. 43, 1439-1447 (1992).

56. Wattenberg, L.W. and Bueding, E. Inhibitory effects of oltipraz on carcinogenesis induced by benzo(a)pyrene and diethylnitrosamine. Carcinogenesis 7, 1379-1381 (1986).

57. Kensler, T.W., Groopman, J.D., Eaton, D.L., Curphey, T. and Roebuck, B. Potent inhibition of aflatoxin induced hepatic tumorigenesis by the monofuctional enzyme inducer 1,2 dithiole-3-thione. Carcinogenesis 13, 95-100 (1992).

58. Eling, T.E. and Curtis, J.F. Hydroperoxide-dependent oxidation. Pharmacol. Ther. 53, 261 (1993).

59. Sarkar, F.H., Radcliff, G. and Callewaert, D.M. Purified prostaglandin synthase activates aromatic amines to mutagenic derivatives. Mutn. Res. 282, 273-281 (1992).

60. Pruess-Schwartz, D., Nimesheim, A. and Marnett, L.J. Peroxyl radical and cytochrome P-450 metabolic activation of 7,8 dihydroxy 7,8 dihydrobenzo(a)pyrene in mouse skin in vitro and in vivo. Cancer Res. 49, 1732-1737 (1989).

61. Zenser, T.V. and Davis, B.B. Arachidonic acid metabolism by human urothelial cells: implications in aromatic amine-induced bladder cancer. Prostaglandins Leuk. Essent. Fatty Acids: Rev. 31: 199-207 (1988).

62. Yamazoe, Y., Zenser, T.V., Miller, D.W. and Kadlubar, F.F. Mechanism of formation and structural characterisation of DNA adducts derived from peroxidative activation of benzidine. Carcinogen. 9, 1635-1641 (1988).

63. Krauss, R.S., Angerman-Stewart, J., Dooley, K.L., Kadlubar, F.F. and Eling, T.E. The formation of 2 aminofluorene-DNA in vivo: evidence for peroxidase-mediated activation. Biochem. Toxicol. 4, 111-117 (1989).

64. Josephy, P.D., Chiu, A.I.H. and Eling, T.E. Prostaglandin H synthase dependent mutagenic activation of benzidine in a S.typhimunium Ames tester strain possesing elevated N-acetyltransferase levels Cancer Res. 49, 853-856 (1989).

65. Majid, S., Kharduja, K., Gandhi, R.K., Kapus, S. and Sharma, R.R. Influence of ellagic acid on antioxidant defense system and lipid peroxidation in mice. Biochem. Pharmacol. 42, 1441-1445 (1991).

66. Marnett, L.J., Reed, G.A. and Johnson, J.T. Prostaglandin synthetase dependent benzo(a)pyrene oxidation: products of the oxidation and inhibition of their formation by antioxidants. Biochem. biophys. Res. Comm. 79, 569-576 (1977).

67. O'Brien, P.J. Radical formation during the peroxidase catalyzed metabolism of carcinogens and xenobiotics: the reactivity of these radicals with GSH, DNA and unsaturated lipid. Free Radic. Biol. Med. 4, 169-183 (1988).

68. Tsuruta, Y., Subrahmanyan, V.V., Marshall, W. and O'Brien, P.J. Peroxidase mediated irreversible binding of aryamine carcinnogens to DNA in intact polymorphonuclear leukocytes activated by a tumor promoter. Chemico-Biol. Interacns. 53, 25-35 (1985).

69. Uetrecht, J.P. The role of leukocyte-generated reactive metabolites in the pathogenesis of idiosyncratic drug reactions. Drug Metab. Rev. 24, 299-366 (1992).

70. Cross, A.R. Inhibitors of the leukocyte superoxide generating oxidase. Free Rad. Biol. Med. 8, 71-93 (1990).

71. Chena, K.C., Cahill, D.S., Kasai, H., Nishimura, S., and Loeb, L.A. 8 hydroxydeoxy-guanosine, an abundant form of oxidative DNA damage causes G-T and A-T substitution. J. Biol. Chem. 267, 166-172 (1992).

72. Kensler, T.W., Bush, D.M. and Kozumbo, W.J. Inhibition of tumor promotion by a biomimetic superoxide dismutase. Science 221, 75-77 (1983).

73. Kasai, H., Okada, Y., Nishimura, S., Rao, M.S., and Reddy, J.K. Formation of 8-hydroxydeoxyguanosine in rat kidney DNA after intraperitoneal administration of ferric nitrilotriacetate. Carcinogenesis 11, 345-347 (1989).

74. Hinrichsen, L.I., Floyd, R.A. and Sudilovsky, O. Is 8 hydroxydeoxyguanosine a mediator of carcinogenesis by choline-devoid diet in the rat liver? Carcinogenesis 11, 1879-1881 (1990).

75. Kasai, H., Nishimura, S., Kurokawa, Y., and Hayoslii, Y. Oral administration of the renal carcinogen, potassium bromate, specifically produces 8-hydroxydeoxyguanosine in rat target organ DNA. Carcinogenesis, 8, 1959-1961 (1987).

76. Umemura, T., Sai, K., Takagi, A., Hasegawa, R. and Kurokawa, Y. Formation of 8-hydroxydeoxyguanosine in rat kidney DNA after intraperitoneal administration of ferric nitrilotriacetate. Carcinogenesis 11, 345-347 (1990).

77. Tchou, J., and Grollman, A.P. Repair of DNA containing the oxidatively damaged base, 8 oxyguanine. Mutn. Res. 299, 277-287 (1993).

78. Slaga, T.J., Fischer, S.M., Weeks, C.E., Nelson, K., Mamrack, M. and Klein-Szanto, A.J.P. Specificity and mechanism(s) of promoter inhibitors in multistage promotion. In: Carcinogenesis, Vol. 7, pp. 19-34, Hecker, E., Kuntz, W., Fusenig, N.E., Marks, F. and Thielmann, H.W. (eds) Raven Press, New York (1982).

79. Athar, M., Raza, H., Bickers, D.M. and Muktar, H. Inhibition of benzoyl peroxide mediated tumor promotion in 1,2 dimethylbenz(a)anthracene-initiated skin of Sencar mice by antioxidants, nordihydroguaiaretic acid and diallyl sulfide. J. Invest. Dermatol., 94, 162- (1990).

80. Huang, M.T., Smart, R.C., Wong, C.Q., Conney, A.H. Inhibitory effect of curcumin, chlorogenic acid, caffeic acid, ferulic acid on tumor promotion in mouse skin by tetradecanoylphorbol acetate. Cancer Res. 48, 5941-5946 (1988).

81. Rao, M.S., Lalwani, N.D., Watanabe, T.K. and Reddy, J.K. Inhibitory effects of antioxidants ethoxyquin and butylhydroxyanisole on hepatic tumorigenesis in rats fed ciprofibrate, a peroxisome proliferator. Cancer Res. 44, 1072-1076 (1984).

82. Ghoshal, A.K., Rushmore, T.H., Calderon, P., Roberfroid, M. and Farber, E. Prevention by a free radical scavenger of prooxidant effects of choline deficiency. Free Rad. Biol. Med. 8, 3-7 (1990).

83. Okada, S., Hamazaki, S., Ebina, Y., Li, J.L. and Midorikawa, O. Nephrotoxicity and its prevention by vitamin E in ferric nitrilotriacetate-promoted lipid peroxidation. Biochim. biophys. Acta 922, 28-33 (1987).

84. Tanaka, T., Kojima, T., Kawamori, T., Wang, A., Suzui, M., Okamoto, K. and Mori, H. Inhibition of 4 nitroquinoline 1 oxide induced rat tongue carcinogenesis by the naturally occurring plant phenolics caffeic, ellagic, chlorogenic and ferulic acids. Carcinogenesis 14, 1321-1325 (1993).

85. Nunoshiba, T., and Demple, B. Potent intracellular oxidative stress exerted by the carcinogen 4 nitroquinoline N oxide. Cancer Res. 53, 3250-3252 (1993).

86. Wargowich, M.J., Imada, O. and Stephens, L.C. Initiation and post initiation chemopreventive effects of diallyl sulfide in oesophageal carcinogenesis. Cancer Lett. 64, 39-42 (1992).

87. Cerutti, P.A. Oxidant stress and carcinogenesis. Eur. J. Clin. Invest. 21, 1-5 (1991).

88. Keller, B.J., Marsman, D.S.O., Popp, J.A., Thyman, R.G. Several nongenotoxic carcinogens uncouple mitochondrial oxidative phosphorylation. Biochim. biophys. Acta 1162, 237 (1992).

89. Monteiro, H.P., Ivaschenko, Y., Fischer, R. and Stern, A. Inhibition of protein tyrosine phosphatase activity by diamide is reversed by epidermal growth factor in fibroblasts. FEBS Lett. 295, 146-148 (1991).

90. Milo, G.E., Kurican, P., Kirsten, E. and Kun, E. Inhibition of carcinogen-induced cellular transformation of human fibroblasts by drugs that interact with the poly(ADP-ribose) polymerase system. FEBS Lett. 179, 332-336 (1985).

91. Yamamoto, S. and Kyoto, R. Inhibitors of the arachidonic acid cascade and their chemoprevention of skin cancer, in Cancer Chemoprevention. Ed Wattenberg, L., Lipkin, M., Boone, C.W. and Keloff, G.J. pp 141-151, CRC press (1992).

92. Marnett, L.J. Aspirin and the potential role of prostaglandins in colon cancer. Cancer Res. 52, 5575-5589 (1992).

93. Reddy, B.S. Inhibitors of the arachidonic acid cascade and their chemoprevention of colon carcinogenesis, in Cancer Chemoprevention. Ed. Wattenberg, L., Lipkin, M., Boone, C.W. and Kelopff, G.J. pp. 153-163, CRC Press (1992).

94. Nguyen, T. and Pickett, C.B. Regulation of rat glutathione S transferase Ya subunit gene expression. J. Biol. Chem. 267, 1535-13539 (1992).

95. Prestera, T., Holtzclaw, Zhang, Y. and Talalay, P. Chemical and molecular regulation of enzymes that detoxify carcinogens. Proc. Natl. Acad. Sci. 90, 2965-2969 (1993).

96. Li, Y. and Jaiswal, A.K. Regulation of human NAD(P)H: quinone oxidoreductase gene: role of AP1 binding site contained within human antioxidant responsive element. J. Biol. Chem. 267, 15097-15104 (1992).

97. Prestera, T., Zhang, Y., Spencer, S.R., Wilczak, C.A. and Talalay, P. The electrophile counterattack response: protection against neoplasia and toxicity. Adv. Enz. Regul. 33, 281-296 (1993).

98. DeLong, M.J., Dolan, P., Santamaria, A. and Bueding, E. Dithiolethione analogs: effects on NAD(P)H quinone reductase and glutathione levels in murine hepatoma cells. Carcinog. 7, 977-980 (1986).

99. Morel, F., Fardel, O., Meyer, D.J., Kensler, T.W., Ketterer, B. and Guillouzo, A. Preferential increase of GS transferase class α-transcripts in cultured human hepatocytes by phenobarbital, methylcholanthrene and dithiolethiones. Cancer Res. 53, 231-234 (1993).

100. Munday, R. and Winterbourn, C.C. Reduced glutathione in combination with superoxide dismutase as an important biological antioxidant defense mechanism. Biochem. Pharmacol. 38, 4349-4352 (1989).

101. Keyse, S.M. and Tyrrell, R.M. Both near ultraviolet radiation and the oxidising agent hydrogen peroxide induced a 32KDa stress protein in normal human skin fibroblasts. J. Biol. Chem. 262, 14821-14825 (1987).

102. Cantoni, A., Guidarelli, O., Sestili, P., Mannello, F., Gazzanelli, G. and Cattabeni, F. Development and characterization of hydrogen peroxide resistant Chinese hamster ovary cell variants. Biochem. Pharmacol. 45, 2251-2257 (1993).

103. Spitz, D.R., Adams, D.T., Sherman, C.M. and Roberts, R.J. Mechanisms of cellular resistance to hydrogen peroxide, hyperoxia and 4 hydroxy 2 nonenal toxicity: the significance of increased catalase activity in H_2O_2 resistant fibroblasts. Arch. Biochem. Biophys. 292, 221-227 (1992).

104. Spitz, D.R., Sullivan, S.J., Malcom, R.R., and Roberts, R.J. Glutathione dependent metabolism and detoxification of 4 hydroxy 2 nonenal. Free Rad. Biol. Med. 11, 415-423 (1991).

105. Chu, F.F., Esworthy, R.S., Aleman, S. and Dorshow, J.H. Modulation of glutathione peroxidase expression by selenium. Nucl. Acid. Res. 18, 1531-1539 (1990).

106. Sandstrom, B., Carlsson, J. and Marklund, S.L. Selenite-induced variation in glutathione peroxidase activity of three mammalian cell lines. Radiation Res. 117, 318-325 (1989).

107. Bannai, S., Sato, H., Ishii, T. and Taketani, S. Enhancement of glutathione levels in mouse peritoneal macrophages. Biochim. biophys. Acta. 1092, 175-179 (1991)

108. Deneke, S.M. and Fanburg, B.L. Regulation of cellular glutathione Am. J. Physiol. 257, L163-L173 (1989).

109. Taketani, S., Sato, H., Yoshinaga, T., Tokunaga, R., Ishii, T., and Bannai, S. Induction in mouse peritoneal macrophages of 34kDa stress protein and heme oxygenase by sulfhydryl-reactive agents. J. Biochem. 108, 111-115 (1990).

110. Mimnaugh, E.G., Dusre, L., Atwell, J. and Myers, C.E. Differential oxygen radical susceptibility of adriamycin-sensitive and -resistant human breast tumor cells. Cancer Res. 49, 8-15 (1989).

111. Sinha, B.K., Minnaugh, E.C., Rajagopalan, S. and Myers, C.E. Adriamycin activation and oxygen free radical formation in human breast tumor cells. Cancer Res. 49, 3844-3848 (1989).

112. Meier-Bratschi, A., Lutz, W.K. and Schlatter, C. Methylation of liver DNA of a rat and mouse by N-nitrosodimethylamine formed in vivo from dimethylamine and nitrite. Food Chem. Toxicol. 21, 285-290 (1983).

113. Astill, B.D. and Mulligan, L.T. Phenolic antioxidants and the inhibition of hepatotoxicity from N-dimethylnitrosamine formed in situ in the rat stomach. Fd. Cosmet. Toxicol. 15, 167-171 (1977).

114. Stich, H.F., Dunn, B.P., Pignatelli, B., Ohshinia, H. and Bartsch, H. Dietary phenolics and betel nut extracts as modifiers of N-nitrosation in rat and man. IARC Sc. Publ. 57, 213-222 (1984).

115. Pignatelli, B., Bereziat J., Descotes, G. and Bartsch, H. Catalysis of nitrosation in vivo in rats by catechin and resorcinol and inhibition by chlorogenic acid. Carcinogenesis 3, 1045-1049 (1982).

116. Kuenzig W., Chan, J., Norkus, E., Newmark, H., Mergens, W. and Conney, A.H. Caffeic acid and ferulic acid as blockers of nitrosamine formation. Carcinogenesis 5, 309-313 (1984).

117. Nguyen, T., Brunson, D., Crespi, C.L., Periman, B.W., Wishnok, J.S. and Tannenbaum, S.R. DNA damage and mutation in human cells exposed to nitric oxide in vitro. Proc. Natl. Acad. Sci. 89, 3030-3034 (1992).

118. Carreras, M.C., Paragament, G.A., Catz, S.D., Poderoso, J.J. and Boveris, A. Kinetics of nitric oxide and hydrogen peroxide production and formation of peroxynitrite during the respiration burst of human neutrophils. FEBS Lett. 341, 65-68 (1994).

119. Michnovicz, J.J., Bradlow, H.L. Induction of estradiol metabolism by dietary indole-1-carbinol in humans. J. Nat'l. Cancer Inst. 82, 947-949 (1990).

120. Baldwin, W.S. and Leblanc, G.A. The anticarcinogenic plant compound indolecarbinol differentially modulates P450-mediated steroid hydroxylase activities in mice. Chem.-Biol. Interacns. 83, 155-169 (1992).

121. Bradlow, H.L., Michnovicz, J.J., Telang, N.T. and Osborne, M.P. Effects of dietary indole-3-carbinol on estradiol metabolism and spontaneous mammary tumors in mice. Carcinogenesis 12, 1571-1574 (1991).

122 Takahashi, S., Hakoi, K., Yaga, H., Hirose, M., Ito, N. and Fukushima, S. Enhancing effects of diallyl sulfide on hepatocarcinogenesis and inhibitory actions of the related diallyl disulfide on colon and renal carcinogenesis in rats. Carcinogenesis 13, 1513-1518 (1992).

123. Bailey, G.S., Hendricks, J.D., Shelton, D.W. Enhancement of carcinogenesis by the natural anticarcinogen indole-3-carbinol. J. Nat'l. Cancer Inst. 78, 931-934 (1987).

124. Ito, N. and Hirose, M. Antioxidants-carcinogenic and chemopreventive properties. Adv. Cancer Res. 53, 247-302 (1989).

125. Farber, E. Clonal adaptation during carcinogenesis. Biochem. Pharmacol. 39, 1837-1846 (1990).

126. Jones, G.R.N. Cancer destruction in vivo through disrupted energy metabolism. Physiol. Chem. Phys. and Med. NMR 24, 181-194 (1992).

127. De Vries, C.E. and Van Noorden, C.J.F. Effects of fatty acids on tumor growth and metastasis. Anticancer Res. 12, 1513-1522 (1992).

128. Corrocher, R., Casaril, M., Bellisola, G., Gabrielli, G.B., Guidi, G.C. and DeSandre, G. Severe impairment of antioxidant systems in human hepatoma. Cancer 58, 1658-1662 (1986).

129. Canuto, R.A., , Muzio, G., Biocca, M.E. and Dianzani, M.U. Lipid peroxidation in rat AH-130 hepatoma cells enriched in vitro with arachidonic acid. Cancer Res. 51, 4603-4608 (1991).

130. Cogrel, P., Morel, I., Lescoat, G., Chevanne, M., Brissot, P., Cillard, P. and Cillard, J. The relationship between fatty acid peroxidation and α-tocopherol consumption in isolated normal and transformed hepatocytes. Lipids 28, 115-119 (1993).

131. Kandaswami, C., Perkins, E., Soloniuk, D.S. and Middleton, E. Differential inhibition of proliferation of human squamous cell carcinoma, gliosarcoma and embryonic fibroblast-like lung cells in culture by plant flavonoids. Anti-Cancer Drugs 3, 525-30 (1992).

132. Khan, S., Ramuvani, J.J. and O'Brien, P.J. The toxicity of mechlorethamine and other alkylating anti-cancer drugs: role of lipid peroxidation. Biochem. Pharmacol. 43, 1963-1967 (1992).

133. Powis, G. and Hacker, M.P. The toxicity of anticancer drugs. Pergamon Press (1991).

NEW DIRECTIONS FOR FREE RADICAL CANCER RESEARCH AND MEDICAL APPLICATIONS

Stephen M. Hahn, C. Murali Krishna, and James B. Mitchell

Radiation Biology Branch
National Cancer Institute
National Institutes of Health
Bethesda, MD 20892

Introduction

The study of free radicals in biomedical research has dramatically increased in recent years. There is evidence that free radicals play a role in aging, chronic inflammatory or autoimmune diseases, atherosclerosis, and ischemia/reperfusion injury[1]. Free radicals may also play an important role in several areas of cancer research including the study of carcinogenesis, mechanisms of radiation and chemotherapy action, and the toxicity of therapies. As our knowledge of free radical mechanisms has grown, we have increased our ability to devise strategies in cancer prevention and in modifying the actions of chemotherapeutic agents and radiation. Concomitant with this increased knowledge base, has been the need for further study of free radicals as they relate to carcinogenesis and therapies based upon free radical damage.

Oxidant stress from a variety of agents will cause DNA damage. Based upon this basic premise, it has been proposed that oxidant stress plays a role in the initiation, promotion, and progression of cancers. Cerrutti has proposed that oxidants may cause DNA damage and lead to alterations in the expression of proto-oncogenes and suppresser genes[2,3]. It has also been suggested that oxidants promote cell growth and that natural antioxidant defenses regulate this effect. Oxidant induction of cell growth may be mediated via the activation of protein kinases or expression of growth-regulatory proto-oncogenes[2]. Investigation of the free radical mechanisms of carcinogenesis has led to the consideration of anti-oxidants as chemopreventive agents. Indeed there are ongoing clinical trials evaluating the effect of beta carotene, vitamin A, vitamin E, selenium and aspirin in the development of cancers.

It has been known for some time that several chemotherapeutic agents used to treat cancer exert their anti-neoplastic effects via free radical mechanisms or via free radical intermediates. Examples of such agents include bleomycin, actinomycin D, and mitomycin C[4-7]. Other compounds such as the anthracyclines are thought to produce normal tissue toxicity through free radical intermediates. Redox cycling of doxorubicin by the microsomal electron transport chain to form reactive semiquinones may be important in the toxicity of this compound. The semiquinone can react with oxygen to form superoxide. The conversion of superoxide or hydrogen peroxide (H_2O_2) to the more damaging species, hydroxyl radical (•OH), may be

Free Radicals in Diagnostic Medicine, Edited by
D. Armstrong, Plenum Press, New York, 1994

accomplished via the formation of a doxorubicin-iron complex. Preclinical studies suggesting a role for iron in the development of doxorubicin cardiotoxicity has led to clinical trials with the metal chelator ICRF-187[8,9]. Initial studies are encouraging and suggest a reduction of cardiotoxicity when the chelator is administered with doxorubicin. It is, therefore, likely that additional study of the free radical mechanisms of chemotherapeutic drugs will lead to the development of newer agents with enhanced therapeutic ratios or compounds that will ameliorate drug toxicity.

Much of radiation cytotoxicity is mediated through oxygen free radicals. Attempts have been made to modulate the effect of radiation upon tumor and normal tissue by enhancing or eradicating the production of free radicals. Several hypoxic cell sensitizers have been developed which appear to enhance the production of free radicals under hypoxic conditions. The mechanism of action of nitroimidazole and benzotriazene compounds is not fully elucidated but may at least partially involve their reduction to toxic radical species. Radiation protectors are compounds which enhance the therapeutic differential of radiation by limiting normal tissue toxicity. The development of radiation protectors has been important not only to increase the effectiveness of cancer treatment but also for the study of the underlying mechanisms of radiation cytotoxicity.

In this chapter, we will describe the development of a class of anti-oxidant compounds, the nitroxides, which highlight many of the features of free radicals as they pertain to cancer research. The nitroxides are stable free radical compounds (Figure 1) which have been investigated as possible contrast agents in magnetic resonance imaging and as spin labels for biophysical studies[10-12]. Several studies have shown that the nitroxides react with a variety of biologically relevant compounds including other free radicals[13-16]. It was the observation that several nitroxides themselves reacted with oxy radicals[17] that led to the investigation of these compounds as radioprotectors and anti-oxidants.

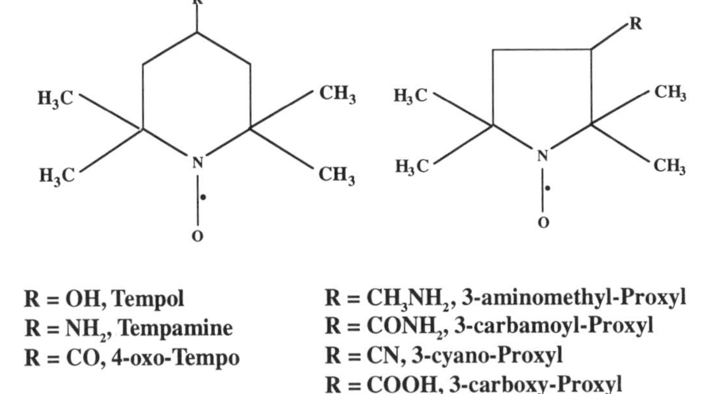

R = OH, Tempol
R = NH_2, Tempamine
R = CO, 4-oxo-Tempo

R = CH_3NH_2, 3-aminomethyl-Proxyl
R = $CONH_2$, 3-carbamoyl-Proxyl
R = CN, 3-cyano-Proxyl
R = COOH, 3-carboxy-Proxyl

Figure 1. Basic structure of the nitroxides. The six- and five-membered ring nitroxides are displayed. Data from reference 29 is used with permission.

THE CHEMISTRY OF NITROXIDES

Initial studies of the nitroxides revealed that superoxide was capable of directly reducing the nitroxide, 2-ethyl-2,5,5-trimethyl-3-oxazolidine-1-oxyl (OXANO·) to the corresponding hydroxylamine OXANOH as measured by a decrease in the EPR signal of the nitroxide[18]. The oxidation of OXANOH and the reduction of OXANO· were found to be pH dependent and resulted in a steady state distribution of [OXANO] and [OXANOH]. The ratio of [nitroxide]/[hydroxylamine] was independent of the initial concentration of either the nitroxide or the hydroxylamine. This cyclic reduction/oxidation of the nitroxide/hydroxylamine pair suggests a metal-independent superoxide dismutase (SOD) activity for the nitroxides[18]. Subsequent evaluation of six-membered piperidine ring nitroxides showed similar results consistent with the observation that SOD-like activity is shared by the nitroxides in general[19]. Piperidine nitroxide derivatives such as Tempo, Tempol, or Tempamine (Figure 1) are oxidized by superoxide but the oxidized intermediate, oxo-ammonium cation[20], is rapidly reduced by another superoxide as shown below for Tempol:

$$\text{O}\!\!-\!\!\text{ring}\;\dot{\text{N}}\text{-O} + {}^{\cdot}\text{O}_2^- + \text{H}^+ \underset{k_{-1}}{\overset{k_1}{\rightleftharpoons}} \text{O}\!\!-\!\!\text{ring}\;\text{N-OH} + \text{O}_2 \qquad (1)$$

$$\text{O}\!\!-\!\!\text{ring}\;\text{N-OH} + {}^{\cdot}\text{O}_2^- + \text{H}^+ \longrightarrow \text{O}\!\!-\!\!\text{ring}\;\dot{\text{N}}\text{-O} + \text{H}_2\text{O}_2 \qquad (2)$$

In a more recent study, it has been shown that nitroxides induce a catalase-like conversion of H_2O_2 to oxygen in the presence of hemoglobin and phenols through the oxoammonium intermediate[21].

$$\text{HO-ring}\;\dot{\text{N}}\text{-O} + {}^{\cdot}\text{O}_2^- + 2\text{H}^+ \longrightarrow \text{HO-ring}\;{}^{+}\text{N=O} + \text{H}_2\text{O}_2 \qquad (3)$$

$$\text{HO-ring}\;{}^{+}\text{N=O} + {}^{\cdot}\text{O}_2^- \longrightarrow \text{HO-ring}\;\dot{\text{N}}\text{-O} + \text{O}_2 \qquad (4)$$

ANTIOXIDANT ACTION

The observation that nitroxides reacted with and could potentially scavenge free radicals led to the investigation of these compounds as anti-oxidants. Initial studies showed that the nitroxides were membrane permeable compounds[22] and therefore could exert their action intracellulary. Pretreatment of mammalian cells with selected nitroxides provided protection against a variety of oxidative stresses[22]. Specifically, the nitroxides afforded complete protection against H_2O_2 cytotoxicity (Figure 2). The cytotoxicity of H_2O_2 was transition metal dependent since desferrioxamine, a metal chelator, also provided complete cytoprotection. Several mechanisms of nitroxide protection were proposed. Nitroxides oxidize reduced metals thereby lessening the metal-catalyzed hydroxyl radical production from H_2O_2[22]. In addition, nitroxides also react with and potentially detoxify free radicals[13,14,16]. The nitroxides themselves exhibited no catalase activity.

As a result of the discovery of antioxidant activity, it was suggested that the nitroxides might have application in protection from postischemic reperfusion injury such as after

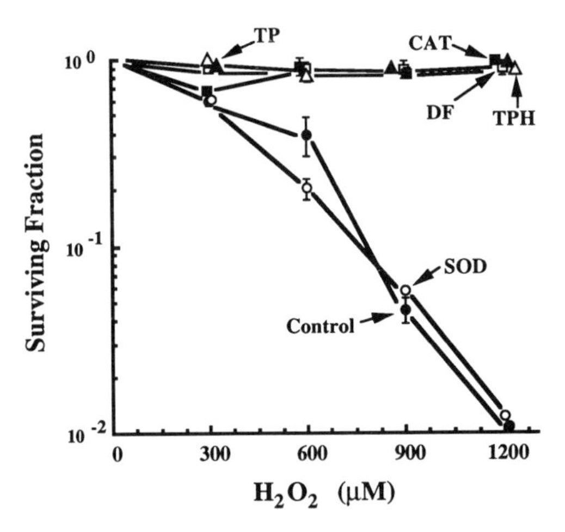

Figure 2. The effect of various agents on clonogenic survival of Chinese hamster V79 cells against H_2O_2 cytotoxicity. (closed circle) control, (closed square) catalase, (open circle) SOD, (open square) Desferrioxamine, (closed triangle) Tempol, (open triangle)Tempol-H. Data from reference 22 is used with permission.

myocardial infarction or cerebrovascular accidents[22]. Samuni et al.[23] isolated newborn rat myocardial cells and exposed them to H_2O_2 providing a model of ischemia-reperfusion injury. H_2O_2 exposure blocked the rhythmic beating of monolayered cardiomyocytes and resulted in cellular injury as manifested by LDH release. The nitroxides provided complete protection against this damage (Figure 3). Gelvan et al.[24] extended these findings to an *in vivo* system showing that the administration of the nitroxide, Tempo, inhibited reperfusion injury using an isolated rat heart model.

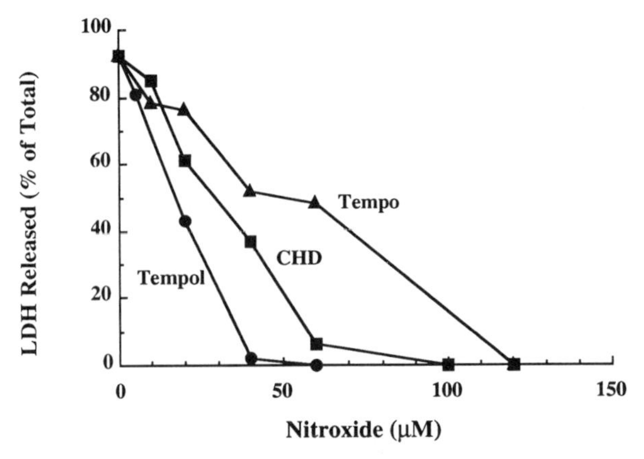

Figure 3. Dose-dependence of nitroxide protection against H_2O_2-induced (500 μM) LDH leakage from cardiomyocytes. Data from reference 23 is used with permission.

RADIOPROTECTION

Protection against radiation-induced cytotoxicity has been afforded by several nitroxides both *in vitro* and *in vivo*. This might be expected given that the production of free radicals (especially oxy-radicals) is an important mode of radiation cytotoxicity[25]. Mitchell et al. first showed that a water soluble nitroxide, Tempol, inhibited radiation-induced cytotoxicity in mammalian cells[26]. Under aerobic conditions, pretreatment of Chinese hamster V79 cells, with Tempol afforded substantial radioprotection in a dose-dependent fashion (Figure 4). The reduced form of Tempol (Tempol-H), did not provide aerobic radioprotection (Figure 4). When cells were exposed to Tempol under hypoxia, no radioprotection was observed. On the contrary, Tempol treatment resulted in modest radiosensitization. Previous studies have also shown that nitroxides are radiosensitizers under hypoxic conditions[27,28].

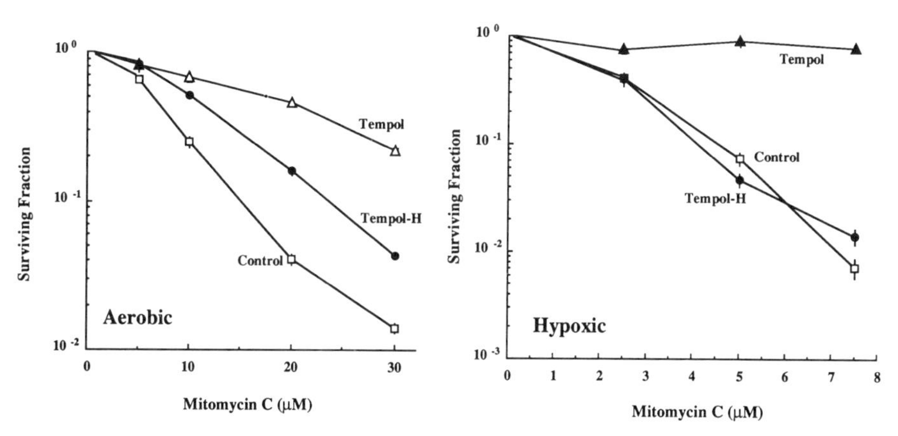

Figure 4. Full x-ray dose survival curves for Tempol and Tempol-H for Chinese hamster V79 cells. Left panel: Increasing aerobic protection is observed with increasing Tempol concentration. Right panel: Tempol-H provided no aerobic radioprotection. Data from reference 26 is used with permission.

The mechanism of differential nitroxide action under aerobic and hypoxic conditions is unclear. Aerobic radioprotection was shown not to be the result of an induction of hypoxia, an increase in intracellular glutathione concentration, or an induction of intracellular SOD mRNA[26]. Radiation under aerobic and hypoxic conditions produces several common radical species including carbon-centered free radicals (R·), ·OH, H_2O_2, H·, and aquated electrons. The nitroxides could directly react with and therefore detoxify radiolytically produced ·OH, H·, or aquated electrons; however, these species are produced under both aerobic and hypoxic conditions and Tempol only protected aerobic cells. Tempol might react with R· under hypoxic conditions, forming a non-repairable lesion and resulting in radiosensitization. Since R· is also produced under aerobic conditions, modest sensitization might also occur but may be overwhelmed by other Tempol-mediated protective mechanisms.

Six additional water soluble nitroxides with different structural features have also been evaluated for radioprotective properties *in vitro*[29]. Two nitroxides, 3-aminomethyl-Proxyl (3AM) and Tempamine (Figure 1), were found to exceed the radioprotection afforded by Tempol when given on an equimolar basis. 3AM and Tempamine provided protection factors of 2.3 and 2.4, respectively, at the 10% survival level versus 1.3 for Tempol (Table 1). One nitroxide, 3-carboxy-Proxyl did not afford radioprotection. The differential radioprotection provided by the various nitroxides was evaluated in this study[29] and likely involves differences in structure (Figure 1). Both 3AM and Tempamine contain amino groups and are more likely

Table 1. The effect of selected nitroxides (10mM concentration) were assessed using full x-ray dose response curves. Protection factors were calculated by determining the ratio of radiation dose of nitroxide-treated cells to that of control cells at 10% and 1% survival. Data from reference 29 is used with permission.

	10% Survival	1% Survival
Tempol	1.3	—
Tempamine	2.3	2.4
3-Aminomethyl Proxyl	2.4	2.6
3-Cyano Proxyl	1.6	1.6
3-Carbamoyl Proxyl	1.5	1.6
4-Oxo Tempo	1.2	1.2

to be positively charged at a neutral pH while 3-carboxy-Proxyl is more likely to be negatively charged. These charge differences could alter the affinity of the nitroxides for DNA, the putative target of radiation. Such differences in DNA affinity were confirmed with nonequilibrium dialysis studies[29].

The direct protection of DNA has also been explored *in vitro*. Using the XPRT forward mutation assay in mammalian cells, DeGraff et al.[30] found that Tempol provided near complete protection against the mutagenic effects of radiation. In addition, Tempol protected against DNA double-strand breaks as measured by field inversion gel electrophoresis[30]. Preliminary studies have also shown that Tempol protects against radiation-induced chromosomal aberrations in human peripheral blood lymphocytes in a dose-dependent fashion.

In vivo studies of the nitroxides have also confirmed the radioprotective properties of these compounds. Tempol was administered to C3H mice intraperitoneally (IP) and the toxicity, pharmacology and radioprotective properties were investigated[31]. A maximally tolerated dose of 275 mg/kg was found for Tempol. Above this dose, restlessness and seizure activity were found. Using EPR spectroscopy, Tempol levels were measured in the whole blood of mice after IP injection. A peak level of 600 μg/ml was found 5-10 minutes after injection, corresponding to a concentration of approximately 3.0 mM (Figure 5). With a peak concentration of nitroxide found 5-10 minutes after injection, C3H mice were given Tempol and then exposed to increasing doses of whole body radiation 5-10 minutes after drug administration. Significant *in vivo* radioprotection was observed[31]. The dose of radiation that caused 50% lethality 30 days after administration (the $LD_{50/30}$ dose) was 9.97 Gy for Tempol-treated mice and 7.84 Gy for control mice. The dose modification factor, DMF, (the $LD_{50/30}$ dose in the Tempol-treated mice divided by the $LD_{50/30}$ dose in the control mice) was calculated to be 1.3 (Figure 5). This data showed that a nonlethal dose of a nitroxide, Tempol, could provide *in vivo* radioprotection. Additional *in vivo* studies are needed to evaluate the protection afforded by the other water soluble nitroxides identified in *in vitro* studies.

The mechanism of *in vivo* radioprotection is not clear. Certainly, reaction with and detoxification of oxy-radicals could be a major mechanism for nitroxide protection as suggested by the *in vitro* studies. However, other possibilities exist. Preliminary studies in pigs have shown that the intravenous administration of Tempol results in decreases in arterial blood pressure. It cannot be excluded that Tempol, through its effect upon blood pressure, produces hypoxia in the bone marrow by shunting blood away from the bone marrow compartment. Indeed, it has been suggested that the bone marrow is normally functioning under marginally oxygenated conditions. A hypoxic environment for bone marrow cells could make them more resistant to radiation thereby providing radioprotection to the whole animal.

The survival of C3H mice 30 days after radiation is essentially a measure of the hematologic toxicity[32]. Although the mechanism is unknown, significant radioprotection of bone marrow

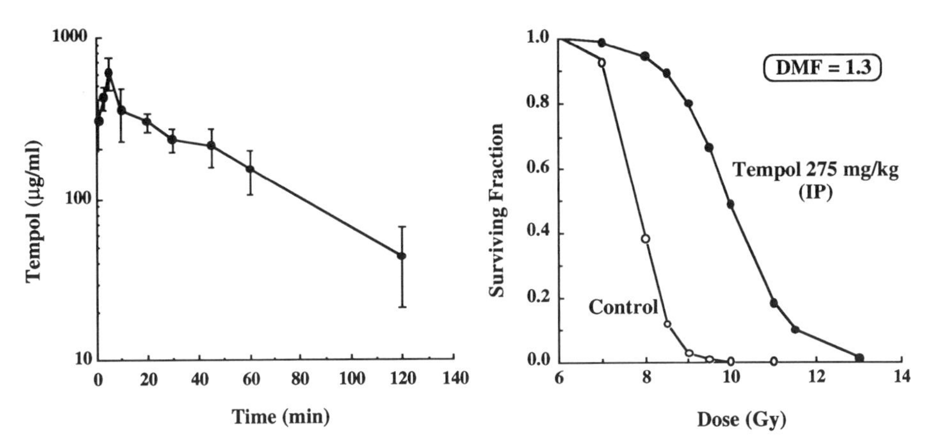

Figure 5. Left panel: Whole blood levels of Tempol in C3H mice with time. Mice were injected with Tempol IP. Total nitroxide (oxidized and reduced) blood levels were measured using EPR spectroscopy. Right panel: Survival of C3H mice, 30 days after radiation with and without Tempol administration. Significant radioprotection is observed with a DMF 1.3. Data from reference 31 is used with permission.

is observed with nitroxide administration. It is unclear, however, if the nitroxides would protect tumors. Simultaneous protection of tumors and bone marrow of similar magnitude would provide no therapeutic advantage for the administration of nitroxides. Preliminary studies of Tempol evaluating the tumor control probability of the radiation-induced fibrosarcoma (RIF-1) in mice, suggest that no protection is observed. Additional studies involving other transplantable tumors will be necessary.

There are several explanations for the possible differential protection of tumor and normal tissues. The discovery that selected nitroxides provide aerobic radioprotection and hypoxic radiosensitization suggests that a differential protection of normal aerated tissues and hypoxic tumor cells might exist. If this were true, one would expect that tumors with the greatest hypoxic fraction would have the most radiosensitization from nitroxide administration. There is also evidence that tumor cells have a greater capacity for bioreduction compared to normal cells[33]. Aerobic Tempol radioprotection is dependent upon the presence of the oxidized form, as Tempol-H does not afford aerobic radioprotection[26]. If it is discovered that tumor cells bioreduce nitroxides to a greater extent, differential radioprotection might be expected.

Tempol has also been shown to protect against radiation-induced alopecia. Goffman et al.[34] demonstrated that the topical administration of Tempol to guinea pig hair and skin prior to single doses of radiation provided a marked increase in the rate and extent of new hair recovery when compared to control animals. Using EPR spectroscopy, Tempol was detected in the treated skin, but not in whole blood samples or in tissues directly underneath the skin[34]. This demonstrated that the topical administration of Tempol could provide cutaneous protection without systemic effects.

NITROXIDES AND THE FREE RADICAL MECHANISMS OF CHEMOTHERAPY

Several chemotherapeutic agents exert their cytotoxicity through the generation of free radicals. These free radicals may be responsible for the toxicity of chemotherapy, the antitumor effects, or both. Examples include doxorubicin, mitomycin C, and bleomycin[35,36]. The nitroxides have been employed to explore the mechanisms of free radical drug cytotoxicity and to provide cytoprotection [7]. Krishna et al.[7] studied the aerobic and hypoxic cytotoxicity of mitomycin C using Tempol. Both Tempol and Tempol-H provided partial cytoprotection to mammalian cells under aerobic conditions (Figure 6). Only Tempol provided protection against hypoxic mitomycin C cytotoxicity[7]. Using EPR spectroscopy, the authors found that mitomycin C was activated by the human NADPH:cytochrome P-450 enzyme system to its one electron product the semiquinone radical (Figure 7). Under aerobic conditions, the semiquinone radical reduced molecular oxygen leading to the production of superoxide (Figure 7). Under hypoxic conditions, the semiquinone radical reduced H_2O_2 to produce ·OH. The semiquinone radical was also found to reduce Tempol[7]. Therefore, Tempol was able to accept an electron from the semiquinone radical and afforded complete protection against hypoxic cytotoxicity. These studies suggested that the one electron reduction of mitomycin C to the semiquinone radical was the mechanism of the selective hypoxic cytotoxicity of this compound. Furthermore, since Tempol provided only partial aerobic cytoprotection, mechanisms of aerobic cytotoxicity in addition to the formation of the semiquinone radical were likely.

The protection afforded by Tempol against mitomycin C cytotoxicity could lead to the use of the nitroxides in ameliorating toxicity in the clinic. Preliminary studies in our laboratory

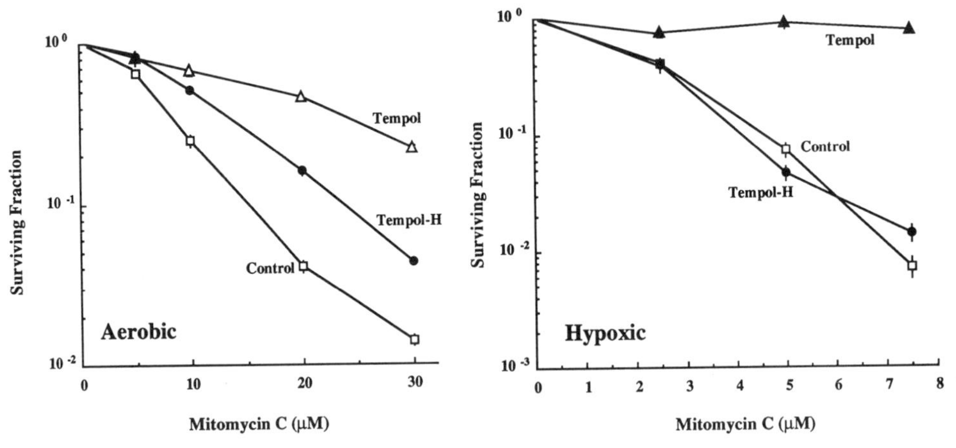

Figure 6. Full dose-response curves of mitomycin C cytotoxicity in Chinese hamster V79 cells with and without Tempol and Tempol-H treatment. Left panel: aerobic conditions. Right panel: hypoxic conditions. Data from reference 7 is used with permission.

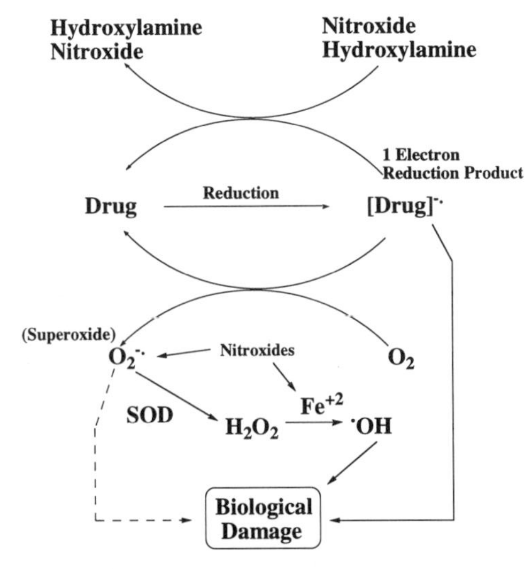

Figure 7. Free radical mechanisms of drug action. Selected drugs undergo a one electron reduction. Under either hypoxic or aerobic conditions, toxic species are subsequently produced.

have investigated the use of selected water soluble nitroxides against the skin extravasation of mitomycin C. A swine model has been used because of the similarity of pig skin with human skin. The initial results of these studies show that one nitroxide, 3-carbamoyl-Proxyl provides protection against the skin necrosis and scarring produced by intradermal and subcutaneous injection of mitomycin C. Additional studies are necessary to fully evaluate this finding.

The mechanism of doxorubicin cytotoxicity has also been evaluated with Tempol. Initial EPR studies have shown that Tempol reacts with the one electron reduction product of doxorubicin, a semiquinone radical as well as with superoxide radicals produced during the redox cycling of this drug in the presence of oxygen. However, Tempol does not protect against the cytotoxicity of doxorubicin in V79 cells. In addition, Tempol administration did not result in inhibition of doxorubicin-induced double stranded DNA breaks (Krishna, MC personal communication). These results suggest that free radical mechanisms have little role in the cytotoxicity of V79 cells by doxorubicin.

A similar study has evaluated the hypoxic cytotoxicity of the benzotriazene SR-4233. SR-4233 is being evaluated presently in clinical trials as a radiation sensitizer and hypoxic cell cytotoxic agent. In evaluating the mechanism of SR-4233 cytotoxicity, it has been found that Tempol protects against the hypoxic cytotoxicity in V79 cells. EPR studies suggest that Tempol reacts with the 1 electron reduction product of SR-4233 and therefore detoxifies it (Herscher, L personal communication). This study is important for understanding the mechanisms of drug cytotoxicity under hypoxic conditions and may help in the development of compounds which reverse the negative effects of hypoxia upon antineoplastic therapies.

SUMMARY

The nitroxides are stable, low molecular weight free radical compounds which are freely membrane permeable. These properties make the nitroxides valuable for the study of and possible protection against oxidative stresses. It is becoming increasingly clear that oxidative stress is important to the pathogenesis of cancer as well as to the development of treatments for cancer. Several nitroxides have been shown to interrupt the toxicity of oxidative stress with the protection against H_2O_2 toxicity and possibly ischemia/reperfusion injury being of primary importance. With respect to radiation, the nitroxides have afforded both *in vitro* and *in vivo* protection. The redox activity of the nitroxides may allow for the differential activity of these agents in normal versus tumor tissues. Further study of these compounds may yield a nitroxide with clinical applications as well as provide insight into the mechanisms of radiation cytotoxicity. Finally, the nitroxides have allowed us to explore the mechanisms of action of several chemotherapeutic agents. Understanding these processes is important to the process of ameliorating the toxicity of therapies and to the rationale design of future agents.

REFERENCES

1. B. Halliwell, and J. M. C. Gutteridge, "Free Radicals in Biology and Medicine," Clarendon Press, Oxford, UK (1989).
2. P. Cerutti, G. Shah, A. Peskin, and P. Amstad, Oxidant carcinogenesis and antioxidant defense, *Ann. N. Y. Acad. Sci.* 663:158 (1992).
3. P. A. Cerutti, Oxidant stress and carcinogenesis, *Eur. J. Clin. Invest.* 21:1 (1991).
4. A. Russo, J. B. Mitchell, S. McPherson, and N. Friedman, Alteration of bleomycin cytotoxicity by glutathione depletion or elevation, *Int. J. Radiat. Oncol. Biol. Phys.* 10:1675 (1984).
5. C. Borek, and W. Troll, Modifiers of free radicals inhibit *in vitro* the oncogenic actions of x-rays, bleomycin, and the tumor promoter 12-O-tetradecanoylphorbol 13-acetate, *Proc. Natl. Acad. Sci. USA* 80:1304 (1983).

6. T. Komiyama, T. Kikuchi, and Y. Sugiura, Generation of hydroxyl radical by anticancer quinone drugs, carbazilquinone, mitomycin C, aclacinomycin A and adriamycin, in the presence of NADPH-cytochrome P-450 reductase, *Biochem. Pharmacol.* 31:3651 (1982).
7. C. M. Krishna, W. DeGraff, S. Tamura, F. Gonzalez, A. Samuni, A. Russo, and J. B. Mitchell, Mechanisms of hypoxic and aerobic cytotoxicity of Mitomycin C in Chinese hamster V79 cells, *Cancer Res.* 51:6622 (1991).
8. B. B. Hasinoff, The interaction of the cardioprotective agent ICRF-187 [+]-1,2-bis(3,5-dioxopiperazinyl-1-yl) propane; its hydrolysis product (ICRF-198; and other chelating agents with the Fe(III) and Cu(II) complexes of adriamycin, *Agents Actions* 26:378 (1989).
9. G. F. Vile, and C. C. Winterbourn, dl-N,N'-dicarboxcamidomethyl-N,N'-dicarboxymethyl-1,2-diaminopropane (ICRF-198) and d-1,2-bis(3,5-dioxopiperazine-1-yl) propane (ICRF-187) inhibition of Fe3+ reduction, lipid peroxidation, and CaATPase inactivation in heart microsomes exposed to adriamycin, *Cancer Res.* 50:2307 (1990).
10. H. F. Bennett, H. M. Swartz, R. D. Brown III, and S. H. Koenig, Modification of relaxation of lipid protons by molecular oxygen and nitroxides, *Invest. Radiol.* 22:502 (1987).
11. H. M. McConnell, "Spin Labelling: Theory and Applications," C.C. Thomas Publ., Springfield, IL (1965).
12. H. M. Swartz, Interactions between cells and nitroxides and their implications for their uses as biophysical probes and as metabolically responsive contrast agents for *in vivo* NMR, *Bull. Mag. Res.* 8:172 (1983).
13. S. Belkin, R. J. Mehlhorn, K. Hideg, O. Hankovsky, and L. Packer, Reduction and destruction of nitroxide spin probes, *Arch. Biochem. Biophys.* 256:232 (1987).
14. J. Chateauneuf, J. Lusztyk, and K. U. Ingold, Absolute rate constants for the reactions of some carbon-centered radicals with 2,2,6,6-tetramethylpiperidine-N-oxyl, J. *Org. Chem.* 53:1629 (1988).
15. R. J. Mehlhorn, and L. Packer, Electron paramagnetic resonance spin destruction methods for radical detection, *Methods in Enzymology* 105:215 (1984).
16. U. A. Nilsson, L. I. Olsson, G. Carlin, and A. C. Bylund-Fellenius, Inhibition of lipid peroxidation by spin labels. Relationships between structure and function, J. *Biol. Chem.* 264:11131 (1989).
17. A. Samuni, C. M. Krishna, P. Riesz, E. Finkelstein, and A. Russo, Superoxide reaction with nitroxide spin-adducts, *Free Radic. Biol. Med.* 6:141 (1989).
18. A. Samuni, C. M. Krishna, P. Riesz, E. Finkelstein, and A. Russo, A novel metal-free low molecular weight superoxide dismutase mimic, J. *Biol. Chem.* 263:17921 (1988).
19. A. Samuni, C. M. Krishna, J. B. Mitchell, C. R. Collins, and A. Russo, Superoxide reaction with nitroxides., *Free Rad. Res. Comms.* 9:241 (1990).
20. M. C. Krishna, D. A. Grahame, A. Samuni, J. B. Mitchell, and A. Russo, Oxoammonium cation intermediate in the nitroxide-catalyzed dismutation of superoxide, *Proc. Natl. Acad. Sci.* 89:5537 (1992).
21. R. J. Mehlhorn, and C. E. Swanson, Nitroxide-mediated H2O2 decomposition by peroxidases and pseudoperoxidases, *Free Radic. Res. Comm.* 17:157 (1992).
22. J. B. Mitchell, A. Samuni, M. C. Krishna, W. G. DeGraff, M. S. Ahn, U. Samuni, and A. Russo, Biologically active metal-independent superoxide dismutase mimics, *Biochemistry* 29:2802 (1990).

23. A. Samuni, D. Winkelsberg, A. Pinson, S. M. Hahn, J. B. Mitchell, and A. Russo, Nitroxide stable radicals protect beating cardiomyocytes against oxidative damage, J. *Clin. Invest.* 87:1526 (1991).
24. D. Gelvan, P. Saultman, and S. Powell, Cardiac reperfusion damage prevented by a stable nitroxide free radical, *Proc. Natl. Acad. Sci. USA* 88:4680 (1991).
25. E. J. Hall, "Radiobiology for the Radiologist," J. B. Lippincott Co., Philadelphia, PA (1994).
26. J. B. Mitchell, W. DeGraff, D. Kaufman, M. C. Krishna, A. Samuni, E. Finkelstein, M. S. Ahn, S. M. Hahn, J. Gamson, and A. Russo, Inhibition of oxygen-dependent radiation-induced damage by the nitroxide superoxide dismutase mimic, Tempol, *Arch. Biochem. Biophys.* 289:62 (1991).
27. B. C. Millar, E. M. Fielden, and C. E. Smithen, Polyfunctional radiosensitizers IV. The effect of contact time and temperature on sensitization of hypoxic Chinese hamster cells in vitro by bifunctional nitroxyl compounds, *Br. J. Cancer* 37:73 (1978).
28. P. T. Emmerson, and P. Howard-Flanders, Preferential sensitization of anoxic bacteria to X-rays by organic nitroxide-free radicals, *Radiat. Res.* 23:54 (1965).
29. S. M. Hahn, L. Wilson, C. M. Krishna, J. Liebmann, W. DeGraff, J. Gamson, A. Samuni, D. Venzon, and J. B. Mitchell, Identification of nitroxide radioprotectors, *Radiat. Res.* 132:87 (1992).
30. W. G. DeGraff, M. C. Krishna, D. Kaufman, and J. B. Mitchell, Nitroxide-mediated protection against x-ray- and neocarzinostatin-induced DNA damage, *Free Radic. Biol. Med.* 13:479 (1992).
31. S. M. Hahn, Z. Tochner, C. M. Krishna, J. Glass, L. Wilson, A. Samuni, M. Sprague, D. Venzon, E. Glatstein, J. B. Mitchell, and A. Russo, Tempol, a stable free radical, is a novel murine radiation protector, *Cancer Res.* 52:1750 (1992).
32. J. M. Yuhas, and J. B. Storer, The effect of age on two modes of radiation death and on hematopoietic cell survival in the mouse, *Radiat. Res.* 32:596 (1969).
33. K. A. Kennedy, B. A. Teicher, S. Rockwell, and A. C. Sartorelli, The hypoxic tumor cell: a target for selective cancer chemotherapy, *Biochem. Pharmacol.* 29:1 (1980).
34. T. Goffman, D. Cuscela, J. Glass, S. Hahn, C. M. Krishna, G. Lupton, and J. B. Mitchell, Topical application of nitroxide protects radiation induced alopecia in guinea pigs, *Int. J. Radiat. Oncol. Biol. Phys.* 22:803 (1992).
35. R. T. Dorr, G. T. Bowden, D. S. Alberts, and J. D. Liddil, Interactions of mitomycin C with mammalian DNA detected by alkaline elution, *Cancer Res.* 45:3510 (1985).
36. N. R. Bachur, S. L. Gordon, M. V. Gee, and H. Kon, NADPH cytochrome P-450 reductase activation of quinone anticancer agents to free radicals, *Proc. Natl. Acad. Sci. USA* 76:954 (1979).

SHOCK AND MULTIPLE ORGAN FAILURE

Patricia A. Abello, Timothy G. Buchman,
and Gregory B. Bulkley

Department of Surgery
The Johns Hopkins University School of Medicine
Baltimore, MD 21287

INTRODUCTION

Multiple organ dysfunction syndrome (MODS), previously known as multiple organ failure, has emerged as the leading cause of mortality in surgical critical care. It is a syndrome of sequential and progressive organ dysfunction, associated with a sustained, massive inflammatory response that is often preceded by insults such as sepsis, hemorrhagic shock, inflammatory states such as pancreatitis, and tissue injury. Classically, organ involvement occurs in a predictable sequence, initially involving the lung, then the liver, then the gut, and other organs. With four organ system involvement, mortality approaches 100%. Once the process is initiated, MODS often progresses despite eradication of the inciting cause.

Our understanding of this complex disease is admittedly limited. Prior to 1970, this disease was essentially non-existent. An injury sufficient enough to result in even one organ failure was universally lethal, since we did not have the technology to support the failing organ. During the Vietnam War, the commonest cause of death after successful resuscitation from traumatic injury was "shock lung," or post-traumatic pulmonary insufficiency, later called the Adult Respiratory Distress Syndrome (ARDS). With better understanding of pulmonary pathophysiology and improvement in ventilatory management, more patients survived this initial organ failure long enough for the syndrome of multiple organ dysfunction to become manifest. In 1973, Tilney and his colleagues first reported sequential system failure in a series of patients after rupture of abdominal aortic aneurysms.[1] In this article, he recognized the relationship between massive acute blood loss with resuscitation and postoperative organ failure, attributing such factors as operative trauma, anesthesia, transfusion, and emboli from surgical vascular clamping in its pathogenesis. By 1980, it was recognized that an almost universal predisposing factor in the development of MODS was infection. It was not uncommon surgical practice at that time to take these critically ill patients from the intensive care unit to the operating room for empiric laparotomy to look for a hidden source of infection in the abdomen when none was otherwise apparent clinically. With the advent of computerized tomography, better culturing techniques, and broad spectrum antibiotics, we became more successful in localizing and eradicating infection. Despite these advances, patients continued

Free Radicals in Diagnostic Medicine, Edited by
D. Armstrong, Plenum Press, New York, 1994

to die of MODS. Certainly, eradication of the infection was necessary, but by itself, insufficient to reverse the process.

1985 was an important year in our understanding of MODS. At that time, investigators successfully isolated and sequenced a plasma-borne factor derived from macrophages which they produced in response to bacterial endotoxin.[2] This factor, called cachectin or tumor necrosis factor-alpha (TNF-∝), when injected into experimental animals, elicited systemic hypotension and tissue injury almost indistinguishable from that seen in septic shock .[3] With this discovery came the recognition that the patient was no longer the passive victim, but rather an active participant in his or her own disease process. The focus in MODS shifted from attempts to improve what we had perceived as a failing immune system, to attempts to control an inappropriate or excessive immune response.

Thus the change in nomenclature from multiple organ failure to multiple organ dysfunction syndrome[4] reflects our recognition of the syndrome as an active and progressive process, characterized by two distinct features: disseminated inflammation and dysregulated metabolism. In 1985, Goris and his colleagues found that only a third of trauma patients that went on to develop MODS had evidence of bacterial sepsis.[5] Generalized inflammation, however, was apparent in all involved organs, even those distant from the site of initial injury. He attributed this generalized inflammatory response to activation of the complement system with subsequent endothelial damage, and suggested that this microvascular injury was primarily responsible for the organ dysfunction that ensued. In addition, it became clear from work by Cerra and his colleagues in the nutrition field that a universal finding in MODS is a hypermetabolic state, with consumption of calories and metabolic substrates disproportionate to the amount of productive work being performed.[6] Thus the organs are not dying, but are in fact, very much alive. Their function, however, is dysregulated.

There remains a certain difficulty in defining this disease which is inherent in all nomenclature. This difficulty lies in the fact that all definitions have assumed etiologies. Herein we discuss features of MODS that reflect our cumulative understanding of the disease process. Our information is descriptive; whether these features have a causal or an associative relationship remains to be discerned.

CLINICAL PICTURE

The typical scenario of a patient with MODS is that of a critically ill patient in the intensive care unit, perhaps after major trauma or emergency surgery, complicated by sepsis or circulatory shock. The patient is intubated and on the ventilator. His or her skin is somewhat yellow and edematous. A nasogastric tube decompresses the stomach and allows gastric pH monitoring and antacid administration. Intravenous nutrients are being infused via a catheter into the subclavian vein. The patient is attached to various cardiac monitors that show his or her heart rhythm and allow the measurement of cardiovascular function. A Foley catheter collects the small amount of urine that the patient is making, while a dialysis machine waits at bedside. The patient is unresponsive. This scenario depicts a case of MODS in which there is dysfunction in almost every organ. The pathologic and pathophysiologic derangements associated with each organ system are well defined.

The first organ characteristically involved is the lung. Dysfunction is manifested as decreased compliance (requiring higher ventilation pressures), impaired gas exchange (requiring an increased concentration of inspired oxygen), and shunting of blood through unventilated areas, further impairing oxygenation of the pulmonary venous blood. This constellation of findings is collectively called the Adult Respiratory Distress Syndrome. Mild hypoxia and respiratory alkalosis are often the first clinical finding. With worsening pulmonary function, mechanical ventilation become necessary to maintain adequate oxygenation and carbon dioxide removal. The pathologic hallmark of early ARDS is the finding of neutrophil-fibrin aggregates within the pulmonary microvasculature. As the disease progresses, one sees extension of the inflammation into the interstitium and alveoli, with the associated exudation of proteinaceous fluid, and ultimately, consolidation.

The next organ characteristically involved is the liver, in which derangements are seen in all three areas of hepatic function: synthesis, detoxification, and immunologic (reticuloendothelial) function. Blood levels of transaminases and lactic dehydrogenase are elevated, reflective of hepatic parenchymal injury. Low levels of albumin and clotting factors are indicative of the liver's decreased capacity to synthesize proteins. Serum ammonia levels are often elevated, indicating inadequate detoxification. Clinically, the patient is often jaundiced, reflecting inadequate biliary excretion, with altered mental status from hepatic encephalopathy. The classic pathologic lesion is a characteristic gradient of parenchymal injury with the center of the lobule (venous end) most disturbed, and with the portal end (arterial) relatively spared. Congestion of the microvasculature associated with frank necrosis of the hepatocytes at the center of the lobule (termed centrolobular necrosis) is characteristic of this injury.

Renal involvement is seen often at seven to ten days, with impairment of renal concentrating ability and decreased excretion. Decreased glomerular perfusion results in low urine output, inability to excrete waste water with consequent edema and systemic hypervolemia, and associated electrolyte abnormalities, including hyperkalemia, which might threaten cardiac function. As renal function worsens, the patient may require dialysis, often for these life threatening electrolyte abnormalities. The pathologic hallmark of early renal failure is sludging of red blood cells and leukocytes in the dilated vasa recta of the loop of Henle, with interstitial edema, swelling of tubular cells with collection of proteinaceous debris in the tubule, and sloughing of tubular cells into the lumen (termed acute tubular necrosis).

The intestine is also involved early in the progression of MODS, and importantly, may contribute to the pathogenesis of the disease. The gut mucosa appears to play an important role as an impermeable barrier, protecting the body from the bacteria, toxins and digestive enzymes that reside in the gut lumen. It is this barrier function that allows us to physiologically coexist with our intestinal flora. In MODS, this epithelial barrier is often disrupted or even denuded from the underlying basement membrane. Translocation of bacteria across the compromised barrier has been implicated as the "hidden gut wound" that may initiate or sustain the inflammatory response characteristic of MODS.[7] Clinically, gut failure is manifested as hemorrhagic stress erosions of the stomach, which can be a significant cause of bleeding, morbidity, and mortality. Malabsorption and ileus are also common, and pancreatitis and acalculous cholecystitis can be serious complications. Impaired absorption of nutrients in the face of high metabolic demands can result in poor nutritional status systemically with subsequent impairment of wound healing and immunologic function. Patients often require total parenteral nutrition, wherein high levels of nutrients are infused via central venous access, which has its attendant risks and complications, including line sepsis. The classic pathologic intestinal injury seen in experimental ischemia and low flow states is a gradient of injury which progresses from the inside of the gut lumen (villus tip) outward.[8] The epithelial layer is initially involved, and one often sees mucosal sloughing after such non-occlusive mesenteric ischemia. Prolonged ischemia may lead to transmural infarction, which can be immediately life-threatening.

The hematologic and immunologic manifestations of MODS include leukocytosis, neutropenia, and thrombocytopenia. Immunologic abnormalities are seen in both the humoral and cell-mediated arms of the immune system, making these patients at greater risk for development of ICU-acquired infections. Infections with opportunistic flora such as candida, staphylococcus epidermidis, and pseudomonas may well be a reflection of this host immunocompromise. Disseminated intravascular coagulation is a serious complication arising from the activation of the clotting cascade within the microvasculature, leading to thrombosis. Paradoxically, the resultant depletion of pro-coagulant factors in the blood can lead to hypocoagulopathy and uncontrolled hemorrhage.

Cardiac dysfunction may be manifested as supraventricular arrhythmias and impairment of myocardial contractility. This primary myocardial dysfunction may be the consequence of circulating cardiotoxins released from the gut after ischemia/reperfusion (termed "myocardial depressant factor"), although these toxins have yet to be clearly defined.[9] Cardiac failure can ensue, leading to maldistribution of volume with worsening pulmonary edema and deterioration of oxygen delivery to the body. This failure may necessitate pharmacologic inotropic support to improve cardiac function, but this may exacerbate arrhythmias. In the face of severe dysfunction, hypotension may persist despite maximal pharmacologic support.

Central nervous system dysfunction is characterized by varying decreased levels of consciousness, which may reflect the cumulative contributions of hepatic encephalopathy, sepsis, electrolyte abnormalities, decreased perfusion, hypoxia, thrombosis, and hemorrhage. Fortunately, if the patient recovers from MODS, there is rarely any permanent cerebral dysfunction recognized.

PATHOPHYSIOLOGY

Again, our descriptions of the clinical picture of MODS lead us back to the recognition that associative relationships do not necessarily establish cause and effect. Such causal relationships have not been well defined in the etiology of MODS. Part of the difficulty in MODS research is that a truly characteristic animal model for MODS has yet to be established. Much of what we know relies heavily on the assumption that organ pathology secondary to clinical events that precede MODS are the precursors to the organ dysfunction syndrome. Consequently, several animal models have been employed that mimic various aspects of sepsis, including infection, inflammation, and circulatory shock (Table 1).

Despite these limitations, and a paucity of definitive interventional experiments, a number of unifying theories have arisen regarding the pathophysiology of MODS. These theories are interrelated and share the common observation that proximal triggers of inflammation secondary to clinical events such as sepsis and circulatory shock can initiate a cascade of events that leads to tissue injury. Whether this tissue injury necessarily progresses to the sustained, aberrant syndrome of inflammation and hypermetabolism that defines MODS remains to be established.

MEDIATOR THEORY

From the identification of TNF as an endogenous link between infection and shock emerged the mediator theory of MODS. This theory attributes the organ dysfunction and disseminated inflammation of MODS to the effects of a multitude of endogenous agents released during stimulation of the immune response. These agents interact to amplify or modulate each other, with continued recruitment of cellular and humoral inflammatory components. Products of activated leukocytes and the complement cascade can effect damage to the microcirculation. Vasoreactive agents can override normal vascular homeostatic mechanisms and impair oxygen delivery to various organs. Resultant tissue injury may then

Table 1. Experimental Models of Sepsis.

Model	Animal	Mechanism	Condition Modelled	References
IV or IP Endotoxin	Babboon, Rat, Mouse	Inflammation	Inflammation	13,14,16,17, 18,23
IP Zymosan	Mouse, Rat	Inflammation	Inflammation	59,61,69
Cecal Ligation and Puncture	Rat	Controlled Peritonitis	Sepsis	65,66,67,68
Wiggers Model	Dog, Pig, Rat	Hemorrhage	Hemorrhagic Shock	47,62,64
Cardiac Tamponade	Pig	Tamponade	Cardiogenic Shock	26,45,63
Gut Ischemia-Reperfusion	Cat, Rat	Ischemia-reperfusion	Shock-Resuscitation with intestinal compromise	9,11,27,28, 31,32,33,60

perpetuate the inflammatory process through a positive feedback loop. This sustained and aberrant activation of the immune response is self-perpetuating, and may lead to the "generalized autodestructive inflammation" that characterizes MODS.[5]

Much of the ongoing research in this field has focused on the various inflammatory responses elicited by lipopolysaccharide (LPS), the active component of bacterial endotoxin. LPS is a component of the bacterial wall of gram-negative organisms and, when injected into animals, results in systemic hypotension, generalized inflammation, and tissue injury even in the absence of intact, viable bacteria. For this reason, LPS is often considered a mediator of sepsis, despite its initially exogenous origin. Its effects are mediated primarily via the cytokine TNF-∝, discussed above.

Growing evidence supports the role of TNF-∝ as a central mediator of endotoxic shock. After its initial isolation and sequencing in 1985[2], Tracey and his colleagues demonstrated that the injection of recombinant TNF-∝ into rats resulted in pulmonary hemorrhage and inflammation, acute tubular necrosis of the kidney, and intestinal ischemia and infarction.[3] Subsequent experiments showed that monoclonal antibodies against TNF-∝ protected baboons against organ injury sustained after infusion of an otherwise lethal dose of live E. coli.[10] Moreover, anti-TNF-∝ antibody has been found to ameliorate pulmonary microvascular injury after intestinal or hepatic ischemia/reperfusion.[11,12] This effect may be due in part to blockade of TNF-mediated neutrophil activation.[11]

Other important inflammatory mediators include interleukin-1-alpha (IL-1), released by endothelial cells, which effects the same hypotension, tachycardia, and lactic acidosis as endotoxin when injected into experimental animals.[13] Platelet activating factor (PAF) has also been implicated as an effector of tissue injury after LPS administration. IV administration of PAF produced necrosis and vascular congestion of the stomach and small intestine in the rat indistinguishable from endotoxin-induced injury.[14] PAF has also been shown to be an important mediator of neutrophil adhesion in the microvasculature after ischemia/reperfusion.[15] Moreover, PAF blockade with receptor antagonists attenuated endotoxin-induced injury and hemodynamic changes.[14,16]

Nitric oxide appears to play a central role in mediating the vascular hyporeactivity induced by endotoxin.[17,18] The nitric oxide free radical, previously known as endothelial derived relaxing factor (EDRF), is a potent vasodilator and second messenger. Inhibition of nitric oxide synthesis reduces the hypotension seen in rats treated with LPS.[17]

Thus there are a growing number of endogenous mediators that fulfill Koch-Dale criteria[19] for a causal relationship to MODS. Clinical trials with HA-1A, the human monoclonal antibody against endotoxin, have shown some beneficial results in reducing mortality.[20] The redundancy and complexity of the immune response, however, may necessitate a multi-inhibitor approach. Blockade of proximal mediators in the inflammatory cascade, such as LPS and TNF, appear to hold the most promise.

GUT TRANSLOCATION THEORY

The question that arises from the mediator theory is, how is this cycle of aberrant immune responsiveness initiated or sustained? It is evident that infection or the byproducts thereof (*ie.* LPS) can initiate the inflammatory cascade. However, often no infectious focus is found, or even when successfully eradicated, the process progresses. There has been continued speculation over the last four decades as to the role that intestinal bacteria play in the pathogenesis of disease. In 1959, Dr. Jacob Fine proposed that absorption of bacterial endotoxin from the GI tract was responsible for death from "irreversible shock" after trauma or hemorrhage.[21] While this concept was later disputed and dismissed as unique to the dog, there has been renewed interest in the alimentary tract as a potential source of bacteria or toxin that initiates or sustains MODS. Marshall and Meakins refer to the gastrointestinal tract as the "motor of inflammation" in multiple organ failure, providing the reservoir of pathogens by which the hyperinflammatory state is sustained.[22]

It is clear that, in addition to the process of nutrient digestion and absorption, the gut performs an equally important task of providing a protective barrier between the internal milieu and the outside world (*ie.* the gut lumen). This barrier function protects us from the caustic digestive juices of the alimentary tract as well as the overwhelming population of bacteria and their toxins that reside within our bowel lumen. At the same time, its digestive role necessitates a complex villus and microvillus architecture to provide maximal functional surface area to facilitate nutrient absorption. It is perhaps more remarkable that this barrier integrity is ordinarily maintained than that it may be compromised in disease.

The anatomic basis for this barrier function is the integrity of the mucosal epithelial layer. In normal intestine, the villi are covered from crypt to tip with an epithelial layer comprised of enterocytes joined to one another by intercellular tight junctions to form a continuous, relatively impermeable monolayer. Disruption of this monolayer, however, leaves denuded villi with exposed basement membrane which could allow the egress of intraluminal bacteria and toxins into the mesenteric lymph nodes and into the blood stream.

The role of bacterial translocation in the evolution of MODS remains controversial. There is clear evidence in animal models of circulatory shock that bacteria can be induced to translocate across a compromised mucosal barrier,[23,24] sometimes even before morphological evidence of injury is apparent. However, studies in severe trauma patients have failed to demonstrate this phenomenon clinically.[25] These negative findings may reflect rapid venous clearance of bacteria from the bloodstream, or inadequate mesenteric lymph node sampling. The clinical significance of this phenomenon remains to be quantitated. Clearly, shock and resuscitation increase intestinal mucosal permeability. Moreover, whether or not it initiates the process, bacterial translocation may serve to sustain the hyperinflammatory state once triggered. However, translocation of intraluminal toxins may have greater importance clinically, for which bacterial translocation serves only as a marker. The gut has also been implicated as a source of cardiotoxin after ischemia and may contribute to the primary myocardial dysfunction seen in MODS (see above).[9]

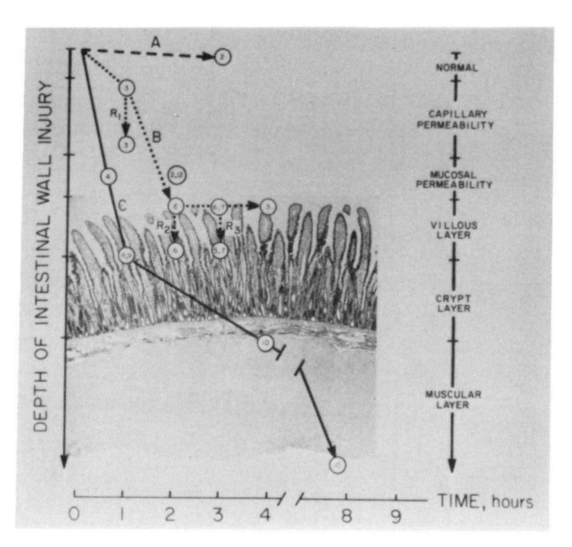

Fig. 1. This figure illustrates the severity gradient of injury after intestinal ischemia with initial involvement of the villus tips followed by progression of injury down the vili and eventual involvement of the muscularis with increasingly severe or prolonged ischemia. Initial injury is manifest as increased capillary and mucosal permeability, even before morphologic changes are evident. R1 to R3 indicate the contribution of subsequent reperfusion to the initial ischemic injury. The encircled numbers within the figures refer to specific references in the original article. (from ref. 8 with permission)

REPERFUSION THEORY

How does the intestine sustain the mucosal injury responsible for disruption of its barrier function? As mentioned previously, the epithelial layer is particularly susceptible to ischemia. In the villus, there is a well-described severity gradient of mucosal injury seen following ischemia/reperfusion, wherein villus tips are initially and primarily involved, with progression of injury down the villus with increasingly severe or prolonged shock states[8] (Figure 1). This gradient has been attributed to the villus countercurrent exchange mechanism in which arteriovenous short-circuiting of oxygen, exacerbated during low flow states, contributes to villus tip hypoxia. Moreover, blood is preferentially redirected away from the intestine during circulatory shock secondary to a profound angiotensin-mediated disproportionate mesenteric vasoconstriction.[26] Thus villus tips can experience a significant decrease in oxygen delivery consequent to even mild episodes of hypotension.

In 1981, Granger and his colleagues examined the postischemic (resuscitation) injury in the feline small intestine. They found that infusion of superoxide dismutase (SOD), a specific scavenger of the toxic oxygen metabolite superoxide, ameliorated the microvascular injury sustained by the small intestine after one hour of portal ischemia and subsequent reperfusion.[27] Moreover, the SOD was effective when given near the end of ischemia, prior to reperfusion. Later experiments found that inhibition of the superoxide-generating enzyme xanthine oxidase by allopurinol would inhibit not only the microvascular injury, but also the intestinal epithelial necrosis seen after longer periods of ischemia.[28] These experiments clearly demonstrated the gut mucosa's susceptibility to reperfusion-mediated injury, and implicated the generation of toxic oxygen metabolites as an important mechanism of injury.

Reperfusion injury can be defined as the damage that occurs to an organ at the time of resumption of blood flow after an episode of ischemia. This injury is distinct from the injury caused by the ischemia *per se*, and must be distinguished from the predetermined evolution of an injury that had been sustained during that ischemic period. The biochemical events leading to reperfusion injury are illustrated in Figure 2. Catabolism of ATP during ischemia leads to an accumulation of the purine metabolites hypoxanthine and xanthine. At the same time, ischemia appears to mediate the (probably proteolytic) conversion of xanthine dehydrogenase to xanthine oxidase. While xanthine dehydrogenase catalyzes the two step oxidation of purines to urate using NAD+ as an electron acceptor, xanthine oxidase can only use molecular oxygen, thereby generating the superoxide radical as a by-product of this oxidation. When oxygen is reintroduced suddenly and in excess at the time of reperfusion, superoxide is generated, which then triggers a free radical chain reaction that ultimately leads to microvascular injury, partially via the upregulation of endothelial adhesion molecules and the consequent arrest, trapping, and activation of circulating neutrophils within the microvasculature.[27,29]

In Granger's feline intestinal model, injury could be ameliorated by (1) scavenging of hydrogen peroxide with catalase,[30] (2) scavenging of hydroxyl radicals with dimethylsulfoxide (DMSO) or mannitol,[31] or (3) preventing formation of hydroxyl radicals by chelating iron with desferoxamine or transferrin.[32] These studies suggest that, while superoxide is the initial trigger of free radical generation, tissue injury is triggered by the highly reactive hydroxyl radical.

Further studies in the same model suggest that the neutrophil plays an important role in the

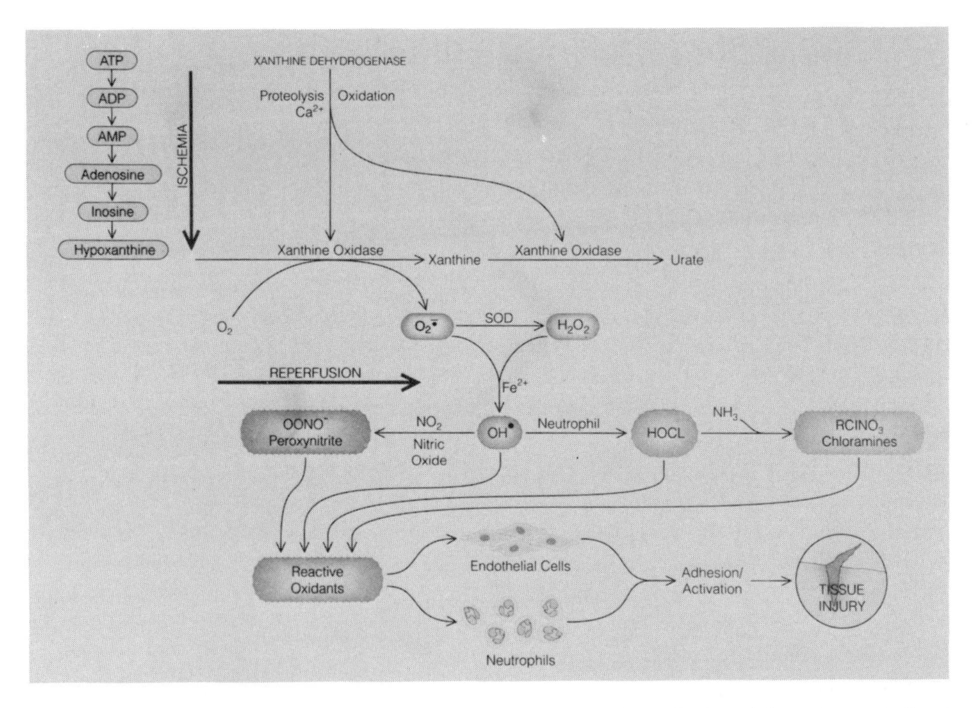

Fig. 2. This figure illustrates the biochemical chain of events that culminates in tissue injury after postischemic resuscitation. During ischemia, the purine metabolites hypoxanthine and xanthine accumulate from the breakdown of ATP. Simultaneously, xanthine dehydrogenase is converted to xanthine oxidase. With the introduction of oxygen at the time of reperfusion, superoxide is generated from the xanthine oxidase-catalyzed oxidation of purines. Subsequent generation of other reactive oxidants as well as neutrophil adhesion and activation result in tissue injury. (from ref. 29 with permission)

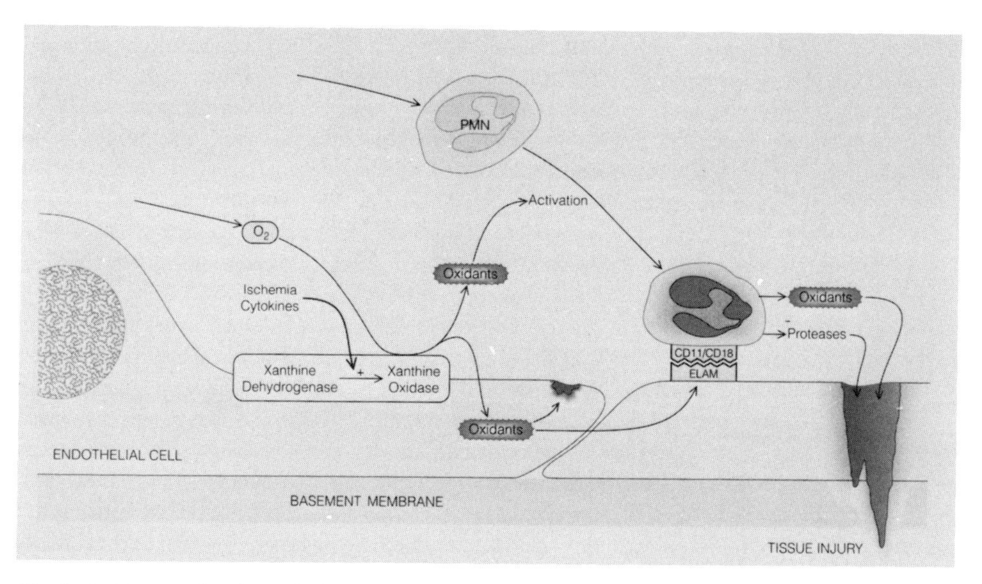

Fig. 3. Oxidant-mediated reperfusion injury appears to start at the microvascular level, at the interface of the endothelium with the bloodstream. While generation of oxidants via the xanthine oxidase pathway may cause substantial microvascular injury by itself, an important component of reperfusion injury appears to be the subsequent activation and adhesion of circulating neutrophils, which contribute to injury by production of oxidants and highly toxic proteases. This latter component constitutes the neutrophil amplifier. (from ref. 29 with permission)

final mediation of reperfusion injury. The injury can be largely prevented by treatment with (1) antineutrophil serum, (2) monoclonal antibody that blocks neutrophil binding to the endothelial cell,[33] or (3) agents that inhibit platelet activating factor, which in part mediates neutrophil-endothelial cell interaction.[15] Thus, it is likely that the initiator of tissue injury is xanthine-oxidase generated superoxide, with the neutrophil as a subsequent amplifier of injury, as illustrated in Figure 3.

While first studied extensively in the intestine, reperfusion injury has subsequently been reported in a large number of organs susceptible to injury after ischemia/reperfusion. The liver, for example, demonstrates hepatocellular necrosis in the pericentral region after ischemia/reperfusion which can be ameliorated by prior infusion of SOD and catalase or allopurinol, at least in some studies.[34,35] The liver is also particularly rich in xanthine oxidoreductase, which has been localized to this same pericentral region by histochemical analysis.[36] Other organs, such as the pancreas, heart, kidney, and central nervous system also manifest tissue injury after postischemic resuscitation that can be blocked by various antioxidant agents.[37,38,39,40] That circulatory insufficiency often precedes multiple organ dysfunction syndrome suggests that widespread reperfusion injury may account for the microvascular injury and sustained inflammation that we see.

The universality of reperfusion injury has lead investigators to examine the endothelial cell as a possible source of xanthine oxidase-generated superoxide. Xanthine oxidase is abundant within the microvascular endothelium of all organs studied (including the human heart and the brain, in which parenchymal levels of xanthine oxidase have been difficult to detect).[41] Both *in vivo* and *in vitro*, the endothelial trigger mechanism can be activated by anoxia via xanthine dehydrogenase to oxygenase conversion.[42] Recently, we have found that the enzyme is expressed constituitively as xanthine dehydrogenase, both in the cytoplasm and on the outer surface of the endothelial cell membrane.[43] This localization would explain the efficacy of circulating (extracellular) high molecular weight antioxidants such as superoxide dismutase in ameliorating reperfusion injury.

COMPETING GENE EXPRESSION THEORY

To this point, we have discussed how clinical events of sepsis and ischemia/reperfusion can give rise to proximal triggers of injury such as inflammatory cytokines and reactive oxygen metabolites. In our laboratory, we have looked specifically at how cellular genetic responses to these inducers might contribute to the dysregulated metabolism and disseminated inflammation of MODS. In 1989, initial studies in a porcine model of cardiogenic shock, focusing on the liver as an organ involved early in the evolution of MODS, revealed that messenger RNA representing two distinct programs of stress gene expression, the acute phase response and the heat shock response, accumulated during resuscitation.[44] The acute phase response is the cell-type specific response to adverse environmental stimuli, and is an integrative response involved in maintaining systemic homeostasis in the multicellular organism. The liver's secretory contribution to this response are the pro-coagulants and the anti-proteases that are collectively called the acute phase reactants. This response is elicited by such mediators as LPS, TNF, and IL-1. Clinically, acute phase reactants can be identified in the blood after trauma, infection, surgery, or even a mild stressor such as anesthesia alone. The heat shock response, on the other hand, is the cell-type generic response to environmental stress, and is involved in maintaining intracellular homeostasis.[45] Heat shock proteins are involved in stabilizing or restoring tertiary protein structure, or eliminating permanently denatured proteins. The clinician is blind to the production of heat shock proteins, since they are not secreted and therefore cannot be measured in the plasma. Importantly, in our porcine model, heat shock gene expression was seen in the pigs that appeared the most ill following shock/resuscitation.

The heat shock response is the primitive stress response, and has been remarkably preserved from E. coli to man. With progression from unicellular to multicellular organisms, the need for a systemic, integrative stress response has resulted in the subsequent evolution of the acute phase response. In the hepatocyte, we have found that these two programs of gene expression are both exclusive and prioritized, both *in vivo* and *in vitro*.[46,47] Induction of the heat shock response effectively preempts both acute phase gene expression, as well as constitutive gene expression. Thus, while these two stress responses are individually beneficial, competition between these two programs may have adverse consequences. For example, if a patient sustained an injury sufficient to induce enough hepatocytes to make heat shock proteins at the expense of acute phase as well as liver-specific proteins, then what we may clinically perceive as liver failure may in fact be the individual hepatocytes executing a genetic program of cellular defense that overrides the needs of the organism as a whole. This could well account for the apparent organ failure in the face of hypermetabolism that characterizes the organ dysfunction of MODS.

Subsequent experiments were performed to identify the clinical inducer responsible for heat shock gene expression. In the isolated, perfused liver subjected to two hours of warm ischemia followed by reperfusion, heat shock gene expression was apparent one hour after the reestablishment of blood flow, but not after prolonged ischemia alone.[48] Moreover, this response could be *completely* ablated with the intravascular infusion of superoxide dismutase at the time of reperfusion (Figure 4). Thus superoxide generated at the time of postischemic reperfusion is here the necessary clinical inducer of this response. Importantly, this superoxide must be generated either at or near the intravascular space, again suggesting a role for endothelial xanthine oxidase as a signal transducer.

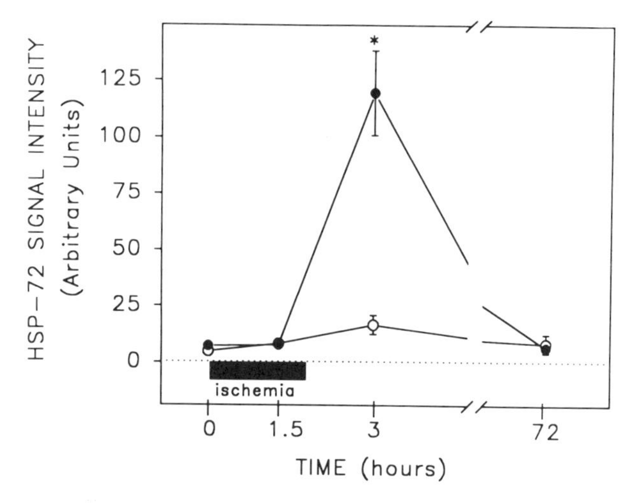

Fig. 4. This figure illustrates the role of superoxide in the induction of heat shock gene expression. When isolated, perfused porcine livers were subjected to 120 minutes of ischemia followed by reperfusion, heat shock gene expression was induced, as evidenced by accumulation of hsp 72 mRNA detected by Northern analysis (closed circles). Ischemia alone was insufficient to induce such a response. Moreover, intravascular infusion of superoxide dismutase (SOD) at the time of reestablishment of blood flow completely ablated heat shock gene expression (open circles). These findings suggest that superoxide, generated at the time of reperfusion and in or near the intravascular space, is an important clinical inducer of the heat shock response. (from ref. 48 with permission)

We next examined the endothelial cell, the putative target of oxidants generated within the microvasculature, and exposed them to various inducers of the acute phase and heat shock responses to see how competition between these two programs were manifest. When porcine endothelial monolayers were treated with either sodium arsenite or heat, standard inducers of the heat shock response *in vitro*, global reduction in other protein synthesis was observed, again demonstrating the ability of the heat shock response to preempt other programs of gene expression. Exposure of these cells to LPS, an acute phase inducer, was well tolerated. However, sequential exposure to LPS followed by heat shock induction aberrantly triggered a genetic program of cell death known as apoptosis.[49] Apoptosis is ordinarily a physiologic mechanism used to remove senescent cells from constantly expanding cell populations, as in intestinal epithelium or the lymphopoietic system, or for remodelling of tissues during embryologic development. It is characterized by morphologic changes that include cellular shrinkage, cytoplasmic blebbing, and formation of apoptotic bodies.[50] Its biochemical hallmark is a distinct pattern of internucleosomal DNA cleavage, wherein DNA is cleaved where it is exposed between nucleosomes, thus generating fragments which are multiples of the internucleosomal distance (160 base pairs).[51] Whereas apoptosis is a physiologic mechanism in certain cell types, endothelial apoptosis is pathologic. This *in vitro* model mimics the two-hit initiation of MODS, where an otherwise tolerable level of low grade sepsis is followed by an equally survivable episode of hypotension and resuscitation. This effect could be mimicked precisely when LPS exposure was followed by protein synthesis inhibition with cycloheximide.[49] This suggests that it may not be the heat shock response *per se* that is deleterious, but rather its preempting of expression of a gene product or products that is necessary to hold this cell suicide program in check. Thus endothelial cell apoptosis may contribute to the microvascular injury and disseminated inflammation of MODS. In reperfusion

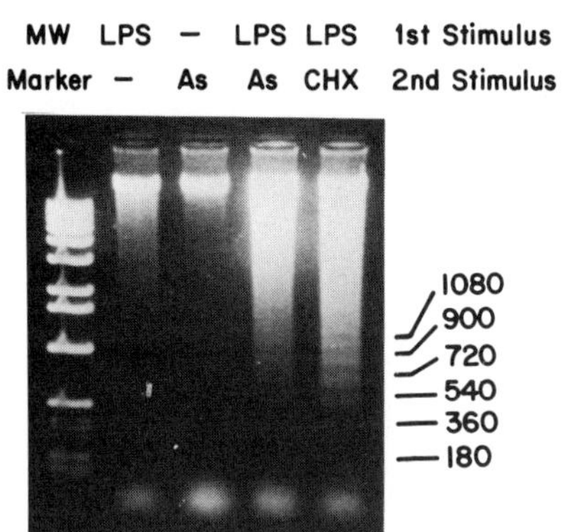

Fig. 5. Programmed endothelial cell death may contribute to the microvascular injury seen in MODS. This figure shows a gel electrophoretic analysis of low molecular weight DNA obtained from porcine endothelial cells after exposure to the stimuli indicated. When cells are treated with either LPS or sodium arsenite (As), a chemical inducer of the heat shock response, no significant fragmentation is observed. However, when cells are treated first with LPS followed by protein synthesis inhibition either by cycloheximide (CHX) or by heat shock induction with As, a characteristic "ladder" pattern of DNA fragmentation is observed, which is the biochemical marker for programmed cell death. Morphologically, there is disruption of the endothelial monolayer, cell shrinkage, and detachment, corresponding to loss of viability. (from ref. 52 with permission)

injury, superoxide is viewed as an initial trigger of indiscriminate free radical endothelial damage. We propose that superoxide may also serve as an inducer of heat shock gene expression resulting in genetically mediated cell death.

Subsequent experiments were performed to elucidate the mechanism by which heat shock induction in LPS-primed endothelial cells leads to apoptosis. When dimethyl sulfoxide, an hydroxyl radical scavenger, or *o*-phenanthroline, an iron chelator, was applied prior to LPS priming, cell killing was ameliorated, suggesting a role for intracellular hydroxyl radical as a non-lethal signal transducer.[52] Extracellular superoxide dismutase and catalase were ineffective in preventing this injury. Moreover, application of the thiol reducing agents dithiothreitol and n-acetylcysteine prior to heat shock induction with sodium arsenite was also protective against cell death, suggesting a role for the redox status of the cell in the execution of apoptosis (unpublished data).

Hockenberry and his colleagues have also found that antioxidants including N-acetyl cysteine and glutathione peroxidase will prevent apoptosis *in vitro*.[53] Moreover, the gene product of the proto-oncogene bcl-2, which has been found to inhibit apoptotic cell death in a variety of cell types,[54] is in fact a naturally occurring antioxidant that ameliorates hydrogen peroxide- and menadione-induced oxidative deaths.[53]

Pathologic apoptosis in circulatory shock may be a more general mechanism contributing to the development of MODS. Recently, we have found that resuscitation from hemorrhagic shock potentiates physiologic apoptosis in rat intestinal epithelium (unpublished data). This pathologic cell death may contribute to the disruption of the epithelial barrier. That resuscitation is a necessary component of this accelerated apoptosis again suggests a role for reactive oxygen species in the induction of this genetic program.

Cell injury, dysfunction, and death in MODS have been traditionally viewed as the passive sequelae of overwhelming infectious and metabolic insults. We offer a novel theory that adverse competition between two otherwise beneficial programs of stress gene expression may be a common mechanism underlying both the dysregulated metabolism and the disseminated inflammation of MODS. That cell death may in fact be a genetically executed process suggests a greater potential for therapeutic manipulation.

OVERVIEW AND PERSPECTIVE

In sum, multiple organ dysfunction syndrome represents a final common aberrant host response to a variety of complex and independent stimuli. The question that follows is, how has such a maladaptive host response evolved over time without an apparent natural selective advantage? Clearly, the most common cause of death of primitive animals and man through the ages has been infection. Today, infectious diarrhea accounts for the greatest number of deaths worldwide. All of the host responses that are elicited in the evolution of MODS are involved in the trapping and killing of bacteria and the control of hemorrhage.[55] In MODS, however, these responses are taken from the beneficial to the pathologic extreme, without the normal negative feedback loops seen in other systems (for example, the endocrine system) to keep the response in check. How did these evolve?

Prior to our ability to intervene, people did not survive severe infection or injury. Thus the pathologic extreme of the activation of the immune response went unopposed because there was no selective pressure to develop a mechanism to counter it. Now, with the advent of paramedics, blood banks, surgeons, and critical care, initial survival after overwhelming infection or massive hemorrhage is seen commonly. We are only now confronted with the challenge of managing the consequences of the pathologic extremes of the biologic responses that were our only armament against such insults. The inflammatory response is beneficial for its ability to localize infection, engulf and kill bacteria, and augment clot formation to control hemorrhage. Hyperinflammation, however, may lead to tissue injury from indiscriminate phagocytosis, microvascular injury, and thrombosis. Mesenteric vasoconstriction in the face of severe blood loss may be helpful in preferentially redirecting blood flow to vital organs such as the brain and the heart. Non-occlusive mesenteric ischemia, or perhaps bacterial translocation, however, may be the attendant morbidity. Heat shock gene expression is important in maintaining the integrity of protein structure within the individual cell. Preemption of other vital programs of gene expression, including genes necessary for cell survival, may lead to cellular dysfunction or *programmed* cell death.

Reperfusion injury, on the other hand, probably does not occur in nature. Why would such an elaborate cascade mechanism therefore exist? We have proposed that this mechanism has been favored by natural selection as a means of initiating reticuloendothelial function via the trapping and killing of circulating pathogens within the microvasculature. Its triggering by the limited proteolysis of xanthine dehydrogenase to xanthine oxidase ("D to O conversion") would be a convenient strategy for more rapid activation than could be achieved by changes in gene expression. In this way, this system would be analogous to activation of the clotting cascade, the complement system, the renin-angiotensin system and protein kinase-activated intracellular control systems, and many other regulation cascades that are characteristically activated by limited proteolysis. Since D to O conversion can be facilitated in endothelial cells by a number of cytokines, complement, proteases, bacteria, endotoxin, and oxidants themselves *in vitro*[56] suggests that this activation of xanthine oxidase on the cell surface may be a common mechanism of inflammatory signal transduction, and thereby constitute a major effector of reticuloendothelial function.[55] In this sense, the free radical superoxide, like the free radical nitric oxide, could act as the second messenger.

FUTURE DIRECTIONS

Since the initial description of multiple organ dysfunction syndrome offered by Tilney and his colleagues in the 1970s, therapeutic strategies (as opposed to tactics) in clinical management have changed remarkably little. It is clear given its multifactorial nature that successful treatment will necessitate a multimodality approach. Clearly, elimination of the provoking stimulus remains the paramount strategy. Preliminary studies have suggested that maintenance of intestinal epithelial integrity may be improved by early enteral feeding with diets enriched with glutamine and arginine.[57] Mediator inhibitors such as monoclonal antibodies to TNF-∝ and LPS have had some success in improving survival in clinical trials.[20,58] Antioxidant therapy in patients at high risk for development of MODS may also prove useful. Lastly, the recognition that cellular dysfunction and death may be the result of an active, genetically executed process allows the potential for modulation of this process through gene therapy. We anticipate that this modality of therapy will become an increasingly useful and familiar tool for the clinician, and hopefully improve survival in the intensive care unit.

REFERENCES

1. N.L. Tilney, G.L. Bailey, and A.P. Morgan, Sequential system failure after rupture of abdominal aortic aneurysms: an unsolved problem in postoperative care, *Ann Surg* 178:117 (1973).
2. B. Beutler, D. Greenwald, J.D. Hulmes, M. Chang, Y.-C.E. Pan, J. Mathison, R. Ulevitch, and A. Cerami, Identity of tumour necrosis factor and the macrophage-secreted factor cachectin, *Nature* 316(8):552 (1985).
3. K.J. Tracey, B. Beutler, S.F. Lowry, et al., Shock and tissue injury induced by recombinant human cachectin, *Science* 234:470 (1986).
4. Members of the American College of Chest Physicians/Society of Critical Care Medicine Consensus Conference Committee, Definitions for sepsis and organ failure and guidelines for the use of innovative therapies in sepsis, *Crit Care Med* 20(6):864 (1992).
5. R.J.A. Goris, T.P.A. te Boekhorst, J.K.S. Nuytinck, and J.S.F. Gimbrere, Multiple-organ failure, *Arch Surg* 120:1109 (1985).
6. F.B. Cerra, Hypermetabolism, organ failure, and metabolic support, *Surgery* 101(1):1 (1987).
7. J.R. Border, Multiple systems organ failure, *Arch Surg* 216(2):111 (1992).
8. U. Haglund, G.B. Bulkley, and D.N. Granger, On the pathophysiology of intestinal ischemic injury, *Acta Chir Scand* 153:321 (1987).
9. U. Haglund and O. Lundgren, Intestinal ischemia and shock factors, *Federation Proc* 37:2729 (1978).
10. K.J. Tracey, U. Fong, D.G. Hesse, et al., Anti-cachectin/TNF monoclonal antibodies prevent septic shock during lethal bacteraemia, *Nature* 330(17):662 (1987).
11. M.G. Cary, K.S. Guice, K.T. Oldham, D.G. Remick, and S.L. Kunkel, Evidence for tumor necrosis factor-induced pulmonary microvascular injury after intestinal ischemia-reperfusion injury, *Ann Surg* 212(6):694 (1990).
12. L.M. Colletti, D.G. Remick, G.D. Burtch, S.L. Kunkel, R.M. Strieter, and D.A. Cambell, Jr., Role of tumor necrosis factor-∝ in the pathophysiologic alterations after hepatic ischemia/reperfusion injury in the rat, *J Clin Invest* 85:1936 (1990).
13. E. Fischer, M.A. Marano, A.E. Barber, et al., Comparison between effects of interleukin-1∝ administration and sublethal endotoxemia in primates, *Am J Physiol* 261:R442 (1991).
14. J.L. Wallace, G. Steel, B.J.R. Whittle, V. Lagente, and B. Vargaftig, *Gastroenterology* 93:765 (1987).
15. P. Kubes, G. Ibbotson, J. Russell, et al., Role of platelet-activating factor in ischemia/reperfusion-induced leukocyte adherence, *Am J Physiol* 259:G300 (1990).

16. S. Chang, C.O. Feddersen, P.M. Henson, and N.F. Voelkel, Platelet-activating factor mediates hemodynamic changes and lung injury in endotoxin-treated rats, *J Clin Invest* 79:1498 (1987).
17. C. Thiemermann and J. Vane, Inhibition of nitric oxide synthesis reduces the hypotension induced by bacterial lipopolysaccharides in the rat in vivo, *Eur J Pharm* 182:591 (1990).
18. G. Julou-Schaeffer, G.A. Gray, I. Fleming, C. Schott, J.R. Parratt, and J.C. Stoclet, Activation of the L-arginine-nitric oxide pathway is involved in vascular hyporeactivity induced by endotoxin, *J Cardiovasc Pharm* 17(Suppl. 3):S207 (1991).
19. E. Neugebauer, W. Lorenz, D. Maroske, and W. Barthlen, Mediators in septic shock: strategies of securing them and assessment of their causal significance, *Chirurg,* 58(7):470 (1987) (English abstract).
20. E.J. Ziegler, C.J. Fisher, Jr., C.L. Sprung, et al., Treatment of gram-negative bacteremia and septic shock with HA-1A human monoclonal antibody against endotoxin, *N Engl J Med* 324:429 (1991).
21. J. Fine, E.D. Frank, H.A. Ravin, S.H. Rutenberg, and F.B. Schweinburg, The bacterial factor in traumatic shock, *N Engl J Med* 260:214 (1959).
22. J.C. Marshall and J.L. Meakins, The gastrointestinal tract: the 'motor' of MOF, *Arch Surg* 121:197 (1986).
23. E.A. Deitch, R.Berg, and R. Specian, Endotoxin promotes the translocation of bacteria from the gut, *Arch Surg* 122:185 (1987).
24. A.J. Sori, B.F. Rush, Jr., T.W. Lysz, S.Smith, and G.W. Machiedo, The gut as source of sepsis after hemorrhagic shock, *Am J Surg* 155:187 (1988).
25. A.B. Peitzman, A.O. Udekwu, J. Ochoa, and S. Smith, Bacterial translocation in trauma patients, *J Trauma* 31(8):1083 (1991).
26. P.M. Reilly, S. MacGowan, M. Miyachi, H.J. Schiller, S. Vickers, and G.B. Bulkley, Mesenteric vasoconstriction in cardiogenic shock in pigs, *Gastroenterology* 102:1968 (1992).
27. D.N. Granger, G. Rutilli, and J.M. McCord, Superoxide radicals in feline intestinal ischemia, *Gastroenterology* 81:22 (1981).
28. D.A. Parks, G.B. Bulkley, D.N. Granger, et al., Ischemic injury in the cat small intestine: role of superoxide radicals, *Gastroenterology* 82:9 (1982).
29. P.M. Reilly, H.J. Schiller, and G.B. Bulkley, Reactive oxygen metabolites in shock, *in:* "Care of the Surgical Patient," D.W. Wilmore, M.F. Brennan, A.H. Harken, J.W. Holcroft, and J.L. Meakins, eds., Scientific American, New York (1991).
30. D.N. Granger, M.E. Hollwarth, and D.A. Parks, Ischemia-reperfusion injury: role of oxygen-derived free radicals, *Acta Physiol Scand Suppl* 548:47 (1986).
31. D.A. Parks and D.N. Granger, Ischemia-induced vascular changes: role of xanthine oxidase and hydroxyl radicals, *Am J Physiol* 245:G285 (1983).
32. L.A. Hernandez, M.B. Grisham, and D.N. Granger, A role for iron in oxidant-mediated ischemic injury to intestinal microvasculature, *Am J Physiol* 253:G49 (1987).
33. L.A. Hernandez, M.B. Grisham, B. Twohig, et al., Role of neutrophils in ischemia-reperfusion-induced microvascular injury, *Am J Physiol* 253:H699 (1987).
34. D. Adkison, M.E. Hollwarth, J.N. Benoit, et al., Role of free radicals in ischemia-reperfusion injury to the liver, *Acta Physiol Scand Suppl* 548:101 (1986).
35. G. Nordstrom, T. Seeman, P.O. Hasselgren, Beneficial effect of allopurinol in liver ischemia, *Surgery* 97:679 (1985).
36. A. Kooij, W.M. Frederiks, R. Gossarau, and C. Van Noorden, Localization of xanthine oxidoreductase activity using the tissue protectant polyvinyo alcohol and final electron acceptor tetranitro BT, *J Histochem and Cytochem* 39(1):87 (1991).
37. H. Sanfey, G.B. Bulkley, and J.L. Cameron, The pathogenesis of acute pancreatitis: the source and role of oxygen-derived free radicals in three different experimental models, *Ann Surg* 201:633 (1985).
38. J.R. Stewart, S.L. Crute, V. Loughlin, et al., Prevention of free radical-induced myocardial reperfusion injury with allopurinol, *J Thorac Cardiovasc Surg* 90:68 (1985).
39. M.S. Paller, J.R. Hoidal, and T.F. Ferris, Oxygen free radicals in ischemic acute renal failure in the rat, *J Clin Invest* 74:1156 (1984).
40. T.H. Liu, J.S. Beckman, B.A. Freeman, et al., Polyethylene glycol-conjugated superoxide dismutase and catalase reduce ischemic brain injury, *Am J Physiol* 256:H589 (1989).
41. S. Vickers, J. Hildreth, F. Kujada, et al., Immunohistoaffinity localization of xanthine oxidase in the microvascular endothelial cells of porcine and human organs, *Circ Shock* 31:87 (1990) (Abstract).
42. R.E. Ratych, R.S. Chuknyiska, and G.B. Bulkley, The primary localization of free radical generation after anoxia/reoxygenation in isolated endothelial cells, *Surgery* 102:122 (1987).
43. H.J. Schiller, S. Vickers, J. Hildreth, I. Mather, F. Kujada, and G.B. Bulkley, Immunoaffinity localization of xanthine oxidase on the outside surface of the endothelial cell plasma membrane, *Circ Shock* 34:A435 (1991).

44. T.G. Buchman, D.E. Cabin, S.Vickers, C.S. Deutschman, E.Delgado, M.M. Sussman, and G.B. Bulkley, Molecular biology of circulatory shock. Part II. Expression of four groups of hepatic genes is enhanced after resuscitation from cardiogenic shock, *Surgery* 108:559 (1990).
45. W.J. Welch, The mammalian stress response: cell physiology and biochemistry of stress proteins, *in*: "Stress Proteins in Biology and Medicine," R.I. Morimoto, A. Tissieres, and C. Georgopoulos, eds., Cold Spring Laboratory Press, Cold Spring Harbor (1990).
46. L.O. Schoeniger, P.M. Reilly, G.B. Bulkley, and T.G. Buchman, Heat-shock gene expression excludes hepatic acute-phase gene expression after resuscitation from hemorrhagic shock, *Surgery* 112:355 (1992).
47. D.E. Cabin and T.G. Buchman, Molecular biology of circulatory shock. Part III. Human hepatoblastoma (HepG2) cells demonstrate two patterns of shock-induced gene expression that are independent, exclusive, and prioritized, *Surgery* 108:902 (1990).
48. L.O. Schoeniger, K.A. Andreoni, G.R. Ott, T.H. Risby, G.B. Bulkley, R. Udelsman, J.F. Burdick, T.G. Buchman, Induction of heat-shock gene expression in postischemic pig liver depends on superoxide generation, *Gastroenterology* 106:177 (1994).
49. T.G. Buchman, P.A. Abello, E.H. Smith, and G.B. Bulkley, Induction of heat shock response leads to apoptosis in endothelial cells previously exposed to endotoxin, *Am J Physiol* 265:H165 (1993).
50. J.F.R. Kerr, A.H. Wyllie, and A.R. Currie, Apoptosis: a basic biological phenomenon with wide ranging implications in tissue kinetics, *Br J Cancer* 26:239 (1972).
51. A.H. Wyllie, Glucocorticoid-induced thymocyte apoptosis is associated with endogenous endonuclease activation, *Nature* 284:555 (1980).
52. P.A. Abello, S.F. Fidler, G.B. Bulkley, and T.G. Buchman, Antioxidants modulate induction of programmed endothelial cell death (apoptosis) by endotoxin, *Arch Surg* 129:134 (1994).
53. D.M. Hockenberry, Z.N. Oltvai, X.M. Yin, C.L. Milliman, and S.J. Korsmeyer, Bcl-2 functions in an antioxidant pathway to prevent apoptosis, *Cell* 75:241 (1993).
54. D. Hockenberry, G. Nunez, C. Milliman, R.D. Schreiber, and S.J. Korsmeyer, Bcl-2 is an inner mitochondrial membrane protein that blocks programmed cell death, *Nature* 348:334 (1990).
55. G.B. Bulkley, Endothelial xanthine oxidase: a radical transducer of inflammatory signals for reticuloendothelial activation, *Br J Surg* 80:684 (1993).
56. H.P. Friedl, G.O. Till, U.S. Ryan, et al., Mediator-induced activation of xanthine oxidase in endothelial cells, *FASEB J* 3:2512 (1989).
57. J.M. Daly, M.D. Lieberman, J. Goldfine, J. Shou, F. Weintraub, E.F. Rosato, and P. Lavin, Enteral nutrition with supplemental arginine, RNA, and omega-3 fatty acids in patients after operation: immunologic, metabolic, and clinical outcome, *Surgery* 112:55 (1992).
58. J. Cohen, A.R. Exley, W. Buutman, R. Own, G. Hanson, J. Lumley, J.M. Aulakh, M. Bodmer, A. Riddle, S. Stephens, and M. Perry, Monoclonal antibody to TNF in severe septic shock, *Lancet* 335:1275 (1990).
59. E.A. Deitch, A.C. Kemper, R.D. Specian, and R.D Berg, A study of the relationships among survival, gut-origin sepsis,and bacterial translocation in a model of systemic inflammation, *J Trauma* 32(2):141 (1992).
60. S. Zhi-Yong, D. Yuan-Lin, and W. Xiao-Hong, Bacterial translocation and multiple system organ failure in bowel ischemia and reperfusion, *J Trauma* 32(2):148 (1992).
61. R.J.A. Goris, W.K.F. Boekholtz, I.P.T. van Bebber, J.K.S. Nuytinck, and P.H.M. Schillings, Multiple-organ failure and sepsis without bacteria, *Arch Surg* 121:897 (1986).
62. I.P.T. van Bebber, C.F.J. Lieners, E.L. Koldewijn, H. Redl, and R.J.A. Goris, Superoxide dismutase and catalase in an experimental model of multiple organ failure, *J Surg Research* 52:265 (1992).
63. R.W. Bailey, G.B. Bulkley, S.R. Hamilton,J.B. Morris, and U.H. Haglund, Protection of the small intestine from nonocclusive mesenteric ischemic injury due to cardiogenic shock, *Am J Surg* 153:108 (1987).
64. C.J. Wiggers. "Physiology of Shock." New York: The Commonwealth Fund, 1950.
65. P. Wang, Z.F. Ba, and I.H. Chaudry, Hepatic extraction of indocyanine green is depressed early in sepsis despite increased hepatic blood flow and cardiac output, *Arch Surg* 126(2):219 (1991).
66. W.J. Schirmer, J.M. Schirmer, G.B. Naff, D.E. Fry, Complement activation in peritonitis. Association with hepatic and renal perfusion abnormalities. *Am Surg* 53(12):683 (1987).
67. M. Sodeyama, K.R. Gardiner, S.J. Kirk, G. Efron, and A. Barbul, Sepsis impairs gut amino acid absorption, *Am J Surg* 165(1):150 (1993).
68. H.P. Madden, R.J. Breslin, H.L. Wasserkrug, G.Efron, and A. Barbul, Stimulation of T cell immunity by arginine enhances survival in peritonitis, *J Surg Res* 44(6):658 (1988).
69. R.J.A. Goris, I.P.T. van Bebber, R.M. Mollen, and J.P. Koopman, Does selective decontamination of the gastrointestinal tract prevent multiple organ failure? An experimental study, *Arch Surg* 126(5):561 (1991).

ROLE OF NUTRIENTS IN THE CAUSE AND PREVENTION OF OXYGEN RADICAL PATHOLOGY

Harold H. Draper and William J. Bettger

Department of Nutritional Sciencees
University of Guelph
Guelph, Ontario, Canada N1G 2W1

INTRODUCTION

Nutrients function in both the generation and catabolism of oxygen radicals, i.e., as oxidants and antioxidants. Experimental animals fed large excesses of prooxidant nutrients, such as polyunsaturated fatty acids and iron, develop pathologies that have been firmly linked to the formation of toxic concentrations of oxygen radicals in the tissues. Similar pathologies have been demonstrated to occur as a result of experimental deficiencies of some antioxidant nutrients, notably vitamin E and selenium. Clinical diseases resulting from dietary deficiencies of these nutrients also have been well documented in farm animals. There is evidence that pharmacological intakes of some antioxidants may reduce the risk of developing some clinical diseases associated with oxygen radical pathology in human subjects consuming a normal diet and ameliorate the tissue damage associated with others. It remains unclear, however, whether oxygen radical pathology occurs in humans as a result of inadequate or excessive intakes of nutrients from the general food supply.

This chapter deals primarily with the relationship between the intake of prooxidant and antioxidant nutrients from diets composed of common foodstuffs and the incidence of diseases associated with oxygen radical pathology. The nutrients involved in the metabolism of oxygen radical fall into two interrelated but distinct categories: those that function in the metabolism of radicals of molecular oxygen and those that function in the metabolism of lipoxy radicals (i.e., in lipid peroxidation). The chapter also deals briefly with the current evidence that pharmacological amounts of lipid antioxidants may be efficacious in disease prevention.

COPPER, ZINC AND MANGANESE

Dietary copper, zinc and manganese are required for the synthesis of enzymes that catalyze the dismutation of superoxide anion radical ($O_2\cdot^-$). Cu,Zn superoxide dismutase (CuZnSOD) occurs in the cytosol and Mn superoxide dismutase (MnSOD) occurs in the

Free Radicals in Diagnostic Medicine, Edited by
D. Armstrong, Plenum Press, New York, 1994

mitochondria of all cell types. An extracellular CuZnSOD occurs on the surface of some cell types[1] and at low concentrations in extracellular fluid.[2] The products of this reaction are hydrogen peroxide and molecular oxygen.

$$O_2^{\cdot -} + O_2^{\cdot -} + 2H^+ \longrightarrow H_2O_2 + O_2$$

The importance of superoxide as a direct cause of free radical pathology (and consequently the clinical significance of the metal-dependent SODs) is still not fully clarified.[3,4] Superoxide is a weak oxidant that can dismutate spontaneously in the presence of trace quantities of transition metals. However, there is substantial evidence for a significant physiological role of SOD enzymes in higher as well as lower species.[5] Mutants of lower organisms that lack SODs are more susceptible to oxidants or have a shortened life span. Species of animals that are adaptable to hyperoxia undergo an increase in SOD activity in the lungs when exposed to a high oxygen environment, whereas those that are unadaptable do not. Administration of bacterial endotoxin to rats results in an increase in SOD activity (as well as in catalase and glutathione peroxidase activity) in the lungs that is associated with the development of resistance to hyperoxia. Interest in the physiological role of SOD has been enhanced by evidence that amyotrophic lateral sclerosis (Lou Gerhig's Disease) is associated with a defect in the gene for this enzyme.[6] The two most important reactions of superoxide from the standpoint of oxygen radical pathology appear to be its reduction of ferric to ferrous ions, which react with H_2O_2 to form the highly destructive hydroxyl radical in the Fenton reaction[7] (see below), and its reaction with the nitric oxide radical NO· to produce $ONOO^-$, which decomposes to $NO_2\cdot$ and hydroxyl radical.[8-11]

Dietary copper deficiency is not a recognized clinical condition in human nutrition. Menkes' disease, an X-linked inborn error of copper metabolism resulting in malabsorption of dietary copper and severe depletion of copper stores, has not been demonstrated to result in oxygen radical pathology. Severe dietary copper deficiency in experimental animals causes a significant depression of Cu,ZnSOD and there is evidence of free radical damage in the tissues.[12-15] However, this damage appears to apply to only selected developmental and degenerative processes.[16,17]

Chronic zinc deficiency has been identified as the cause of dwarfism in children in the Middle East[18] and subclincal zinc deficiency has been reported in some North American children. However, neither of these conditions is associated with clear indications of oxygen radical pathology. This is also true of acrodermatitis enteropathica. a condition of severe zinc deficiency in humans caused by a metabolic error in the metabolism of this element. Dietary zinc deficiency has been found to increase the susceptibility of experimental animals to oxidative stress,[19-21] even though there is no decrease in CuZnSOD activity in the tissues.[22] Moreover, zinc deficiency pathology is rapidly reversible,[23] in contrast to the developmental abnormalities seen in dietary copper deficiency in experimental animals.

There is no evidence of manganese deficiency in humans consuming normal diets. Severe manganese deficiency in experimental animals causes a decrease in MnSOD activity in some tissues and an increase in free radical pathology.[24-26] The decrease in MnSOD activity in the liver of manganese deficient rats and chickens is accompanied by an increase in CuZnSOD activity.[5] There is still little information on the physiological importance of this enzyme in the prevention of oxygen radical damage.

Pharmacological supplements of copper, zinc and/or manganese do not produce a supraphysiological elevation of SOD activity in animals. Excess Zn can cause Cu deficiency and a decrease in CuZnSOD activity in the tissues.[27] Large intakes of copper and/or zinc cause an elevation in the tissue concentration of metallothioneins. Both Cu[28] and Zn[29] metallothioneins have been reported to have antioxidant activity.

SELENIUM

In contrast to copper, zinc and manganese, the importance of dietary selenium in the prevention of oxygen radical pathology has been clearly demonstrated in laboratory and domestic animals as well as in human subjects. Several prevalent clinical conditions in domestic animals, notably cardiac and striated muscle dystrophy in cattle and sheep and necrotic liver degeneration in swine, which previously had been attributed to vitamin E deficiency, have been shown to be related primarily to differences in the selenium content of soils and consequently in the selenium content of animal feeds grown on these soils. These conditions frequently yield to supplements of either selenium or vitamin E. The perplexing interrelationship between dietary selenium and vitamin E appeared to be explained by the findings that selenium is a constituent of a peroxidase enzyme (Se-GP_X) that can catalyze the reduction of lipid hydroperoxides to their corresponding alcohols by glutathione, whereas vitamin E serves as a hydrogen donor to lipoxy radicals, resulting in their conversion to hydroperoxides and thereby preventing their propagation of a lipid free radical chain reaction. Subsequently it was determined that a non-Se-dependent glutathione peroxidase (GP_X) is the enzyme primarily involved in the reduction of lipid hydroperoxides. The protective effect of the classical cellular Se-containing enzyme (now called Se-GP_X-1) against lipid peroxidative tissue damage is attributable mainly to its reduction of H_2O_2 and thereby the formation of hydroxyl radicals, which are prime initiators of lipid peroxidation.[30]

$$H_2O_2 + 2GSH \longrightarrow 2H_2O + GSSG + 2H^+$$

There are at least four different genes that code for Se-dependent glutathione peroxidases.[31,32] Each gene product has its own substrate specificity and cellular or subcellular location. In addition, animal tissues contain a non-Se-dependent glutathione peroxidase (GP_X) with a high affinity for organic hydroperoxidases but not for H_2O_2. In addition, Se-dependent glutathione peroxidases catalyze the destruction of free fatty acid hydroperoxides (FAOOH).

$$FAOOH + 2GSH \longrightarrow FAOH + GSSG + H_2O$$

The reduction of phospholipid hydroperoxides (PLOOH) to their corresponding alcohols (PLOH) by glutathione (GSH) is catalyzed by a recently discovered phospholipid-hydroperoxide Se-GP_X[32] and by non-selenium glutathione peroxidases.

$$PLOOH + 2GSH \longrightarrow PLOH + GSSG + H_2O$$

A further Se-GP_X isozyme recently has been found in the gastrointestinal tract of rodents,[31] where it may have a function in the reduction of lipid hydroperoxides in the diet. This would explain the observation that oral administration of ^{14}C-linoleic acid hydroperoxide to the rat results in the appearance of its analog alcohol and other metablic products in the lymph, but not to the appearance of unmetabolized peroxide.[33]

Evidence for cooperativity between Se-GP_X-1 and catalase in the prevention of excess H_2O_2 accumulation in animal tissues has been reviewed by Halliwell and Gutteridge.[5] There is marked variability in the concentrations of the two enzymes at different sites, but the sum of their concentrations is relatively uniform. In erythrocytes, H_2O_2 is metabolized mainly by Se-GP_X-1, as indicated by the fact that humans who have an inborn error in the gene for catalase suffer no serious consequences unless they are given a drug that increases H_2O_2 generation. In the liver, catalase is confined mainly to the peroxisomes, whereas Se-GP_X occurs mainly in the cytosol. The distribution of GSH is similar. Thus, H_2O_2 produced by the cytosolic SOD is metabolized mainly by Se-dependent peroxidases. Considering the weak oxidizing capacity of superoxide compared to that of hydroxyl radicals, preventing the conversion of the former to the latter by the removal H_2O_2 is of central importance in the prevention of oxygen radical pathology. In animals fed a selenium deficient diet, Se-GP_X-1 can undergo extensive depletion before signs of oxidative damage appear, indicating that there is normally excess combined Se-dependent glutathione peroxidase and catalase activity to prevent H_2O_2 accumulation.

The relative amounts of Se-GP_X and non-Se-GP_X activity vary markedly among tissues. The latter enzyme accounts for nearly all the GP_X activity in the rat testis but is absent from the spleen, lung and heart.[5] It is a major component of total activity in the liver of both humans and animals. The relative importance of these enzymes in the prevention of oxidative damage in different tissues is still obscure.

In nutritional terms, vitamin E and selenium have a mutually "sparing effect", i.e., each has a modifying effect on the requirement for the other. Most clinical cases of so-called "selenium-vitamin E deficiency" encountered in veterinary practice signify a low intake of both nutrients and respond to a supplement of either one. The first unambiguous demonstration that selenium is an essential nutrient for humans arose out of field studies on a severe cardiomyopathy in children living in the Keshan District of China (Keshan Disease), which was found to be responsive to selenium but not to vitamin E. This disease is caused by an extremely low level of Se in local soils and in the foodstuffs grown on these soils that constitute most of the diet of the indigenous population.[34]

Commerce in the foodstuffs that comprise the food supply of industrialized countries compensates for the low Se content of food items prepared from plants grown on Se deficient local soils. However, in countries such as New Zealand and Finland, where low Se soils are widespread, human plasma Se-GP_X-1 levels are significantly lower than those in other countries. Extensive clinical observations on New Zealand adults have failed to reveal any unusual disease epidemiology that could be associated with their lower plasma level of this enzyme.[35] A higher incidence of cancer has been reported among Finlanders living in areas where the soil is unusually low in selenium, and a program of soil fertilization with selenium salts has been instituted. However, no unusual incidence of cancer has been observed in the Keshan District of China, where the Se content of the soil is the lowest ever reported and where there is clinical evidence of Se deficiency in the population. Inhibition of chemically induced tumorigenesis in rodents fed a high selenium diet is attributable to the cytotoxic action of this element.

Use of selenium as a dietary supplement in human nutritiion is constrained by its narrow margin of safety. Endemic selenium intoxication caused by contamination of the soil with high-selenium coal dust has been observed at an intake of 1000 μg per day.[36] Some composite North American diets contain more than 200 μg per day.[37] Pharmacological intakes of selenium do not result in supraphysiological levels of Se-GPx in the tissues of animals.

IRON

Although iron has an antioxidant role in metabolism by virtue of its presence in catalase, the predominant interest in this element from the standpoint of oxygen radical pathology lies in its catalytic action in both hydroxyl radical and lipoxy radical generation. Iron salts are highly toxic when administered parenterally to animals and free iron exists only in soluble pool from which it is drawn for the synthesis of iron proteins. About two-thirds of the iron in the body is present in hemoglobin and about 10% in myoglobin. Iron is stored mainly in the liver, spleen and bone marrow in the form of ferritin and hemosiderin. As there is no means of disposing of excess absorbed iron, the absorption of this element from the intestine is tightly regulated so as to prevent iron toxicity at high levels of intake and iron deficiency at low levels.

Catalysis of hydroxyl radical generation by iron is perceived in terms of a two-step process commonly referred to as the iron-catalyzed Haber Weiss reaction. The first step involves the reduction of ferric to ferrous ions by superoxide and the second the reduction of hydrogen peroxide by ferrous ions to form hydroxyl radicals.

$$O_2\cdot^- + Fe^{3+} \longrightarrow Fe^{2+} + O_2$$

$$H_2O_2 + Fe^{2+} \longrightarrow HO\cdot + OH^- + Fe^{3+} \text{ (Fenton reaction)}$$

$$\text{Sum: } O_2\cdot^- + H_2O_2 \longrightarrow O_2 + HO\cdot + OH^-$$

Catalytic iron is released from heme and non-heme proteins in the presence of oxidants (e.g., from ferritin by $O_2\cdot^-$ and from hemoglobin by H_2O_2) and there is evidence that it is released from bound forms as a result of tissue injury. Kozlov *et al.*[38] observed that the amount of Fe(III) bound by the iron-binding agent desferal was nearly twice as great for liver homogenate as for whole liver. Homogenization of animal tissues (Polytron) as opposed to mincing with scissors doubles the values obtained for their apparent content of malondialdehyde (MDA), a product of the oxidative decomposition of polyunsaturated fatty acid hydroperoxides.[39] Parenteral iron administration is the most effective means of stimulating lipid peroxidation in experimental animals, as judged by the resulting increases in pentane exhalation and MDA excretion.

Neither iron deficiency nor iron toxicity appears to be a significant cause of oxygen radical pathology in humans over the range of concentrations present in normal diets. In many countries the diet is enriched with iron for the prevention of anemia. However, iron overload occurs as a result of accidental ingestion of vitamin-mineral tablets by small children, use of iron vessels for brewing alcoholic beverages (so-called African siderosis), and overconsumption of iron supplements. Excess iron absorption also occurs in some individuals as a result of an inborn error of metabolism. Chronic iron overload can lead to hemochromatosis, a syndrome marked by cirrhosis, diabetes, hyperpigmentation of the skin and other symptoms. To what extent these pathologies are due to iron catalysis of oxygen radical generation is conjectural.

POLYUNSATURATED FATTY ACIDS

Polyunsaturated fatty acids (PUFA) constitute the main prooxidant load imposed on the body by the diet. These acids are concentrated in the phospholipids of cell membranes, where they have a vital role in transport phenomena and in metabolic processes catalyzed by membrane-bound enzymes. In addition, they are the precursors of a group of hormone-like substances, the prostanoids, that have pervasive regulatory effects on many metabolic pathways. Prostanoids are produced from specific n-6 and n-3 PUFA by enzymatic peroxidation. Recent reports that various forms of tissue trauma are associated with an increase in the non-enzymatic peroxidation of these acids have fostered an interest in the possible effect of tissue PUFA composition on any pathology that may occur as a result.

Lipid peroxidation is initiated by H abstraction at the α-position allylic to the olefinic groups of PUFA where bond energy is weakest. The main initiators appear to be hydroxyl and lipoxy radicals (see below), though superoxide, singlet oxygen and iron can also act as initiators. Bond strength at the α position decreases with the number of double bonds in the fatty acid chain. The relative peroxidation rates of fatty acids containing from one to six double bonds have been estimated to increase in the order 0.025, 1, 2, 4, 6, 8.[40] Hence eicosapentaenoic acid (EPA, 20:5 n-3) and docosahexaenoic acid (DHA, 22:6 n-3), highly unsaturated fatty acids commonly associated with fish oils, are several times more susceptible to peroxidation than linoleic acid (18:2 n-6), the main PUFA present in cereal oils. Linoleic acid is the most prevalent PUFA in most diets and in total body lipids, but some phospholipid fractions of the brain and retina, for example, contain a preponderance of more highly unsaturated fatty acids. These tissues exhibit an age-dependent increase in fatty acid hydroperoxides.[41] Variability in the PUFA content of different tissues may be a factor in the sites at which lipid peroxidative damage occurs.

$RH + HO\cdot \longrightarrow R\cdot + H_2O$	initiation by hydroxy radical
$R\cdot + O_2 \rightarrow ROO\cdot$	oxygen uptake reaction
$RH + ROO\cdot \longrightarrow R\cdot + ROOH$	catalysis by peroxy radical
$RH + RO\cdot \longrightarrow R\cdot + ROH$	catalysis by alkoxy radical

Alkoxy radicals are formed by the metal-catalyzed decomposition of preformed peroxides.

$$ROOH + Fe^{2+} \longrightarrow RO\cdot + OH^- + Fe^{3+}$$

An analogous reaction results in the generation of peroxy radicals.

$$ROOH + Fe^{3+} \longrightarrow ROO\cdot + H^+ + Fe^{2+}$$

Hence iron catalyzes the initiation of lipid peroxidation by increasing the generation of hydroxyl radicals from hydrogen peroxide in the Fenton reaction and promoting the decomposition of peroxides to form propagating peroxy and alkoxy radicals. Lipid peroxides are relatively stable in the absence of transition metals but are highly unstable in their presence. These considerations underscore the critical importance of catalase and the glutathione peroxidases in removing excess H_2O_2 and preformed lipid peroxides in the control of *in vivo* lipid peroxidation. Release of free iron (or some low molecular weight iron compound) from bound forms may account for the increase in lipid peroxidation associated with tissue injury. Oxidative rancidity of fat caused by iron and other trace element contaminants derived from processing machinery is a major problem in the food industry.

The lipid free radical chain reaction is terminated by the quenching action of vitamin E on lipoxy radicals and by interactions among products of the reaction to form polymers and other non-oxidative compounds.

$$ROO\cdot + EOH \longrightarrow ROOH + EO\cdot$$

$$RO\cdot + EOH \longrightarrow ROH + EO\cdot$$

Alkoxy radicals are reduced by vitamin E to their corresponding alcohols, which are metabolized through established fatty acid oxidation pathways. Studies on the metabolism of ^{14}C-α-tocopherol *in vitro* and *in vivo* have shown that the tocopheroxy radical undergoes further oxidation to form a p-quinone and a dimer formed as a result of an intramolecular shift of electrons to form a carbon-centered radical at the 5-methyl position.[42]

Lipoxy radicals, peroxides and certain of their decomposition products have been shown to react with nucleic acids, proteins and various other macromolecules. These observations have raised two important nutritional questions: does a high intake of PUFA increase the rate of lipid peroxidation *in vivo* and, if so, does this increase constitute a risk factor for oxygen radical pathology? The answer to both questions is obscure. The first is made difficult by lack of a practical, non-invasive method of assessing *in vivo* lipid peroxidation in human subjects.[43]

Breath analysis for pentane, a product of the oxidative decomposition of n-6 fatty acids, has been used to monitor lipid peroxidation in patients maintained by intravenous feeding,[44] but it is impractical for free living subjects, requires special equipment and is prone to errors of estimate. Urinalysis for an adduct of lysine with MDA, a product of the oxidative decomposition of PUFA with three or more double bonds, is compromised as a measure of lipid peroxidation *in vivo* by its presence in food digesta.[45] By feeding MDA-free diets to experimental animals, it has been demonstrated that an increase in the excretion of MDA compounds formed *in vivo* occurs in response to a high intake of fish oil, a low intake of vitamin E, treatment with lipolytic hormones, and administration of iron and certain drugs.[46] Most of this increase is due to two conjugates of MDA with lysine, one acetylated and the other unacetylated, that constitute most of the MDA excreted in human and rat urine. These compounds reflect reactions of MDA with proteins in the tissues and in the diet. MDA-modified proteins, like proteins modified by hydroxyl radicals,[47] are rapidly degraded by a cellular protease (unpublished results). MDA adducts with other amino compounds, including the phospholipid bases serine and ethanolamine and the nucleic acid bases guanine, guanosine and deoxyguanosine also have been identified in human and rat urine.[48] An MDA adduct with deoxyguanosine has been found in a hydrolysate of DNA isolated from rat tissues,[49] but neither its frequency in DNA nor its excretion in the urine appears to be affected by vitamin E deficiency or the administration of lipid oxidants (unpublished results).

These studies on MDA metabolism indicate that lipid peroxidation in the body is a normal physiological event that is related to metabolic rate. Like the excretion of 8-hydroxy deoxyguanosine,[50] the excretion of endogenous MDA adducts is much lower on a body weight basis in humans than it is in rats (unpublished results). Lipolysis, whether induced by epinephrine, ACTH. exercise or fasting, results in an increase in the excretion of MDA adducts in the urine. Free MDA is a minor and inconsistent component of human urine, and constitutes only a small fraction of total MDA in the tissues, most of which is bound in Schiff's base linkages to proteins. Methods of tissue analysis that do not provide for hydrolysis of its bound forms grossly underestimate the MDA content of animal tissues.[39]

Evidence that MDA-modified proteins can result in overt disease has been obtained by Haberland and coworkers[51] in the case of familial hypercholesterolemia in humans and its counterpart in the Watanabe rabbit. In these conditions there is an accumulation of lipoprotein-derived lipid in the macrophage-related foam cells of the arterial wall, arising from a functional deficiency in the LDL receptor that normally regulates the intracellular level of cholesterol. Modification of more than 15% of the lysine residues of the apoB-100 protein of LDL by MDA results in rapid clearance of the modified particles by macrophage scavenger receptors. These receptors are genetically distinct from the LDL receptors. Modification of more than 12% of the lysine residues abolishes lipoprotein uptake by the LDL receptor. It has been proposed that chronic production of modified LDL and its unrestrained cellular clearance accounts for the rapid accumulation of lipids in foam cells that marks this disease. Products of reactions with MDA and a more highly reactive aldehydic product of lipid peroxidation, 4-hydroxynonenal (HNE), have been identified among the proteins in the arterial lesions. These observations suggest the possibility that decomposition products of lipid peroxidation generated in the arterial wall may initiate the development of atheroma.

The cytotoxicity and genotoxicity of lipid oxidation products have been discussed by Esterbauer.[52] The likelihood of toxicity caused by the amounts of these products normally present in the diet appears to be low, but feeding large quantities to experimental animals has been reported to produce various adverse effects, including an increase in the frequency of atherosclerosis and tumors. The two compounds about which there is most information are HNE and MDA. Both have been found to be toxic to a variety of cell types at low concentrations in the culture medium and to react *in vivo* with DNA and proteins. Whether the reactions of these compounds in human tissues have any clinical implications, whether their formation is increased by high intakes of PUFA, and whether large vitamin E supplements reduce their formation, are important unresolved questions. It is of interest in this connection, however, that the decrease in the U.S. death rate from cardiovascular disease associated with an increased intake of PUFA and a decreased intake of saturated fat during the past two decades has been offset by an increase in the death rate from other diseases including cancer.[53] This was a consideration in the shift in emphasis in current diet recommendations for the prevention of cardiovascular disease from a high intake of PUFA to a lower intake of saturated fatty acids.

Recent reports of high levels of prostanoid-like compounds being formed by non-enzymatic peroxidation of arachidonic acid requires further investigation.[54] The physiological impact of non-enzymatic peroxidation products as agonists or antagonists of prostanoid receptors is unknown.

VITAMIN E

The strong natural association between PUFA and vitamin E in foods of both plant and animal origin tempers the risk of vitamin E deficiency on high PUFA diets, despite the increased requirement for this vitamin imposed by such intakes. An exception to this general association prevails in the case of the oils of fish and marine mammals, which contain enough vitamin E to stabilize their highly unsaturated fatty acids at the low temperature and oxygen tension of the aquatic environment, but not when they are exposed to the high oxygen environment of land animals and humans. Consumption of fish oil preparations for the prevention of heart attacks therefore results in an increased vitamin E requirement that should be countered by fortification of these products with vitamin E.[55]

Except in metabolic disorders of lipid absorption and transport such as abetalipoproteinemia, dietary vitamin E deficiency is not a recognized clinical problem in the nutrition of any segment of the human population. The absorption, transport and metabolism of vitamin E recently have been reviewed by Kayden and Traber.[56] The plasma concentration of this vitamin is maintained within relatively narrow limits by several factors, including the action of a tocopherol-binding protein in the liver that regulates its secretion into the plasma in VLDL, the limited carrying capacity of plasma lipoproteins, and the low efficiency with which megadoses of the vitamin are absorbed from the intestine. Such doses result in only a doubling of the normal level of vitamin E in the plasma. These factors also explain the relative lack of toxicity of the vitamin. The hepatic tocopherol-binding protein appears to be responsible for a discrimination against non-α-tocopherol forms of the vitamin, including γ-tocopherol, which occurs at three times the concentration of α-tocopherol in the U. S. diet but constitutes only about 10% of the total vitamin E in the plasma. Tissue concentrations of vitamin E are only slowly depleted by low intakes in the diet and slowly increased by high intakes. In the only experimental study on dietary vitamin E deficiency in human adults,[57] it was found necessary to feed a vitamin E deficient diet high in PUFA for two to four years before the incipient signs of vitamin E deficiency (consisting mainly of mild creatinuria) were detectable.

Evidence that lipid peroxidation *in vivo* may be a risk factor for cardiovascular disease (CVD) and that this risk is reduced by pharmacological intakes of vitamin E has come from two recent studies. In a study on 87,245 women aged 34 to 59 years who were initially free of CVD and cancer, the consumption of large vitamin E supplements for two to eight years was associated with a risk of major coronary disease of 0.59 relative to that in a oontrol group that consumed either no vitamin supplement or a low potency multivitamin preparation that contained vitamin E.[58] The protective effect of vitamin E occurred only at a pharmacological level of intake (median ~200 IU per day) and then only if it had been consumed for at least two years. No relationship was found between the incidence of coronary disease and the intake of dietary vitamin E, even when it was augmented by a multivitamin supplement containing this vitamin. This finding is noteworthy because the close association of vitamin E with PUFA in foodstuffs dictates that those subjects who had a high intake of dietary vitamin E also had a high intake of PUFA, which have a well documented lowering effect on the level of plasma cholesterol. Whether the protection afforded by vitamin E against coronary disease is related to the modification of LDL by MDA[51] is unknown. Large intakes of vitamin E did not protect against breast in this study.[59]

Similar results were obtained in a large prospective study on U.S. men.[60] After controlling for age and other relevant factors, men consuming more than 60 IU of vitamin E per day for four years were found to have a relative risk for coronary heart disease of 0.64 compared to men consuming less than 7.5 IU per day from the diet. The relative risk for those who had consumed 100 IU per day or more for at least two years was similar to those who consumed 60 IU per day.

This reported response to pharmacological, but not to physiological, intakes of vitamin E is consistent with the unique function of this vitamin in metabolism. Most vitamins belong to the B complex group and function as coenzymes in a variety of enzymatic reactions. These reactions are not accelerated by an excess of cofactors. Consequently, the efficacy of the B complex vitamins is not increased at pharmacological intakes unless there is a defect in their metabolism (as in the prevention of neural tube defects by folic acid supplements). Large supplements of vitamins A, D and K are precluded by their toxicity. In contrast, vitamin E functions non-enzymatically *in vivo* in much the same way that it functions in foods and other natural products, i.e., by scavenging lipoxy radicals that catalyze the peroxidation of PUFA and other unsaturated lipids. This function is not absolutely specific, as shown by the fact that certain other structurally unrelated synthetic lipid antioxidants can substitute for vitamin E in the prevention of its classical deficiency diseases, including sterility in rodents and muscular dystrophy in most other species.[61] The common food antioxidants, BHA and BHT, however, are biologically inactive.

These considerations suggest that the dose response curve for vitamin E with respect to the prevention of oxygen radical pathology *in vivo* follows chemical rather than enzyme kinetics. If so, its efficacy may increase beyond the normal range of intakes in the diet and the normal concentrations in the tissues. The apparent efficacy of large, but not small, vitamin E supplements in the prevention of CVD supports this possibility.

Large vitamin E supplements also have been reported to be useful in modulating tissue damage associated with major surgery and other forms of trauma. The evidence that free radicals are mediators of tissue injury has been reviewed by Kehrer[62] and by Rice-Evans and Diplock[63]. It is strongest in the case of myocardial reperfusion injury, although attempts to prevent this injury by administration of superoxide dismutase, catalase and iron chelators have had indifferent success. There is an increase in serum lipid peroxide levels following coronary by-pass surgery that is probably secondary to the generation of superoxide and hydroxyl radicals during reperfusion.[64] The concentration of breath pentane is elevated following myocardial infarction[65] and is reduced by administration of the superoxide scavenging drug captopril.[66] Excretion of MDA has been reported to increase 10- to 20-fold over a period of 9 days following burn injury.[67] The clinical use of antioxidants in the treatment of burn and smoke injury has been discussed by Youn *et al.*[68]

The increase in superoxide and hydroxyl radical formation following tissue trauma probably reflects a release of catalytic iron from bound forms (see above) as an initiating event. Although these radicals are the primary oxidants implicated in free radical pathology associated with tissue injury, there is evidence for a concomitant increase in the generation of lipoxy radicals, signifying the occurrence of peroxidative damage to cell membranes. Babies with rhesus hemolytic disease have higher plasma ferritin levels, lower iron-binding capacity and high concentrations of lipid peroxidation products.[69] Hereditary hemochromatosis is associated with high levels of serum iron and TBARS and with low levels of vitamin E.[70] The ability of pharmacological amounts of vitamin E to modulate iron toxicity in animals[71] suggests that iron-catalyzed lipid peroxidation following tissue injury also may be ameliorated by this vitamin.

Although increased lipid peroxidation has been reported in a number of metabolic abnormalities (for example, non-insulin-dependent diabetes),[72] in some such conditions it is difficult to determine whether peroxidation is a cause or a result of pathology. Breath pentane and urinary MDA are increased by hyperventilation, fasting, fever, lipolysis and other conditions commonly caused by disease. In the case of surgery and other forms of trauma, peroxidation is clearly the result of tissue damage. The level of lipid peroxides in the tissues may be of clinical importance in critically ill patients, and the question has been raised whether they should be provided with systematic antioxidant therapy.[73]

Uncertainty still surrounds the long-standing question whether vitamin E supplements can protect humans against the pathological effects of ozone in smog.[74] Prolonged exposure of animals to the concentrations of ozone present in heavy smog increases the formation of lipoxy radicals and causes inflammation and edema in the lungs.[75] Large intakes of vitamin E are protective against these effects in animals. Increased serum concentrations of lipid oxidation products have been observed in patients with asthma, an inflammatory disorder of the airways aggravated by exposure to smog.[76] Whether this common disorder can be modulated by a high intake of vitamin E is a significant clinical question that deserves investigation in a prospective study of the type that has been carried out with reference to coronary disease.

Intakes of vitamin E up to 1,000 IU per day are generally well tolerated, though the low fractional absorption of such doses may result in intestinal disturbances. They may also cause some mild impairment of the absorption of vitamins A and K.

β-CAROTENE

β-carotene is an avid quencher of singlet oxygen ($^1O_2$), a non-radical form of oxygen that is capable of carrying out many of the reactions associated with oxygen radicals. It is an important stabilizer of PUFA and other unsaturated lipids in foods of plant origin, particularly under conditions of exposure to light or ionizing radiation, which catalyze singlet oxygen generation. It has been reported to be effective in the treatment of erythropoietic propoporphyria and other diseases of hyperphotosensitivity[77] and may be important in the prevention of oxygen radical damage in tissues such as the skin and the eye that are exposed to light. There is evidence that β-carotene (as well as vitamin E) acts as a scavenger of oxyradicals that damage the ocular lens and thereby increase the risk of cataracts.[5]

Analysis of plasma samples taken from adults consuming the Canadian mixed diet indicated that β-carotene was the second most important lipid antioxidant present next to vitamin E.[78] However, the plasma level of β-carotene and its associated carotenoids varies enormously among different diet cultures, as illustrated by the orange color of plasma taken from consumers of "soul food" in the southern regions of the United States and the colorless plasma of the inhabitants of its most northern regions who still pursue the carnivorous aboriginal diet culture. The experience of these Arctic inhabitants serves to confirm in humans what has been established for experimental animals, i.e., that β-carotene is not essential for life or for general health and wellbeing.

Other explanations therefore must be sought for an inverse association between β-carotene intake and the incidence of several diseases, including cancer, heart disease and diabetes, that has been observed in epidemiological studies. The negative correlation between β-carotene intake and cancer incidence is stronger than that for any of the other so-called antioxidant nutrients,[79] but this association appears to be largely, if not entirely, associative rather than causative. The fact that it is stronger for β-carotene than for vitamin A[80] indicates that it is not due to the conversion of the former to the latter. A further indication is that there is an equally strong correlation with the intake of carotenoids that have no vitamin A activity.[80]

The lower incidence of cancer associated with diets high in β-carotene is most probably attributable to the fact that this compound serves as a marker for diets high in fruits and vegetables and low in fat of the type currently recommended for the reduction of cancer incidence in the populations of industrialized countries.[81,82] It may also serve as a marker for a generally healthier lifestyle on the part of consumers of this type of diet. If so, no benefit should accrue in terms of cancer incidence from the consumption of β-carotene supplements. The efficacy of such supplements in the prevention of cancer is being addressed in ongoing studies analogous to those carried out on the effect of large supplements of vitamin E.

Administration of 120 mg of β-carotene per day to adults previously fed a carotene-free diet for two weeks lowered pentane exhalation, indicating that a large supplement is capable of decreasing lipid peroxidation in depleted subjects.[83] However, it was not determined whether β-carotene depletion increased pentane exhalation above the prestudy level. No decrease was produced by a moderate supplement of 15 mg per day. It was concluded that the results supported previous evidence that there is a weak association between β-carotene intake and lipid peroxidation *in vivo*.

VITAMIN A

Although vitamin A is frequently included in a list of dietary antioxidants, its susceptibility to oxidation has been of greater interest in nutrition than its antioxidant activity. As an unsaturated lipid, it is vulnerable to trace metal-catalyzed oxidation in the diet unless it is protected by vitamin E or a synthetic food antioxidant such as BHA or BHT. Co-oxidation of vitamin A and PUFA occurs in vitamin E deficient animals, but the low concentration of vitamin A in the tissues imposes a negligible stress on the vitamin E requirement. The mechanism of its antioxidant action presumably is similar to that of β-carotene and likewise is probably unrelated to its essential biochemical functions.

Interest in the possible efficacy of retinol in cancer prevention was stimulated in the 1980s by evidence for a role of retinoic acid in cell differentiation[84] and for an inhibitory effect of synthetic retinoids, notably 13-*cis*-retinoic acid, on the growth of tumor cells.[85]

However, the efficacy these synthetic retinoids in the prevention of carcinogen-induced tumors in animals is unrelated to their vitamin A activity, they are highly toxic, and they stimulate the growth of some tumors.[80] Further, the evidence for a specific involvement of retinoic acid in cell differentiation has been challenged.[86] Epidemiological and case-control studies on the relationship between vitamin A intake and cancer incidence have yielded equivocal results. In a recent prospective study, evidence was obtained for an increased incidence of breast cancer in a large cohort of women whose intake of retinol equivalents (vitamin A plus β-carotene) was in the lowest of five percentiles.[59] The same quintile had the lowest intake of β-carotene, an indication that their diets were low in fruits and vegetables, which have been found to be protective against cancer. This possible explanation of the results is strengthened by the fact that the only significant correlation between the incidence of breast cancer and the intake of specific food groups was an inverse relationship with the consumption of vegetables. Hence the results may reflect an increased risk of breast cancer on diets low in vitamin A activity or a reaffirmation of previous evidence for a modulating effect of vegetarian and semi-vegetarian diets on cancer incidence. There was no relationship between the frequency of cancer and a high intake of vitamin A from supplements. Present evidence indicates that there is no reason to change the conclusion drawn from a conference on vitamin A and cancer prevention convened by the U. S. National Cancer Institute a decade ago: "There is no basis now for proposing that vitamin A protects against human cancer...".[87]

Unlike vitamin C and E supplements, the use of large vitamin A supplements is constrained by considerations of toxicity. Toxic effects in adults have been reported at intakes as low as 50,000 IU per day (15 times the RDA)[88] and in pregnant women at 25,000 IU per day.[89]

VITAMIN C

Vitamin C has an ambivalent image with respect to its effect on the metabolism of oxygen radicals. Its capacity to reduce ferric to ferrous ions and thereby to stimulate hydroxyl radical generation in the Fenton reaction has been widely used to increase lipid peroxidation in liver microsomes. However, there is no indication , based on MDA excretion or other measures of lipid peroxidation, that megadoses of vitamin C have a prooxidant effect *in vivo*.

In addition to some specific metabolic function relating to the prevention of scurvy, vitamin C clearly acts *in vivo* as a water soluble reductant. Its reducing action on dietary iron increases the absorption of this element from the intestine. Its ability to scavenge superoxide, hydroxyl radicals and singlet oxygen[90] may account for its mild inhibitory effect on lipid peroxidation in experimental animals. Exposure of plasma to aqueous peroxy radicals results in rapid oxidation of ascorbic acid and the appearance of fatty acid hydroperoxides.[91] The oxidants in cigarette smoke increase the intake of vitamin C required to maintain the normal plasma level of ascorbate by about 40%.[92] The ability of vitamin C to reduce oxidized thiols has led to the proposal that its antioxidant action arises from the regeneration of reduced glutathione consumed in peroxidase reactions.

It has also been proposed that vitamin C reduces the α-tocopheroxy radical *in vivo,* thereby effecting the recycling of α-tocopherol.[93] While this reaction has been demonstrated under chemical conditions, the conditions necessary are unlikely to prevail *in vivo*. Its occurrence *in vivo* is also inconsistent with the results of animal experiments. In guinea pigs chronically labelled with deuterated α-tocopherol, the turnover rate of this vitamin in the tissues was similar whether they were subsequently fed a diet deficient, normal or high in vitamin C.[94] Feeding high levels of vitamin C to rats has been reported to aggravate the symptoms of vitamin E deficiency.[95] Conversely, the level of vitamin E in the diet has no influence on the vitamin C content of rat tissues or on the ratio of ascorbic acid to dehydroascorbic acid.[96] Further, it is conceptually improbable that normal vitamin E status in humans is dependent upon another vitamin that can vary in intake by two orders of magnitude, from the 10 mg per day necessary for the prevention of scurvy to the 1,000 mg per day consumed by some subjects, with no apparent effect on the plasma level of vitamin E.

An integrative concept of the role of vitamin C in the metabolism of oxygen radicals remains elusive. Indeed, there no satisfactory proof of its proposed role in the hydroxylation of proline and lysine that has been put forward as an explanation of its antiscorbutic activity.[97]

Gross claims for the efficacy of megadoses of vitamin C in the prevention of cancer made several decades ago have been disproved and there is little evidence that variations in its intake from the diet are related to cancer incidence. In a recent large prospective study on women,[59] it was concluded that neither diets high in vitamin C nor vitamin C supplements provided protection against breast cancer. The recommended intakes of vitamin C in affluent societies reflect the abundance of this vitamin in the food supply. The average intake of U.S. and Canadian adults is in excess of 100 mg per day, well above the recommended intakes of 60 and 40 mg per day, respectively, and far above the 10 mg per day required to prevent clinical signs of deficiency.

SULFUR AMINO ACIDS

Sulfur amino acids (SAA) have a mild "sparing effect" on the vitamin E requirement of experimental animals. They afford partial protection against the muscular dystrophy of vitamin E deficiency in laboratory and domestic animals, and in birds this disease develops only on a diet low in SAA as well as vitamin E. The antioxidant action of SAA is attributed to their role as precursors of glutathione, the reductant employed in the metabolism of lipid peroxides and H_2O_2 by the glutathione peroxidase enzymes. Protein or SAA deficiency in experimental animals results in depletion of reduced glutathione in the tissues and increased susceptibility to oxygen stress.[98,99] The glutathione pool is of key importance in the defence against oxygen radical pathology.[100,101]

The high protein diet of industrialized countries contains a surfeit of SAA. Even low protein cereal based diets are unlikely to cause SAA deficiency, as cereals proteins contain adequate amounts of methionine plus cysteine to meet nutritional requirements, provided they are available in amounts sufficient to meet energy requirements. The likelihood that dietary protein/SAA deficiency is a significant cause of oxygen radical pathology in the general population therefore is remote. However, SAA/glutathione status should be considered in patients with protein and/or energy malnutrition.[98]

NIACIN

Maintenance of normal levels of glutathione peroxidase activity depends on the recycling of oxidized glutathione by reductases that have a specific requirement for reducing equivalents in the form of niacin coenzymes generated in the pentose shunt.

$$GSSG + NADPH^+ + H^+ \rightarrow 2GSH + NADP^+$$

Niacin deficiency is no longer a significant problem in human nutrition. On high protein diets the supply of niacin can be augmented by synthesis from tryptophan. However, some forms of oxidative damage may result in an inadequate supply of NADPH in specific tissues. Large numbers of people in tropical areas have an inborn error in the gene for glucose-6-phosphate dehydrogenase, which is necessary for the generation of NADPH, and this results in susceptibility to hemolysis induced by drugs and other agents that increase H_2O_2 production.

FLAVONOIDS

Flavonoids are polyphenolic compounds with antioxidant properties that are present in vegetables, fruits and beverages such as tea and wine. The most important groups of flavonoids are the anthocyanins, flavonols, flavones, catechins and flavanones. Flavanols are scavengers of superoxide, singlet oxygen and lipid peroxy radicals as well as metal ions. They have been found to inhibit the oxidation of LDL *in vitro* and to reduce thrombogenic processes by inhibiting the activity of cyclooxygenase. In the first investigation of their possible effect on human health, the relationship between flavonoid intake and coronary heart disease was investigated in a 5-year prospective study on men with an initial age of 65-84 years.[102] After adjustment for known risk factors and for differences in the intake of vitamins C and E, β-carotene, fiber and energy, the relative risk of coronary heart disease mortality in the highest versus the lowest tertile of flavonoid intakes was 0.32. This finding adds to uncertainty about the cause of the lower rate of atherosclerotic heart disease associated with the consumption of fruits, vegetables and wine.

OVERVIEW

The foregoing findings indicate that dietary deficiencies of the so-called antioxidant nutrients are not an important cause of oxygen radical pathology among consumers who are in general compliance with current diet recommendations. An inverse relationship between β-carotene intake and the incidence of cancer and heart disease is probably attributable to the fact that this compound serves as a marker for diets high in fruits and vegetables and low in fat that have been widely observed to provide protection against these diseases. If so, β-carotene supplements are unlikely to provide such protection. There is persuasive recent evidence that large vitamin E supplements reduce the incidence of atherosclerotic heart disease. These findings lend support to current recommendations that the intake of polyunsaturated fatty acids, the main determinants of the vitamin E requirement, be limited to 10% of energy intake. The results of similar studies on the efficacy of vitamin E in the prevention of other diseases in which oxygen radicals have been implicated, particulartly cancer, will be awaited with interest. Large vitamin A and selenium supplements are precluded for reasons of safety. The evidence that large vitamin C supplements are efficacious in the prevention of cancer, colds and other diseases is weak. The use of large supplements of antioxidant nutrients for disease prevention is a pharmacological measure and should not be addressed by nutritionists in terms of increases in their RDAs or of massive food fortification.

REFERENCES

1. K. Karlsson, J. Sandström, A. Edlund, T. Edlund, and S.L. Marklund, Pharmacokinetics of extracellular-superoxide dismutase in the vascular system, *Free Rad. Biol. Med.* 14:185 (1993).

2. T. Adachi, H. Ohta, H. Yamada, A. Futenma, K. Kato, and K. Hirano, Quantitative analysis of extracellular-superoxide dismutase in serum and urine by ELISA with monoclonal antibody, *Clin. Chim. Acta* 212:89 (1992).

3. D.A. Peterson and J.W. Eaton, Electron transfer facilitated by superoxide dismutase: a model for membrane redox systems?, *Biochem. Biophys. Res. Commun.* 165:164 (1989).

4. M.B. Yim, P.B. Chock, and E.R. Stadtman, Enzyme function of copper,zinc superoxide dismutase as a free radical generator, *J. Biol. Chem.* 268:4099 (1993).

5. B. Halliwell and J.M.C. Gutteridge, "Free Radicals in Biology and Medicine" (2nd ed.), Clarendon Press, Oxford (1989).

6. D.R. Rosen, T. Siddique, D. Patterson, D.A. Figlewicz, P. Sapp, A. Hentati, D. Donaldson, J. Goto, J.P. O'Regan, H.-X. Deng, Z. Rahmani, A. Krizus, D. McKenna-Yasek, A. Cayabyab, S.M. Gaston, R. Berger, R.E. Tanzi, J.J.. Halperin, B. Herzfeldt, R. Van den Bergh, W.-Y. Hung, T. Bird, G. Deng, D.W. Mulder, C. Smyth, N.G. Laing, E. Soriano, M.A. Pericak-Vance, J. Haines, G.A. Rouleau, J.S. Gusella, H.R. Horvitz, and R.H. Brown, Jr. Mutations in Cu/Zn superoxide dismutase gene are associated with familial amyotrophic lateral sclerosis, *Nature* 362:59 (1993).

7. S.I. Liochev and I. Fridovich, The role of $O_2^{\cdot -}$ in the production of HO· in vitro and in vivo, *Free Rad. Biol. Med.* 16:29 (1994).

8. J.S. Stamler, D.J. Single, and J. Loscalzo, Biochemistry of nitric oxide and its redox-activated forms, *Science* 258:1898 (1992).

9. G. Yang, T.E.G. Candy, M. Boaro, H.E. Wilkin, P. Jones, N.B. Nazhat, R.A. Saadalla-Nazhat, and D.R. Blake, Free radical yields from the homolysis of peroxynitrous acid, *Free Rad. Biol. Med.* 12:327 (1992).

10. S.A. Lipton, Y.-B. Choi, Z.-H. Pan, S.Z. Lei, H.-S.V. Chen, N.J. Sucher, J. Loscaizo, D.J. Singel, and J.S. Stamler, A redox-based mechanism for the neuroprotective and neurodestructive effects of nitric oxide and related nitroso-compounds, *Nature* 364: 626 (1993).

11. Y. Henry, M. Lepoivre, J.-C. Drapier, C. Ducrocq, J.-L. Boucher, and A. Guissani, EPR characterization of molecular targets for NO in mammalian cells and organelles, *FASEB J.* 7:1124 (1993).

12. Y. Rayssiguier, E. Gueux, L. Bussiere, and A. Mazur, Copper deficiency increases the susceptibility of lipoproteins and tissues to peroxidation in rats, *J. Nutr.* 123:1343 (1993).

13. M. Fields, J. Ferretti, J.C. Smith, and S. Reiser, Interaction between dietary carbohydrate and copper nutriture on lipid peroxidation in rat tissues, *Biol. Trace Elem. Res.* 6:379 (1984).

14. M. Fields, C.G. Lewis, M. Lure, and W.E. Antholine, The influence of gender on developing copper deficiency and on free radical generation of rats fed a fructose diet, Metabolism 41:989 (1992).

15. S.K. Nelson, C-J. Huang, M.M. Mathias, and K.G.D. Allen, Copper-marginal and copper-deficient diets decrease aortic prostacyclin production and copper-dependent superoxide dismutase activity, and increase aortic lipid peroxidation in rats, *J. Nutr.* 122:2101 (1992).

16. J.R. Prohaska, Biochemical changes in copper deficiency, *J. Nutr. Biochem.* 1:453 (1990).

17. R.G. Allen, Oxygen-reactive species and antioxidant responses during development: the metabolic paradox of cellular differentiation, *Proc. Soc. Exp. Biol. Med.* 196:117 (1991).

18. A.S. Prasad, A. Miele, Jr., Z. Farid, H.H. Sandstead, A.R. Schubert, and W.J. Darby, Biochemical studies on dwarfism, hypogonadism and anemia, *Arch. Intern. Med.* 111:407 (1963).

19. R.A. DiSilvestro and G.P. Carlson, Effects of mild zinc deficiency, plus or minus acute phase response, on CCl_4 hepatotoxicity, *Free Rad. Biol. Med.* 16:57 (1994).

20. J.D. Hammermueller, T.M. Bray, and W.J. Bettger, Effect of zinc and copper deficiency on microsomal NADPH-dependent active oxygen generation in rat lung and liver, *J. Nutr.* 117:894 (1987).

21. J.F. Sullivan, M.M. Jetton, H.K. Hahn, and R.E. Burch, Enhanced lipid peroxidation in liver microsomes of zinc-deficient rats, *Am. J. Clin. Nutr.* 33:51 (1980).

22. T.M. Bray and W.J. Bettger, The physiological role of zinc as an antioxidant, *Free Rad. Biol. Med.* 8:281 (1990).

23. W.J. Bettger, and B.L. O'Dell, Physiological roles of zinc in the plasma membrane of mammalian cells, *J. Nutr. Biochem.* 4:194 (1993).

24. S. Zidenberg-Cherr, C.L. Keen, B. Lönnerdal, and L.S. Hurley, Superoxide dismutase activity and lipid peroxidation in the rat: developmental correlations affected by manganese deficiency, *J. Nutr.* 113:2498 (1983).

25. K.H. Thompson, D.V. Godin, and M. Lee, Tissue antioxidant status in streptozotocin-induced diabetes in rats. Effects of dietary manganese deficiency, *Biol. Trace Elem. Res.* 35:213 (1992).

26. K.H. Thompson and M. Lee, Effects of manganese and vitamin E deficiencies on antioxidant enzymes in streptozotocin-diabetic rats, *J. Nutr. Biochem.* 4:476 (1993).

27. S. Samman, Dietary versus cellular zinc: the antioxidant paradox, *Free Rad. Biol. Med.* 14:95 (1993).

28. K.T. Tamai, E.B. Gralla, L.M. Ellerby, J.S. Valentine, and D.J. Thiele, Yeast and mammalian metallothioneins functionally substitute for yeast copper-zinc superoxide dismutase, *Proc. Natl. Acad. Sci. USA* 90:8013 (1993).

29. M. Sato and I. Bremner, Oxygen free radicals and metallothionein, *Free Rad. Biol. Med.* 14:325 (1993).

30. T.C. Stadtman, Specific occurrence of selenium in enzymes and amino acid tRNAs, *FASEB J.* 1:375 (1987).

31. F.-F. Chu, J.H. Doroshow, and R.S. Esworthy, Expression, characterization and tissue distribution of a new cellular selenium-dependent glutathione peroxidase, GSHPx-GI, *J. Biol. Chem.* 268:2571 (1993).

32. R.A. Sunde, J.A. Dyer, T.V. Moran, J.K. Evenson, and M. Sugimoto, Phospholipid hydroperoxide glutathione peroxidase: full-length pig blastocyst cDNA sequence and regulation by selenium status, *Biochem. Biophys. Res. Commun.* 193:905 (1993).

33. J.G. Bergan and H.H. Draper, Absorption and metabolism of 1-^{14}C-hydroxy octadecadienoate in the rat, *Lipids* 5:983 (1970).

34. G. Yang, J. Chen, Z. Wen, K. Ge, L. Zhu, X. Chen, and X. Chen, The Role of selenium in Keshan disease, *Adv. Nutr. Res.* 6:203 (1984).

35. C.D. Thompson and M.F. Jones, Selenium in human health and disease with emphasis on those aspects peculiar to New Zealand, *Amer. J. Clin. Nutr.* 33:303 (1980).

36. G. Yang, S. Wang, R. Zhou, and S. Sun, Endemic selenium intoxication of humans in China, *Amer. J. Clin. Nutr.*37:872 (1983).

37. J.N. Thompson, P. Erdody, and D.C. Smith, Selenium content of food consumed by Canadians, *J. Nutr.*105:274 (1975).

38. A.V. Kozlov, D.Y. Yegorow, Y.A. Vladimirov, and O.A. Azizova, Intracellular free iron in liver tissue and liver homogenate: studies with electron paramagnetic resonance on the formation of paramagnetic complexes with desferal and nitric oxide, *Free Rad. Biol. Med.* 13:9 (1992).

39. H.H. Draper, E.J. Squires, H. Mahmoodi, J. Wu, and M. Hadley, A comparative evaluation of thiobarbituric acid methods for the determination of malondialdehyde in biological materials, *Free Rad. Biol. Med.* 15:353 (1993).

40. L.A. Witting, Lipid peroxidation *in vivo*, *J. Amer. Oil Chem.* 42:908 (1965).

41. T. Miyazawa, T. Suzuki, and K. Fujimoto, Age-dependent accumulation of phosphatidyl hydroperoxide in the brain and liver of the rat, *Lipids* 28:789 (1993).

42. H.H. Draper, Antioxidant role of vitamin E, *in*: Atmospheric Oxidation and Antioxidants, Vol. III, G. Scott, ed., Elsevier, Amsterdam (1993).

43. W.A. Pryor and S.S. Godber, Noninvasive measures of oxidative stress status in humans, *Free Rad. Biol. Med.* 10:177 (1991).

44. K.N. Jeejeebhoy, In vivo breath alkane as an index of lipid peroxidation, *Free Rad. Biol. Med.* 10:191 (1991).

45. L.A. Piché, P.D. Cole, M. Hadley, R. van den Bergh, and H.H. Draper, Identification of N-ε-(2-propenal)lysine as the main form of malondialdehyde in food digesta, *Carcinogenesis* 9:473 (1988).

46. S.N. Danakoti and H.H. Draper, Response of urinary malondialdehyde to factors that stimulate lipid peroxidation *in vivo*, *Lipids* 22:643 (1987).

47. K.J.A. Davies and A.L. Goldberg, Proteins damaged by oxygen radicals are rapidly degraded in extracts of red blood cells, *J. Biol. Chem.* 262:8227 (1987).

48. H.H. Draper and M. Hadley, A review of recent studies on the metabolism of exogenous and endogenous malondialdehyde, *Xenobiotics* 20:901 (1990).

49. S. Agarwal and H.H. Draper, Isolation of a malondialdehyde-deoxyguanosine adduct from rat liver DNA, *Free Rad. Biol. Med.* 13:695 (1992).

50. R. Adelman, R.L. Saul, and B.N. Ames, Oxidative damage to DNA: relation to species metabolic rate and life span, *Proc. Natl. Acad. Sci. USA* 85:2706 (1988).

51. M.E. Haberland, D. Fong, and L. Cheng, Malondialdehyde, modified lipoproteins, and atherosclerosis, *Europ. Heart J.* 11 (Suppl. E):100 (1990).

52. H. Esterbauer, Cytotoxicity and genotoxicity of lipid-oxidation products, *Amer. J. Clin. Nutr.*57: (Suppl):7795 (1993).

53. C.C. Seltzer, Letter to the editor, *New Engl. J. Med.* 323:1705 (1990).

54. J.D. Morrow, K.E. Hill, R.F. Burk, T.M. Nammour, K.F. Badr, and L.J. Roberts, II, A series of prostaglandin F_2-like compounds are produced *in vivo* in humans by a non-cyclooxygenase, free radical-catalyzed mechanism, *Proc. Natl. Acad. Sci. USA* 87:9383 (1990).

55. M. Meydani, F. Natiello, B. Goldin, N. Free, M. Woods, E. Schaefer, J.B. Blumberg, and S.L. Gorbach, Effect of long-term fish oil supplementation on vitamin E status and lipid peroxidation in women, *J. Nutr.* 121:484 (1991).

56. H.J. Kayden and M.G. Traber, Absorption, lipoprotein transport, and regulation of plasma concentrations of vitamin E in humans, *J. Lipid Res.* 34:343 (1993).

57. M.K. Horwitt, Vitamin E and lipid metabolism in man, *Amer. J. Clin .Nutr.*8:451 (1960).

58. M.J. Stampfer, C.H. Hennekens, J.E. Manson, G.A. Colditz, B. Rosner, and W.C. Willett, Vitamin E consumption and the risk of coronary disease in women, *New Engl. J. Med.* 328:1444 (1993).

59. D.J. Hunter, J.E. Manson, G.A. Colditz, M.J. Stampfer, B. Rosner, C.H. Hennekens, F.E. Speizer, and W.C. Willett, A prospective study of the intake of vitamins C, E and A and the risk of breast cancer, *New Engl. J. Med.* 329: 234 (1993).

60. E.B. Rim, M.J. Stampfer, A. Ascherio, E. Giovannucci, G.A. Colditz, and W.C. Willett, Vitamin E consumption and the risk of coronary heart disease in men, *New Engl. J. Med.* 328:1450 (1993).

61. H.H. Draper, Nutrient interrelationships, *in*: "Vitamin E. A Comprehensive Treatise", L.J. Machlin, ed., Marcell Dekker, New York (1980).

62. J.P. Kehrer, Free radicals as mediators of tissue inury and disease, *Critical Rev. Toxicol.* 23:21 (1993).

63. C.A. Rice-Evans and A.T. Diplock, Current status of antioxidant therapy, *Free Rad. Biol. Med.* 15:77 (1993).

64. T. Tihan, P. Chiba, O. Krupicka, M. Fritzer, R. Seitelbergen, and M.M. Muller, Serum lipid peroxide levels in the course of coronary by-pass surgery, *Eur. J. Clin. Chem. Clin. Biochem.* 30:205 (1992).

65. Z.W. Weitz, A.J. Birnbaum, P.A. Sobolka, E.J. Starling, and J.L. Skosey, High breath pentane concentrations during acute myocardial infarction, *Lancet* 337:933 (1991).

66. P.A. Sobolka, M.D. Brottman, Z. Weitz, A.J. Birnbaum, J. Skosey, and E.J. Zarling, Elevated breath pentane in heart failure reduced by free radical scavenger, *Free Rad. Biol. Med.* 14:643 (1993).

67. D.L. Gee and R.E. Litov, Lipid peroxidation and antioxidant status in burn patients, *Amer. J. Clin. Nutr.* 31:P-31 (1991).

68. Y.-K. Youn, C. Lalonde, and R. Demling, Oxidants and the pathophysiology of burn and smoke inhalation injury, Free Rad. Biol. Med. 12:409 (1992).

69. H.M. Berger, J.H.N. Lindeman, D. van Zoeren-Grobben, E. Houdkamp, J. Schrijver, and H.H. Kanhai, Iron overload, free radical damage, and rhesus haemolytic disease, *Lancet* 335:933 (1990).

70. I.S. Young, T.G. Trouton, J.J. Forney, D. McMaster, M.E. Callender, and E. Trimble, Antioxidant status and lipid peroxidation in hereditary haemochromatosis, *Free Rad. Biol. Med.* 16:393 (1994).

71. H.H. Draper, L. Polensek, M. Hadley, and L.G. McGirr, Urinary malondialdehyde as an indicator of lipid peroxidation in the diet and in the tissues, *Lipids* 19:836 (1984).

72. G. Paolisse, A. D'Amote, D. Giuliano, A. Ceriello, M. Varrichio, and F. D'Onofrio, Pharmacological doses of vitamin E improve insulin action in healthy subjects and non-insulin-dependent diabetic patients, *Amer. J. Clin. Nutr.* 57:650 (1993).

73. A.B. Baouali, H. Aube, V. Maupoil, B. Blettery, and L. Rochette, Plasma lipid peroxidation in critically ill patients: importance of mechanical ventilation, *Free Rad. Biol. Med.* 16:223 (1994).

74. W.A. Pryor, Can vitamin E protect humans against the pathological effects of ozone in smog?, *Amer. J. Clin. Nutr.* 53:702 (1991).

75 C.K. Chow, CG. Plopper, and D.L. Dungworth, Influence of dietary vitamin E on the lungs of ozone-exposed rats, *Envir. Res.* 20:309 (1977).

76. S. Owen, D. Pearson, V. Suarez-Mendez, R. O'Driscoll, and A. Woodcock, Evidence of free radical action in asthma, *New Engl. J. Med.* 325:586 (1991).

77. M.M. Mathews-Roth, Beta carotene therapy for erythropoietic protoporphyria and other photosensitivity diseases, *Biochimie* 68:875 (1986).

78. G.W. Burton and K.V. Ingold, Beta carotene: an unusual type of lipid antioxidant, *Science* 224: 569 (1984).

79. B.A. Underwood, The diet-cancer conundrum, *Public Health Rev.* 14:191 (1986).

80. J.A. Olson, Carotenoids, vitamin A and cancer, *J. Nutr.* 116:1127 (1986).

81. C. La Vecchia, S. Franceschi, A. Decarli, A. Gentile, M. Fasoli, S. Pampallona, and G. Tognoni, Dietary vitamin A and the risk of invasive cervical cancer, *Int. J. Cancer* 34:319 (1984).

82. K. Katsouyami, W. Willett, D. Trichopoulos, P. Boyle, A. Trichopoulu, S. Vasilaros, J. Papadiamantis, and B. MacMahon, Risk of breast cancer among Greek women in relation to nutrient intake, *Cancer* 61:181 (1988).

83. K. Gottlieb, E.J. Zarling, S. Mobarhan, P. Bowen, and S. Sugerman, β-carotene decreases markers of lipid peroxidation in healthy volunteers, *Nutr. Cancer* 19:207 (1993).

84. J.M.W. Slack, We have a morphogen!, *Nature* 327:553 (1987).

85. M.B. Sporn and A.B. Roberts, Role of retinoids in differentiation and carcinogenesis, *Cancer Res.* 43:3034 (1983).

86. J. Brockes, We may not have a morphogen, *Nature* 350:15 (1991).

87. G. Kolata, Does vitamin A prevent cancer?, *Science* 233:1161 (1984).

88. J.C. Bauernfeind, "The Safe Use of Vitamin A", International Vitamin A Consultative Group, The Nutrition Foundation, Washington, D.C. (1980).

89. F.W. Rosa, A.L. Wilk, and F.O. Kelsey, Teratogen update: vitamin A congeners, *Teratology* 33:355 (1986).

90. R.C. Rose and A.M. Bode, Biology of free radical scavengers: an evaluation of ascorbate, *FASEB J.* 7:1135 (1993).

91. B. Frei, R. Stocker, and B.N. Ames, Antioxidant defenses and lipid peroxidation in human blood plasma, *Proc. Natl. Acad. Sci. USA* 85:9748 (1988).

92. A.B. Kallner, D. Harbmann, and D.H. Hornig, On the requirements of ascorbic acid in man: steady state turnover and body pool in smokers, *Amer. J. Clin. Nutr.* 34:1347 (1981).

93. A.C. Chan, K. Tran, T. Raynor, P.R. Ganz, and C.K. Chow, Regeneration of vitamin E in human platelets, *J. Biol. Chem.* 266:17290 (1991).

94. G.W. Burton, U. Wronska, L. Stone, D.O. Foster, and K.U. Ingold, Biokinetics of dietary RRR-α-tocopherol in the male guinea pig at three dietary levels of vitamin C and two levels of vitamin E. Evidence that vitamin C does not "spare" vitamin E *in vivo*, *Lipids* 25:199 (1990).

95. L. Chen, An increase in vitamin E requirement induced by high supplementation of vitamin C in rats, *Amer. J. Clin. Nutr.* 34:1036 (1981).

96. W.A. Behrens and R. Madère, Ascorbic and dehydroascorbic acid status in rats fed diets varying in vitamin E levels, *Intern. J. Vit. Nutr. Res.* 59:360 (1989).

97. S. England and S. Seifter, The biochemical functions of ascorbic acid, *Ann. Rev. Nutr.* 6:365 (1986).

98. T.M. Bray and C.G. Taylor, Enhancement of tissue glutathione for antioxidant and immune functions in malnutrition, *Biochem. Pharmacol.* (in press).

99. C. Huang and M. Fwu, Degree of protein deficiency affects the extent of depression of the antioxidative enzyme activities and the enhancement of tissue lipid peroxidation in rats, *J. Nutr.123*:803 (1993).

100. C.C. Winterbourne, Superoxide as an intracellular radical sink, *Free Rad. Biol. Med.* 14:85 (1993).

101. T.M. Bray and C.G. Taylor, Tissue glutathione, nutrition and oxidative stress, *Can. J. Physiol. Pharmacol.* 71:746 (1993).

102. M.G.L. Hertog, E.J.M. Feskens, P.C.H. Hollman, M.B. Katan and D. Kromhout, Dietary antioxidant flavanoids and risk of coronary heart disease: the Zutphen elderlystudy, *Lancet* 342:1007 (1993).

INVOLVEMENT OF FREE RADICAL MECHANISM IN THE TOXIC EFFECTS OF ALCOHOL: IMPLICATIONS FOR FETAL ALCOHOL SYNDROME

C.Guerri,[1] C.Montoliu[1] and J.Renau-Piqueras[2]

[1]Inst. Invest. Citológicas (FIB)
[2]Ctr. Invest. Hospital La Fe
Valencia, Spain

INTRODUCTION

The role of free-radicals in alcohol-induced organ damage is still a matter of debate, in spite of decades of research. Moreover, during the last few years, great advances have been made in our understanding of the mechanisms of ethanol-induced enhancement of oxidative damage in liver and certain extrahepatic tissues. These advances have led to promising therapeutic approaches for the prevention of ethanol-induced oxidative stress. The present review focuses on the ethanol-induced production of pro-oxidant reactive species in different organs, an on the factors involved in ethanol metabolism that contribute to the generation of excess amounts of free radicals. Finally, the possible contribution of free radical generation to alcohol-induced fetal effects (fetal alcohol syndrome) are also discussed.

FREE-RADICAL MECHANISMS IN ALCOHOL-INDUCED LIVER DAMAGE

The first evidence of free radical involvement in liver injury came from Di Luzio[1] in the early 1960s. Di Luzio observed that pretreatment with antioxidants alleviated the liver fat accumulation of rats treated with ethanol. These investigators also showed[2] that "in vitro" ethanol addition to liver homogenates results in an increase in lipid peroxidation and that this can also be prevented by the simultaneus addition of antioxidants[2]. An enhancement of lipid peroxidation was also reported by Comporti et al.[3] after 1 to 12 h of an ethanol intubation. These observations suggested that an acute dose of ethanol induced free radical disturbances in the liver that resulted in lipid peroxidation. However, other investigators questioned the ethanol-induced lipid peroxidation since they did not observe any increase in thiobarbituric acid-reacting

Free Radicals in Diagnostic Medicine, Edited by
D. Armstrong, Plenum Press, New York, 1994

substances (TBArs) and/or no increase in conjugated dienes in the liver after acute ethanol administration[4]. These controversial results, in part due to the methodology used, have been reviewed by several authors[5,6].

Moreover, the early observation have been confirmed in many laboratories, where it was found that acute and chronic ethanol intoxication is associated with an increase in lipid peroxidation, in conjugate dienes and chemiluminescence, increased exhalation of pentane and ethane (see reviews[6,7]). More recently, researchers using Electron Spin Resonance (ESR) spectroscopy in combination with a spin trapping technique, have also found direct evidence that lipid-derived free radicals are generated after the acute[8,9] and chronic administration[10] of ethanol to experimental animals.

One of the main questions concerning ethanol-induced free radical formation is whether ethanol metabolism is involved in the free radical generation and whether this mechanism is implicated in alcohol-induced liver or tissue damage. In this context three main pathways have been proposed to explain the peroxidative effects of alcohol (Fig. 1).

1) Increases production of oxygen and ethanol-derived free radicals (hydroxyethyl radicals), generated during ethanol metabolism by the microsomal mono-oxygenase system and particularly by the ethanol-inducible form of cytochrome P-450 (CYP2E1).

It is well documented that ethanol can be metabolized in the endoplasmic reticulum by the microsomal ethanol oxidizing systems (MEOS) involving cytochrome P-450[11]. This enzymatic system oxidizes ethanol to acetaldehyde using NADPH as a cofactor. Chronic ethanol consumption produced an induction of this enzymatic system (MEOS)[11] and recently it has been demonstrated that one isoform of the cytochrome P-450, namely CYP2E1, is specifically induced by ethanol (see review[12]). This particular isoform has a high potential of production of both superoxide anion and hydrogen peroxide through the NADPH oxidase activity[13,14]. In fact, the enhancement in the microsomal mono-oxygenase system, which occurs after chronic ethanol treatment, has been associated with the increased production of superoxide and hydroxyl radicals found in isolated liver microsomes of ethanol-fed rats[15].

Recent findings obtained using electron spin resonance (ESR) spectroscopy indicate that free radical species, namely hydroxyethyl radicals, are generated during ethanol metabolism in isolated microsomes[10,16,17] and that these ethanol derived radicals are generated by the alcohol--inducible form of cytochrome P-450 (CYP2E1)[18]. The formation of the hydroxyethyl radicals has also been confirmed in vivo in deer-mice and rats intoxicated with ethanol[19,8].

It has been proposed that hydroxyethyl radicals can be generated in the microsomes by two separate metabolic pathways. The first is mediated by hydroxyl radicals originating from a Fenton-type reaction from endogenously formed hydrogen peroxide in presence of trace amounts of iron[19]. The second is $OH^{\cdot}$-independent[16,18] and can occur at the active site of cytochrome P-450. Although, both pathways can be partially responsible for the formation of hydroxyethyl radicals, some studies demonstrate that in the absence of contaminating iron, CYP2E1 exhibits a high specificity of ethanol oxidation[20], and that antibodies against CYP2E1 inhibit the spin trapping of alcohol-derived radicals[18].

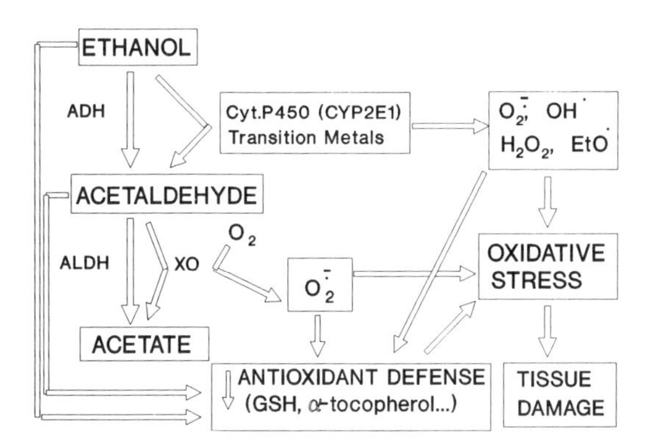

Figure 1. Proposed pathways leading to the formation of reactive oxygen species and oxidative stress during alcohol metabolism. ADH: alcohol dehydrogenase; ALDH: aldehyde dehydrogenase; XO: xanthine oxidase; $EtO^{\cdot}$: hydroxyethyl radicals.

Therefore, it seems that the hydroxyethyl radicals are mainly generated during ethanol metabolism by the alcohol-inducible form of cytochrome P-450 (CYP2E1) and that human liver microsomes are similarly capable of producing ethanol-derived radicals[21]. Indeed, in humans a direct relationship has been demonstrated between the level of CYP2E1 in different microsomal preparations and the capacity of these preparations to form hydroxyethyl radicals[21]. On the basis of these observation it has been postulated that by inducing a CYP2E1, ethanol consumption might lead to the formation of hydroxyethyl radical and reactive oxygen species, which may be responsible for the stimulation on lipid peroxidation detectable in liver biopsies and in the blood of patients suffering from alcohol-related liver diseases[21]. This might be particularly important in patients who are heavy drinkers, in whom the content of CYP2E1 in the centrilobular areas of the liver is greatly enhanced[22]. Therefore, the induction of CYP2E1 with the concomitant free radical formation, which occurs after chronic alcohol consumption, might play an important role in the pathogenesis of alcohol-induced liver damage.

2) Acetaldehyde, the first product of ethanol metabolism, might enhance free radical generation by its oxidation via aldehyde oxidase and/or xantine oxidase (XO) with the concomitant generation of superoxide anions.

The evidence that acetaldehyde may play a role in the ethanol-induced increase in free radical formation comes from several studies on perfused rat liver and hepatocytes that demonstrated that acetaldehyde enhances the production of alkanes[23], causes an increase in the antioxidant-sensitive respiration, and is more powerful than ethanol in stimulating lipid peroxidation and in decreasing glutathione levels[5]. "In vivo" acetaldehyde administration to rats has also been shown to increase hepatic lipid peroxidation an decrease glutathione levels[24].

Two different mechanisms by which acetaldehyde could stimulate peroxidative reactions have been proposed: first, by increasing the formation of oxygen-derived free radicals and hydrogen peroxide by its metabolism through aldehyde oxidase and XO[25] and, second, by decreasing the content of reduced glutathione in the cell[26].

Regarding the contribution of XO and aldehyde oxidase, Shaw and Jayatilleke[27] have reported that in isolated hepatocytes incubated with ethanol or acetaldehyde, the addition of allopurinol, an inhibitor of XO, or of menadione, an inhibitor of aldehyde oxidase, decreased lipid peroxidation by 35-50% and 80-95%, respectively. Allopurinol administration "in vivo" similarly decreased ethanol-induced lipid peroxidation in liver homogenates of rats treated acutely with alcohol[28], although these results have not been confirmed by others[29].

However, the "in vivo" contribution of the acetaldehyde metabolism by XO is questionable. Aldehydes have been reported to be inefficient substrates for XO, in which the Km of the enzyme for acetaldehyde is too high (~30mM)[30] as compared to the concentration of acetaldehyde present in the liver during alcohol metabolism. Therefore, it is not yet clear to what extent XO and aldehyde oxidase might contribute to an enhance production of free radicals following ethanol administration.

Acetaldehyde has also been implicated in the depletion of glutathione (GSH) found after alcohol intoxication. It has been shown that the alcohol dehydrogenase inhibitor, pyrazole, prevents the decrease in GSH levels produced by ethanol "in vivo"[31] or in isolated hepatocytes[5,26] while inhibition of aldehyde dehydrogenase by disulfiram[26,31] potentiates the effect. The formation of an adduct between acetaldehyde and glutathione[26] or its precursor L-cysteine may explain this mechanism. However, it has been postulated that unphysiological high acetaldehyde concentrations are required to form significant amounts of the adduct[32].

3) Ethanol consumption depletes antioxidant defenses and causes oxidative stress.

Due to the critical role of glutathione (GSH) in the detoxication processes of cells, including hydroperoxide catabolism, conjugation with electrophiles and direct interception of free radicals, the relationship beween GSH levels and alcohol-induced free radical production has been extensively explored during the past few years.

Acute ethanol administration has been shown to be associated with a decrease in the hepatic GSH content (see reviews[7,33]). The mechanisms involved in this diminution are the formation of GSH-acetaldehyde conjugates (as discussed above) and an enhanced hepatic GSH oxidation into GSSG as a consequence of increased generation of pro-oxidants free radical and lipid peroxides induced by ethanol. The latter interpretation is supported by the finding that antioxidants, given prior to ethanol, abolished both liver GSSG accumulation and the increase in lipid peroxidation induced by ethanol[34]. However, some investigators did not observe any changes in the hepatic oxidized glutathione (GSSG)[31]. Changes in the glutathione release into the bile and/or blood or in the synthesis of the tripeptide[32] have been also suggested.

Conflicting results on the hepatic content of GSH have been obtained from chronic ethanol administration to experimental animals. Some investigators have reported no change in this content, while others found it either decresed or increased (see review[33]). The discrepancies observed could be ascribed to differences in the experimental designs and the different procedures used to measured GSH.

Concomitant with the assessment of the effect of chronic ethanol intake on hepatic GSH levels, enhancement of the activities of the enzymes involved both in the synthesis (γ- glutamylcysteine and GSH synthetases)[32] and degradation (γ- glutamyl transferase, γ-GT)[31] of the tripeptide has been reported. It has been suggested that if alcohol treatment increases both liver GSH synthesis and sinusoidal GSH efflux[35] to a similar extent, a lower steady-state level would be established in the hepatocyte. Moreover, the increases in the activity of hepatic glutathione-S-transferases[36], glutathione peroxidase[36], γ-GT[31] and lipid peroxidation reported after chronic ethanol intake would also contribute to a diminished GSH level in the liver. In addition, Fernandez-Checa et al.[37] have recently demonstrated that chronic ethanol feeding in rats induces an important decrease in the hepatic mitochondria GSH pool size due to an impaired transport of GSH from cytosol into mitochondria. They further show that the mitochondrial GSH depletion is associated with mitochondrial lipid peroxidation and progession of liver damage[38].

Studies of liver biopsy samples from alcoholic patients indicate a decreased content of hepatic GSH[39,40,41,42] when compared to values obtained in nonalcoholics. This change was found to be unrelated to the nutritional status of the patients and influenced by the length of abstinence and the presence of liver necrosis[40,41]. Liver GSH depletion observed in alcoholic patients occurs concomitantly with an enhanced free-radical prooxidative activity, with elevated indexes of lipid peroxidation in the liver tissue[39,40], serum[43,44,45] and breath[46,47]. Reduction of hepatic GSH and enhancement of cellular free-radical activity are evidenced by the development of an oxidative stress conditions in the liver of alcoholic subjects, and this may constitute an important mechanism in the pathogenesis of alcoholic liver cell necrosis and in the progression and persistence of alcoholic liver disease. The oxidative stress phenomenon could be exacerbated by contributory factors related to prolonged alcohol consumption such as 1) induction of the CYP2E1 associated with the production of hydroxyethyl free radicals and reactive oxygen species (O_2^-, H_2O_2)[21], 2) hepatic accumulation of iron and the role of iron mobilization and microsomal induction[48], and 3) nutritional depletion. This last factor could involve a decreased ingestion and/or absorption of essential nutrients, which could limit the availability of the cellular antioxidant vitamins E[44] and A[49], the precursors for GSH biosynthesis, trace elements such as selenium[44,50,51,52] (constituent of glutathione peroxidases) and zinc[51] (constituent of superoxide dismutase), and possibly other molecules needed to repair cellular components altered by oxidative damage.

ETHANOL-INDUCED FREE RADICALS AND OXIDATIVE DAMAGE IN EXTRAHEPATIC TISSUES

Although most of the experimental and clinical data has been focused on the involvement of free radicals and lipid peroxidation in alcohol-induced liver damage, recent data indicate that this mechanism may also be involved in the toxic effect of ethanol on various extrahepatic tissues. Most of the data concern the gastric mucosa, testes, heart and central nervous system (see review[7]).

Ethanol-induced free radicals in gastric mucosa

An enhancement of lipid peroxidation and a decrease in nonprotein sulfhydryl levels have been reported in gastric mucosa of rats after oral administration of ethanol. Moreover, administration of antiperoxidative substances reduce the severity of mucosa damage and prevent the increase in lipid peroxidation[53]. These results were further confirmed in cultured mucosal cells, in which ethanol induces both the generation of superoxide anions and cellular damage. Furthermore, cell incubation with N-acetyl-L-cysteine, a substrate for glutathione synthesis, prevented the ethanol-induced damage to gastric mucosal cells[54]. These results suggest the involvement of a free radical mecanism in ethanol-induced gastric injury. In fact, Loguercio et al.[55] recently demonstrated that parental administration of glutathione prevents ethanol-induced depletion of gastric sulfhydryl compounds and decreases gastric mucosal injury in humans.

Concerning the mechanism, it has been postulated that the metabolism of acetaldehyde by the xanthine oxidase increases free radical production, which may contribute to the gastric injury observed after ethanol exposure[56]. Other authors suggest that ethanol-induced injury to gastric mucosa is neurophil-mediated and does not involve oxy-radicals[57].

Ethanol-induced free radicals in the testes

It is known that alcohol abuse induces testicular damage leading to endocrine and reproductive dysfuctions. Rosenblum et al.[58] postulated that free radical generation and lipid peroxidation may be an important mechanism of the toxic effects of ethanol on the testes. In fact, these authors show an increase in testicular lipid peroxidation with a decrease in peroxidizable polyunsaturated fatty acid in rats fed alcohol chronically. A direct correlation was also found between testicular atrophy and reduced glutathione levels. They further reported that supplementation with vitamin A, which can potentially fuction as a free radical scavenger, attenuates both the testicular atrophy and the depletion of testicular glutathione observed in rats fed with alcohol[59].

Involvement of free radical mechanism in ethanol-induced heart injury

It has been postulated that free radicals play important role in the production of mycardial injury, and this mechanism has also been implicated in the pathogenesis of alcoholic myopathy[60]. Moreover, using spin trapping and ESR spectroscopy, Reinke et al.[10] provided direct evidence that lipid radicals are generated in the hearts of rats after chronic and acute administration of ethanol.

It has been postulated that ethanol-induced lipid peroxidation in the heart is mediated by an increase in the conversion of xanthine dehydrogenase to xanthine oxidase[61] and an enhancement of the activity of acyl CoA-oxidase[62].

Chronic ethanol intake also induces changes in the components of the myocardial antioxidant defense[60] system, such as glutathione levels[31], and in the cytosolic and membranous protein thiols[63], which could contribute to the increase in free radical formation in this tissue. Changes in the levels of some antioxidant enzymes has also been observed after chronic alcohohol treatment, suggesting an adaptive mechanism acting against the prooxidative effect[63].

Ethanol-induced free radical formation in the central nervous system

The brain possess certain characteristics which may make it especially prone to oxyradical injury: its membranes are rich in polyunsaturated fatty acids, and the levels of some antioxidant enzymes (e.g. catalase, superoxide dismutase) and physiological scanvengers (e.g. glutathione, ascorbate and α-tocopherol) are relatively low. In fact, free radicals and lipid peroxidation appear to play a major role in the pathogenesis of some neurodegenarative diseases.

Regarding the contribution of this mechanism to alcohol-induced free radical formation in the nervous system, an enhancement of lipid peroxide and a decrease in α-tocopherol, ascorbate and glutathione levels have been reported in brain and cerebellum of rats after acute alcohol administration (see review[7]).

Chronic ethanol administration is also able to induce oxidative stress in the brain by increasing lipid peroxidation in some brain regions[64] and enhacing the generation of oxygen radical species in synaptosomes of chronic alcohol-fed rats[65]. This increase in oxygen radical generation was not observed when the rats were fed a ethanol diet supplemented with vitamine E (unpublished results).

Chronic ethanol intake also affects the antioxidant defense of the nervous system, as shown by the decrease in the levels of brain glutathione[31,65] and α-tocopherol content in cerebellum[66]. In addition, it is associated with adaptive changes in the anti-oxidant defense enzymes such as increased levels of superoxide dismutase and catalase in brain[65].

The mechanism involved in ethanol-induced oxygen radical formation in brain has not been fully identified. It has been suggested that an increase in the low-molecular-weight iron content, which has been observed in the cerebellum of ethanol-treated rats, and its putative links to local ethanol metabolism may contribute to the enhanced prooxidant radicals in cerebellum[7,66]. In this respect, although low alcohol metabolism was belived to occur in the brain, recent studies have indicated that ethanol could be metabolized in rat brain via catalase[67], and an ethanol-inducible form of cytochrome P-450 (CYP2E1) has been observed in various regions of the rat brain[68]. Moreover, we have observed that chronic ethanol intake, as in liver, induced an increase in the brain activity of total cytochrome P-450 (230%) as well as CYP2E1[65]. These results suggest that the induction of CYP2E1 could increase the generation of pro-oxidant free radical species (e.g. hydroxyethyl radicals), thereby decreasing the GSH and α- tocopherol levels leading to an oxidative stress in the brain of chronic alcohol-fed rats.

It has been postulated that an enhanced lipid peroxidation could disturb synaptosomal membrane fluidity as well as the metabolism of neurotransmitters and could contribute to the development of ethanol tolerance and dependence(see review[7]). At the same time, increased free radical generation could be involved in the ethanol-induced brain alterations as well as in the alcohol-induced acceleration of the aging process in the brain of alcoholics.

IMPLICATIONS OF FREE RADICALS IN ALCOHOL-INDUCED ALTERATIONS DURING CENTRAL NERVOUS SYSTEM DEVELOPMENT

It has been well established that maternal alcohol consumption during pregnancy can produce embryo-toxic effects that, in humans, culminate in a variety of birth anomalies designated as "Fetal Alcohol Syndrome"(FAS). This syndrome is characterized by facial dysmorphism, central nervous system (CNS) dysfunctions, mental retardation, growth deficiencies, and additional organ abnormalities[69]. Despite considerable research to elucidate the potential mechanism of alcohol teratogenicity, the primary cause is still unknown.

An increase in free radical formation has been implicated, by some authors, in the pathogenesis of alcohol-induced CNS dysfunctions in the FAS. Davis et al.[70], using neural crest cell cultures, demonstrated that ethanol severely depressed cell viability while simultaneously inducing the generation of reactive oxygen intermediates such as superoxide, hydrogen peroxide and hydroxyl anions. Addition of free radical scavenging enzymes, such as SOD, to the culture medium significantly reversed these effects. Since craniofacial malformations associated with FAS have been related to the alterations of the structure derived from neural crest ecto-mesenchyma, the authors suggest that free radicals are, at least partly, involved in the ethanol--induced injury of neural crest cells that can lead to developmental craniofacial anomalies associated with the fetal alcohol syndrome.

Table 1. Effect of Vitamin E supplementation on the cerebral level of GSH and on body and brain weights in 5-day-old offspring exposed "in utero" to alcohol[1].

	Body wt. (g)	Brain wt. (g)	Brain GSH (μmol/g)
Control	11.5±0.8	0.53±0.03	1.9±0.2
Alcohol	8.5±0.6*	0.43±0.04*	1.5±0.2*
Alcohol+Vit.E	11.6±0.7	0.54±0.03	1.8±0.2

[1]Offspring were from dams fed, during the entire gestation, control or alcohol-liquid diet with or without vitamin E supplementation (145mg/l). Brain GSH was determined as described[65]. Values are mean ± SD of 10-12 determination. * P<0.01.

In our laboratory we have demonstrated that the free radical mechanism could play a role in the pathogenesis of brain alterations observed after "in utero" exposure to alcohol. Thus, using an experimental model which reproduces in the rat most of the characteristics of the FAS (growth deficiencies, alterations in brain and neural development, etc.)[71,72] we have observed a decrease in the brain levels of

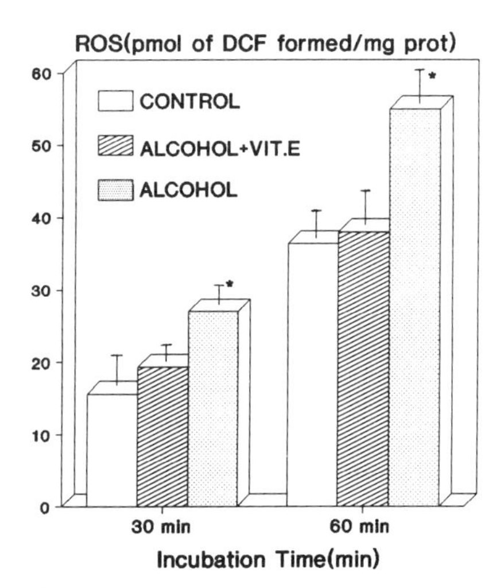

Figure 2. Rate of oxygen radical generation [measured as dichlorofluorescein (DCF) formed[65]] in synaptosomes from the offspring of dams fed control or ethanol-diet with or without vitamin E supplementation, as described in table 1. Data are the mean ± SD of 2 different synaptosome preparations. * P<0.01. v.s. control.

glutathione and an enhancement of the activities of SOD, catalase and glutathione transferase in 15-day and 21-day old fetuses, compared to controls.In addition, we have found that "in utero" exposure to ethanol, significantly increased the formation of reactive oxygen species (ROS) in synaptosomes from 5-day old rats (Fig. 2). Supplementation of dietary vitamin E to alcohol-fed pregnant rats during the entire gestation prevents the increase in ROS, the reduction in brain and body weights and the decrease in the cerebral level of GSH in their offspring (Table 1,Fig.2).Furthermore, using primary culture of astrocytes as a model to study the effect of alcohol on astroglial development, we have demonstrated that alcohol dramatically alters the development of astrocytes[73]. We can, therefore, postulate that derangement of astrogliogenesis could be an important mechanism involved in the effects of alcohol on brain development (see review [74]). Therefore, we have investigated the possible role of ethanol-induced free radical generation in the alterations observed in astrocytes. We have observed that alcohol added to the culture of astrocytes induces: an increase in ROS, a decrease in the levels of glutathione, and cleavage of some cytoskeletal proteins such as glial fibrillary acidic protein and vimentin. This last effect was prevented by the addition of scavengers such as GSH and SOD (unplublished results).

These results suggest that reactive free radicals may be involved in the pathogenesis of brain alterations that are the hallmark of FAS.

In conclusion, there is ample evidence for the involvement of free radical in the toxic effects of alcohol. However, more clinical studies are needed to stablish the quantitative importance of free radicals in the pathogenesis of alcohol-induced organ damage in humans. Nevertheless, the present knowledge suggests that the use of pharmacological agents which could prevent oxidative stress may have therapeutical potential against the clinical manifestations of ethanol toxicity in humans and perhaps, to try to prevent of some of the CNS dysfunctions of the FAS.

ACKNOWLEDGMENTS

The research work of the authors was supported by CICYT, grant SAL 91-0020-C02 (Spain).

REFERENCES

1. N.R.Di Luzio. Prevention of the acute ethanol-induced fatty liver by antioxidants. Physiologist 6:169 (1963).
2. N.R.Di Luzio and A.D. Hartman. Role of lipid peroxidation in the pathogenesis of ethanol-induced fatty liver. Fed. Proc. 26:1436(1967).
3. M.Comporti, A. Hartman and N.R. Di Lucio. Effect of in vivo and in vitro ethanol administration on liver lipid peroxidation. Lab. Invest. 16:616(1967).
4. S.Hashimoto and R.O. Recknagel. No chemical evidence of hepatic lipid peroxidation in acute ethanol toxicity. Exp. Mol. Pathol. 8:225(1968).
5. M.U.Dianzani. Lipid peroxidation in ethanol poisoning:A critical reconsideration. Alcohol and Alcoholism 20: 161(1985).
6. E.Albano, M. Ingelman-Sundberg, A. Tomasi and C. Poli.

Free radical mediated reactions and ethanol toxicity: Some considerations on the methodological approaches, in:Alcoholism: A Molecular Perspective (T.N.Palmer, ed.), pp. 45-56. Plenum, New York(1991).
7. R.Nordman, C. Ribière and H. Rouach. Implication of free radical mechanisms in ethanol-induced cellular injury. Free Radical Biology and Medicine 12: 219(1992).
8. L.A.Reinke, Y. Kotake, P.B. McCay and E.G. Janzen. Spin-trapping studies of hepatic free radicals formed following the acute administration of ethanol to rats: in vivo detection of 1-hydroxyethyl radicals with PBN. Free Rad. Biol. Med. 11:31(1991).
9. K.T.Knecht, B.U. Bradford, R.P. Mason and R.G. Thurman. In vivo formation of a free radical metabolite of ethanol. Mol. Pharmacol. 38:26(1990).
10. L.A.Reinke, E.K. Lai, C.M. DuBose and P.B. McCay. Reactive free radical generation in vivo in heart and liver of ethanol-fed rats: correlation with radical formation in vitro. Proc. Natl. Acad. Sci. USA 84: 9223(1987).
11. C.S.Lieber. Microsomal ethanol oxidizing system. Enzyme 37:45(1987).
12. C.S.Lieber. Mechanism of ethanol induced hepatic injury. Pharm. Ther.46:1(1990).
13. G.Ekstrom and M. Ingelman-Sundberg. Rat liver microsomal NADPH-supported oxidase activity and lipid peroxidation dependent on ethanol-inducible cytochrome P-450 (P-450 II E1). Biochem. Pharmacol. 38:1313(1989).
14. J.O.Persson, Y. Terelius and M. Ingelman-Sundberg. Cytochrome P450-dependent formation of oxygen radicals. Isoenzyme-specific inhibition of P450-mediated reduction of oxygen and carbon tetrachloride. Xenobiotica 20:887(1990).
15. J.Rashba-Step, N.J. Turro and A.I. Cederbaum. Increased NADPH- and NADH-dependent production of superoxide and hydroxyl radical by microsomes after chronic ethanol treatment. Arch. of Biochem. and Biophys. 300:401(1993).
16. E.Albano, A. Tomasi, L. Goria-Gatti, G. Poli, V. Vannini and M.U. Dianzani. Free radical metabolism of alcohols by rat microsomes. Free Rad. Res. Commun. 3:243(1987).
17. L.A.Reinke, J.M. Rau and P.B. McCay. Possible roles of free radicals in alcoholic tissue damage. Free Rad. Res. Commun. 9:205(1990).
18. E. Albano, A. Tomasi, J.O. Persson, Y. Terelius, L. Goria-Gatti, M. Ingelman-Sundberg, and M.U. Dianzani. Role of ethanol-inducible cytochrome P450 (P450IIE1) in catalysing the free radical activation of aliphatic alcohols. Biochem. Pharmacol. 41:1895 (1991b).
19. A.I.Cederbaum. Oxygen radical generation by microsomes: role of iron and implications for alcohol metabolism and toxicity. Free Radical Biol. and Med. 7:559(1989).
20. G.Ekstrom, T. Gronholm and H. Ingelman-Sundberg. Hydroxyl-radical production and ethanol oxidation by

liver microsomes isolated from ethanol-treated rats. Biochem. J. 233:755(1986).
21. E.Albano, P. Clot, M. Tabone, S. Aricò and M. Ingelman-Sundberg. Oxidative damage and human alcoholic liver diseases. Experimental and clinical evidence. In: Free Radicals: From Basic Science to Medicine (G. Poli, E. Albano, M.U. Dianzani, eds.) pp.310-322(1993).
22. R.Bühler, K.O. Lindros, K. Von Boguslawsky, P. Karkkainen, J. Makinen and M. Ingelman-Sundberg. Perivenous expression of ethanol-inducible cytochrome P450 IIE1 in livers from alcoholics and chronically ethanol-fed rats. Alcohol Alcohol. (Suppl.) 311 (1991).
23. A.Muller and H. Sies. Ethane release during metabolism of aldehydes and monoamines in perfused rat liver. Eur. J. Biochem. 134:599(1983).
24. L.A.Videla, V. Fernandez and A. de Marinis. Liver lipoperoxidative pressure and glutathione status following acetaldehyde and aliphatic alcohols pretreatment of rats. Biochem. Biophys. Res. Commun. 104:965(1982).
25. E.W.Kellogg and I. Fridovich. Superoxide hydrogen peroxide and singlet oxygen in lipid peroxidation by a xanthine oxidase system. J. Biol. Chem. 250:8812 (1975).
26. J.Viña, J.M. Estrela, C. Guerri and F.J. Romero. Effect of ethanol on glutathione concentration in isolated hepatocytes. Biochem. J. 188:549(1980).
27. S.Shaw and E. Jayatilleke. The role of aldehyde oxidase in ethanol-induced hepatic lipid peroxidation in the rat. Biochem. J. 268:579(1990).
28. S.Kato, T. Kawase, J. Alderman, N. Inatomi and C.S. Lieber. Role of xanthine oxidase in ethanol-induced lipid peroxidation in rats. Gastroenterol. 98:203 (1990).
29. Y.Kera, Y. Ohbora and S. Komura. The metabolism of acetaldehyde and not acetaldehyde itself is responsible for in vivo ethanol-induced lipid peroxidation in rats. Biochem. Pharmacol. 37:3633 (1988).
30. I.Fridovich. The mechanism of the enzymatic oxidation of aldehydes. J. Biol. Chem. 241:3126(1966).
31. C.Guerri and S. Grisolia. Changes in glutathione in acute and chronic alcohol intoxication. Pharmacol. Biochem. Behav. 13:53(1980).
32. H.Speisky, A. MacDonald, G. Giles, H. Orrego and I. Israel. Increase loss and decresed synthesis of hepatic glutathione after acute ethanol administration. Biochem. J. 225:565(1985).
33. L.A.Videla and C. Guerri. Glutathione and alcohol, in: Glutathione: Metabolism and Physiological Functions (J. Viña, ed.) CRC Press, pp.57-67(1990).
34. A.Valenzuela, C. Lagos, K. Schmidt and L.A. Videla. Silymarin protection against hepatic lipid peroxidation induced by acute ethanol intoxication in the rat. Biochem. Pharmacol. 34:2209(1985).
35. J.C.Fernandez-Checa, M. Ookhtens and N. Kaplowitz. Effect of chronic ethanol feeding on rat hepatocytic glutathione. Compartmentation, effux and response to

incubation with ethanol. J. Clin. Invest. 80:57 (1987).
36. G.Aycak, M. Uysal, A.S. Yalcin, N. Kocak-Toker, A. Sivas and OzH. The effect of chronic ethanol ingestion on hepatic lipid peroxide glutathione, glutathione peroxidase and glutathione transferase in rats. Toxicology 36:71(1985).
37. J.C.Fernandez-Checa, C. Garcia-Ruiz, M. Ookhtens and N. Kaplowitz. Impaired uptake of glutathione by hepatic mitochondria from chronic ethanol-fed rats. J. Clin. Invest. 87:397(1991).
38. T.Hirano, N. Kaplowitz, H. Tsukamoto, S. Kamimura and J.C. Fernandez-Checa. Hepatic mitochondrial glutathione depletion and progression of experimental alcoholic liver disease in rats. Hepatology 16:1423 (1992).
39. T.Suematsu, T. Matsumura, N. Sato, T. Miyamoto, T. Ooka, T. Kamada and H. Abe. Lipid peroxidation in alcoholic liver disease in humans. Alcoholism: Clin. Exp. Res. 5:427 (1981).
40. S.Shaw, K. Rubin and C.S. Lieber. Depressed hepatic glutathione and increased diene conjugates in alcoholic liver disease. Evidence of lipid peroxidation. Dig. Dis. Sci. 28:585(1983).
41. L.A.Videla, H. Iturriaga, M.E. Pino, D. Bunout, A. Valenzuela and G. Ugarte. Content of hepatic reduced glutathione in chronic alcoholic patiens: influence of the length of abstinence and liver necrosis. Clin. Sci. 66: 283(1984).
42. S.A.Jewell, D. DiMonte, A. Gentile, A. Guglielmi, E. Altomare and O. Albano. Decreased hepatic glutathione in chronic alcoholic patiens. J. Hepatol. 3:1(1986).
43. R.Fink, M.R. Clemens, D.H. Marjot, P. Patsalos, P. Cawood, A.G. Norden, S.A. Iversen and Domandy. Increased free-radical activity in alcoholics. Lancet 2:291(1985).
44. A.R.Tanner, I. Bantock, H.B. Lloyd, N.R. Turner and R. Wright. Depressed selenium and vitamin E levels in alcoholic population. Possible relationship to hepatic injury thuogh increased lipid peroxidation. Dig. Dis. Sci. 31:1307(1986).
45. M.Uysal, H. Bulur, S. Erdine-Demirelli and C. Demiroglu. Erythrocyte and plasma lipid peroxides in chronic alcoholic patiens. Drud. Alc. Dep. 18:385(1986).
46. S.Moscarella, G. Laffi, C. Buzzelli, R. Mazzanti, L. Caramelli and P. Gentillini. Expired hydrocarbons in patiens with chronic liver disease. Hepato--Gastroenterol. 31: 60(1984).
47. P.Lettéron, V. Duchatelle, A. Berson, B. Fromenti, C. Fisch, C. Degot and J.P. Benhamou. Increased ethane exhalation, an in vivo index of lipid peroxidation, in alcohol-abusers. Gut 34:409(1993).
48. S.Shaw, E. Jaytilleke and C.S. Lieber. Lipid peroxidation as a mechanism of alcoholic liver injury: role of iron movilization and microsomal induction. Alcohol: 135 (1988).
49. A.M.Leo and C.S. Lieber. Hepatic vitamine A depletion in alcoholic liver injury. N. Eng. J. Med. 307:597 (1982).

50. B.Dworkin, W.S. Rosenthal, R.H. Jankowski, G.G. Gordon and D. Haldea. Low blood selenium levels in alcoholics with and without advanced liver disease. Correlations with clinical and nutritional status. Dig. Dis. Sci. 30:838(1984).
51. G.E.A.Bjorneboe, J. Johnsen, A. Bjorneboe, J. Morland and C.A. Drevon. Effect of heavy alcohol consumption on serum concentrations of fat-soluble vitamins and selelium. Alcohol and Alcoholism. Suppl.1:533(1987).
52. C.Girre, E. Hispard, P. Therond, S. Guedj, R. Bourdon and S. Dally. Effect of abstinence from alcohol on the depression of glutathione peroxidase activity and selenium and vitamin E levels in Chronic alcoholic patiens. Alcoholism: Clin. Exp. Res. 14:909(1990).
53. T.Mizui, H. Sato, F. Hirose and M. Doteuchi. Effect of antiperoxidative drugs on gastric damage induced by ethanol in rats. Life Sci. 41:755(1987).
54. H.Mutoh, H. Hiraishi, S. Ota, H. Yoshida, K.J. Ivey and A. Terano. Protective role of intracellular glutathione against ethanol-induced damage in cultured rat gastric mucosal cells. Gastroenterology. 96:14452(1990b).
55. C.Loguercio, D. Taranto, F. Beneduce, C. del Vecchio Blanco, A. de Vincentiis, G. Nardi, and M. Romano. Glutathione prevents ethanol induced gastric mucosal damage and depletion of sulfhydryl compounds in humans. Gut 34:161(1993).
56. S.Shaw, V. Hebert, N. Colman and E. Jayatilleke. Effect of ethanol-generated free radicals on gastric intrinsic factor and glutathione. Alcohol 7:153(1990).
57. P.R.Kvietys, B. Twohig, J. Danzell and R.D. Specian. Ethanol-induced injury to the rat gastric mucosa. Role of neurophils and xanthine oxidase-derived radicals. Gastroenterology 98:909(1990).
58. E.R.Rosemblum, J.S. Gavaler and D.H. Van Thiel. Lipid peroxidation: a mechanism for alcohol-induced testicular injury. Free Rad. Biol. Med.7:569(1989).
59. E.R.Rosemblum, J.S. Gavaler and D.H. Van Thiel. Vitamin A at pharmacologic doses ameliorates the membrane lipid peroxidation injury and testicular atrophy that occurs with chronic alcohol feeding in rats. Alcohol and Alcoholism 22:241(1987).
60. I.Edès, A. Toszegi, M. Csanady and B. Bozoky. Myocardial lipid peroxidation in rats after chronic alcohol ingestion and the effects of different antioxidants. Cardiovasc. Res. 20: 542(1986).
61. H.H.Oei, W.E. Stroo, K.P. Burton, S.W. Schaffer. A possible role of xanthine oxidase in producing oxidative stress in the heart of chronically ethanol-treated rats. Res. Commun. Chem. Pathol. Pharmacol. 38: 453(1982).
62. L.F.Panchenko, S.V. Pirozhkov, S.V. Popova, V.D. Antonenkov. Effect of chronic ethanol treatment on peroxisomal acyl-CoA oxidase activity and lipid peroxidation in rat liver and heart. Experientia 43:580(1987).
63. C.Ribiere, I. Hininger, H. Rouach and R. Nordmann. Effects of chronic ethanol administration on free

radical defence in rat myocardium. Biochem. Pharmacol. 44:1495(1992).
64. H.A.Nadiger, S.R. Marcus, M.V. Chandrakala. Lipid peroxidation and ethanol toxicity in rat brain:Effect of vitamin E deficiency and supplementation. Med. Sci. Res. 16:1273(1988).
65. C.Montoliu, S. Vallés, J. Renau-Piqueras, C. Guerri. Ethanol-induced oxygen radical formation and lipid peroxidation in rat brain.Effect of chronic alcohol consumption.(1994) J. Neurochem. (in press).
66. H.Rouach, P. Houzé, M.T. Orfanelli, M. Gentil, R. Nordmann. Effects of chronic ethanol intake on some anti- and pro-oxidants in rat cerebellum. Alcohol and Alcoholism 26:257(1991).
67. C.M.G. Arago, F. Rogan and Z. Amit. Ethanol metabolism in rat brain homogenates by a catalase-H_2O system. Biochem. Pharmacol. 44:93(1992).
68. T. Hansson, N. Tindberg, M. Ingelman-Sundberg, and C. Kühler. Regional distribution of ethanol-inducible cytochrome P450 IIEI in the rat central nervous system. Neuroscience 34:451(1990).
69. S.K. Clarren, D.W. Smith. The fetal alcohol syndrome. N. Engl. J. Med. 298:1063(1978).
70. W.L.Davis, L.A. Crawford, O.J. Cooper, G.R. Farmer, D.L. Thomas, B.L. Freeman. Ethanol induces the generation of reactive free radicals by neural crest cells in vitro. J. Craniofac. Genet. Biol. 10: 277(1990).
71. R.Sanchis, M. Sancho-Tello, M. Chirivella, C. Guerri. The role of maternal alcohol damage on the ethanol teratogenicity in the rat. Teratology 35:199(1987).
72. C.Guerri. Synaptic membrane alterations in rats exposed to alcohol. Alcohol and Alcoholism. Suppl. 1:467 (1987).
73. C.Guerri, R. Sáez, M. Sancho-Tello, E. Martín de Aguilera, J.Renau-Piqueras. Ethanol alters astrocyte development: A study of critical periods using primary cultures. Neurochem. Res. 15:559(1990).
74. C.Guerri, R. Sáez, M. Portolés, J. Renau-Piqueras. Derangement of astrogliogenesis as a possible mechanism involved in alcohol-induced alterations of central nervous system development. Alcohol and Alcoholism, Suppl.2:203(1993).

21-AMINOSTEROIDS ("LAZAROIDS")

Eugene D. Means

Clinical Research II
The Upjohn Company
526 Jasper St.
Kalamazoo, MI 49007

INTRODUCTION

Acute central nervous system (CNS) injury, eg, ischemia, trauma, results in the almost immediate death of neurons. Surrounding tissue is at risk for further cell death. This circumjacent tissue has been labeled *tissue-at-risk* or the *ischemic penumbra*. A complex interrelated series of pathophysiological events occurs in this vulnerable tissue which, if not interrupted, will inexorably progress to tissue necrosis (secondary injury).[1] A discussion of the various factors involved in this secondary injury cascade are beyond the scope of this chapter. However, one of the pathophysiological events believed to play a crucial role in this injury cascade is free radical-mediated lipid peroxidation.

The Upjohn Company has developed a series of synthetic, nonhormonal (nonglucocorticoid) 21-aminosteroid compounds to provide therapy for the damaging effects of CNS trauma and ischemia. One of these compounds, tirilazad mesylate, is a novel lipid peroxidation inhibitor/cytoprotective agent which is currently in trials for the treatment of CNS trauma, subarachnoid hemorrhage (SAH), and ischemic stroke. The purpose of this report is to briefly describe the mechanism of action of the 21-aminosteroids and report on some of the early and recently completed trials with this compound.

MECHANISM OF ACTION

One of the 21-aminosteroid compounds, tirilazad mesylate, is a potent inhibitor of radical-induced, iron-catalyzed lipid peroxidation. It is a highly lipophilic compound and is preferentially distributed to the lipid layer of cell membranes. The compound seems to

Free Radicals in Diagnostic Medicine, Edited by
D. Armstrong, Plenum Press, New York, 1994

exert its anti-lipid peroxidation effect through a dual mechanism of action: a free radical scavenging effect (antioxidant action) and a physiochemical interaction with the cell membrane to decrease membrane fluidity (ie, membrane stabilization).

Tirilazad has been reported to scavenge lipid peroxyl and phenoxy radicals, to slow the oxidation of vitamin E and potentiate vitamin E's antioxidant efficacy.[2] Tirilazad can also interact with the highly reactive hydroxyl radicals generated during in vitro Fenton reactions.[3] In vivo studies utilizing the salicylate trapping method for measuring the hydroxyl radical have demonstrated that tirilazad administration decreases brain hydroxyl radical levels in models of traumatic brain injury in mice[4,5] and global cerebral ischemia-reperfusion injury in gerbils.[6]

The 21-aminosteroids are highly lipophilic reflecting its membrane interaction. Experiments to characterize the membrane effects of the drug showed that tirilazad induced changes in the molecular packing order in membrane hyprophalic domains in the endothelial cells of brain microvessels.[7] It has been hypothesized that the pyramidine amine moiety of the molecule should help compress membrane phospholipid head groups by interacting with the phosphate-containing head groups of the hydrophilic portion of the membrane. These membranes stabilizing effects of tirilazad may help to inhibit the propagation of lipid peroxidation by inhibiting the movement of peroxyl and alkalyl radicals in the membrane. Thus, tirilazad blocks oxygen-radical lipid peroxidation, apparently by a chemical antioxidant action and membrane stabilizing effect.

CLINICAL PHARMACOLOGY OF TIRILAZAD MESYLATE*

Central nervous system trauma and stroke are a major cause of death and disability and are unmet medical needs.[8,9] There has been a general pessimism in the past among practitioners about current or potential therapies for the treatment of these diseases. There are several reasons for this sentiment which include: 1) the complexity of the pathophysiology of the tissue injury, 2) getting the patient to a setting where treatment can be administered quickly (time to treatment),[10] adequate trial design to delineate the efficacy of the therapy,[11,12,13] and 4) the cost to society and industry to develop drugs for the treatment of these diseases. There is reason, however, to be more optimistic about the possibilities for the development of new therapies for CNS trauma and stroke. There is currently a much better understanding of the pathophysiology of the secondary injury following the initial traumatic or ischemic episode. Animal research has contributed a number of potential categories of drug candidates for the treatment of CNS trauma and stroke. Included among the more common drug categories are calcium channel blockers, excitatory amino acid antagonists, thrombolytics, and lipid peroxidation inhibitors like the 21-aminosteroids.

---

* Taken from E.D. Hall, J.M. McCall, E.D. Means, Therapeutic potential of the lazaroids (21-aminosteroids) in acute CNS trauma, ischemia and subarachnoid hemorrhage, *Advances in Pharmacology*, 28:221-268 (1994).

A worldwide development program has been undertaken to study tirilazad mesylate in acute head and spinal trauma, aneurysmal subarachnoid hemorrhage (SAH), and ischemic stroke. To date, a number of Phase I safety and pharmacokinetic studies have been completed in normal volunteers, as well as Phase II safety studies in patients with head and spinal trauma, aneurysmal SAH, and ischemic stroke. Currently, Phase III registration studies are underway using tirilazad mesylate for the treatment of acute head trauma, acute spinal trauma (NASCIS III), aneurysmal SAH, and ischemic stroke. A concerted attempt was made in these registration studies to treat the patient as early after the onset as possible, even at the risk of reduced enrollment.

It should be mentioned that the registration study using tirilazad mesylate in patients with acute spinal cord injury (NASCIS III) is a successor to NASCIS II. The NASCIS II study showed that patients treated with MP, when given in large doses (30 mg/kg i.v. bolus followed by infusion at 5.4 mg/kg/hr for 23 hours), had significant improvement in motor function and sensation of pinprick and light touch within 8 hours of injury compared to placebo.[14,15] Patients treated with methylprednisolone after 8 hours did not differ in their neurological outcome compared to placebo. The NASCIS III study will contain 3 arms. Each arm will initially receive an i.v. bolus of MP (30 mg/kg). Two arms will receive either an infusion of 5.4 mg/kg/hr for 24 or 48 hours. The third arm will receive tirilazad mesylate, 2.5 mg/kg every 6 hours for 48 hours. It is anticipated that this study will be complete in 3-4 years.

Phase I Studies

Phase I studies with tirilazad mesylate included both single- and multiple-dose regimens, and interaction with several marketed drugs. Safety and pharmacokinetics of tirilazad mesylate were first tested in single- and multiple-dose studies in normal healthy volunteers. Single-dose tolerance and pharmacokinetics of tirilazad mesylate were assessed in 47 healthy male subjects.[16] Subjects were randomized to receive citrate vehicle or tirilazad mesylate, 0.25, 0.5, 1.0, or 2.0 mg/kg by 0.5-hour infusion.

Injection site pain was observed, with approximately equal frequency in both vehicle- and tirilazad mesylate-treatment groups. No statistically significant effects of tirilazad mesylate on blood pressure, heart rate, electrocardiogram, liver enzymes or renal function were apparent. Tirilazad mesylate did not significantly affect measures of glucocorticoid activity (blood glucose, adrenocorticotropic hormone, cortisol, eosinophil, or lymphocyte levels). Maximal plasma concentrations of tirilazad mesylate increased linearly with dose. The apparent elimination half-life at the higher doses was 3.7 hours. Clearance of tirilazad mesylate approached liver blood flow. Results indicated that intravenous infusions at these doses were well tolerated and devoid of glucocorticoid effects. Tirilazad mesylate appears to be efficiently cleared by the liver, and its pharmacokinetics are apparently linear over the dosage range studied.

The multiple-dose tolerability and pharmacokinetics of tirilazad mesylate were assessed in 50 healthy male volunteers.[17] Volunteers were randomized to receive intravenous normal saline placebo, citrate vehicle placebo, or 0.5, 1.0, 2.0, 4.0, or 6.0 mg/kg/day tirilazad mesylate in divided doses every 6 hours for 5 days for a total of 21 doses. Drug was infused over 10 or 30 minutes. All tirilazad mesylate treatment groups and the citrate vehicle group had significantly more frequent and more intense pain at the injection site than did the saline group, but the pain intensity did not require interruption of dosing. Three episodes of clinical thrombophlebitis were observed. No statistically significant effects of tirilazad mesylate on blood pressure, heart rate, electrocardiograms, or renal function were apparent. Moderate and transient increases in serum alanine transaminase were observed in several subjects. In the 6 mg/kg/day group, 50% of the subjects exhibited increased alanine transaminase. Tirilazad mesylate did not significantly affect measures of glucocorticoid activity (blood, glucose, adrenocorticotropic hormone, cortisol, eosinophil, or lymphocyte levels). Tirilazad mesylate pharmacokinetics were linear over the dosage range studied. Steady state appeared to be achieved by the fifth day of dosing. After the last dose, a mean terminal half-life of 35 hours was observed. This terminal phase was not observed after single dosing, but the portion of the area under the plasma tirilazad concentration-time curve represented by this terminal phase is estimated to account for less than 5% of the total area after a single dose. Thus, single-dose estimates of clearance may be predictive of clearance on multiple dosing, but volume of distribution estimates are not.

The interaction of tirilazad mesylate with nimodipine was studied in normal healthy volunteers.[18] Subjects received 60 mg nimodipine p.o., 2 mg/kg tirilazad mesylate as a 10-minute i.v. infusion, and a combination of the two treatments according to a balanced 3-way crossover design. Plasma tirilazad mesylate concentrations were determined by HPLC; nimodipine plasma concentrations were determined by gas chromatography with electron-capture detection. Blood pressure, heart rate, respiration, and cardiac rhythm were also monitored. No significant effects of nimodipine on tirilazad pharmacokinetic parameters were observed ($p>0.05$). Values for tirilazad mesylate clearance (34.9 ± 8.96 L/hr) and $t^{1/2}$ (29 ± 7.83 hr) were consistent with previous studies. Nimodipine pharmacokinetic parameters exhibited substantial variability, and clearance estimates were approximately 3 times those previously reported. However, no significant differences in nimodipine pharmacokinetics were observed between treatments. No clinically significant changes were observed in the cardiac parameters assessed. Thus, no significant interaction between tirilazad mesylate and nimodipine was detectable after single dose administration.

Phase II Study of Subarachnoid Hemorrhage

A Phase II study was recently completed using tirilazad mesylate for the treatment of patients with aneurysmal SAH. This was a double-blind, dose-escalating safety trial in which the patients were treated within 72 hours with either placebo (vehicle) or with tirilazad mesylate, i.v. (0.6, 2.0, or 6.0 mg/kg/day) for 10 days post-hemorrhage. All patients received nimodipine, 60 mg every 4 hours p.o. for up to 21 days. The data analysis included all patients who were randomized (i.e. intent to treat). The placebo group (vehicle) and each active group were well matched for age, sex, neurological and medical conditions on admission. The amount of blood on the admission CT scan suggested a worse prognosis for the Tier 2 (2/0 mg/kg/day) group compared to vehicle or the other two active drug groups. There were no clinically meaningful differences between the vehicle and drug groups for any of the safety parameters measured. The mortality for the 6 mg/kg/day tier was the highest (20%), while the lowest mortality was observed in the 2 mg/kg/day tier (5%) (Table 1). The difference in mortalities between the 6 mg/kg/day group and the vehicle group did not reach statistical significance. The slightly higher mortality for the 6 mg/kg/day compared to placebo was thought to be secondary to the study design (consecutive randomization) and to small numbers of patients, but the possibility of an inverted U-shaped dose response cannot be ruled out. This is presently under investigation in a Phase III study in which patients with aneurysmal SAH will be concurrently randomized to three doses of tirilazad mesylate. Symptomatic vasospasm between admission and day 14, and favorable outcome (good recovery and moderate disability) on the 3-month Glasgow Outcome Scale (GOS) were used as primary outcome measures. Symptomatic vasospasm was about the same for the vehicle and 6 mg/kg/day groups (Table 2), but was reduced from 41% in the vehicle group to 31% and 19% in the 0.6 and 2 mg/kg/day groups, respectively. The 2 mg/kg/day dosing group had the lowest rate of symptomatic vasospasm despite having the greatest amount of blood on the admission CT scan. Favorable outcome on the 3-month GOS was best for the 2 mg/kg/day tier (90%), while favorable outcome for the vehicle and 6 mg/kg/day groups was 72% (Table 3). The value for the 0.6 mg/kg/day was 81%. These data show that tirilazad mesylate is a safe drug in patients with aneurysmal SAH. The dose-related reductions in symptomatic vasospasm for the 0.6 and 2 mg/kg/day groups compared to the vehicle group and a dose-related increase in the 3-month GOS for the 0.6 and 2 mg/kg/day groups compared to the vehicle group suggest that tirilazad mesylate may be effective in the treatment of patients with aneurysmal SAH.

In summary, tirilazad mesylate has been tested in over 1700 patients and has proven to be safe except for pain at the injection site, and phlebitis in a few volunteers and patients. These complications have largely been overcome by diluting the drug, frequent changes of the catheter site, and the use of a central line when appropriate. There was no difference between placebo and tirilazad mesylate for cardiac or hepatotoxicity in the Phase I-II studies. In a Phase II safety study of 245 patients with aneurysmal SAH, symptomatic vasospasm was reduced and favorable outcome on the 3-month GOS was improved by tirilazad mesylate compared to placebo. Studies to date demonstrate that tirilazad mesylate is safe in normal volunteers and patients with acute head and spinal cord injury, aneurysmal SAH, and ischemic stroke. In addition, data suggest that tirilazad mesylate is a promising agent for the treatment of aneurysmal SAH.

REFERENCES

1. B.K. Siesjo, C.-D. Agardh, and F. Bengtsson, Free radicals and brain damage, *Cerebrovasc. Brain Metab. Rev.* 1:165-211 (1989).
2. J.M. Braughler and J.F. Pregenzer, The 21-aminosteroid inhibitors of lipid peroxidation: reactions with lipid peroxyl and phenoxyl radicals, *Free Rad. Biol. Med.* 7:125-130 (1989).
3. J.S. Althaus, C.M. Williams, P.K. Andrus, P.A. Yonkers, G.J. Fici, E.D. Hall, and P.F. Vonvoigtlander, *In vitro* and *in vivo* analysis of tirilazad (U-74006F) as a hydroxyl radical scavenger, *Neurosci. Abs.* 17:164 (1991).

4. E.D.Hall, P.A. Yonkers, P.K. Andrus, J.W. Cox, and D.K. Anderson, Biochemistry and pharmacology of lipid antioxidants in acute brain and spinal cord injury, *J. Neurotrauma* 9 (Suppl 2):425-442 (1992).
5. E.D. Hall, P.K. Andrus, and P.A. Yonkers, Brain hydroxyl radical generation in acute experimental head injury, *J. Neurochem.* 60:588-594 (1993).
6. P.K. Andrus, E.D. Hall, B.M. Taylor, L.M. Sam, and F.F. Sun, Effects of the 21-aminosterioid tirilazad mesylate (U-74006F) on the eicosanoid levels in gerbil brain following ischemia and reperfusion, *Neurosci. Abs.* 18:1952 (1992).
7. P.K. Andrus, E.D. Hall, J.S. Althaus, C.M. Williams, and P.F. VonVoigtlander, Post-ischemic detection of hydroxyl radicals in gerbil brain: attenuation by the 21-aminosteroid tirilazad mesylate (U-74006F), *Neurosci. Abs.* 17:1086 (1991).
8. M. Goldstein, The decade of the brain, challenge and opportunities in stroke research. *Stroke* 21:373-374 (1990).
9. C.L. Harrison, and M. Dijkers, Traumatic brain injury registries in the United States: An overview, *Brain Injury* 6:203-212 (1992).
10. W.G. Barsan, J. Marler, T. Brott, E.C. Haley, and D. Levy, Delays in seeking medical attention in patients with acute stroke, *Neurology* 40 (Suppl. 1): 146 (1990).
11. L.F. Marshall, D.P. Becker, S.A. Bowers, C. Cayard, H. Eisenberg, C.R. Gross, R.G. Grossman, J.A. Jone, S.C. Kunitz, R. Rimel, K. Tabaddor, and J. Warren, The National Traumatic Coma Data Bank: Part 1. Design, purpose, goals and results, *J. Neurosurg.* 59:276-284 (1983).
12. A.B. Sterman, A.J. Furlan, M. Pressin, C. Kase, L. Caplan, and G. Williams, Acute stroke therapy trials, an introduction to reoccurring design issues, *Stroke* 18: 524-527 (1987).
13. H.P. Adams, W.K. Amery, Clinical trial methodology in stroke, *Stroke* 20:1276-1278 (1989).
14. M.B. Bracken, M.J. Shepard, W.F. Collins, T.R. Holford, D.S. Baskin, H.M. Eisenberg, E.S. Flamm, L. Leo-Summers, J.C. Maroon, L.F. Marshall, P.L. Perot, J. Piepmeier, V.K.H. Sonntag, F.C. Wagner, J.L. Wilberger, H.R. Winn, and W. Young, A randomized controlled trial of methylprednisolone or naloxone in the treatment of acute spinal cord injury, *New Engl. J. Med.* 322: 1405-1411 (1990).
15. M.B. Bracken, M.J. Shepard, W.F. Collins, T.R. Holford, D.S. Baskin, H.M. Eisenberg, E.S. Flamm, L. Leo-Summers, J.C. Maroon, L.F. Marshall, P.L. Perot, J. Piepmeier, V.K. H. Sonntag, F.C. Wagner, J.L. Wilberger, H.R. Winn, and W. Young, Methylprednisolone or naloxone treatment after acute spinal cord injury: 1 year follow-up data, *J. Neurosurg.* 76:23-31 (1992).
16. J.C. Fleishaker, G.R. Peters, and K.S. Cathcart, Evaluation of the pharmacokinetics and tolerability of tirilazad mesylate, a 21-aminosteroid free radical scavenger. I. Single Dose Administration, *J. Clin. Pharmacol.* 33:175-181 (1993a).
17. J.C. Fleishaker, G.R. Peters, and K.S. Cathcart, Evaluation of the pharmacokinetics and tolerability of tirilazad mesylate, a 21-aminosteroid free radical scavenger. II. Multiple Dose Administration, *J. Clin. Pharmacol.* 33:182-190 (1993b).
18. J.C. Fleishaker, L.K. Hulst, and G.R. Peters, Evaluation of the possible interaction between nimodipine and tirilazad in healthy volunteers, *Clin. Pharm. and Therap.* 53:162 (1993c).

THE POTENTIAL OF GLICLAZIDE, A SULPHONYLUREA TO INFLUENCE THE OXIDATIVE PROCESSES WITHIN THE PATHOGENESIS OF DIABETIC VASCULAR DISEASE

Paul E. Jennings

York Diabetes Centre
The York District Hospital
Wigginton Road
York
YO3 7HE
United Kingdom

INTRODUCTION

The major consequence of diabetes mellitus is that it predisposes to serious vascular disease which affects both the quality and quantity of life. The vascular disease affects both the small vessels (microangiopathy) and the large vessels (macroangiopathy) [1]. Specific diabetic microangiopathic changes are characterised by increased basement membrane thickening, increased capillary permeability and microthrombus formation. These changes are generalised, found throughout the vascular tree, having major clinical effects in the kidney, retina, peripheral nerve, myocardium and skin. Diabetic patients have a 25 fold increase in the risk of blindness, and diabetes is the major cause of renal failure in the UK and USA [2].

Macroangiopathy is manifest by cardiovascular disease (CVD) and cerebrovascular disease and is the commonest cause of death for diabetic patients who have a 2-5 fold increased risk of developing these conditions when compared to the general population. The risk of amputation due to peripheral vascular disease is increased by 40 fold. The extent of the problem is such that 50% of insulin dependent patients, diagnosed before the age of 30 will die before the age of 50 and will have suffered the consequences of vascular disease for many years before their death. Macroangiopathy is due to accelerated atherosclerosis - the atheromatous process developing earlier and progressing more rapidly in diabetic than non-diabetic populations. In Type II diabetes over 80% of patients die of macroangiopathy which reduces life expectation by 5-10 years as other major risk factors for atherosclerosis co-exist such as hypertension, dyslipidaemia, hyperinsulinaemia and increased thrombotic tendency. These factors result in accelerated ageing of the vascular tree to the extent that it is related to the duration and severity of hyperglycaemia.

The aetiology of diabetic complications remains controversial although the development of vascular disease is positively associated with duration of diabetes and poor long term blood glucose control. Improving diabetic control may retard or prevent the progression of diabetic microangiopathy in the kidney, retina and delay damage to the

Free Radicals in Diagnostic Medicine, Edited by
D. Armstrong, Plenum Press, New York, 1994

autonomic and peripheral nerves. Although these associations are well established they do not explain the mechanism of the development of microangiopathy. The pathogenesis reflects many factors including functional abnormalities within the microcirculation, the physiological consequences of enhanced glucose metabolism via different pathways of non-glycolytic metabolism, and genetic susceptibility. Both micro-and macroangiopathy share some of the factors now thought to be important such as, changes in endothelial cells and platelets and an increase in oxidative stress due to excess free radical activity. The association of these abnormalities with hyperglycaemia has suggested that they may be caused by a common pathological process induced by non-enzymatic protein glycosylation. This factor is dependent on levels of glycaemia and duration of diabetes, and therefore appears to be crucial in the development of vascular disease.

PROTEIN GLYCOSYLATION

Glycosylation is a non-enzymatic process which is initially reversible resulting in the formation of Ketoamines such as frustosamines and HbA1c which can be assesssed to determine short term diabetic control. On proteins having low physiological turnover rates glucose causes further rearrangements and dehydrations to form advanced glycosylation end (AGE) products [3]. The process produces and is augmented by free radicals. The formation of these products is irreversible and the resulting protein is characterised by a brown pigment, fluorescence and extensive cross-linking. AGE of collagen increases normally with ageing but is greatly increased in diabetes particularly in association with microangiopathic complications. In diabetes, the arteries and joints are prematurely stiff and elasticity is lost, a feature attributed to extensive glycosylation and cross-linking of collagen. Many proteins are modified adversely by glycosylation[4]. For example albumin undergoes conformational alterations which affect its binding characteristics, superoxide dismutase (SOD) loses activity, lens crystallins aggregate, oxidise and opacify, low density lipoprotein (LDL) becomes poorly recognised by fibroblasts and preferentially accumulated by macrophages. This has been suggested to contribute to LDL accumulation and the accelerated atherosclerosis of diabetes. The pathological role of AGE are therefore related to conformational changes of the protein, cancellation of the positive charges, blocking critical amino groups, loss of hydrogen bonding capacity, loss of cellular recognition and the formation of complex products capable of cross-linking. In addition the decreased susceptability of AGE to degradation by proteases leads to further accumulation and accelerated capillary basement membrane thickening and arteriosclerosis. Fibrin, for example, is trapped in capillary basement membrane and becomes resistant to fibrinolytic activities generating a potential for microvascular thrombosis.

Vascular permeability is not only dependent on cellular integrity but also to ionic charge barriers. The basement membrane contains negatively charged proteoglycan molecules associated with collagen and glycoproteins such as fibronectin and laminim. In diabetes there is markedly decreased anionic proteoglycans allowing leakage of negatively charged proteins such as albumin across the membrane into extra-vascular spaces such as the urine with resultant proteinuria. Non-enzymatic glycosylation reduces the binding of anionic proteoglycans to proteins such as collagen and basement membrane. Furthermore the increased formation of cross-linked AGE within basement membranes decreases vascular elasticity and traps extravasated plasma proteins to the vessel wall. Albumin and IgG are trapped by reactive groups generated by non-enzymatic glycosylation. Accumulation of plasma proteins trapped by this mechanism over the years of diabetes contributes to the basement membrane thickening and progressive microvascular occlusion [5]. Binding of IgG and albumin to basement membranes is characteristically seen in diabetic extravascular membranes such as the glomerular structures. Their presence may also contribute to additional tissue damage by immunological mechanisms. Similarly AGE forming on collagen traps LDL and this may contribute to the accelerated atherosclerosis in diabetic patients even at normal levels of plasma LDL.

Normally as long-lived extra-cellular proteins such as collagen and basement membrane age they are degraded and replaced at a constant slow rate throughout adult life [6]. This removal appears to be dependent on macrophage function and the efficacy of the process will determine the rate at which vascular damage, specifically atherosclerosis, develops. crophages and possibly also endothelial cells have specific, high affinity receptors which mediate the uptake and degradion of AGE. This receptor is distinct from the scavenger receptor for oxidised LDL. Therefore the net accummulation of AGE - proteins reflects an equilibrium between hyperglycaemia accelerated AGE formation and macrophage mediated removal. In diabetes it appears that the macrophage removal mechanism is saturated or that the activity is significantly decreased by genetic and/or metabolic factors unique to each individual.

Oxidative Stress

In addition to basement membrane thickening and increased vascular permeability in diabetes there is evidence for reduced endothelial cell production of prostacyclin and the activators of fibrinolysis, together with increased platelet reactivity and factor VII:vW production. This produces a thrombotic tendency which will lead to microthrombus formation and small vessel occlusion, contributing both to the abnormalities of blood flow in the small vessels and arterial occlusion [2]. Damage to endothelial cells, modification of platelet reactivity and the arachidonic acid cascade are all properties of free radicals and their reaction products lipid peroxides. Both free radicals and their lipid peroxides are directly cytotoxic to vascular endothelial cells [7]. They will also modulate the arachidonic acid cascade, reducing the synthesis of prostacyclin (vasodilator and anti-platelet aggregating agent) while stimulating cycloxygenase to promote platelet production of thromboxane A_2 (vasoconstrictor and platelet aggregating agent). This is the concept of "peroxide tone" which relates exogenous peroxides to the arachidonic acid cascade [8]. There is, therefore, a relationship between oxidative stress and the thrombotic tendency in diabetic patients with complications.

Determination of free radical activity in patients is difficult. Studies in Type II diabetes have reported increased lipid peroxidation and oxidative stress in diabetes particularly in the presence of microvascular disease. The possible anti-oxident defences in diabetic patients are found to be generally impaired. Normally, and as demonstrated in other diseases free radicals are rapidly eliminated by antioxidants such as reduced glutathione, and vitamins C and E. In non-diabetic patients free radical activity has been implicated in the enhancement of atherosclerosis [9]. In diabetic patients ischaemic heart disease occurs earlier than in non-diabetic patients and it is suggested that altered free radical processes are involved. Cellular levels of reduced glutathione [10] and serum vitamin C both primary antioxidants are reduced in diabetes in part due to poor glycaemic control. Furthermore diabetic patients even without evidence of large vessel disease are under increased oxidative stress particularly in association with microangiopathy [11, 12].

The reasons why diabetic patients should be under oxidative stress is unclear. Peroxides appear to have a positive but poorly defined relationship with hyperglycaemia; i.e. complicated patients have had poorer control. Peroxides could act synergistically with hyperglycaemia, through monosaccharide auto-oxidation and oxidative/simple glycosylation in causing damage. The origin of peroxides is unclear; are they the result of excessive production or inadequate elimination?

At present there are three possibilities for consideration:

1. They are a consequence of cell damage and destruction due to other mechanisms

and due to their ability to participate in chain reactions they amplify any damage.

2. They are produced from protein glycosylation and monosaccharide auto-oxidation.

3. They are the result of a reduction in antioxidant reserve.

There is considerable evidence that diabetic patients are under increased oxidative stress [13]. Oxidative stress occurs when there is an imbalance between the production of free radicals (oxidants) and the body's ability to reduce (or scavenge) free radicals Elaborate defence mechanisms against damaging oxidants are widespread both intra-and extracellularly. However, SOD for which the superoxide radical is the specific substrate is widespread but it's activity is impaired by glycosylation. Reduced glutathione detoxifies organic peroxides producing oxidised glutathione which is rapidly reduced back to it's active form by reactions utilising NADPH generated from "redox cycling". NADPH may be lacking in hyperglycaemia as it is a co-factor for aldose reductase in the polyol pathway [14]. The polyol pathway is of particular relevance in insulin insensitive tissues (retina, lens, nerve etc.) where hyperglycaemia drives the over expression of the enzyme aldose reductase which converts glucose to sorbitol. Consumption of NADPH by increased flux through this pathway leaves insufficient remaining to generate antioxidants such as glutathione or vitamin C and leaves the tissues (those classically affected by diabetic complications) susceptible to free radical mediated oxidant damage. There is evidence in diabetes for an increase in free radical production and a decline in antioxidant capacity preceeding the development of complications.

IMPLICATIONS FOR PREVENTING VASCULAR DISEASE WITH GLICLAZIDE

Both the end products of glycosylation in the form of AGE and the free radicals generated in this process can be implicated in the accelerated atherosclerosis and the vascular and prothrombotic microangiopathic changes typified by diabetes. The rate of formation of free radicals is dependent on the rate of protein glycosylation and therefore the level and duration on hyperglycaemia. AGE are cross-linked and heavily oxidised. Glycation and oxidation are inextricably linked.

The importance of the demonstration of the mechanisms whereby hyperglycaemia contributes to vascular damage opens exciting therapeutic options. It is now conceivable to use specific agents to block protein glycosylation or scavenge free radicals which will have effects independent of improving diabetic control. There is reduced free radical scavenging in diabetes and a compound which could directly scavenge free radicals as well as reduce blood sugar levels may be more beneficial than a pure hypoglycaemic agent alone.

FREE RADICAL SCAVENGING BY GLICLAZIDE IN-VITRO

Gliclazide is a second generation sulphonylurea that lowers blood glucose . The technique used to assess the scavenging ability of gliclazide and glibenclamide was that described by Misra and Fridovich [15]. In this assay free radicals are generated by the aerobic photo-oxidation of o-dianisidine sensitised by riboflavin. The photo-oxidation comprises of a complex series of free radical chain reactions involving the superoxide ion ($O_2^{.-}$) as the propagating species (**table 1**). This assay can be used to determine whether a compound is a general free radical scavenger or a scavenger specific for $O_2^{.-}$ A general free radical scavenger will inhibit the assay, a $O_2^{.-}$ scavenger will augment the assay and a compound with both features will also inhibit the assay. A substance with no free radical scavenging activity will have no effect on the assay. This assay has been validated in other studies [16].

Table 1
Photo-oxidation of o-dianisidine

Riboflavin absorbs a phototon and becomes electronically excited **(step 1)**. The excited state oxidises the dianisidine, yielding the flavin semiquinone and a dianisidine radical **(step 2)** which in the absence of competing reactions would dismute to form the divalently oxidised dianisidine **(step 5).** However the flavin semiquinone RbH can reduce oxygen to $O_2^{\cdot-}$ **(step 3)** and this radical can in turn reduce the dianisidine radical $DH^{\cdot}$ preventing its oxidation **(step 4)**. A general free radical scavenger will remove $DH^{\cdot}$ from **step 2** leading to a decrease in oxidised dianisidine. In contrast any compound which specifically scavenges $O_2^{\cdot-}$ will remove the $O_2^{\cdot-}$ from **steps 3** and **4**, thus increasing the amount of oxidised dianisidine and hence will have an augmentary effect on the assay.

$Rb + hv$	-	Rb^*	(1)
$Rb^* + DH_2$	-	$RbH + DH^{\cdot}$	(2)
$RbH + O_2$	-	$Rb + O_2^{\cdot-} + H^+$	(3)
$DH + O_2^{\cdot-} + H^+$	-	$DH_2 + O_2$	(4)
$DH^{\cdot} + DH^{\cdot}$	-	$D + DH_2$	(5)
$O_2^{\cdot-} + O_2^{\cdot-} + 2H^+$	SOD	H_2O_2	(6)

Rb, riboflavin: RbH flavin semiquinone; hv. energy of photon of light: Rb*, excited riboflavin; DH_2 o-dianisidine; $DH^{\cdot}$, dianisidine radical: $O_2^{\cdot-}$ superoxide anion; O_2 molecular oxygen; D, product formed by photo-oxidation measured at 460 nm.

Materials

Ribofavin and o-dianisidine were purchased from Sigma Chemicals. Riboflavin solution (1.3 x 10.5 M) was prepared in 0.01 M potassium phosphate buffer pH 7.5 and o-dianisidine solution (1 x 10.2 M) was prepared in ethanol. Gliclazide (Servier Laboratories) was dissolved in phosphate buffer, glibenclamide (Hoechst U.K.) in white fluorescnt tubes mounted 6 inches apart in an aluminium foil lines open ended box. These tubes provided a constant source of side-band radiation.

Assay procedure

The procedure for the assay was as follows: 60 ul of ethanol was added to a cuvette containing 2.94 ml of riboflavin solution. Thereafter 60 ul of o-dianisidine solution was added. Absorbance of light was measured at 460 nm using a Pye Unicam PU 8600 UV spectrophotometer. The cuvette was then transferred to the box, illuminated for 4 min and the absorbtion was remeasured at 460 nm. The change in absorbtion after 4 min served as the control leading and referred to as zero per cent inhibition on the assay 60 ul of gliclazide or glibenclamide solution then added to 2.88 ml of riboflavin solution followed 60 ul of o-ianisidine and the measurements carried out as above. The assay was repeated on several occasions at different final concentrations gliclazide 0.5, 1.0, 2.5, 5.0 mg/ml and control olution alone. Percentage inhibition on the assay was calculated at each concentration. The intra-assay variation for the assay is 4.3% [17]

Table 2
Free radical scavenging assay.

Percentage inhibition on the assay by sulphonylureas

Concentration of drug in final soluion (ug/ml) n = 7	% Inhibition of control (mean + SD) Gliclazide	Glibenclamide
0.5	11.05 + 2.5	-
1.0	20.8 + 2.9	-
2.5	31.4 + 2.2	- 0.4 + 3.3

Glibenclamide has no scavenging effects at concentrations up to 25 mg/ml.

Results

The results, as shown in **table 2** demonstrated that gliclazide is a general free radical scavenger as demonstrated by the inhibition of the photo-oxidation dianisidine at concentrations as low as 0.5 ug/ml glibenclamide has no effect on the assay even at concentrations as high as 25 mg/ml. The pharmacological concentration of Gliclazide is 4ug/ml 4 hours after a dose of 80 mg [18]. therefore the effect is seen well below the expected therapeutic level. Glibenclamide however had no free radical scavenging effect even at concentrations as high as 25 mg/ml. This study does not demonstrate which part of the Gliclazide molecule is reponsible for the free radical scavenging but it is likely to be due to the azabicyclo-octyl ring which is not found in other sulphonylureas.

The effects of Gliclazide as a general free radical scavenger could explain the previously documented effects on prostanoid release and platelet function both in animal [19]. and human platelets [20]. These appear to be independent of any effects on glycaemic control and are not necessarily mimicked by other oral hypoglycaemic agents. A clinical study documented a decline in the vasoconstrictor prostanoid TXA_2 together with a decline in lipid peroxides which was independent of glycaemic control [21]. It is suggested that the effect may be due to direct free radical scavenging. This was further investigated in a Glibenclamide-controlled study over a 6 month period [22].

CLINICAL STUDY PROTOCOL FOR GLICLAZIDE EFFECTS

Thirty non-insulin-dependent diabetic patients were studied. Twenty males aged 38 - 65 years (mean age, 58.1 years) were recruited into the study. All patients had been treated for diabetes for more than 2 years (mean 8 years) and had been established on glibenclamide for more than 2 years with or without adjunctive metformin therapy. The dosage of glibenclamide had not been altered at two consecutive visits before enrolment in the trial, and all subjects at had postprandial blood glucose values below 15 mmol/l Patients taking additional drugs known to influence platelet aggregation (eg. nonsteroidal anti-inflammatory agents) or to have free radical scavenging properties (eg, vitamins C or E, probucol, and Captopril) were not studied.

Patients were only studied if they had retinopathy and/or incipient nephropathy. However, they were excluded if they require laser therapy or had undergone laser treatment in the previous 6 months, or if the serum creatinine was greater than 200 mmol/l Patients were also excluded if they had clinically significant large vessel disease, a history of intermittent claudication, myocardial infarction, angina or cerebrovascular ischemia, or signs of vascular disease on examination, such as absent foot pulses or ECG evidence of ischemia.

To assess the degree of oxidative stress found in the patients studied, two non diabetic control groups were compared . The degree of plasma oxidation as lipid peroxides and plasma thiols was assessed. These were age - and sex-matched nondiabetic subjects; one group were healthy controls taking no medication with no hisitory of vascular disease; the second group were patients with proven ischemic heart disease awaiting coronary artery bypass surgery.

Patients were studied at -1, 0, 1, 3, and 6 months and randomized at time point 0 to continue either their present dose of glibenclamide or to be converted to an equivalent dose of gliclazide. Five milligrams of glibenclamide was considered to be equivalent to 80 mg of gliclazide. Randomization details were not available to the scientific staff until all biochemical and platelet invesitgations were completed.

Venous samples were taken through a 21-gauge butterfly cannula, without venous stasis after 30 minutes recumbancy, 2 hours after tablets and a standard breakfast. Diabetic control was asssessed by HbA_1 and blood glucose measurement at the same time.

Measurements were taken of hemostatic variables, the oxidative status of the plasma, and the redox status, both extracellularly as plasma albumin-thiols (PSH) and intracellularly as red blood cell superoxide dismutase activity (SOD). The following standard assays were performed within 2 hours of the venepuncture, and samples were coded so that the clinical status of the patient or control was unknown to the investigator performing the assay.

Assays

For PSH concentration, the plasma was separated by centrifugation at 1500 x g for 15 minutes at 4 C. PSH levels were measured using a spectrophotometric technique. Lipid peroxides were determined as malondialdehyde-like material by the spectrophotometric method of Aust [23]. SOD was detected in erythrocytes that had been separated, washed, and then lysed. The red blood cell SOD was measured by the method of Misra and Fridovich [15]. Platelet aggregation to collagen in whole blood was quantified by measuring the decrease in single platelet count in response to 1 ug/ml collagen using a whole blood platelet counter (Clay Adams Ultra Flo 100, Becton-Dickinson, Cowley, Oxford, England. Glycosylated hemoglobin was measured by agar gel electrophoresis (Corning Medical, Halstead, UK) (normal range 6.0% to 8.0%). Plasma glucose was measured on an autoanalyzer using the glucose oxidase method (Cobas Fara. Roche Diagnostics, Basel, Switzerland).

Statistical Analysis

ANOVA was performed together with the Mann-Whitney U test comparing the results obtained between patients continuing on glibenclamide and those changing to gliclazide.

Results

One subject randomized to continue on glibenclamide was withdrawn after 6 weeks, after having sustained a nonfatal myocardial infarction. His results are not included in the analysis. When subjects are under oxidative stress, lipid peroxides increase, while PSH and red blood cells SOD decreases. **Table 3** demonstrates that the diabetic patients, on entry into the trial, were under oxidative stress, as their lipid peroxides were greater than that seen in the age-matched control groups and their PSH were reduced.

Table 3

Comparison of Oxidative Status in Diabetic Patients, Healthy Subjects and Non Diabetic Controls With Ischemic Heart Disease.

	Normal Controls (n = 29)	Controls with IHD (n = 30)	Diabetic Patients (n = 29)
Age (yr)	59.5 (40 - 68)	60 (45 - 68)	58.1 (38 - 65)
PSH (umol/L)	506 (433 - 462)	435 (361 - 462)*	463 (420 - 490)*
Lipid peroxides (umol/L)	7.0 (5.4 - 9.6)	8.4 (6.5 - 10.1)*	8.7 (6.6 - 11.3)*

NOTE: Values are the mean, with ranges in parentheses.
Abbreviations; IHD, ischemic heart disease.
Significant differences between patient groups and normal controls.
*P < 0.05 for all values. No differences between diabetic patients and IHD patients.

There were no differences between the values obtained for any of the measurements at - 1 month, baseline, and 1 month after randomization. For subsequent analysis, comparisons were made between baseline values and 3 - and 6-month values. Baseline characteristics were similar between patients continuing on glibenclamide and those randomized to change to gliclazide **(Table 4).**

Table 4
Baseline Characteristics of the Two Patient Groups.

	Gliclazide Treated (n = 15)		p.	Glibenclamide Treated (n = 14)	
Blood glucose (mmol/L)	8.6	(3.1)	.16	10.6	(4.7)
HbA_1 (%)	8.2	(2.3)	.23	9.1	(2.0)
Lipid peroxides (mmol/L)	8.3	(1.1)	.11	9.0	(1.2)
PSH (umol/L)	458	(42)	.70	451	(63)
SOD (ug/mL)	135	(21)	.75	132	(19)
Platelet aggregation (%)	65.1	(14)	.97	70.2	(14)

NOTE: Results are presented as the mean, with the SD in parentheses.
Significance tested by Mann-Whitney U test comparing gliclazide-treated to glibenclamide treated patients.

Table 5
Three-Month Characteristics of the Two Patient Groups

	Gliclazide-Treated (n = 15)		p.	Glibenclamide-Treated (n = 14)	
Blood glucose (mmol/L)	9.7	(2.8)	.72	10.1	(3.8)
HbA_1(%)	8.1	(1.9)	.21	9.2	(2.1)
Lipid peroxides (umol/L)	7.0	(0.6)	.0002	8.3	(0.8)
PSH (umol/L)	458	(38)	.004	414	(34)
SOD (ug/mL)	152	(36)	.016	123	(16)
Platelet aggregation (%)	50.8	(24)	.006	72.3	(15)

NOTE: Results are presented as the mean (SD)Significance tested by Mann-Whitney U test comparing gliclazide-treated with glibenclamide-treated patients.

At 3 months, diabetic control was unaltered, but there were significant improvements in the oxidative status of the gliclazide-treated patients. Lipid peroxides decreased (8.3 + 1.1 to 7.0 +0.6 umol/ $P < 0.01$) and red blood cell SOD increased (135 + 21 to 152 + 36 ug/mL, $P<0.05$). PSH levels were unaltered at 458 + 38 umol/L, while they had decreased significantly in the glibenclamide patients (414 + 34 umol/L, $P <0.05$), resulting in a significant difference between the two treatment groups ($P<0.004$; **Table 5**). Platelet reactivity to collagen also improved in the gliclazide-treated patients, decreasing from 65.1% + 14% to 50.8% + 24% ($P<0.01$). The reactivity of the platelets remained unaltered in the glibenclamide patients. At 6 months, the significant differences between the two treatment groups remained, although there were no further improvements in the gliclazide patients **(Table 6).**

Table 6
Six-Month Characteristics of the Two Patient Groups

	Gliclazide-Treated (n = 15)		p.	Glibenclamide-Treated (n = 14)	
Blood glucose (mmol/L)	8.8	(2.9)	.63	9.6	(4.5)
HbA_1 (%)	8.8	(2.3)	.51	9.3	(2.3)
Lipid peroxides (umolL)	7.2	(0.7)	.009	8.8	(1.9)
PSH (umol/L)	449	(42)	.007	418	(30)
SOD (ug/mL)	158	(33)	.003	117	(27)
Platelet aggregation (%)	49.3	(22)	.004	72.4	(19)

NOTE: Results are presented as the mean (SD)
Significance tested by Mann-Whitney U test comparing gliclazide-treated to glibenclamide treated patients.

SUMMARY

The clinical studies have demonstrated that patients showed significant improvements in their oxidative status after 3 months treatment . This improvement in oxidative status which was associated with a reduction in platelet reactivity, remained constant for the rest of the study period. The effect was independent of glycaemic control.

The demonstration of a benefit to clinical vascular disease has proved difficult to achieve in all studies of non-insulin dependent diabetes. This is due to the multifactorial nature of complications and the long duration of disease required before microvascular complications such as retinopathy became apparent.

The Japanese Diabetic Retinopathy Program [24] studied the progression of retinopathy over a 5 year period comparing Gliclazide with other sulphonylureas and with placebo. The study suggested that with equivalent metabolic control there was a trend towards a lower rate of deterioration of retinopathy and a significantly lower incidence of pre-proliferative retinopathy in the group receiving Gliclazide compared with patients receiving other sulphonylureas or placebo.

There is little comparative evidence on the effect of specific sulphonylureas on large vessel disease. Although improvement in parameters of hyperglycaemia is associated with an improvement in morbidity from large vessel disease. In Type II diabetes atherosclerosis co-exists in the majority of patients and often predates the clinical diagnosis of diabetes. The presence of atherosclerosis which often determines the ultimate fate of the patient, further increases the level of lipid peroxidation of oxidative stress, amplifying the effects of hyperglycaemia and potentiating vascular damage. In diabetes, therefore where increased glycation and oxidation are fundemental in the pathogenesis of diabetic vascular disease agents such as Gliclazide with anti-oxidant activities may have an enhanced therapeutic role.

Acknowledgments

The studies reported here were performed with Dr. J.J.F. Belch in the University of Dundee, Scotland.

References

(1) P.J. Palumbo L.R. Elveback C.P. Chu et al. Diabetes mellitus; incidence, prevalence, survivorship, and causes of death in Rochester, Minnesota, 1945-1970 Diabetes 1970; Diabetes 25:566-73 (1976)

(2) P.E. Jennings A.H. Barnett . New approaches to the problem of diabetic complications Diabetic Med 5: 111-117. (1988)

(3) V.W. Monnier . Toward a Maillard theory of ageing. In Baynes J.W; Monnier V.M eds. The Malliard Reaction in ageing, diabetes and nutrition. New York; Alan R. Liss inc; 1-22 (1989)

(4) M. Brownlee H.Vlassara and A. Cerami . The pathogenetic role of non-enzymatic glycosylation in diabetic complications. In Crabbe M.J.C. Ed. Diabetic complications; Scientific and clinical aspects. Edinburgh; Churchill Livingstone, 94-139 (1987)

(5) M. Brownlee S. Ponger and A.Cerami . Covalent attachment of soluble proteins by non enzymatically glycosylated collagen; role of in situ formation of immune complexes. J Ex Med 158; 1739-44 (1983)

(6) S.P. Wolff Z.Y. Jiang and J.V. Hunt . Protein glycation and oxidative stress in diabetes and ageing. Free Radic Biol Med. 10: 339-352 (1991)

(7) P.H. Procter and E.S. Reynolds. Free radicals and disease in man. Physiol Chem Phys 16; 175-195 (1984)

(8) M.A. Warso and W.E.M. Lands. Lipid Peroxidation in relation to Prostacyclin and thromboxane physiology. Br Med Bull 39: 277-280 (1983)

(9) J.J.F. Belch M. Chopra S . Hutchinson et al; Free radical pathology in chronic arterial disease. Free Radical Biol Med 6:375-378, (1989)

(10) S.W. Chari N. Nath and A.B. Rathi . Glutathione and its redox system in diabetic polymorphonuclear leucocytes. Am. J. Med. Sci. 287.14 (1984)

(11) A.F. Jones P.E. Jennings A. Wakefield J.W. Winkles J. Lunec and A.H. Barnett . The fluoresence of serum proteins in diabetes mellitus. Relationship to microangiopathy Diabetic Medicine 5; 547-551 (1988)

(12) P.E. Jennings A.F. Jones C.K.M. Florkowski J. Lunec and A.H. Barnett Increased diene conjugates in diabetic subjects with microangiopathy. Diabetic Medicine 5; 111-117 (1988)

(13) S.P. Wolff. The potential role of oxidative stress in diabetes and its complications; novel implications for theory and therapy. In diabetic Complications; Scientific and clinical aspects. Crabbe M.J.C. (Ed) Churchill Livingstone 167-221 (1987)

(14) D.G. Cogan Aldose Reductase and complications of diabetes. Annal Int Med 101; 82-91 (1984)

(15) H.P. Misra and I. Fridovich. Superoxide radical and free radicals in clinical chemistry Ann Clin Biochem. 13: 393 - 398. (1979)

(16) M. Chopra N. Scott W.E. Smith and J.J.F. Belch. Captopril a free radical scavengar Br. J. Pharmacol. 27, 396 (1989)

(17) N.A. Scott P.E. Jennings J. Brown and J.J.F. Belch. Gliclazide a Free Radical Scavenger. Eur. J. Pharm 208 175-177 (1991)

(18) D.B. Campbell. P. Adrigenssens Y.W. Hopkins B. Gordon and J.R.B. Williams Pharmacokinetics and metabolism of gliclazide; a review in; Gliclazide and the Treatment of Diabetes (Proceedings Internatinal Symposium, London) eds. Keen et al. p.71 (1979)

(19) T. Tsuboi B. Fuijitani T. Maeda . et al; Effect of gliclazide on prostaglandin and thromboxane synthesis in guinea-pig platelets. Throm Res 21; 103-110 (1981)

20) O. Ponari E. Givardi S. Megta. et al. Anti-platelets effects of long term treatment with gliclazide on platelet function in patients with diabetes mellitus. Throm Res 16; 191-203. (1979)

21) C.M. Florkowski M.R. Richardson C. Le Guen P.E. Jennings M.J. O'Donnell. A.F. Jones J. Lunec and A.H. Barnett Effect of gliclazide on thromboxane B2, parameters of haemostasis, fluorescent 1gG and lipid peroxides in non-insulin dependent mellitus. Diabetes Research 9; 87-90 (1988)

(22) P.E. Jennings N.A. Scott A.R. Saniabadi and J.J.F. Belch . Effects of Gliclazide on Platelet Reactivity and Free Radicals in Type II Diabetic Patients Clinical Assessment. Metabolism 41 (5) 36-39 (1992)

(23) Austs: Lipid peroxidation. In Greenwald R.A. (ed): Handbook of Methods for oxygen Free Radicals. Research, Boca Raton, FL, LRL, pp 203-207 (1987)

(24) Y. Akanuma K. Kosaka Y. Kanazawa K. Kasugam M. Fukuda and A. Aoki. Long . Term Comparison on Oral Hypoglycaemic Agents in Diabetic Retinopathy. Diabetes Research and Clinical Practice. 5; 81-90 (1988)

INTERACTIONS BETWEEN VITAMIN E, FREE RADICALS, AND IMMUNITY DURING THE AGING PROCESS

Jeffrey B. Blumberg
Antioxidants Research Laboratory
USDA Human Nutrition Research Center on Aging at Tufts University
Boston, MA 02111

INTRODUCTION

There is a substantial amount of experimental evidence indicating a role for oxygen free radicals in the aging process and the development of several chronic diseases common among the elderly. Comparisons between mammalian species reveal a strong correlation between their life span, metabolic generation of free radicals, and tissue concentration of specific antioxidants including α-tocopherol, carotenoids, uric acid, and superoxide dismutase.[1] Experimental, clinical, and epidemiological studies have been largely consistent in demonstrating a reduced risk of atherosclerosis, cancer, and other chronic diseases in groups with the highest status of antioxidant vitamins C, E, and/or ß-carotene.[2]

The free radical theory of aging proposed by Harman[3] hypothesizes that the degenerative changes associated with aging might be produced by the accumulation of deleterious side reactions of free radicals produced during cellular metabolism. Oxygen free radicals (e.g., superoxide, hydroxyl, and peroxyl radicals) are the most common radicals generated during metabolism and could contribute to aging via several mechanisms. Oxyradicals can also be generated *in vivo* by environmental exposure to prooxidant toxicants like cigarette smoke and smog.

Although supporting evidence keeps the free radical theory of aging both reasonable and attractive, several basic predictions are not observed, particularly the fact that supplementary antioxidants do not lengthen the maximum life span of mammals appreciably. Pryor[4] has suggested that the theory be reproposed with free radicals considered as involved in the etiology and development of chronic diseases that are most life-limiting, i.e., with radicals implicated in processes that shorten life below the maximum life span possible even if dampening radical reactions cannot lengthen maximum life span.

FREE RADICALS AND IMMUNITY DURING AGING

Vitamin E and other dietary antioxidants may play an important role in reducing the age-related decline of physiological function and the risk of several chronic diseases via their influence on the immune system. An immunological basis for many age-associated diseases such as amyloidosis, atherosclerosis, and cancer has been proposed by Walford[5] and others. Free radical formation associated with aging may be

Free Radicals in Diagnostic Medicine, Edited by
D. Armstrong, Plenum Press, New York, 1994

an underlying factor in the depressed immune response observed in the elderly. Oxygen metabolites, especially H_2O_2, produced by activated macrophages depress lymphocyte proliferation but vitamin E has been demonstrated to decrease H_2O_2 formation in polymorphonuclear leukocytes (PMN). Aging is also associated with an increased production of prostaglandin (PG) E_2, an inhibitor of lymphocyte proliferation, but vitamin E can depress its rate of synthesis and enhance cellular immunity.

It is not surprising that the immune system might require a different level of tocopherol for its optimal function than other organ systems. White blood cells are particularly rich in polyunsaturated phospholipids prone to oxidative destruction. The mounting of an immune response requires membrane bound receptor-mediated communication between cells as well as between protein and lipid mediators which can be directly or indirectly affected by tocopherol status. However, although several hypotheses have been proposed, the exact nature of the action of vitamin E on immunity has yet to be fully elucidated.

Arachidonic acid metabolism via a peroxidative casade to PG, leukotrienes (LT), and hydroxyeicosatetranoic acid (HETE) has been demonstrated in PMN. In addition to PGE_2, LT and HETE have also been shown to inhibit lymphocyte proliferation.[6] These eicosanoids appear to act via decreasing T helper and increasing T suppressor/cytotoxic cell proliferation. In the spleen, macrophages are the major contributors of PGE_2 and lipoxygenase products, although arachidonic acid released from lymphocytes can be used by macrophages for PG synthesis and lymphocytes can synthesize lipoxygenase products.

Vitamin E could enhance immune function through its antioxidant function by decreasing reactive oxygen metabolites such as H_2O_2 and/or by altering formation of arachidonic acid metabolites such as PG. However, the immunostimulatory effect of α-tocopherol cannot be fully explained by its antioxidant function as other antioxidants do not produce similar actions. Interestingly, cyclooxygenase inhibitors such as aspirin and indomethacin have been shown to enhance several parameters of cell-mediated immunity and, as noted, vitamin E has been shown to decrease PG production in immune cells. Vitamin E-induced changes in cell surface glycoconjugates suggest another possible mechanism of action, i.e., changes in membrane receptor molecules involved in the immune response.

EXPERIMENTAL MODELS

In experimental models almost every aspect of the immune system, including resistance to infection, specific antibody responses, splenic plaque forming cells, *in vitro* mitogenic responses of lymphocytes, reticuloendothelial system clearance, and phagocytic index, have been shown to be impaired by vitamin E deficiency and enhanced by increases in α-tocopherol intake.[7]

Rat peritoneal macrophages contain relatively high concentrations of vitamin E which are markedly depleted upon exposure to prooxidant agents.[8] Vitamin E has been reported to effectively protect against membrane damage caused by oxidative products of neutrophils in liposome model cell systems and human umbilical vein endothelial cells.[9,10] Vitamin E may also play an indirect protective role in endothelial cells via enhancing PGI_2 (prostacyclin) synthesis.[11] *In vitro* studies also reveal that vitamin E may enhance the biosynthesis of glycoconjugates which are important in cell-cell interactions and can serve as antigenic determinants.[12]

Nutritional requirements for vitamin E are dependent on the criteria selected for this determination. Several studies indicate that the dietary α-tocopherol requirement for the maintenance of optimal immune responsiveness in rodents, chickens, and large farm animals is considerably higher than the levels required for normal growth and reproduction. For example, Bendich[13] demonstrated that 15 mg/kg diet of vitamin E was adequate to prevent myopathy in SHR rats but optimal lymphocyte proliferation was manifest only at intakes of 50-200 mg/kg. Meydani[14] observed that intakes of 500 ppm dietary vitamin E to senescent C57BL/6j mice decreased PGE_2 synthesis and increased splenocyte proliferation, interleukin (IL)-2 production, and delayed cutaneous hypersensitivity (DCH) relative to control animals fed a standard chow with 30 ppm vitamin E. Vitamin E supplementation has also been shown to increase survival against infection with *Diplococcus pneumonia* Type I in mice and against *E. Coli* in chickens.[15,16] This immunostimulatory action of vitamin E and resultant decreased infectivity has also been reported in dogs, pigs, calves, and rams. It has been suggested that vitamin E supplementation may present a cost effective way to enhance immunological performance and decrease stress in farm animals.[17]

Vitamin E has been noted in animal studies to have a beneficial action in reducing the pathogenesis of autoimmune disorders, including kidney amyloidosis and adjuvant arthritis.[18,19] Harman[20] has also reported an increase in the average life span of short-lived autoimmune-prone NZB/NZW mice receiving vitamin E supplements.

CLINICAL TRIALS

Studies of the effects of vitamin E on immune response in humans are limited. An early report by Baehner[21] showed that *in vivo* supplementation with vitamin E could affect the bactericidal activity of leukocytes without affecting other blood chemistry and hematological parameters. They hypothesized that directed movement and phagocytosis by PMN are attenuated by autooxidative damage to the cell membrane by endogenously derived H_2O_2 and that administration of vitamin E may prevent this damage via scavenging of reduced oxygen radicals. They administered 1600 I.U. daily for 7 days to 3 volunteers and found their isolated PMN were hyperphagocytic but killed *Staphylococcus aureus* 502A less effectively than controls suggesting that less H_2O_2 was available to damage PMN or kill bacteria. H_2O_2-dependent stimulation of the hexose monophosphate shunt, H_2O_2 release from phagocytizing PMN, and fluoresceinated concanavalin A (ConA) cap formation promoted by H_2O_2 damage to microtubules were all diminished, but the release of superoxide from phagocytizing PMN was not reduced in the vitamin E group. It is difficult to assess the dichotomy between the depressant effect on PMN killing *in vitro* and the reduced PMN autotoxicity leading to improved phagocytosis. Boxer[22] suggested these results indicate that the stable reduction product of oxygen, H_2O_2, may be responsible for modulating PMN motile functions.

Prasad[23] studied the effect of a daily, 3 week treatment of 300 mg dl-α-tocopherol acetate on DCH in 5 teenage boys and on bactericidal activity of peripheral leukocytes and cell-mediated immunity in 13 young men. The vitamin E supplementation decreased leukocyte bactericidal activity, the release of acid phosphatase activity, and phytohemagglutinin (PHA)-stimulated lymphocyte proliferation but did not alter DCH to an intradermal injection of PHA. The results of this trial reveal a discrepancy between the *in vivo* and *in vitro* results of cell-mediated immunity parameters. Interestingly, 2 subjects showed clinical improvement in the symptoms of asthmatic attacks and nasal allergy, respectively.

Vitamin E deficiency has been suggested to contribute to alterations in neonatal PMN function through peroxidative damage to the cell's membrane. Newborn infants, especially premature babies, have a low vitamin E status and neonatal PMN have been shown to be deficient in phagocytosis, bactericidal activity, and chemotaxis[24] These defects may play a role in determining the high susceptibility to infections of the newborn infant by impairment of host defenses. Chirico[25] administered a total of 120 mg/kg dl-α-tocopherol in divided doses to 10 of 20 healthy premature infants and assessed neutrophil phagocytosis. In the treated and untreated infants no differences were found in PMN function before treatment with vitamin E although phagocytosis, bactericidal activity, and chemotaxis were lower than in 30 adult controls. During their first week of life, the untreated infants maintained a low index and frequency of phagocytosis while these parameters were significantly increased in the group receiving vitamin E. However, after 2 weeks of age, phagocytosis was normal in both groups of infants with no differences being noted in bactericidal activity, chemotaxis or metabolic activity. Thus, vitamin E accelerated the normalization of phagocytic function during the first week of life in these premature infants.

Zimlanski[26] supplemented 20 institutionalized elderly women (63-93 years) with 100 mg dl-α-tocopherol acetate twice daily and assessed serum proteins and immunoglobulin concentrations after 4 and 12 months. Vitamin E increased total serum protein with the principal effect on α_2- and β_2-globulin fractions occurring at 4 months. No significant effects were noted on levels of immunoglobulins and complement C3 although another group administered vitamin C (400 mg daily) with vitamin E displayed significant increases in IgG and complement C3 levels.

Harman and Miller[27] supplemented 103 elderly patients from a chronic care facility with 200 or 400 mg α-tocopherol acetate daily but did not see any beneficial effect on antibody development against influenza virus vaccine. Unfortunately, data on the health status, medication use, antibody levels, and other relevant parameters were not reported.

Meydani[28] have described a residential 30 day placebo-controlled, double-blind vitamin E supplementation (800 IU daily) trial of 32 healthy elderly subjects. *In vivo* and *in vitro* indices of immune response were evaluated in each subject before and after supplementation. Vitamin E supplementation significantly improved the DCH response to 7 recall antigens as well as the mitogenic response of lymphocytes to ConA. Further, the vitamin E supplementation markedly reduced plasma lipid peroxides and the production of PGE_2 by PMN.

Penn[29] studied 30 geriatric patients who had been hospitalized for over 3 months and assessed cell-mediated immune function before and after daily antioxidant supplementation (50 mg vitamin E, 100 mg vitamin C, 8000 IU vitamin A) or placebo treatment for 28 days. Following vitamin supplementation, cell-mediated immune function improved as indicated by a significant increase in the absolute number of T cells, T4 subsets, T4:T8 ratio, and PHA-stimulated lymphocyte proliferation. In contrast, no signifcant changes were noted in the placebo group.

Long-duration or damaging exercise can initiate reactions that resemble the "acute phase response" to infection and offers a model of immune responsiveness to stress conditions. Cannon[30,31] subjected a group of 21 young (<30 years) and older (>55 years) men receiving placebo or vitamin E (800 IU daily) treatment for 48 days to a bout of intense eccentric exercise. This exercise is associated with mobilization and activation of neutrophils, proteolysis of skeletal muscle, and increased hepatic production of certain plasma proteins. Both placebo and supplemented groups of younger men responded to the exercise with a significantly greater neutrophilia and higher plasma creatine kinase concentration than the older placebo subjects. However, vitamin E supplementation of the older men significantly reduced the differences in values between the two age groups by increasing their acute phase immune responses. The concentration of vitamin E decreased and conjugated dienes increased in muscle tissue biopsied from the young subjects following the eccentric exercise. These and other alterations in antioxidant and lipid peroxide status following the exercise bout were consistent with the concept that vitamin E provides protection against exercise-induced oxidative injury[32] Vitamin E was also noted to blunt the exercise-induced production of the pro-inflammatory mediators IL-1 and IL-6 in both groups. Thus, vitamin E appears to restore more youthful immune responses to eccentric exercise in older individuals. The effect of vitamin E on IL production suggests an action that may be exploited in future studies for determining the fundamental role of these cytokines and the application of vitamin E supplementation in non-infectious inflammatory diseases as well as muscle damage and repair.

Chandra[33] enrolled 96 apparently healthy, free-living men and women older than 65 years in a double-blind, placebo-controlled trial to investigate the effect of an antioxidant-enriched nutrient multivitamin/mineral supplement (containing 44 mg vitamin E and 16 mg ß-carotene) on infection-related illness and immune responsiveness. After 12 months of the supplemented group showed a significant improvement in immune function, including the number of T cells, mitogen-stimulated lymphocyte proliferation, production of IL-2, number and activity of natural killer cells, and antibody response to influenza vaccination. Importantly, the supplemented group had 48% fewer days of infection-related illness than the control subjects and 56% fewer days of antibiotic use.

Meydani[34] have reported preliminary evidence from a randomized, double-blind, placebo-controlled trial in which they provided 60 healthy young (<30 years) and older (>65 year) men and women with 400 IU d-α-tocopherol for 6 months. Both the young and older treated groups displayed significant increases in the total number and induration of positive DCH responses (58 and 91% increases, respectively) to 7 different recall antigens. These data suggest that the immune responses of even apparently healthy younger adults may be suboptimal and can be elevated by vitamin E supplementation.

EPIDEMIOLOGICAL STUDIES

The use of vitamin E supplements in high doses by healthy individuals is widely promoted and, along with vitamin C supplements, represents one of the most popular single nutrient supplements.[35] Goodwin and Garry[36] studied a population of 270 healthy older adults (65-94 years) consuming megadoses of vitamin supplements but did not see any correlation between vitamin E intake and DCH, mitogen stimulation, serum antibodies or circulating immune complexes. Persons taking more than 150 IU vitamin E or any of several B vitamins had lower absolute circulating lymphocyte counts than the rest of the population. However, the study was complicated by the fact that several vitamin supplements at megadose levels were used by each subject and the interaction between different nutrients present confounding variables. It has been suggested that at high doses some nutrients may act as non-specific adjuvants, the effects of which diminish with time.

Chavance[37,38] conducted a community-based survey on the relationship between nutritional and immunological status in 100 healthy subjects over 60 years of age. They reported that plasma vitamin E levels were positively correlated with positive DCH responses to diphtheria toxoid, candida, and trichophyton. In men only, positive correlations were also observed between vitamin E levels and the number of positive DCH responses. Subjects with tocopherol levels greater than 135 mg/l were found to have higher helper-inducer/cytotoxic-suppressor cell ratios. Blood vitamin E concentrations were also negatively correlated with the number of infectious disease episodes in the three preceding years. Thus, this epidemiological study is consistent with the results of the positive *in vitro* and animal model experiments as well as the controlled clinical trials of vitamin E supplementation.

IMMUNOLOGICAL ROLE OF VITAMIN E IN CHRONIC DISEASE

Evidence has been accumulating that oxidant damage to tissues underlies the pathology of several diseases including many associated with altered immune function, e.g., amyloidosis, rheumatoid arthritis, and cancer. A beneficial effect of vitamin E supplementation has been demonstrated in animal models and clinical trials in reducing the pathogenesis of autoimmune disorders including activated arthroses and osteoarthritis.[19,39] The precise role played by free radicals and H_2O_2 in such disorders and in immune injury to the kidney and lung is unresolved although they are formed and interact with PG, IL, LT, and other modulators of immune function. Vitamin E may be beneficial to host defense mechanisms by affecting the infection-induced alterations in tissue eicosanoid production from arachidonic acid.[14,16,40]

Vitamin E could affect PG production by interfering with its synthesis. PG production requires an active oxygen species and lipid peroxides (e.g., the PG intermediate PGG_2) can stimulate synthesis by providing an oxygen species to enhance the activity of the rate limiting enzyme, cyclooxygenase. Swartz[41] administered 800 IU vitamin E or placebo daily to 30 healthy adults for 8 weeks and observed a dramatic decline in plasma 6-keto-$PGF_{1\alpha}$, the principal stable degradation product of PGI_2, in the treatment group. Vitamin E may reduce PGI_2 production through its antioxidant activity via diminishing the levels of peroxide needed for activation of arachidonic acid. A similar inverse relationship between serum vitamin E and other PG, e.g., PGE_2 and $PGF_{2\alpha}$, has been noted in animal studies.[14,42,43] However, Lauritsen[44] found oral supplements of 1920 IU α-tocopherol daily for 2 weeks in 8 patients with active ulcerative colitis did not affect the disease-induced elevations of PGE_2 or LTB_4 in the lumen. Goetzl[45] reported that vitamin E bidirectionally modulates the activity of the lipoxygenase pathway of human neutrophils *in vitro*. He further noted that normal plasma concentrations of vitamin E enhanced the lipoxygenation of arachidonic acid, whereas higher concentrations exerted a suppressive effect consistent with α-tocopherol's role as a hydroperoxide scavenger. While these data generally support an influence of vitamin E on the arachidonic acid cascade, less evidence is available to explain the involvement of superoxide in neutrophil chemotaxis, the modulating action of platelet-activating factor and tumor necrosis factor on oxidant production by phagocytes, and oxidant involvement in T-lymphocyte activation.[46]

Abnormal PMN function in diabetics, including defective phagocytic uptake and chemotactic responses and excess superoxide anion production, has been observed[47] Hill[48] administered 25 IU α-tocopherol/kg/day orally for several weeks to 7 diabetic patients with consistently depressed monocyte chemotactic responses. The vitamin E treatment doubled monocyte random motility and chemotactic responsiveness to levels comparable to normal controls suggesting this defective function may partially be a result of autooxidative membrane damage.

Free radical damage is thought to be involved in the initiation and promotion of many cancers. The increased incidence of cancer among older adults has been postulated to be due, in part, to the increasing level of free radical reactions with age and the diminishing ability of the immune system to eliminate the altered cells. Although controlled human studies on vitamin E and cancer are limited, epidemiological data suggest that high intakes of vitamin E (and other dietary antioxidants) may decrease the risk for certain cancers, particularly cancers of the breast, colon, lung, and stomach.[49] Knekt[50] assessed blood vitamin E levels and subsequent cancer incidence in a longitudinal study of 21,172 men in Finland. Vitamin E was measured from stored blood samples of 453 subjects who developed cancer during the 6-10 year study period and 841 matched controls. Adjusted relative risks in the two highest quintiles of blood vitamin E concentrations compared to all other quintiles were 0.7 for all cancers and 0.6 for cancers unrelated to smoking. Blot[51] conducted a five-year prospective, randomized trial on the effect of antioxidant supplementation (30 mg vitamin E, 15 mg ß-carotene, 50 μg selenium) and other micronutrient combinations on 29,584 residents of Linxian, China. Death from stomach cancer and esophageal cancer were 21% and 10% lower, respectively, while death from all causes of cancer were reduced by 13% and overall mortality by 9% in the antioxidant supplemented group; the other vitamin/mineral supplements did not significantly alter the risk of cancer mortality.

REFERENCES

1. R.G. Cutler, Aging and oxygen radicals, *in*: "Physiology of Oxygen Radicals," A.E. Taylor, S. Matalon and P. Ward eds., American Physiological Society, Bethesda. pp. 251-285 (1986).
2. T.F. Slater and G. Block, eds., Antioxidant Vitamins and ß-carotene in Disease Prevention, *Am. J. Clin. Nutr.* 53:189S-396S (1991).

3. D. Harman, The aging process, *Proc. Natl. Acad. Sci. USA*. 78:7124-7128 (1980).
4. W.A. Pryor, The free-radical theory of aging revisited: a critique and a suggested disease-specific theory, *in*: "Modern Biological Theories of Aging," H.R. Warner, R. N. Butler, R.L. Sprott, and E.L. Schneider, eds., Raven Press, New York. pg. 89-112 (1987).
5. R.L. Walford, S.R. Gottesman, and R.H. Weindruch, Immunophathology of Aging, *Ann. Rev. Gerontol. Geriatr.* 2:3-15 (1981).
6. M. Rola-Pleszczynski, Immunoregulation by leukotrienes and other lipoxygenease metabolites, *Immunology Today* 6:302-307 (1985).
7. S.N. Meydani and J.B. Blumberg, Vitamin E and the immune response. *in:* "Nutrient Modulation of the Immune Response," S. Cunningham-Rundles, ed., Marcel Dekker, Inc., New York, pp. 223-238 (1993).
8. A. Coquette, B. Vray, and J. Vanderpas, Role of vitamin E in the protection of the resident macrophage membrane against oxidative damage, *Arch. Int. Physiol. Biochem* 94:529-534 (1986).
9. A. Sepe and R.A. Clark, Oxidant membrane injury by the neutrophili myeloperoxidase system. II: Injury by stimulated neutrophils and protection by lipid-soluble antioxidants, *J. Immunol.* 134:1896-1901 (1985).
10. M.A. Boogaerts, J. Van De Broeck, H. Deckmyn, C. Roellant, J. Vermylen, and R.L. Verwilghen, Protective effect of vitamin E on immune triggered granulocyte mediated endothelial injury, *Thromb. Haemost.* 51:89-92 (1984).
11. J.E. Triau, S.N. Meydani, and E.J. Schaefer, Oxidized low density lipoproteins stimulate prostacyclin production by adult human vascular endothelial cells, *Arteriosclerosis* 8:810-818 (1988).
12. G. Yogeeswaran and I.N. Mbawuike, Altered metabolism and cell surface expression of glycosphingolipids caused by vitamin E in cultured murine (K 3T3) reticulum sarcoma cells, *Lipids* 21:643-647 (1986).
13 A. Bendich, E. Gabriel, and L.J. Machlin, Dietary vitamin E requirement for optimum immune response in the rat, *J. Nutr.* 116, 675-681 (1986).
14. S.N. Meydani, M. Meydani, C.P. Verdon, A.C. Shapiro, J.B. Blumberg and K.C. Hayes, Vitamin E supplementation suppresses prostaglandin E_2 synthesis and enhances the immune response in aged mice, *Mech. Ageing Dev.* 34, 191-201, (1986).
15. R.H. Heinzerling, R.P. Tengerdy, L.L. Wicks, and D.C. Lueker, Vitamin E protects mice against Diploccoccus pneumonia type I infection, *Infect. Immun* 10:1292-1295 (1974).
16. R.P. Tengerdy, R.H. Heinzerling and M.M. Mathias, Effect of vitamin E on disease resistance and immune responses, in: "Tocopherol, Oxygen, and Biomembranes," C. de Duve and O. Hayaishi, eds. Elsevier/North-Holland Biomedical Press, Amsterdam, pp. 191-200 (1978).
17. P.G. Reddy, J.L .Morrill, H.C. Minocha, and J.S. Stevenson, Vitamin E is immunostimulatory in calves, *J. Dairy Sci.* 70:993-999 (1987).
18. S.N. Meydani, E.S. Cathcart, R.E. Hopkins, M. Meydani, K.C. Hayes, and J.B. Blumberg, *in:* "Fourth International Symposium of Amyloidosis," G.G. Glenner, E.P. Asserman, E. Benditt, E. Calkins, A.S. Cohen, and D. Zucker-Franklin, eds., Plenum Press, New York, pp. 683-692 (1986).
19. K.D. Pletsityi, E.V. Nikushkin, M. Askerov, and L. Ponomareva, Inhibitory effects of vitamin E on the development of adjuvant arthritis in rats, *Byull Ekp Med* 103, 43-45 (1987).
20. D. Harman, Free radical theory of aging: Beneficial effect of antioxidants on the life span of male NZB mice; role of free radical reactions in the deterioration of the immune system with age and in the pathogenesis of systemic lupus erythematosus, *Age* 3:64-73 (1980).
21. R.L. Baehner, L.A. Boxer, J.M. Allen, and J. Davis, Autooxidation as a basis for altered function by polymorphonuclear leukocytes, *Blood* 50,327-335 (1977).

22. L.A. Boxer, R.E. Harris, and R.L. Baehner, Regulation of membrane peroxidation in health and disease, *Pediatrics* 64: S713-S718 (1979).
23. J.S. Prasad, Effect of vitamin E supplementation on leukocyte function, *Am. J. Clin. Nutr.* 33:606-608 (1980).
24. M.E. Miller, Phagocytic function in the neonate: Selected aspects, *Pediatrics* 64:5709-5712 (1979).
25. G. Chirico, M. Marconi, A. Colombo, A. Chiara, G. Rondini, and A.G. Ugazio, Deficiency of neutrophil phagocytosis in premature infants: Effect of vitamin E supplementation, *Acta. Paediatr. Scand.* 72:521-524 (1983).
26. S. Zimlanski, M. Wartanowicz, A. Klos, A. Raczka, and M. Klos, The effects of ascorbic acid and alpha-tocopherol supplementation on serum proteins and immunoglobulin concentrations in the elderly, *Nutr Intl.* 2:1-5 (1986).
27. D. Harman and R.W. Miller, Effect of vitamin E on the immune response to influenza virus vaccine and incidence of infectious disease in man, *Age* 9:21-23 (1986).
28. S.N. Meydani, M.P. Barklund, S. Liu, M. Meydani, R.A. Miller, J.G. Cannon, F.D. Morrow, R. Rocklin, and J.B. Blumberg, Vitamin E supplementation enhances cell-mediated immunity in healthy elderly subjects, *Am. J. Clin. Nutr.* 52:557-563 (1990).
29. N.D. Penn, L. Purkins, J. Kelleher, R.V. Heatley, B.H. Manscie-Taylor, P.W. Belfield, The effect of dietary supplementation with vitamins A, C, and E on cell mediated immune function in elderly long stay patients: a randomized controlled trial, *Age Ageing* 20:169-174 (1991).
30. J.G. Cannon, S.F. Orencole, R.A. Fielding, M. Meydani, S.N. Meydani, M.A. Fiatarone, J.B. Blumberg, and W.J. Evans, Acute phase response in exercise: interaction of age and vitamin E on neutrophils and muscle enzyme release, *Am. J. Physiol.* 259:R1214-R1219 (1990).
31. J.G. Cannon, S.N. Meydani, R.A. Fielding, M.A. Fiatarone, M. Meydani, M. Farhangmehr, S.F. Orencole, J.B. Blumberg, and W.J. Evans, Acute phase response in exercise. II. Associations between vitamin E, cytokines, and muscle proteolysis, *Am. J. Physiol.* 260:R1235-R1240.
32. M. Meydani, W. Evans, G. Handelman, L. Biddle, R.A. Fielding, S.N. Meydani, J. Burrill, M.A. Fiatarone, J.B. Blumberg, and J.G. Cannon, Protective effect of vitamin E on exercise-induced oxidative damage in young and older adults, *Am. J. Physiol.* 264:R992-R998 (1993).
33. R.K. Chandra, Effect of vitamin and trace element supplementation in immune responses and infection in elderly subjects, *Lancet* 340:1124-1127 (1992).
34. M. Meydani, S.N. Meydani, L. Leka, J. Gong and J.B. Blumberg, Effect of Long-Term Vitamin E Supplementation on Lipid Peroxidation and Immune Response of Young and Old Subjects, *FASEB J.* 7:A415, 1993.
35. S. Hartz and J. Blumberg, Use of vitamin and mineral supplements by the elderly, *Clin. Nutr.* 5:130-136 (1986).
36. J.S. Goodwin and P.J. Garry, Relationship between megadose vitamin supplementation and immunological function in a healthy elderly population, *Clin. Exp. Immunol.* 51:647-653 (1983).
37. M. Chavance, G. Brubacher, B. Herbeth, G. Vernhes, T. Mikstacki, F. Dete, C. Fournier and C. Janot, Immunological and nutritional status among the elderly, *in:* "Lymphoid Cell Functions in Aging," A.L. de Wick, ed., Eurage, Paris, France, pp. 231-237 (1984).
38. M. Chavance, G. Brubacher, B. Herberth, G. Vernhes, T. Mikstacki, F. Dete, C. Fournier and C. Janot, Immunological and nutritional status among the elderly, *in:* "Nutrition, Immunity, and Illness in the Elderly," R.K. Chandra, ed., Pergamon Press, New York, pp. 137-142 (1985).

39. G. Blankenhorn, Clinical efficacy of spondyvit (vitamin E) in activated arthroses: a multicenter, placebo-controlled, double-blind study, *Z. Orthop. Ihre. Grenzgeb.* 124:340-343 (1986).
40. R.O. Likoff, M.M. Mathias and C.P. Neckles, Vitamin E enhancement of immunity mediated by the prostaglandins, *Fed. Proc.* 37:829 (1978).
41. S.L. Swartz, W.C. Willett, and C.H. Hennekens, A randomized trial of the effect of vitamin E on plasma prostacyclin (6-keto-PGF$_{1\alpha}$) levels in healthy adults, *Prost. Leukotr. Med.* 18:105-111 (1985).
42. W.C. Hope, C. Dalton, L.J. Machlin, R.J. Filipski, and F.M. Vane, Influence of dietary vitamin E on prostaglandin synthesis in rat blood, *Prostaglandins* 10:557-561 (1975).
43. L. Machlin, Vitamin E and prostaglandins, *in:* "Tocopherol, Oxygen, and Biomembranes," C. de Dure and O. Hayaishi, eds., Elsevier/North-Holland Biomedical Press, Amsterdam, pp. 208-215 (1978).
44. K. Lauritsen, L.S. Laursen, K. Bukhave, and J. Rask-Madsen, Does vitamin E supplementation modulate in vivo arachidonate metabolism in human inflammation? *Pharmacol. Toxicol.* 61:246-249 (1987).
45. E.J. Goetzel, Vitamin E modulates lipoxygenation of arachidonic acid in leukocytes, *Nature* 288:187-193 (1980).
46. B. Halliwell, Oxidants and human disease: Some new concepts, *FASEB J.* 1:358-364 (1987).
47. M. Kitahara, H.J. Eyre, R.E. Lynch, M.L. Rallison, and H.R. Hill, Metabolic activity of diabetic monocytes, *Diabetes* 29:251-256 (1980).
48. H.R. Hill, N.H. Augustine, M.L. Rallison, and J.I. Santos, Defective monocyte chemotactic responses in diabetes mellitus, *J. Clin.Immunol. 3:*70-77 (1983).
49. H.B. Stähelin, K.F. Gey, M. Eichholzer, E. Lüdin, F. Bernasconi, J. Thurneysen, and G. Brubacher, Plasma antioxidant vitamins and subsequent cancer mortality in the 12-year follow-up of the Prospective Basel Study, *Am. J. Epidemiol.* 133:766-775 (1991).
50. P. Knekt, R. Järvinen, R. Seppänen, A. Rissanen, A. Aromaa, O.P. Heinonen, D. Albanes, M. Heinonen, E. Pukkala, and L. Teppo, Dietary antioxidants and the risk of lung cancer, *Am. J. Epidemiol.* 134:471-479 (1991).
51. W.J. Blot, J-Y Li, P.R. Taylor, W. Guo, S. Dawsey, G-Q Wang, C.S. Yang, S-F Zheng, M. Gail, G-Y Li, Y. Yu, B. Liu, J. Tangrea, Y. Sun, F. Liu, J.F. Fraumeni, Y-H Zhang, and B. Li, Nutrition intervention trials in Linxian, China: supplementation with specific vitamin/mineral combinations, cancer incidence, and disease-specific mortality in the general population, *J. Natl. Cancer Inst.* 85:1483-1492 (1993).

VITAMINS AND CAROTENOIDS - A PROMISING APPROACH TO REDUCING THE RISK OF CORONARY HEART DISEASE, CANCER AND EYE DISEASES

Wolfgang Schalch and Peter Weber

Human Nutrition Research
Vitamins and Fine Chemicals Division
Hoffmann - La Roche Inc., Nutley, NJ, USA

INTRODUCTION

Free radicals, both externally generated and created internally by either normal or aberrant metabolism, play an important role in the etiology of many diseases. In specific disease conditions radicals may lead to tissue, cell or protein injury or the alteration of the genetic material. Thus, radicals can alter DNA, which in turn can lead to aberrant gene products and ultimately to cancer. In addition, radicals and oxidative processes can modify the lipoproteins in the blood with the effect of leading to the formation of foam cells in the arteries, a process, which unless controlled or stopped, will ultimately lead to atherosclerosis. Oxidative modification of the lens proteins mediated by light leads to cataract, while oxidative damage of the photoreceptors in the macula is one factor which is thought to cause the development of age-related macular degeneration.

To ward off or control these detrimental effects of radicals, nature has designed antioxidant defense mechanisms which comprise enzymatic systems such as glutathione peroxidase, superoxide dismutase, and catalase as well as non-enzymatic molecules such as the antioxidant micronutrients vitamins E and C, beta-carotene and selenium. The present paper focusses on the relationship of carotenoid and vitamin intake to the risk reduction of some common diseases. We have selected cancer, cardiovascular diseases and eye diseases, because there is considerable scientific documentation for the involvement of radicals in these diseases, and also because these diseases present major public health problems. Cancer and cardiovascular diseases are the leading causes of death in the western world. However, because of time and space constraints this review is not all-inclusive, but is limited to discussion of more recent data and the fat-soluble vitamins.

Free Radicals in Diagnostic Medicine, Edited by
D. Armstrong, Plenum Press, New York, 1994

RISK REDUCTION OF CARDIOVASCULAR DISEASE

LDL-Oxidation: Current Hypothesis For Atherogenesis And The Possible Role Of Antioxidant Vitamins

Atherosclerotic lesions can be found even in early childhood[1]. Mostly, they remain undetected for several decades until they become manifest with aging at different sites in the vascular system. So far, the possible underlying mechanisms leading to atherosclerosis are only partly understood. Elevated plasma levels of LDL cholesterol are considered to be one of the major risk factors contributing to the development of atherosclerosis[2].

During recent years there has been increasing evidence that oxidatively modified LDL (ox-LDL) may play a pivotal role in this process, because oxidative damage of LDL strongly amplifies its atherogenicity. Ox-LDL, but not native LDL, is highly cytotoxic. It not only is a potent chemo-attractant for the circulating monocytes, but at the same time inhibits the migration of the resident macrophages from within the arterial wall back to the plasma. These properties of ox-LDL effect an accumulation of macrophages in the subendothelial space[3]. Further, the clearance of ox-LDL is mediated by a special receptor, the so-called "scavenger receptor[3]". Unlike the uptake of LDL by the native receptor, uptake of ox-LDL by the scavenger receptor is not down regulated with increasing ox-LDL levels. The increased rate of ox-LDL uptake results in an accumulation of ox-LDL in the macrophage, thus, leading to an increased generation of foam cells and formation of fatty streaks, the precursors of atherosclerotic plaques.

It has been demonstrated in vitro[4] that vitamin E and carotenoids are consumed when LDL is oxidized, and supplementation with beta-carotene and vitamin E in vivo can inhibit the oxidative modification of LDL[5,6] . Further, several studies provide evidence that these results from cell culture studies may apply in humans. It has been reported, that ox-LDL were found to be higher in men with an increased risk for developing atherosclerosis such as diabetics, smokers and patients with coronary heart diseases[7]. Taken together these data provide a strong rationale that antioxidant vitamins like beta-carotene and vitamin E and other antioxidant micronutrients may play an important role in reducing the risk of developing various atherosclerotic diseases.

Epidemiological Studies

During the past decade, a number of epidemiological studies on antioxidant vitamins and cardiovascular disease have been published. Our focus here is mainly on the outcome of recently published trials, which are summarized in Table 1.

The Nurses' Health Study[8] is a prospective epidemiological study utilizing a population of 87,245 female nurses, aged 34 to 59 years. Their dietary intakes of nutrients was estimated at the beginning of the study based on a dietary questionnaire documenting the average frequency of intake of 61 food items. The use of supplements containing antioxidant vitamins was also recorded. From 1980 to 1988, 437 nonfatal myocardial infarctions and 115 deaths caused by coronary disease were documented. After adjusting for age and smoking status, the risk of a coronary event was significantly lower (0.66, $p<0.001$) in the highest quintile of vitamin E intake, as compared to those in the lowest quintile. The apparent benefit was attributable mainly to the use of vitamin E supplements, because high levels of intake from dietary sources alone were not associated with significant reductions in risk. The use of vitamin E supplements for two or more years was associated with a decrease in risk of 41 percent.

The Health Professionals Follow-up Study[9] was started in 1986. This study investigated 39, 910 male health professionals 40 to 75 years of age, who were free of diagnosed coronary heart disease, diabetes, and hypercholesterolemia. The dietary intake of vitamin C, vitamin E and beta-carotene was estimated at baseline from a food questionnaire inquiring about the average frequency of intake of 131 foods during the previous year. The use of supplements was also documented. During four years of follow-up, 667 coronary end points were documented (360 bypass grafts or angioplasties, 201 nonfatal myocardial infarctions, and 106 fatal coronary events). As compared with the lowest quintile for vitamin E intake, the age-adjusted relative risk for coronary heart disease was significant lower (0.59, $p<0.001$) in the highest quintile . Again, the maximal reduction in risk was seen among men consuming 100 to 249 mg vitamin E per day for two or more years, with no further decrease at higher doses.

Table 1. Effects of vitamin E and/or beta-carotene on coronary heart disease (CHD) and peripheral arterial disease (PAD) - results of the most recently published epidemiological studies (DQ=dietary questionnaire, ATL=adipose tissue levels, PL=plasma level, ABPI=ankle brachial pressure index, MI=myocardial infarction).

Author	Population	Parameters	Endpoint/Event	Main Results
Stampfer MJ et al, 1993	87,245 female nurses	vitamin E by DQ	CHD	RR 0.66↓
Rimm EB et al, 1993	39,910 male health professionals	vitamin E beta-carotene by DQ	CHD	RR 0.56↓ RR 0.71↓
Donnan PT et al, 1993	1,592 men and women	vitamin E beta-carotene by DQ	PAD	associated with ABPI not associated
Kardinaal AFM et al, 1993	683 subjects after MI, 727 controls	vitamin E beta-carotene by ATL	acute MI	RR +/- RR 0.56↓
Gey KF et al, 1993	2,947 male volunteers	vitamin E carotene by PL	CHD	RR +/- RR 0.65↓
Hense HW et al, 1993	2,023 men and women	vitamin E by PL	CHD	no difference

For beta-carotene the age-adjusted relative risk for coronary heart disease in the highest quintile was 0.71 as compared to the lowest quintile group. The association between dietary carotene and the risk of coronary heart disease was mainly attributed to baseline smoking status. Among current smokers, the relative risk was 0.30 ($p<0.02$) and among former smokers 0.60 ($p<0.04$) when the highest and lowest intake groups were compared, whereas among men who had never smoked, no association was seen.

The Edinburgh Artery study[10] carried out in 1988 consisted of a cross-sectional survey of 1,592 men and women aged 55 to 74 years. One aim was to examine the relationship between peripheral arterial disease, as measured by the ankle brachial pressure index, and dietary factors. Dietary intake of nutrients was assessed by a food questionnaire including 60 different food items. Dietary vitamin E intake was significantly associated with the ankle brachial pressure index, indicating a lesser extent of peripheral arterial disease.

In the EURAMIC[11] a multicenter case-control study, α-tocopherol and beta-carotene concentrations were measured in adipose-tissue samples collected in 1991 and 1992 from 683 people with acute myocardial infarction and 727 controls. The age-adjusted and center adjusted odds ratio for risk of myocardial infarction in the fifth quintile of beta-carotene was significantly lower (0.56, $p<0.001$) than in the first quintile. Once again the risk reduction was largely attributed to the effect of smoking. The relative risk for a coronary event among smokers in the lowest quintile was more than twofold higher compared to people who had never smoked. For α-tocopherol no significant risk reduction was observed.

After a 12-year follow-up the Basel Prospective Study[12] also showed a lower risk (RR 0.65; $p<0.02$) of mortality from ischemic heart disease in the group with the highest concentrations of the carotene fraction. High vitamin E levels were not associated with a decreased risk for coronary heart disease, most likely because of a good vitamin E supply (median vitamin E plasma levels: 35 µmol/l). In the population-based MONICA Augsburg cohort, in which 2023 men and 1999 women were observed for three to four years, serum vitamin E levels also were not associated with a risk reduction of myocardial infarction. Vitamin E serum levels in cases were 33.9 µmol/l and 32.8 µmol/l in matched controls, respectively. Gey et al[13] recently calculated that vitamin E serum levels above 27.5 - 30.0 µmol/l may be indicative of a decreased risk for coronary heart disease.

In summary, the results of the recently published epidemiological studies report consistently, that a sufficient supply of the dietary antioxidants beta-carotene and vitamin E was inversely associated with the incidence of coronary heart disease. The results derived from the Nurses' Health and the Health Professionals Follow-up Study suggest, in particular, that the use of vitamin E supplements may reduce the risk of coronary heart disease. In subjects who had a good vitamin E supply, as measured by vitamin E plasma levels,[12,13] the risk of coronary heart disease was not associated with vitamin E levels in the plasma. However, the risk for coronary heart disease was decreased in these persons if they had higher carotene plasma levels. In addition, beta-carotene seems to reduce the risk for coronary heart disease particularly in patients who are heavy smokers. Finally, it can be assumed that a good dietary vitamin E intake may also reduce the risk of peripheral artery disease.

Intervention Studies

The promising data from the epidemiological studies on the potential role of antioxidant vitamins such as vitamin E and beta-carotene need to be confirmed by intervention studies, in which people are randomized either to active treatment or placebo groups. At present, data derived from large-scale intervention studies on antioxidant vitamins and cardiovascular disease are currently not available. Only a limited number of intervention studies applying different doses of antioxidant supplements in a small number of subjects are available. The studies can be roughly subdivided into three groups: Studies focussing i) on such risk factors for coronary heart disease as oxidative susceptibility of LDL, ii) on such clinical endpoints as intermittent claudication and myocardial infarction, and iii) on coronary atherosclerosis as measured by percutaneous transluminal coronary angioplasty (PTCA).

Very recently, three papers demonstrated that short-[15] and long-term[5,6] administration of large doses of vitamin E may strongly protect LDL against oxidative modification. A treatment of 1,000 IU D,L-α-tocopherol/d for 7 days resulted in an increase in the oxidative resistance of LDL of 41%, and the resistance times were highly correlated ($r=0.89$, $p=0.014$) with the increase in tocopherol content of the LDL[15]. The other studies using 800 IUα-tocopherol/d and 1,600 mg vitamin E/d, respectively, for several months also reported an increased protection of LDL, whereas beta-carotene given at a level of 60mg/d for 3 months did not change the susceptibility of LDL to oxidation.

Two earlier studies published in 1958 and in 1974 comprising a small number of patients reported that a treatment consisting of 300 mg d-α-tocopheryl acetate and 600mg vitamin E for 3 to 10 months respectively, improved intermittent claudication as estimated by a walking test.[16,17] In another study in patients with stable angina pectoris 400 IU d-α-tocopherol succinate given for 6 months failed to show a beneficial effect on the clinical course of this disease[18].

In 1982, a large randomized placebo-controlled, double-blind intervention trial among 22,701 male physicians from the U.S., aged 40 - 84 years was started at Harvard University. Subjects were randomly assigned to four different treatment groups: 325 mg aspirin alone, 325 mg aspirin and 50 mg beta-carotene (on alternate days), 50 mg beta-carotene alone and a placebo group. Although persons with prior myocardial infarction or stroke were ineligible to enroll, 333 physicians with chronic stable angina or a prior coronary revascularization did enter the study.

An analysis of this risk group demonstrated that those subjects receiving 50 mg beta-carotene on alternate days experienced a statistically significant reduction of 49% in the relative risk of all major vascular events (myocardial infarction, cardiovascular death, coronary revascularization, stroke) as compared to those receiving placebos.[19] The protective effect of beta-carotene supplementation became significant after the second year of treatment and increased further with time.

In a double-blind, placebo-controlled trial DeMaio et al[20] investigated the effects of α-tocopherol on restenosis following PTCA. 52 patients received 1,200 IU dl-α-tocopherol per day and 48 an inactive placebo. Restenosis was defined as the presence of a lesion with ≥ 50% stenosis in a previously dilated artery segment. Patients receiving α-tocopherol had a 35.5% restenosis angiographically documented vs 47.5% restenosis in the placebo group ($p<0.06$).

The data from intervention studies on antioxidant vitamins in humans with accelerated atherosclerosis are limited and preliminary. In summary, there is good evidence that beta-carotene and vitamin E may prevent the oxidation of LDL, which is thought to play an important role in atherogenesis. The data from earlier studies in patients with intermittent claudication as well as the first results from an ongoing trial on the effect of antioxidant vitamins on such clinical endpoints as coronary heart disease and stroke, although convincing, have to be considered as preliminary only. First data on the effects of vitamin E on coronary atherosclerosis are promising, but need also to be confirmed. The outcome of several ongoing studies may provide more information through the course of this decade.

RISK REDUCTION OF CANCER

Potential Mechanisms By Which Beta-Carotene May Exert A Beneficial Role On Carcinogenesis

Carcinogenesis is considered to be a multistage process. So far, the mechanisms involved in the conversion of a normal cell to a one with a malignant phenotype are not fully understood. Briefly and simplified carcinogenesis may be summarized as process of initiation, promotion, and progression.

Tumor initiation is caused by spontaneous or carcinogen-induced genetic (DNA) damage. Some important initiators are mutagens, genotoxic prooxidants, and radiation. At this stage beta-carotene may exert its protective efforts by detoxifying, quenching, or trapping the active intermediates which react with the DNA. In a recently published study it was shown that plasma beta-carotene levels were inversely correlated with the frequency of micronuclei appearance in human lymphocytes. Following six day of supplementation with 30 mg/d beta-carotene, the micronuclei in lymphocytes, indicative of DNA damage, were significantly lower in the supplement group following x-ray irradiation [21].

Promotion means growing and proliferation of the initiated abnormal cells. Promotion may be reversible for a long time. For example, prostatic cancer in humans may have a latent period of up to 40 years.[22] The potency of many promoters (including granulocytes activated by inflammation) to produce active oxygen species provides a rationale for the use of beta-carotene, a known singlet oxygen quencher. In addition, it has been reported that in tobacco chewers beta-carotene decreased the formation of micronuclei in oral mucosa, reflecting a reduced scission of DNA.[23] Further, beta-carotene has been shown to enhance gap junction communications, irrespective of its provitamin A activity.[24] Gap junctions of cell membranes are considered to be part of cell-to-cell communication, and it is believed that increased communication between normal and initiated cells may restrict the rate of initiated cells expressing the neoplastic phenotype. The upregulation of gap junctions by beta-carotene is considered to be caused by the increased expression of connexin 43, a junctional protein[25].

When the cells become more autonomous, a conversion occurs to a malignant tumor, invading healthy tissue and developing metastatic foci at various sites of the body, known as clinical cancer manifestation. Animal studies indicate that beta-carotene may also protect against tumor progression. It was reported that beta-carotene protected in 45% of the cases against skin cancer in mice, when administered after irradiation, but was without any effect when given before irradiation.[26] In another study beta-carotene failed to protect against MNNG-induced gastric preneoplastic lesions in rats, however, it reduced the rate of dysplasia to early and infiltrating carcinomas by 65%.[27]

Epidemiological Studies

During the last 25 years more then 200 epidemiological studies on dietary carotenoids and cancer have been published indicating that intake of dietary carotenoids may reduce the risk of developing cancer at various sites in humans[28]. Despite several drawbacks of retrospective epidemiological studies, for example the difficulty of assessing dietary intakes, different sample sizes and populations, different smoking behavior etc. dietary intake of carotenoids seems most likely to be inversely associated with the risk of developing lung cancer. These results have been confirmed in a number of prospective studies on dietary intake or blood levels of carotenoids and cancer. As recently reviewed by van Poppel,[29] 11 out of 15 studies carried out in different countries around the world reported a significant association of carotenoid intake and the risk of experiencing lung cancer.

These results have been confirmed by the recently published evaluation of the ongoing Basel Prospective Study.[30] After a 12-year follow-up of 2,974 participants, 204 cancer cases have been observed. The relative risk of bronchus and stomach cancer, which was calculated after exclusion of mortality during the first two years, was significantly increased in subjects with low beta-carotene levels. Low plasma levels of beta-carotene and vitamin A were associated with an increased risk for all cancer sites.

Recently, the largest epidemiological study focussing on antioxidant status and mortality from various cancer sites was published. In 130 rural communes widely distributed in the populated areas of 24 provinces in China, fasting blood samples were obtained from 3,225 3,225 randomly selected male and 3,225 female subjects aged 35-64 years[31]. The main biochemical variables determined were several biochemical indicators of reactive oxygen reactivity (glutathione peroxidase, lipid peroxides,catalase, superoxide dismutase), the status of the antioxidant vitamins (beta-carotene, alpha-tocopherol, ascorbic acid) and several minerals (selenium, copper, iron, and zinc). The results of the biochemical measurements were correlated to mortality rates by age and sex for each cancer site as published in the Atlas of Cancer Mortality in China[32]. The results of multiple regression analysis by sex and cancer site showed a significant inverse association of beta-carotene for stomach cancer in males and females. The incidence of lung cancer was negatively associated with ascorbic acid in males.

In summary, a large number of epidemiological studies carried out in different countries around the world have shown that dietary intake of antioxidant vitamins, and in particular of beta-carotene, is associated with risk reduction in various cancer sites. This seems to be true especially for lung and stomach cancer as recently confirmed by the results of the Basel Prospective Study[30] and a study on antioxidant status and mortality in China[31].

Intervention Studies

Results of intervention studies in humans on cancer and antioxidant vitamins can be subdivided into three main groups: i) studies measuring early biomarkers for cancer risk (e.g., immunological response), ii) studies investigating the clinical course of premalignant lesions (e.g., leucoplakia) iii) studies focussing on clinical endpoints of cancer (e.g., incidence, mortality).

Immunological studies have demonstrated a beneficial effect of beta-carotene on several lymphocyte subpopulations, and an enhancement of tumor necrosis factor as summarized by

Bendich et al[33]. A study in elderly subjects who received a micronutrient supplement, the vitamins most likely to improve several determinants of the immune response were beta-carotene, and vitamins A, E, B6 and C[34]. These results are in line with a previous study showing a significant and dose-dependent increase of cells with surface markers for T-helper and natural killer cells and cells with interleukin 2 and transferrin receptors following two-months administration of ≥30mg/d beta-carotene[35].

As recently reviewed by Garewal,[36] it has been shown in seven clinical trials, that the administration of beta-carotene alone, as well as in combination with vitamins E and C, can produce a substantial regression of oral leukoplakia, a premalignant lesion for oral cancer. In addition, it has been reported that discontinuation of the vitamin supplementation resulted in a rapid recurrence of the leukoplakia.

The Linxian dysplasia trial[37] investigated whether supplementation with multiple micronutrients (14 vitamins, 12 minerals) may reduce esophageal and/or gastric cardia cancer among persons with esophageal dysplasia, which is considered as a precancerous lesion of the esophagus. In this study, 3,318 persons with cytological evidence of dysplasia were randomly assigned to receive supplements from 1985 through 1991.In the supplemented group, the risk for total mortality was 7% lower, for total cancer 4% lower, and the cumulative esophageal/gastric cardia death rates were 8% lower ($p<0.1$). It was concluded, that perhaps a longer follow-up would be more informative about the effectiveness of supplementation on this kind of cancer.

A recently published intervention study provides evidence that the use of antioxidant vitamin supplements can lead to a reduction of stomach cancer. The Linxian Nutrition Intervention Trial included 29,584 adults, aged 40-69 years from Linxian County in China[38]. The main purpose of this study was to determine if dietary supplementation with specific vitamins and minerals can lower the mortality or incidence of cancer. Between 1986 and 1991 the subjects where randomly assigned to the following four combinations by a one-half replicate of a 2^4 factorial design (eight intervention groups AB, AC, AD, BC,BD, CD, ABCD, and placebo): 5,000 IU retinol and 22.5 mg zinc (A), 3.2 mg riboflavin and 40 mg niacin (B), 120 mg vitamin C and 30 μg molybdenum (C), and 15 mg beta-carotene, 30 mg vitamin E and 50 μg selenium (D). A total of 2,127 deaths occurred among the trial participants during the intervention period. Cancer was the leading cause of death, with 32% of all deaths due to esophageal or stomach cancer. Persons receiving beta-carotene, vitamin E and selenium showed a reduction in overall mortality of 9% ($p<0.03$).The reduction was mainly because of lower cancer rates, particularly stomach cancer. The risk reduction started to be relevant one to two years following the start of the supplementation. The mortality from noncardiac stomach cancer was reduced by 41% ($p<0.02$).

Another recently published intervention trial investigated the effect of beta-carotene and vitamin E on the incidence of lung cancer and other cancers[38a]. The Alpha-Tocopherol, Beta-Carotene (ATBC) Cancer Prevention Study was a randomized, double-blind, placebo-controlled prevention trial which included 29,133 male smokers 50 to 69 years old from 14 different study sites of southwestern Finland. The participants received a supplementation of either 50mg/d dl-alpha-tocopheryl acetate or 20mg/d beta-carotene, both, or placebo as randomly assigned by a two-by-two factorial design. At study entry the men had a mean age of 57.2 years and had smoked an average of 20.4 cigarettes over a mean period of 35.9 years. The study reported an inverse association between dietary intake of alpha-tocopherol and beta-carotene at baseline and the risk of lung cancer during the trial. Furthermore, the incidence of lung cancer was lower among the subjects who at the beginning of the study had serum alpha-tocopherol and beta-carotene levels in the highest quartile group. During the trial 876 newly diagnosed cases of lung cancer and 564 deaths due to lung cancer were documented in the entire population. Unexpectedly, the patients receiving beta-carotene had a higher incidence of lung cancer (n=474) than those who received placebo (n= 402, $p<0.01$). In the men taking alpha-tocopherol a small reduction of lung cancer incidence was seen (n=433 vs n=443, ns). Overall mortality was also higher among the subjects given beta-carotene ($p<0.02$). It was also noted that the men taking alpha-tocopherol had a lower incidence of prostate cancer (n=99 vs n=151).

Taken together, there are promising results on the effects of antioxidant vitamins in reducing cancer risks. Recently, several studies reported that beta-carotene and vitamin E may improve the immunological response. The use of beta-carotene resulted in a regression of premalignant lesions such as leukoplakia. Finally, one intervention study demonstrated that within a short period of 5 years the use of antioxidant vitamins and several other micronutrients can reduce significantly overall mortality and death from certain stomach cancers. However, the results of the other intervention study carried out in heavy smokers provided no evidence that a beta-carotene or a alpha-tocopherol supplementation may be beneficial for the prevention of lung cancer. Thus, the ATBC-Study findings are in strong contrast to the results of more than 40 epidemiological papers published during the last two decades reporting consistently a beneficial effect particularly of a high dietary beta-carotene intake on the incidence of lung cancer. Thus, it may be assumed that in a population of such heavy smokers being at a significant higher risk to develop lung cancer a supplementation of 20 mg/d beta-carotene may not affect anything within 5 years, but it remains an open question why in the beta-carotene group the incidence of lung cancer was higher. Beta-carotene is considered as safe, and so far there are no plausible mechanisms to explain the higher incidence of lung cancer in the ATBC-Study of the subjects recieving beta-carotene. Thus, further research on this issue is needed to find out if this finding of the ATBC-Study is due to chance as the investigators and other researchers suggested[38b]. More light will be shed on this issue by several ongoing intervention studies, which should be published during the course of this decade.

RISK REDUCTION OF EYE DISEASES

Generation Of Radicals Along The Path Of Light Through The Eye

As indicated earlier free radicals play an important role in the etiology of many diseases[39]. Among all the organs involved the eye is special. Because the eye is exposed to and focusses substantial quantities of visible energy, it is one of the most susceptible organs to damage by free radicals and reactive oxygen species that are produced via photosensitation processes[40]. Through its pathway to the retina, light has the potential to generate reactive molecules mainly at three locations. In the cornea, the outermost part of the eye, exposure to short wavelength UV light can generate free radicals and cause phototokeratitis or "snow-blindness." Less energetic, long wavelength UV light passes through the cornea and is absorbed in the lens. There it can initiate an irreversible, radical-mediated oxidation of lens fibers and lens epithelium leading to the opacification of the lens known as cataract. At the retina, light, particularly the more energetic blue wavelenghts, also has the potential for radical generation. This potential is higher than in the cornea and the lens because of the simultaneous additional presence of oxygen and a large number of photosensitizers. The oxidative damage induced is one of the important factors in the multifactorial process of age-related macular degeneration (AMD). Photoprotection can be achieved by absorption of the sensitizing radiation or by the quenching of the generated reactive molecules. These are processes in which vitamin E[41], vitamin C[42], beta-carotene[43] and other carotenoids[44] may be involved. Based on this, we will focus on a discussion of the role these micronutrients may play in risk reduction of two important eye diseases: cataract, a disease of the lens, and AMD, a disease of the retina.

CATARACT

Description, Incidence, And Etiology

Cataract is an opacification of the lens which occurs in most human lenses as a result of aging. The Framingham Study[45] investigated the epidemiology of cataract in different age groups, and reported an incidence of approximately 10% in 65-74 years old and a marked increase to greater than 25% in individuals over 75 years of age, thus clearly demonstrating that cataract is primarily a disease of old age. If untreated, cataract generally leads to blindness. The only effective treatments are various forms of cataract surgery, which is very expensive and together with related costs consumes $3.2 billion per year of the Medicare budget of the U.S. alone.[46] In the absence of any pharmaceutical treatment this emphasizes the importance of cataract prevention.

The etiology of cataract is not fully understood. One important process appears to be the continuous generation of radicals and highly reactive oxygen species in the lens by a variety of external insults, such as exposure to UV light and sunlight.[47,48] The reactive oxygen species generated can damage the epithelial and fiber-cell membranes of the lens and lens enzymes responsible for the repair processes as has recently been reviewed.[49] In order to neutralize or limit this damage, the lens has developed enzymatic systems,[50] and also uses non-enzymatic defense systems such as the antioxidant vitamins.

Epidemiological Studies

An early observation that vitamin C may be involved in light-related protection mechanisms was put forth.[51] The authors observed that the aqueous humor of diurnal animals, including humans, has a much higher vitamin C content than nocturnal animals. The general role of antioxidant vitamins including beta-carotene in the reduction of the risk for cataract has been reviewed.[52,53,54] All the authors concluded that supplementation, particularly with the antioxidant vitamins C and E, may be one method of delaying cataract development. Definitive proof will have to await the outcome of double-blind, placebo-controlled, intervention trials.

Jacques et al.[55,56] were among the first who reported on carotenoids and cataract risk. In their case-control study of 112 subjects aged 40 to 70 years they found that patients with high plasma levels of at least two of the three antioxidant micronutrients vitamins C, E and beta-carotene were at reduced risk for the development of cataract. In another study,[57] 493 cataract extractions within a prospective cohort study involving more than 50,000 female registered nurses were investigated. The intake of various micronutrients including carotenoids was assessed. The authors reported that the intake of spinach, rather than carrots, was most consistently associated with a lower relative risk for cataract. Furthermore, the risk of cataract was 45% lower among women who used vitamin C supplements for 10 or more years. While this may illustrate the importance of a longer history of vitamin C intake, the findings associated with spinach intake indicate that lutein, the predominant carotenoid in this vegetable, and one of the macular carotenoids may have some effects on the health of the lens. This action may be an indirect one, explained by a carotenoid-mediated reduction of lipid oxidation in the retina resulting in presumably less cataractogenic lipid peroxides being produced and passively transported from the retina through the vitreous to the posterior pole of the lens.[58] A hospital-based case-control study in 1,441 patients investigated cataract risk in relation to some physiologic, behavioral, environmental, and biochemical variables. This study reported a higher cataract risk with lower levels of an antioxidant index based on plasma concentrations of vitamins C and E in addition to other parameters[59]. Another case-control study[60] investigated plasma antioxidant vitamins in 47 patients who underwent cataract extractions and compared the results to 94 controls. Low plasma concentrations of vitamin E and beta-carotene were reported to be risk factors for end-stage, senile cataract in this study. Three epidemiological studies on nutrition and cataract conducted in India, Italy and the United States were reviewed recently[61]. While the authors concluded that it is very difficult to compare parallel studies of the same disease, the use of multiple vitamin supplements and the dietary intake of micronutrients with antioxidant potential consistently emerged as one means to reduce cataract risk. But inconclusive results have also been reported by Vitale et al[62]. These authors investigated 660 subjects and correlated the plasma concentrations of beta-carotene and vitamins E and C with the results of nuclear and cortical lens photographs. While higher plasma levels of vitamin E were associated with a lower risk of nuclear cataract and middle levels with a reduced risk of cortical cataract, no relation regarding cortical cataract was observed for high plasma levels of vitamin E. An index of overall antioxidant status also was not correlated with the risk of cataract. An investigation of 685 Hong Kong fishermen 55 to 74 years old also did not find a relation between antioxidant status and cataract risk[48]. The authors conclude that blood levels of antioxidants measured at only one point in time, may not reflect a subject's past long-term status. An alternative explanation for these data may be an intrinsic difficulty of epidemiological studies, which measure vitamin and carotenoid concentrations in the blood. These concentrations may not necessarily be relevant to their true concentration in the lens epithelium as was shown for some enzyme parameters and vitamin E.[63]

Intervention Studies

The epidemiological evidence so far has only provided data demonstrating that there are statistically significant associations between intake of vitamins and carotenoids or their concentration in the serum and the risk of cataract. These associations do not elucidate whether there is a causal relationship between intake and risk reduction. Such evidence can be provided by intervention studies. Recently, the first double-blind, placebo-controlled, intervention trial assessing the development and progression of cataract in individuals supplemented with vitamins and carotenoids has been published.[64] It indicates that the associations measured by epidemiological studies may be more than chance findings. The authors have reported the results of cataract assessment in two large nutrition intervention trials recently concluded in China. In one of these trials, with 1,241 subjects a statistically significant 36% reduction in the prevalence of nuclear cataract was found for persons aged 65 to 74 years who received a multivitamin/mineral, combination including a daily dose of 15 mg of beta-carotene for 5 years. In the second trial involving 3,249 subjects, a 44% reduction of risk was found in the same age group for those who received a riboflavin/folic acid combination for 6 years. Cortical cataract was not significantly affected by either treatment. Nuclear cataract is most detrimental to vision in elderly people, thus, this is a relevant result regarding the elderly population. However, the nutritional status of the trial population was not optimal. Therefore, it is difficult to extrapolate the findings of these trials to better-nourished western populations. Some ongoing controlled trials have the potential to yield results which may be applicable to these populations. Among them are the REACT: Roche-European-American-Cataract Trial (L.T. Chylack et al., Validation of methods for the assessment of cataract progression in the Roche, European-American Anticataract trial (REACT) submitted to Invest. Ophthalmol. Vis. Sci., 1994) initiated by Hoffmann La Roche and the AREDS: Age Related Eye Disease Study[65] initiated by the National Eye Institute (NEI).

AGE-RELATED MACULAR DEGENERATION (AMD)

Description, Incidence, and Etiology

In patients with cataract the diseased lens can be removed by a fairly routine surgical procedure and vision can be almost completely restored by the insertion of an intraocular lens. However, the diseased retina cannot be removed and surgically replaced. Therefore any damage done to it which can not be repaired by the retina's own internal repair system is irreversible. One of the more important diseases of the retina which leads to an irreversible loss of sight is AMD. Because of its irreversible nature and the fact that there is no treatment available, preventive measures for AMD are even more important than for cataract. AMD is characterized by a degenerative process involving the pigment epithelium and the macula, the central part of the retina. It is this part of the eye which is responsible for the clear and sharp vision necessary to perform tasks such as reading. While the etiological processes leading to AMD are not completely understood, there seems to be agreement that among other factors, short wavelength blue light plays an important role in damage to the retina[66] and the pigment epithelium,[67] the layer just underneath the outer retinal layer. The reason for the potential danger of blue light to the retina is that the retina is special compartment because it is a highly oxygenated tissue. Due to its very high content of docosahexaenoic acid,[68] a highly unsaturated fatty acid, it is vulnerable to oxidative damage by reactive molecules generated by the simultaneous presence of light and oxygen. One of these reactive species is singlet oxygen, which in vitro can effectively be quenched by carotenoids.[69] Therefore, the role of the carotenoids in the retina is twofold. First, they can prevent the energetic blue light from reaching the phototosensitizers, and second, through absorption they can quench reactive molecules once they are generated. With the potential hazard from blue light, it is not surprising to find a yellow pigment in the macula, which absorbs blue light.[70] This yellow pigment is composed of two out of the more than 600 naturally-occurring carotenoids. One is lutein, the yellow pigment of the marigold flower which also occurs in spinach. The other is zeaxanthin, the yellow color of corn. The occurrence of both carotenoids in the macula provides its yellow color and is responsible for the coining of the expression "macula lutea." While it may seem striking to find carotenoids in the retina, the principle of using these substances in systems exposed to light and oxygen is not new. A carotenoid is present in the photosynthetic reaction center of a purple bacterium indicating an evolutionary use of carotenoids at these locations for many thousands of years[71].

Experimental background

That the presence of the macular carotenoids indeed may have some function in the retina was demonstrated in monkeys, the only other mammalian species which has a macula lutea with the same carotenoids in the yellow spot as humans.[72] If monkeys are raised on a carotenoid-free diet, their yellow spot disappears. Concomitant with this, more drusen, pathological manifestations in the retina, and more hyperfluorescence in the fluorescein angiogram of the retina are seen. Interestingly, these are the primary symptoms of developing AMD in humans.

There may be an important additional clue to the importance of carotenoids in the retina. The retina contains only two out of the 10 carotenoids occuring in the blood. In the serum there is a 10 fold greater concentration of lutein than zeaxanthin, conversely in the retina there is more zeaxanthin than lutein. A recent stereochemical analysis of lutein and zeaxanthin in the macula has indicated that the reason for this may be an intraretinal transformation of lutein into zeaxanthin.[73] This would be plausible, because zeaxanthin is a better singlet oxygen quencher than lutein. Furthermore, the yellow color of zeaxanthin is more intensive. As stated above, the yellow color of the macular carotenoids is an important element in their potential to protect against blue light damage. In this context it is important to note that the lens gradually becomes yellow during the normal aging process absorbing increasing amounts of blue light[74]. Up to approximately 30 years of age, the lens is virtually colorless, transmitting all colors equally, including the potentially damaging blue light. Therefore, we can conclude that the presence of macular yellow is particularly important during the early years before the lens builds up any considerable yellowing. Indeed it has been observed[75] that the yellow spot is already present during late intrauterine life while the neonatal lens is clear. On the other hand, the macular carotenoids may also be important in later life. It has been demonstrated that the antioxidant defense systems in the retina deteriorate with age.[76] Therefore, the antioxidant properties of the macular carotenoids could counteract this age-related loss of antioxidant potential in the retina. Whether this reasoning is clinically relevant will have to be demonstrated by prudently designed investigations.

Two hundred years after the discovery of the macular yellow pigments[77] our knowledge regarding the role of the macular carotenoids is considerable. The role of carotenoids in the retina in preventing damage caused by light and oxygen has been extensively reviewed recently.[78] More data have been collected, particularly in epidemiological studies, which will be the subject of the following chapter.

Epidemiological Studies

Vitale et al.[62] have investigated the use of vitamin supplements in 520 persons and correlated it to the presence of AMD. They found weak evidence for a protective effect, while Drews et al.[79] did not find any effect regarding AMD in a case-control study examining the dietary antioxidant intake of 60 patients. These results are contrasted to the findings of Mares-Perlmann et al.[80] who analyzed the intake of vitamins and minerals in a population of 2,000 people. They concluded that intake of zinc and antioxidant nutrients could protect against early AMD. Tsang et al. [81] have measured the plasma levels of vitamin E and selenium in 80 AMD patients and compared them to the values of 86 controls. On the basis of these micronutrients they could not find any significant differences between the groups. However, they did not measure carotenoid levels. Carotenoid plasma levels have been measured in the Eye Disease Case-Control Study[82] which to date is probably the most powerful study undertaken in this field. This case-control study investigated 421 patients with the neovascular form of AMD and compared them to 615 controls. While they did not find any significant effect of vitamin C, vitamin E or selenium individually, an antioxidant index combining vitamins, selenium and beta-carotene showed a significant risk reduction for AMD. When specifically analyzing various carotenoids, a highly significant ($p<0.001$) trend towards a lower risk for AMD with higher plasma concentrations was evident for lutein/zeaxanthin and beta-carotene. Within the framework of the same study intake evaluations were also conducted[83]. These indicated that the consumption of spinach was very strongly correlated with a lower risk of AMD. This result is interesting, because this vegetable is a very good source of lutein, one of the macular carotenoids.

To date no major controlled intervention trial investigating the role of vitamins and carotenoids in AMD prevention has been published. However, based on the evidence from epidemiological studies and on biological plausibility, the NEI has decided to include beta-carotene in the AREDS (Age Related Eye Disease Study) a major controlled clinical study to investigate factors preventing AMD and other eye diseases[65].

CONCLUSIONS

A large body of evidence has been accumulated from the results of recently published epidemiological studies consistently reporting an association of dietary intake of vitamin E and/or beta-carotene and a risk reduction for coronary heart disease. The results from the recently published Nurses' Health Study[8] and from the Health Professional Follow-up Study[9] strongly suggest that the vitamin E dosage required to reduce the risk of coronary heart disease may be 100-250 IU per day. This amount is much higher than the current RDA for vitamin E and cannot be provided by a prudent diet. Thus, the use of vitamin E supplements may be part of the efforts to decrease the risk for coronary heart disease.

A large number of epidemiological studies have consistently shown that beta-carotene may be beneficial in reducing the risk of lung and stomach cancer in particular. In addition, several intervention studies in patients suffering from leukoplakia demonstrated that beta-carotene can induce the regression of premalignant lesions. The recently published results from the Linxian Nutrition Intervention Trial provide further evidence that the use of antioxidant vitamins can reduce overall mortality and the mortality of stomach cancer, whereas another intervention trial, the so-called ATBC-Study, found no beneficial effect of alpha-tocopherol on the incidence of lung cancer in heavy male, finnish smokers. Whether the increased incidence of lung cancer in the men receiving beta-carotene was really due to chance as suggested by the authors requires further investigation.

Based on the natural occurrence of carotenoids in the retina and of vitamins C and E in the lens, AMD and cataract, two of the more important blinding diseases, are emerging as diseases the risk of which may be reduced through dietary adjustment or supplementation with carotenoids and vitamins. The epidemiological evidence available is promising and major controlled intervention trials are underway to test the preliminary findings.

The last decade has clearly shown that vitamins may exert many exciting effects beyond their classical role of avoiding deficiency. In this context the recent data provide further evidence that vitamin E and beta-carotene, which are generally recognized as safe as nutrients and as supplements,[84,85,86,87] may be promising candidates to improve preventive health care.

REFERENCES

1. H.C. Stary, Macrophages, macrophage foam cells, and eccentric intimal thickening in the coronary arteries of young children, Atherosclerosis 64:91 (1987.)
2. S.B. Hulley, J.M.B. Walsh, and T.B. Newman, Health policy on blood cholesterol: time to change directions, Circulation 86:1026 (1992).
3. M.S. Brown and J.L. Goldstein, Lipoprotein metabolism in the macrophage: implications for cholesterol deposition in atherosclerosis, Ann. Rev. Biochem. 52:223 (1983).
4. H. Esterbauer, M. Dieber-Rotheneder, M. Striegel, and G. Waeg, Role of vitamin E in preventing the oxidation of low-density lipoprotein, Am. J. Clin. Nutr. 53:314S (1991)
5. P.D. Reaven, A.Khouw, W.F. Beltz, S. Parthasarathy, and J.L. Witzum, Effect of dietary antioxidant combinations in humans, Arteriosclerosis and Thrombosis 13:590 (1993).
6. I. Jialal and S.M. Grundy, Effect of dietary supplementation with alpha-tocopherol on the oxidative modification of low density lipoprotein, J. Lipid Res. 33:899 (1992).
7. B.C. Blackman, P. White, W. Tsou, and D. Finkel, Peroxidation of plasma and platelet lipids in chronic cigarette smokers and insulin-dependent diabetics, Ann. N Y Acad. Sci. 435:385 (1984).
8. M.J. Stampfer, C.H. Hennekens, J.E. Manson, G.A. Colditz, B. Rosner, and W.C. Willett, Vitamin E consumption and the risk of coronary disease in women, N. Engl. J. Med. 328:1444 (1993).
9. E.B. Rimm, M.J. Stampfer, A. Ascherio, E. Giovannucci, G.A. Colditz, and W.C. Willett, N. Engl. J. Med. 328:1450 (1993).
10. P.T. Donnan, M. Thomson, F.G.R. Fowkes, R.J. Prescott, and E. Housley, Diet as a risk factor for peripheral arterial disease in the general population: The Edinburgh Artery Study, Am. J. Clin. Nutr. 57:917 (1993).
11. A.F.M. Kardinaal, F.J. Kok, J. Ringstad, J. Gomez-Aracena, V.P. Mazaev, L. Kohlmeier, B.C. Maetin, A. Aro, J.D. Kark, M. Delgado-Rodriguez, R.A. Riemersma, P. van t'Veer, J.K. Huttunen, and J.M. Martin-Moreno, Antioxidants in adipose tissue and risk of myocardial infarction:the EURAMIC study, Lancet 342:137 (1993).
12. K.F. Gey, H.B. Staehlin, and M. Eichholzer, Poor plasma status of carotene and vitamin C is associated with higher mortality from ischemic heart disease and stroke: Basel prospective study, Clin. Investig. 71:3 (1993).
13. H.W. Hense, M. Stender, W. Bors, and U. Keil, Lack of an association between serum vitamin E and myocardial infarction in a population with high vitamin supply, Atherosclerosis 103:21 (1993).
14. K.F. Gey, K.U. Moser, P. Jordan, H.B. Staehlin, M. Eichholzer, and E. Luedin, Increased risk of cardiovascular disease at suboptimal plasma concentrations of essential antioxidants: an epidemiological update with special attention to carotene and vitamin C, Am. J. Clin. Nutr. (Suppl.) 57:787S (1993).
15. H.M.G. Princen, G van Poppel, C.Vogelzang, R. Buytenhek, and F.J. Kok, Supplementation with vitamin, Arteriosclerosis and Thrombosis 12:554 (1992).
16. P.D. Livingston and C. Jones, Treatment of intermittent claudication with vitamin E, Lancet 2:602 (1958).
17. K. Haeger, Long-time treatment of intermittent claudication with vitamin E, Am. J. Clin. Nutr. 27:1179 (1974).
18. R.E. Gillilan, B. Mandell, and J.R. Warbasse, Quantitative evaluation of vitamin E in the treatment of angina pectoris, Am. Heart J. 93:444 (1977).
19. J.M. Gaziano, J.E. Manson, P.M. Ridker, J.E. Buring, and C.H. Hennekens, Beta Carotene therapie for chronic stable angina, Circulation 82 (Suppl 4):III-201 (1990).
20. S.J. DeMaio, S.B. King, N.J. Lembo, G.S. Roubin, J.A. Hearn, H.N. Bhagavan, and D.S. Sgoutas, Vitamin E supplementation, plasma lipids and incidence of restenosis after percutaneous transluminal coronary, J. Am. Coll. Nutr. 11:68 (1992).
21. K. Umegaki, S. Ikegami, K. Inoue, T. Ichikawa, S. Kobayashi, and N. Soeno, Beta-carotene prevents x-ray induction of micronuclei in human lymphocytes, Am. J. Clin. Nutr. 59:409 (1994).
22. V.N. Singh and S.K. Gaby, Premalignant lesions: role of antioxidant vitamins and beta-carotene in risk reduction and prevention of malignant transformation, Am. J. Clin. Nutr. 53:386 (1991).
23. H. Stich and M.P. Rosin, Micronuclei in exfoliated human cells as a tool for studies in cancer risk and cancer intervention, Cancer Lett. 22:241 (1984).

24. L.X. Zhang, R.V. Cooney, and J.S. Bertram, Carotenoids enhance gap junctional communication and inhibit lipid peroxidation in C3H/10 T1/2 cells: relationship to their cancer chemopreventive action, Carcinogenesis 12:2109 (1991).
25. J.S. Bertram, Inhibition of chemically induced neoplastic transformation by carotenoids, NY Acad. Sci. 686:161 (1993).
26. M.M. Mathews-Roth and N.I. Krinsky, Carotenoids affect development of UV-B induced skin cancer, Photochem. Photobiol. 46:507 (1987).
27. L. Santamaria, A. Bianchi, A. Arnaboldi, C. Ravetto, L. Bianchi, R. Pizzala, L. Andreoni, G. Santagati, and P. Bermond, Chemoprevention of indirect and direct chemical carcinogenesisby carotenoids as oxygen radical quenchers, Ann. NY Acad. Sci. 534:585 (1988).
28. E.J. Rousseau, A.J. Davidson, and B. Dunn, Protection by beta-carotene and related compounds against oxygen-mediated cytotoxity and genotoxity: implications for carcinogenesis and anticarcinogenesis, Fr. Rad. Biol. & Med. 13:407 (1992).
29. G. Poppel, Carotenoids and cancer: an update with emphasis on human intervention studies, Eur. J. Cancer 29A:1335 (1993).
30. M. Eichholzer, H.B. Staehlin, and K.F. Grey, Inverse correlation between essential antioxidants in plasma and subsequent risk to develop cancer, ischemic heart disease and stroke respectively: 12-year follow-up of the Prospective Basel Study, Fr. Rad. and Aging , 398 (1992).
31. J. Chen, C. Geissler, B. Papria, J. Li, and T.C. Campell, Antioxidant status and cancer mortality in China, Int. J. Epidemiol. 21:625 (1992).
32. J.-Y. Li, .B-Q. Liu, and G.-Y. Li, Atlas of cancer mortality in the people's Republic of China: an aid for cancer control and research, Int. J. Epidemiol. 10:127 (1981).
33. A. Bendich, Carotenoids and immunity, Clin. Appl. Nutr. 1:45 (1991).
34. R.K. Chandra, Effect of vitamin and trace-element supplementation on immune responses and infection in elderly subjects, Lancet 340:1124 (1992).
35. R.R. Watson, R.H. Prabhala, P.M. Plezia, and D.S. Alberts, Effect of beta-carotene on lymphocyte subpopulations in elderly humans: evidence for a dose-response relationship, Am. J. Clin. Nutr. 53:90 (1991).
36. H.S. Garewal, Beta-carotene and vitamin E in oral cancer, J. Cell. Biol. 17F, Suppl:262 (1993).
37. J.Y. Li, P.R. Taylor, B. Li, S. Dawsey, G.-Q. Wang, A.G. Ershow, W. Guo, S.-F. Liu,C.S. Yang, Q. Sheng, W. Wang, S.D. Mark,X.-N. Zou, P. Greenwald, Y.-P. Wu, and W.J. Blot, Nutrition intervention trials in Linxian, China: multiple viatamin/mineral supplementation, cancer incidence, and disease-specific mortality among adults with esophagal dysplasia, J. Natl. Cancer Inst. 85:1492 (1993).
38. W.J. Blot, J.-Y. Li, P.R. Taylor, W. Guo, S. Dawsey, G.-Q. Wang, C.S. Yang, S.-F. Zheng, M. Gail, G.-Y. Li, Y. Yu, B.-Q. Liu, J. Tangrea, Y.-H. Sun, F. Liu, J.F. Fraumeni, Y.-H. Zhang, and B. Li, Nutrition Intervention trials in Linxian, China: supplementation with specific vitamin/mineral combinations, cancer incidence, and disease-specific mortality in the general population, J. Natl. Can. Inst. 85:148 (1993).
38a. The Alpha-Tochopherol, Beta Carotene Cancer Prevention Study Group, The effect of vitamin E and beta carotene on the incidence of lung cnacer and other cancers in male smokers, N. Engl. J. Med. 330:1029 (1994).
38b. C.H. Hennekens, J.E. Buring, and R. Peto, Antioxidant vitamins - benefits not yet proved, N. Engl. J. Med. 330:1080 (1994)
39. J. M.C. Gutteridge, Invited review, free radicals in disease processes, a compilation of cause and consequence, Free Rad. Res. Comms. 19:141-158 (1993).
40. P. Dayhaw-Barker, Ocular photosensitation, Photochem. Photobiol. 46:1051-1055 (1987).
41. M.J. Fryer, Review article, evidence for the photoprotective effects of vitamins E, Photochem. Photobiol. 58:304-312 (1993).
42. A.J. Augustin, W. Breipohl, T. Boker, and A. Wegener, Evidence for the prevention of oxidative tissue damage in the inner eye by vitamins E and C, Ger. J. Ophthalmol. 1:394-398 (1992).
43. P. Palozza, S. Moualla, and N. Krinsky, Effects of β-carotene and alpha-tocopherol on radical initiated peroxidation of microsomes, Free Rad. Biol. Med. 13:127-136 (1992).
44. K. Kirschfeld, Carotenoid pigments: their role in protecting against photooxidation in eyes and photoreceptor cells, Proc. R. Soc. London (B) 216:71-85 (1982).
45. M.M. Kini, H.M. Leibovitz, T. Colton, and R.J. Nickerson, Prevalence of senile cataract, diabetic retinopathy, senile macular degeneration and open-angle glaucoma in the Framingham Eye Study, Am. J. Ophthalmol. 85:28-34 (1978).
46. J. McBride, Avoiding cataracts, Agricultural Res. 38:22-23 (1990).
47. The Italian-American cataract study group risk factors for age-related cortical, nuclear, and posterior subcapsular cataracts, Am. J. Epidemiol. 133: 541-552 (1991).
48. L. Wong, S.C. Ho, D. Coggon, A.M. Cruddas, C.H. Hwang, C.P. Ho, A.M. Robertshaw, and D.M. MacDonald, Sunlight exposure, antioxidant status, and cataract in Hong Kong fishermen, J. Epidemiol. Community Health 47:46-49 (1993).
49. J.V. Ferrer, J. Sastre, F.V. Pallardo, M. Asensi, V. Anton, J.M. Estrela, J. Vina, and J. Miquel, Senile cataract: a review on free radical related pathogenesis and antioxidant prevention, Arch. Gerontol. Geriatr. 13:51-59 (1991).

50. D. Armstrong, G. Santangelo, and E. Connole, The distribution of peroxide regulating enzymes in canine eye, Curr. Eye Res. 1:225-242 (1981).
51. G.R. Reiss, P.G. Werness, P.E. Zollmann, and R.F. Brubaker, Ascorbic acid levels in the aqueous humor of nocturnal and diurnal mammals, Arch. Ophthalmol. 104:753-755 (1986).
52. H. Gerster, Antioxidant vitamins in cataract prevention, Z Ernaehrungwiss 28:56-75 (1989).
53. L.J. Alexander, Ocular vitamin therapy, Optom-Clin. 2:1-34 (1992).
54. A. Taylor, Cataract: relationships between nutrition and oxidation, J. Am. Coll. Nutr. 2:138-146 (1993).
55. P.F. Jacques, S.C. Hartz, L.T. Chylack Jr., R.B. McGandy, and J.A. Sadowski, Nutritional status in persons with and without senile cataract: blood vitamin and mineral levels, Am. J. Clin. Nutr. 48: 152-158 (1988).
56. P.F. Jacques, L.T. Chylack Jr, R.B. McGandy, and S.C. Hartz, Antioxidant status in persons with and without senile cataract, Arch. Ophthalmol. 106:337-340 (1988).
57. S.E. Hankinson, M.J. Stampfer, J.M. Seddon, G.A. Colditz, B. Rosner, F.E. Speizer, and W.C. Willett, Nutrient intake and cataract extraction in women: a prospective study, Br. Med. J. 305:335-339 (1992).
58. H. Hess, and J.S. Zigler, Retina-lens interaction in genesis of cataracts in RCS rats and prevention by dietary supplementation with beta-caroptene and vitamin E, Invest. Ophthalmol. Vis. Sci. (Suppl.) 32:1100 (1991).
59. J. Mohan, R.D. Sperduto, S.K.K. Angra, et al, India-US case-control study of age-related cataracts, Arch. Ophthalmol. 107: 670-676 (1989).
60. P. Knekt, M. Heliovaara, A. Rissanen, A. Aromaa, and R.K. Aaran, Serum antioxidant vitamins and risk of cataract, Brit. J. Med. 305:1392-1394 (1992).
61. A.R. Schoenfeld, M.C. Leske, and S-Y Wu, Recent epidemiological studies on nutrition and cataract in India, Italy and the United States, J. Am. Coll. Nutr. 12:521-526 (1993).
62. S. Vitale, S. West, J. Hallfrisch, C. Alston, F. Wang, C. Moorman, D. Muller, V. Singh, and H.R. Taylor, Plasma antioxidants and risk of cortical and nuclear cataract, Epidemiol. 4: 195-203 (1993).
63. M. Belpoliti, G. Maraini, G. Albert, R. Corona, and S. Crateri, Enzyme activities in human lens epithelium of age-related cataract, Invest. Ophthalmol. Invest. Sci. 34:2843-2847 (1993).
64. R.D. Sperduto, T-S Hu, R.C. Milton, J-L Zhao, D.F. Everett, Q-F, Cheng, W.J. Blot, L. Bing, P.R. Taylor, L. Jun-Yao, S. Dawsey, and W-D Guo, The Linxian cataract studies: two nutrition intervention trials, Arch. Ophthalmol. 111:1246-1253 (1993).
65. AREDS, Manual of procedures, National Eye Institute, Bethesda, MD (1992).
66. W.T. Ham and H.A. Mueller, The photopathology and nature of the blue light and near-UV retinal lesions produced by lasers and other optical sources, *in*: " Laser Applications in Medicine and Biology, "M.L. Wolbarsht, ed., Plenum Publishing Co., New York (1989).
67. B.J. Putting, R.C.V.J. Zweypfennig, G.F.J.M. Vrensen, J.A. Oosterhuis, and J.A. Best, Blood-retinal barrier dysfunction at the pigment epithelium induced by blue light, Invest. Ophthalmol. Vis. Sci. 33:3385-3393 (1992).
68. F.J.G.M. Kujik and P. Buck, Fatty acid composition of the human macula and peripheral retina, Invest. Ophthalmol. Invest. Sci. 33:3493-3496 (1992).
69. P.F. Conn, W. Schalch, and T.G. Truscott, The singlet oxygen and carotenoid interaction, J. Photochem. Photobiol. B: Biol. 11:41-47 (1991).
70. R.A. Bone, J.T. Landrum, and A. Cain, Optical density spectra of the macular pigment in vivo and in vitro, Vision Research 32:105-110 (1992).
71. J. Deisenhofer and H. Michel, The photosynthetic reaction center from the purple bacterium Rhodopseudomonas viridis, Chemica. Scripta. 29:205-220 (1989).
72. M.R. Malinow, L. Feeney-Burns, L.H. Peterson, M.L. Klein, and M. Neuringer, Diet-related macular anomalies in monkeys, Invest. Ophthalmol. Vis. Sci. 19:857-863 (1980).
73. R.A. Bone, J.T. Landrum, G.W. Hime, A. Cains, and J. Zamor, Stereochemistry of the human macular carotenoids, Invest. Optham. Vis. Sci. 34:2033-2040 (1993).
74. S. Lerman, An experimental and clinical evaluation of lens transparency and aging, J. Gerontol. 38:293-301 (1983).
75. R.A. Bone, J.T. Landrum, L. Fernandez, and S.L. Tarsis Analysis of the macular pigment by HPLC: retinal distribution and age study, Invest. Ophthalmol. Vis. Sci. 29: 843-849 (1988)
76. C. Castorina, A. Campisi, C. di Giacomo, V. Sorrenti, A. Russo, and A. Vanella, Lipid peroxidation and antioxidant enzymatic systems in rat retina as a function of age, Neurochem. Res. 17: 599-604 (1992).
77. J.J. Nussbaum, R.C. Pruett, and F.C. Delori, Historic perspectives, macular yellow pigment, the first 200 years, Retina 1:296-310 (1981).
78. W. Schalch, Carotenoids in the retina, a review of their possible role in preventing or limiting damage caused by light and oxygen free radicals and aging, *in:* "Free Radicals and Aging," I. Emerit and B. Chance, eds., Birkhäuser, Basel (1992).
79. C.D. Drews, P. Sternberg, P.S. Samiec, D.P. Jones, R.L. Reed, E. Flagg, A. Boddie, and R. Tinkelman, Dietary antioxidants and age related macular degeneration (ARMD), Invest. Ophthalmol. & Vis. Sci. (Suppl.) 34:1158 (1993).
80. J.A. Mares-Perlman, R. Klein, B.E.K. Klein, and L.L. Ritter, Relationships between age-related maculopathy and intake of vitamin and mineral supplements, Invest. Ophthalmol. & Vis. Sci. (Suppl.) 34:1133 (1993).

81. N.C.K. Tsang, P.L. Penfold, P.J. Snitch, and F. Billson, Serum levels of antioxidants and age-related macular degeneration, Doc. Opthalmol. 81:387-400 (1992).
82. Eye Disease Case-Control Study Group, Antioxidant status and neovascular age-related macular degeneration, Arch. Opthalmol. 111:104-109 (1993).
83. J. Seddon, U. Ajani, R. Sperduto, L. Yannuzzi, T. Burton, J. Haller, N. Blair N., M. Farber, D. Miller, E. oudas, W. Willet, and the EDCCS Group, Dietary antioxidant status and age related macular degeneration: a multicenter study, Invest. Ophthalmol. Vis. Sci. (Suppl) 34:1134 (1993).
84. A. Bendich and L.J. Machlin, The safety of oral intake of vitamin E: data from clinical studies from 1986-1991, *in*: "Vitamin E in Health and Disease," L. Packer and J. Fuchs, eds., Marcel Dekker, Inc., New York, (1993).
85. H. Gerster, Anticarcinogenic effect of common carotenoids, Internat. J. Vit. Nutr. Res. 63:93 (1993).
86. A. Bendich and L. Langseth, Safety of vitamin A, Am. J. Clin. Nutr. 49:358 (1989).
87. M.M. Mathews-Roth, Lack of genotoxiticity with beta-carotene, Toxicology Letters 41:185 (1988).

FREE RADICAL SCAVENGING AND ANTIOXIDANT ACTIVITY OF PLANT FLAVONOIDS

Chithan Kandaswami and Elliott Middleton, Jr.

Department of Medicine
School of Medicine and Biomedical Sciences
State University of New York at Buffalo
Buffalo, NY 14203

INTRODUCTION

The occurrence of reactive oxygen species (ROS), termed as prooxidants, is a characteristic of normal aerobic organisms. The term "reactive oxygen species" collectively denotes oxygen-centered radicals such as superoxide ($O_2^{\bullet -}$)and hydroxyl ($^{\bullet}OH$), as well as nonradical species derived from oxygen, such as hydrogen peroxide (H_2O_2), singlet oxygen ($^1\Delta gO_2$) and hypochlorous acid (HOCl). Radical reactions are central to the maintenance of homeostasis in biological systems. Radical species perform a cardinal role in many physiological processes such as cytochrome P450-mediated oxidative transformation reactions, a plethora of enzymic oxidation reactions, oxidative phosphorylation, regulation of the tone of smooth muscle, and killing of microorganisms.[1-3] Excessive generation of free radicals can have deleterious biological consequences.[4-6] Organisms are equipped with an armamentarium of defense systems, termed antioxidants in order to safeguard them against the onslaught of ROS.[1-3,7] When the generation of prooxidants overwhelms the capacity of antioxidant defense systems oxidative stress ensues. This can cause tissue damage leading to pathophysiological events. ROS play a pivotal role in the action of numerous foreign compounds (xenobiotics). Their increased production seems to accompany most forms of tissue injury.[4,5] Whether sustained and increased production of ROS is a primary event in human disease progression or a secondary consequence of tissue injury has been discussed.[5,6] Whatever may be the case, the formation of free radicals has been implicated in a multitude of disease states ranging from inflammatory/immune injury to myocardial infarction and cancer.

Some of the well known detrimental effects of excessive generation of ROS in biological systems include peroxidation of membrane lipids, oxidative damage to nucleic acids and carbohydrates, and the oxidation of sulfhydryl and other susceptible groups in proteins.[1-4,5] Oxygen-derived free radicals appear to possess the propensity to initiate as well as to promote carcinogenesis. Currently there is heightened interest in the role of ROS in atherosclerosis, stroke, myocardial infarction, trauma, arthritis, ischemia/reoxygenation injury and cancer. [4,5] The involvement of ROS in aging and in many chronic diseases has been considered. The defense provided by antioxidant systems is crucial to the survival of organisms. Detoxification of ROS in the cell is provided by both enzymic and nonenzymic systems which constitute the antioxidant defense systems. Enzymic systems include extensively-studied enzymes such as superoxide dimutase (SOD), catalase, glutathione peroxidases, D-T diaphorase, and glutathione-regenerating enzyme systems, among others.[1,2,7] Some enzymic systems such as SOD and catalase act specifically against ROS,

Free Radicals in Diagnostic Medicine, Edited by
D. Armstrong, Plenum Press, New York, 1994

while certain other enzyme systems reduce thiols. Nonenzymic antioxidants are less specific and can also scavenge other radicals, both organic and inorganic. These antioxidants can be classified as water-soluble or lipid-soluble, depending on whether they act primarily in the aqueous phase or in the lipophilic region of the cell membranes. Hydrophilic antioxidants include ascorbic acid and urate. Ubiquinols, retinoids, carotenoids, flavonoids and tocophenols (vitamin E) are some of the lipid-soluble antioxidants. Plasma proteins,glutathione (GSH), urate, etc. are some of the endogenous antioxidants, while ascorbic acid, carotenoids, retinoids, flavonoids and tocopherols constitute some of the dietary antioxidants. These compounds possess the potential to scavenge and quench various radicals (oxygen-centered, carbon-centered, alkoxyl-, peroxyl- or phenoxyl radicals) and ROS. Certain radical scavengers are not recyclable while others are recycled through the intervention of a series of enzyme systems or other nonenzymic antioxidant systems.

FLAVONOIDS AND THEIR BIOLOGICAL EFFECTS

The flavonoids (phenylchromones) are an unusually large group of naturally occurring phenolic compounds ubiquitously distributed in the plant kingdom.[8-12] These aromatic compounds are formed in plants from the aromatic amino acids, phenylalanine and tyrosine, and acetate units.[13,14] Phenylalanine and tyrosine are converted to cinnamic acid and *p*-coumaric acid which condense with acetate units to form the cinnamoyl structure (cinnamoyl fragment) of the flavonoids. A variety of phenolic acids such as caffeic acid, ferulic acid, and chlorogenic acid are cinnamic acid derivatives. Biotransformation of flavonoids in the gut can release these cinnamic acid derivatives (phenolic acids).[15] The flavonoids are phenylchromones (benzo-γ-pyrones) (Fig. 1). A phenyl group is usually substituted at the 2-position of the pyrone ring. In isoflavonoids, the substitution is at the 3-position. Flavonoids are complex and highly evolved molecules with intricate chemical variation. In plants, they generally occur as glycosylated derivatives. Flavonoids and tocopherols (vitamin E) share a common structure, namely the chromane ring. As of 1985, over 4000 chemically unique flavonoids have been identified in plants. This is only a fraction of the total number that are likely to be present in nature, since only a few percent of plant species have been systematically examined for their flavonoid constituents.[12] Unlike the alkaloids which are restricted to about 20 percent of the flowering plant species, flavonoids occur universally in vascular plants.[12]

The flavonoids are generally known to be plant, flower, leaf and fruit pigments.[16] They are responsible for the brilliant shades of blue, scarlet, orange, etc. in flowers and fruits, and leaves. They are found in various fruits, vegetables, nuts, seeds, grains, spices and herbs, as well as in beverages such as tea, cocoa and wine.[17-19] Dietary exposure to flavonoids is significant. The average diet in the U.K. and U.S.A. may contain up to 1 gram of mixed flavonoids per day.[18,20] Their dietary intake far exceeds that of vitamin E, a monophenolic antioxidant, and of β-carotene.[21]

The past two decades have witnessed a renewed interest in the study of flavonoids and of their biological interactions with mammalian cells and tissues. Flavonoids are known to display a bewildering array of pharmacological and biochemical actions.[22-23] They have long been recognized to possess antiinflammatory, antiallergic, antimicrobial, antihelminthic, hepatoprotective, antihormonal, antithrombotic, antiviral, antimutagenic/anticarcinogenic and antineoplastic activities. The pleiotropic and potentially health-promoting effects of several flavonoids have come to be appreciated in many experimental situations. The flavonoids are typical phenolic compounds and therefore act as potent metal chelators and free radical scavengers.[19,24-27] They are powerful chain-breaking antioxidants. The biological advantages of the ability of flavonoids to safely sequester and thus reduce the activity of the deleterious oxidant-inducing metals such as iron and copper cannot be overemphasized. Interestingly, flavonoids are known to possess vitamin C stabilizing and antioxidant-dependent vitamin C-sparing activities.[19,25,28-30] They are also known to increase the absorption of vitamin C.

Flavonoids are known to modify the activities of a host of enzyme systems including protein kinase C and various other kinases, protein tyrosine kinase, aldose reductase, myeloperoxidase, NADPH oxidase, xanthine oxidase, phospholipase A_2, phospholipase C,

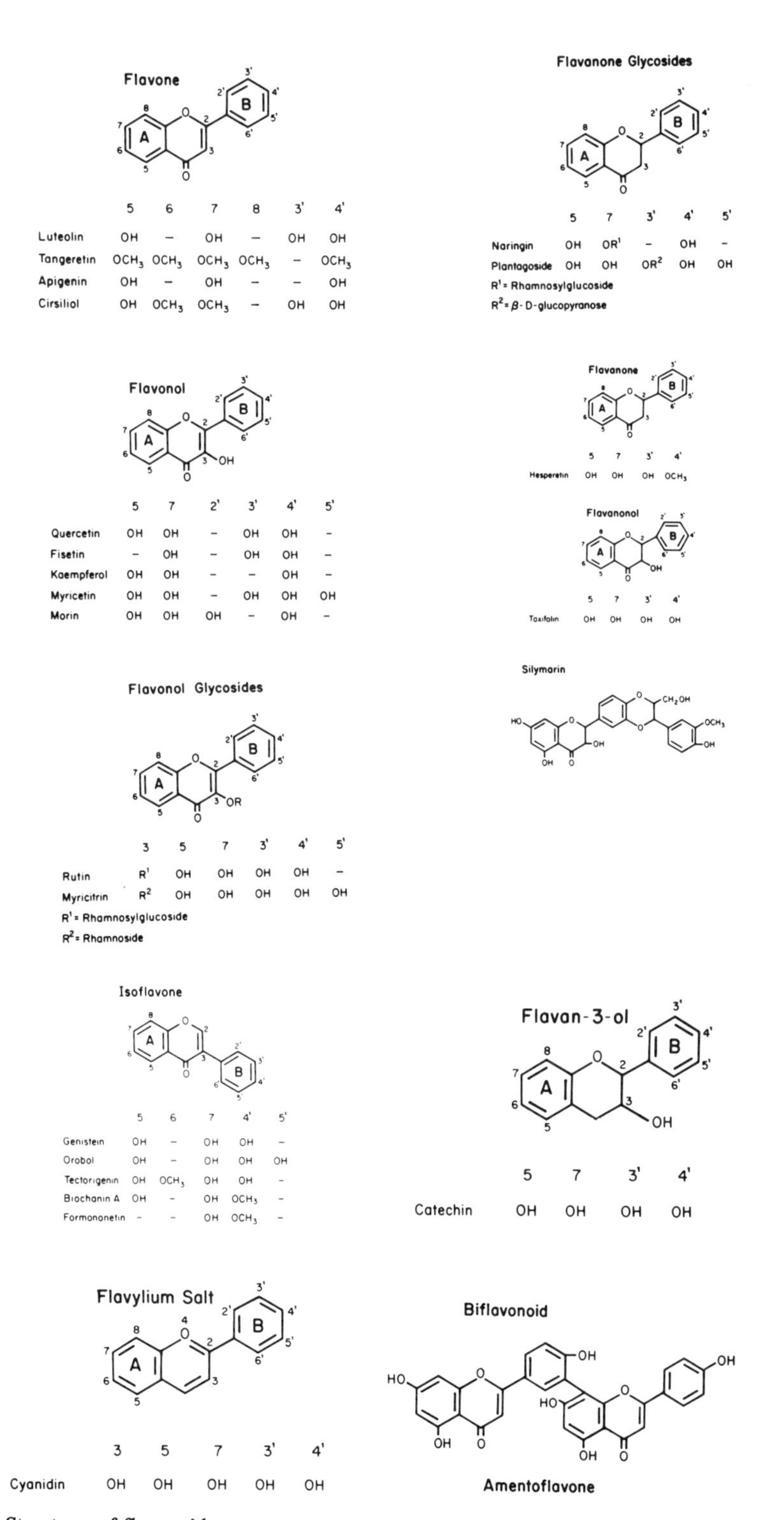

Figure 1. Structures of flavonoids

reverse transcriptases, ornithine decarboxylase, sialidase, several ATPases, nucleotide phosphodiesterases, lipoxygenases, cyclooxygenase, cytochrome P450-dependent mixed-function oxidases, epoxide hydrolase, glutathione-S-transferase, and aromatase, amongst others.[22] Some of these enzyme systems are critically involved in immune function, carcinogenesis, cellular transformation, and tumor growth and metastasis. Studies with mammalian cell systems indicate that different flavonoids can alter the function of mast cells, basophils, neutrophils, eosinophils, macrophages/monocytes, B and T lymphocytes, platelets, nerve, smooth muscle and various cancer cells. The physiologic and pathologic processes affected by flavonoids are diverse and numerous, and include secretion, mitogenesis, platelet aggregation and adhesion to endothelial surfaces, cell motility and malignant cell proliferation, cancer metastasis and, the function/expression of adhesion molecules in various mammalian cell types.[22] The antioxidant function and enzyme-modifying actions of flavonoids could account for many of their pharmacological activities.

With this general background the antioxidant activity of flavonoids will be discussed below. ROS that can be scavenged or whose formation can be inhibited by flavonoids are shown in Table 1.

Table 1. Reactive oxygen species that can be scavenged or whose formation can be inhibited by flavonoids

$O_2^{\bullet-}$, superoxide anion	One electron reduction product of O_2. Produced by phagocytes. Formed in autoxidation reactions (flavoproteins, redox cycling). Generated by oxidases (heme proteins).
$HO_2^{\bullet}$, perhydroxy radical	Protonated form of $O_2^{\bullet-}$
H_2O_2, hydrogen peroxide	Two-electron reduction product of O_2. Formed from $O_2^{\bullet-}$($HO_2^{\bullet}$) by dismutation or directly from O_2. Reactivity of $O_2^{\bullet-}$ and H_2O_2 amplified in the presence of heme proteins.
$^{\bullet}OH$, hydroxyl radical	Three-electron reduction product of O_2. Generated by Fenton reaction, transition metal (iron, copper) - catalyzed Haber-Weiss reaction. Formed by decomposition of peroxynitrite produced by the reaction of $O_2^{\bullet-}$ with $NO^{\bullet}$ (nitric oxide radical).
$RO^{\bullet}$, alkoxyl radical	Example: Lipid radical ($LO^{\bullet}$)
$ROO^{\bullet}$ peroxyl radical	Example: Lipid peroxy radical ($LOO^{\bullet}$). Produced from organic hydroperoxide (e.g. lipid hydroperoxide, LOOH), ROOH by hydrogen abstraction.
$^1\Delta gO_2$(also $^1O_2$)	Singlet molecular oxygen.

INFLUENCE OF FLAVONOIDS ON ROS PRODUCTION BY PHAGOCYTIC CELLS

Phagocytosis is an important physiological process which accompanies the functional production of $O_2^{\bullet-}$. Activated phagocytic cells such as monocytes, neutrophils, eosinophils and macrophages generate $O_2^{\bullet-}$.[31,32] Radical production by phagocytes is extremely important for their bacteriocidal and tumoricidal functions . Phagocytosis is accompanied by a dramatic increase in oxygen consumption (respiratory burst) with the attendant production of $O_2^{\bullet-}$, catalyzed by a membrane-bound NADPH oxidase system.[31,32]

$O_2^{\bullet-}$ generated by phagocytes dismutates to H_2O_2, a fairly unreactive molecule, which in turn gives rise to $^{\bullet}OH$ by reaction with transition metal ions.[5,6] This radical is extremely reactive and is one of the strongest oxidizing agents. The enzyme myeloperoxidase (MPO)

provides another bacterial killing mechanism in neutrophils.[33] This enzyme catalyzes the oxidation of chloride ions by H_2O_2 resulting in the formation of HOCl, a powerful bacteriocidal agent.

Even though $O_2^{\bullet-}$ is far less reactive than $^{\bullet}OH$, it can attack a number of biological targets. It can react with nitric oxide ($NO^{\bullet}$), a reactive free radical produced by cells such as phagocytes and vascular endothelial cells, to yield an even more reactive species, peroxynitrite,[34] which can decompose to form $^{\bullet}OH$ in a reaction independent of transition metal ions.[35] $NO^{\bullet}$ is considered to be a smooth muscle relaxing factor. Endothelium-derived relaxing factor (EDRF), an important mediator of vasodilator responses, has been identified to be $NO^{\bullet}$.[36,37] $O_2^{\bullet-}$ has been reported to react with $NO^{\bullet}$ and inhibit its action.[38] By impairing the physiological function of $NO^{\bullet}$, $O_2^{\bullet-}$ can act as a vasoconstrictor which could have deleterious consequences in some clinical situations.[39]

While ROS generated by phagocytes play an important physiological function, they can also cause cellular damage. These are generated sequentially starting with $O_2^{\bullet-}$ production by a membrane-bound NADPH oxidase.[31,32] The highly reactive oxygen metabolites along with other mediators elaborated by neutrophils and macrophages can promote inflammation and cause tissue damage.[40] Busse *et al*[41] showed that flavonoids inhibited ROS release (as assayed by the production of luminol-dependent chemiluminescence) by human neutrophils. Quercetin and other flavonoids were quite effective inhibitors of $O_2^{\bullet-}$ production by the cells. 'T Hart *et al*[42] recently reported a similar inhibitory effect of different flavonoids on ROS production by activated human neutrophils using the chemiluminescence method. Four selected flavonoids inhibited MPO release, while two of these strongly inhibited this activity. Considering luminol-dependent chemiluminescence production by neutrophils to be an MPO-dependent process, these authors contend that these effects might mask the effects of flavonoids on ROS production. Using the luminescent probe lucigenin for the exclusive detection of $O_2^{\bullet-}$ release, 'T Hart *et al*[42] showed that the release of this species by human neutrophils was inhibited by flavonoids. Essential determinants for inhibition of $O_2^{\bullet-}$ release appear to be the OH groups located in the B-ring of the flavonoid molecule. The formation of $O_2^{\bullet-}$ is dependent on the activation of NADPH oxidase localized in the plasma membrane, which is subject to flavonoid inhibition.[43] The inhibition of protein kinase C by flavonoids can also be implicated in the impairment of the NADPH oxidase activation.[44]

In addition to inhibiting the activity of a purified human neutrophil MPO, quercetin was also found to depress this activity in a system employing intact human neutrophils.[45] In this case, in inhibiting the activity of the purified MPO, quercetin was significantly more potent than methimazole, a specific inhibitor of MPO.[46] In addition, quercetin was found to have an ability to directly scavenge HOCl, a highly reactive chlorinated species generated by the MPO-H_2O_2-Cl system.[45] Flavonoids can inhibit the formation of $O_2^{\bullet-}$ and the generation of $^{\bullet}OH$ radicals, in addition to their radical-scavenging effects. The inhibition of neutrophil MPO activity by flavonoids can result in the impairment of ROS production. Such impairment could diminish the formation of highly toxic HOCl and the hypochlorite ion (OCl^-). A consequence of this would be a decrease in the inactivation of alpha-1-antitrypsin, which could result in turn in the enhanced inactivation of neutrophil-derived and other tissue-damaging enzymes.[47] Quercetin was found to be a potent inhibitor of human neutrophil degranulation and $O_2^{\bullet-}$ production induced by different secretogogues.[48] Quercetin also inhibited the phosphorylation of neutrophil proteins accompanying neutrophil activation by phorbol myristic acetate. Phosphorylation of a specific neutrophil protein (MW 67,000) was reported to be particularly sensitive to quercetin at concentrations that also diminished neutrophil degranulation and $O_2^{\bullet-}$ production suggesting thereby that the phosphorylation of this particular protein is an important intracellular event associated with neutrophil activation.[48]

Ogasawara *et al*[49] described inhibition of anti-IgE-induced H_2O_2 generation and of human basophil histamine release by quercetin, apigenin and taxifolin. All three flavonoids inhibited the generation of H_2O_2 but only quercetin and apigenin inhibited anti-IgE-induced histamine release. The results suggest that quercetin and apigenin possess the structural features necessary for inhibition of histamine secretion whereas all three compounds possess structural features required for inhibition of H_2O_2 generation, suggesting a scavenging activity of the flavonoids.

EFFECT OF FLAVONOIDS ON LIPID PEROXIDATION AND OXY-RADICAL PRODUCTION

Oxidative stress can damage many biological molecules. Proteins and DNA appear to be some significant targets of cellular injury. Another target of free radical attack in biological system is the lipids of cell membranes.[5,50] Highly reactive radicals such as •OH have the propensity to attack biological molecules by abstracting hydrogen. The most widely studied oxidative damage caused by •OH is its capacity to initiate the free radical chain reaction, lipid peroxidation. This damage readily ensues when •OH radicals abstract a hydrogen atom from a methylene carbon of a fatty acid or fatty acid side chain of a lipid, for instance. The lipids initially attacked by free radicals consequently become oxidized to lipid peroxides. Lipid peroxides are potentially toxic and possess the capacity to damage most cells.[4-6,50] Accumulation of lipid peroxides has been reported in altherosclerotic plaques, in brain tissues damaged by trauma or oxygen deprivation, and in tissues poisoned by toxins. The idea that lipid peroxidation is often a secondary event consequent to primary cell damage induced by oxidative stress has been discussed.[50] Rises in intracellular "free" Ca^{2+}, protein and DNA damage, and abnormalities in cellular metabolism produced by oxidative stress have been considered to be more important in causing cellular injury than is the peroxidation of membrane lipids.[50]

Whether lipid peroxidation is a primary event produced by oxidative stress or a consequence of tissue damage, it can still be biologically important in exacerbating tissue injury, in view of the potential cytotoxicity and genotoxicity of the end products of lipid peroxidation.[51] Lipid peroxidation products originating from dying cells could exert a cancer promotional effect.[52] Great emphasis has recently been placed on the significant contribution of lipid peroxidation to the development of atherosclerosis, stroke and myocardial infarction, and to the deterioration of the brain or spinal cord that occurs following trauma or ischemia.[4] Lipid peroxidation has also been implicated in several pathologic conditions including aging, hepatotoxicity, hemolysis, cancer, tumor promotion, inflammation and iron toxicity.[53-56]

Several flavonoids have been reported to inhibit either enzymic or nonenzymic lipid peroxidation. Flavonoids such as quercetin (flavonols) can suppress lipid peroxidation in model systems[57] as well as in several biological systems, such as mitochondria and microsomes,[58,59] chloroplasts[60] and erythrocytes.[61,62] Several studies have reported the inhibitory effects of (+) catechin, quercetin and other flavonoids on *in vitro* lipid peroxidation,[63-67] generally assessed by measuring colorimetrically the formation of thiobarbituric acid-reactive substance.

Bindoli *et al*[58] demonstrated that silymarin, a 3-OH flavanone present in *Silybum marianum* (the European milk thistle), protected rat liver mitochondria and microsomes from lipid peroxide formation induced by Fe^{2+}- ascorbate and NADPH-Fe^{3+}- ADP systems. Its antiperoxidative action was 10-fold higher than that of β-tocopherol at micromolar concentrations. While the impairment of enzymic lipid peroxidation by the flavone might involve its effect on the cytochrome P450 system, inhibition of non-enzymic lipid peroxidation has been considered to involve interaction of silymarin with free radical species responsible for lipid peroxidation.[58] Silymarin and silybin are some flavonoid compounds isolated from the European milk thistle. Cavellini *et al*[59] reported that the inhibitory activity of silybin was superior to that of flavonoids even with *o*- dihydroxy or trihydroxy substitution patterns. Soybean isoflavones have been examined for their antioxidative potency by measuring the extent of inhibition of soybean lipoxygenase action and by the ability to prevent peroxidative hemolysis of sheep, rat and rabbit erythrocytes.[68] The extent of inhibition of the enzyme activity was positively correlated with the number of hydroxyl groups in the isoflavone nucleus. A number of isoflavones and their reduced derivatives (isoflavanones and isoflavans) were examined for inhibitory effects on lipid peroxidation in rat liver microsomes.[69] The parent isoflavones and the isoflavans were by far the most potent inhibitors. Some isoflavans (6,7,4'-trihydroxy- and 6,7-dihydroxy-4-

methoxyisoflavones) excelled α-tocopheral and butylated hydroxyanisole (a synthetic antioxidant) in terms of its inhibitory effect. The 6,7-dihydroxylated isoflavans were 80 times stronger than α-tocopherol in inhibiting lipid peroxidation. Methylation of the C_7-OH of the isoflavones did not reduce the inhibitory effect, while methylation of the C_6-OH group or both hydroxyl groups (C_6 and C_7) resulted in lowered inhibition. The position of the single phenolic group (OH) in the chromane ring of α-tocopherol corresponds to the 6-OH group of the isoflavonoids. A common feature of the active isoflavonoids is an ortho-dihydroxybenzene or catechol structure, which is considered to be important for antioxidative effectiveness of flavonoids.[70-73]

Kimura *et al*[74] reported that flavonoids such as wogonin, orxylin A, chrysin, skull capflavone II, baicalein and baicalin isolated-from the roots of *Scutellaria baicalensis* Georgi, inhibited lipid peroxidation induced by ADP-NADP and Fe^{2+}-ascorbate in rat liver homogenates. The dried roots of *Scutellaria baicalensis* have been used for the treatment of suppurative dermatitis, diarrhea, inflammatory diseases, hyperlipidemia and atherosclerosis in Chinese and Japanese traditional medicine. Another flavonoid isolated from these roots by Kimura *et al*, 2',5,5',7'-tetrahydroxy-6', 8-dimethoxy-flavone, was found to be a very potent inhibitor of lipid peroxidation.[75] It exhibited over 90% inhibition toward lipid peroxidation induced by both ADP plus ascorbate and ADP plus NADPH in rat liver mitochondria and microsomes at a concentration of 100 μM. Wogonin, at the same concentration, inhibited the ADP plus NADPH- induced lipid peroxidation of rat liver microsomes by 90%, where as it inhibited the ADP plus ascorbate-induced lipid peroxidation of rat liver mitochondria by only 19%. It is worth noting that wogonin does not possess any hydroxyl substitution in its B-ring.

It was reported that lipid peroxidation can be inhibited by flavonoids possibly acting as strong $O_2^{\bullet -}$ scavengers[76] and $^1O_2$ quenchers.[61] Although $O_2^{\bullet -}$ itself does not appear to be capable of initiating lipid peroxidation $HO_2^{\bullet}$ (protonated form of $O_2^{\bullet -}$) appears to do so in isolated polyunsaturated fatty acids (PUFAs).[4] The role of $^1O_2$ in lipid peroxidation appears to be minor. The initiation of lipid peroxidation can be induced by $^{\bullet}OH$ and metal-ion-free radical complexes (such as perferryl and ferryl).[4] The scavenging of $^{\bullet}OH$ by flavonoids can impair lipid peroxidation. The induction of lipid peroxidation is shown below.

Initiation: $LH + {}^{\bullet}OH \longrightarrow H_2O + L^{\bullet}$

Propagation: $L^{\bullet} + O_2 \longrightarrow LOO^{\bullet}$

$LOO^{\bullet} + LH \longrightarrow LOOH + L^{\bullet}$

Termination: $LOO^{\bullet} + LOO^{\bullet} \longrightarrow$ Inert Product

$L^{\bullet} + L^{\bullet} \longrightarrow$ Inert Product

$LOO^{\bullet} + L^{\bullet} \longrightarrow$ Inert Product

Lipid peroxidation may be prevented at the initiation stage by free radical scavengers, while the chain propagation reaction can be intercepted by peroxy-radical scavengers such as phenolic antioxidants.[77] The chain-breaking antioxidant action of the flavonoids can be represented as shown below:

$LOO^{\bullet} + FL\text{-}OH \longrightarrow LOOH + FL\text{-}O^{\bullet}$
FL-OH, Flavonoid

Termination of lipid ($L^{\bullet}$) radical, lipid peroxyl radical ($LOO^{\bullet}$) and alkoxyl ($LO^{\bullet}$) radical (formed by reinitiation of lipid peroxidation induced by metal ions) by phenolic antioxidants is shown below:

$$LOO^{\bullet}/L^{\bullet}/LO^{\bullet} + A\text{-}OH \longrightarrow LOOH/LH/LOH + AO^{\bullet}$$

A-OH, Phenol {e.g. α-tocopherol, flavonoids}

$AO^{\bullet}$, Phenoxyl radical

It has also been proposed that flavonoids react with lipid peroxyl radicals ($LOO^{\bullet}$) leading to the termination of radical chain reactions. The oxidation of quercetin and rutin by lauroyl peroxide radicals is suggestive of such a mechanism.[60] The autoxidation of linoleic acid and methyl linoleate was inhibited by flavonoids such as fustin, catechin, quercetin, rutin, luteolin, kaempferol, and morin.[27] Morin and kaempferol were the most inhibitory for the autoxidation of linoleic acid. From the inhibition of the formation of trans, trans hydroperoxide isomers of linoleic acid by flavonoids, inhibition of the autoxidation of fatty acids by radical chain reaction termination was discerned.[27]

Ratty and Das[78] showed that a number of flavonoids inhibited both ascorbic acid- and ferrous sulfate- induced lipid peroxidation in rat brain mitochondria. The concentrations of the flavonoids tested were rather high (0.1 mM-4.0 mM). Structural requirements for antiperoxidative activity included a 3- OH substitution, a 4-keto group, a 2-3 double bond and OH substitutions on rings A and B. Vicinal OH groups in the B ring (3',4'-OH) had no particular effect in increasing the inhibitory potency.

The mechanism of antiradical action of quercetin and its glycoside, rutin, has been evaluated Afanas'ev *et al* using NADPH- and carbon tetrachloride (CCl_4)- dependent lipid peroxidation of rat liver microsomes and iron ion-induced peroxidation of lecithin liposomes.[79] Both flavonoids were significantly more effective inhibitors of the iron-ion-dependent lipid peroxidation system due to their chelation of iron ions. The chelating mechanism of inhibition was more important for rutin than for quercetin. Both flavonoids did not impair the activity of cytochrome P450 as assessed by their influence on microsomal aminopyrine demethylase. It is surprising that no effect of quercetin was found on this mixed-function oxidase activity. The inhibitory action of rutin and quercetin was demonstrated in all the peroxidation (iron ion-dependent and non-dependent) systems studied. This action was explained by both chelating and antioxidative properties of the flavonoids. The inhibitory effects of both quercetin and rutin were more pronounced on NADPH-dependent than on CCl_4-dependent lipid peroxidation in rat liver microsomes. Microsomal NADPH-dependent lipid peroxidation is known to be catalyzed by NADPH cytochrome P450 reductase and proceeds in the presence of iron ions.[80] On the other hand, the activation of CCl_4 involves cytochrome P450 and does not require iron ions.[81] A much stronger inhibitory effect of the flavonoids on NADPH-dependent peroxidation was ascribed to their metal chelating properties. The flavonoids were reported to chelate iron ions and to form inert complexes unable to initiate lipid peroxidation, yet retaining their free radical scavenging properties. At this juncture, it may be recalled that ascorbate can exhibit antioxidant activity only in the absence of transition metal ions.[6] A stronger total inhibitory effect of quercetin in both peroxidation system was thought to be due to its additional phenolic group (3-OH). Quercetin was also found to be oxidized by radicals generated in the decomposition of linoleic acid hydroperoxide in the presence of cytochrome c. The authors surmised that quercetin and rutin were able to suppress free radical processes by inhibiting the formation of $O_2^{\bullet-}$, $^{\bullet}OH$ and lipid peroxyl radicals.

In a recent study, baicalein was found to be a strong inhibitor of lipid peroxidation in rat forebrain homogenates.[82] Its IC_{50} (0.42 μM) was much lower than that of quercetin (1.2 μM). Flavone was found to be inactive. Baicalein also showed free radical scavenging action against 1,1-diphenyl-2-picrylhydrazyl (DPPH). This flavone also inhibited phorbol ester- induced ear edema in mice. Lipid peroxidation was considered to play an important role in the pathogenesis of phorbol ester-induced ear edema in mice.

Polymethoxylated flavones and C-glycosyl derivatives of flavones isolated from medicinal plants were studied for their influence on lipid peroxidation induced by $FeSO_4$+ cysteine in rat liver microsomes.[83] Several hydroxylated flavones, C-glycosyl flavones, methoxyflavones and flavonols, as well as the flavanol, leucocyanidol, and the biflavone, amentoflavone, inhibited inhibitory activity at a concentration of 100 μM. Some hydroxyflavones were as effective as hydroxylated flavonols in inhibiting lipid peroxidation. The same was the case with C-glycosylflavonols (e.g. rutin) and C-glycosylflavones (e.g.

orientin and isoorientin). Some methoxyflavones were also quite potent in inhibiting lipid peroxidation, although their IC_{50} values were much higher than those of hydroxyflavones. The flavanone, naringin, displayed no inhibition at the stated concentration (100 μM). However, the corresponding flavone apigenin (with a 2-3 double bond) was potent inhibitor. Galangin, a flavonol possessing no B-ring hydroxyl groups, was as effective as quercetin in inhibiting lipid peroxidation. Crisiliol and sideritoflavone, potent lipoxygenase inhibitors,[84] showed no inhibitory activity, indicating that the inhibition of arachidonic acid metabolism by these compounds is dependent on flavonoid-enzyme interactions and not related to possible antioxidant properties. A similar conclusion was also made by Laughton *et al*[85] who investigated the ability of various flavonoids to inhibit 5-lipoxygenase and cyclooxygenase activities of rat peritoneal leukocytes, and lipid peroxidation induced by $FeCl_3$+ ascorbate in rat liver microsomes. Several flavonols were potent inhibitors of lipid peroxidation in this system. Rutin, was far less potent than quercetin. The lipid peroxidation inhibitory capacity of the flavonoids was not significantly correlated with their ability to inhibit lipoxygenase or cyclooxygenase activity, suggesting that their mode of inhibition of 5-lipoxygenase/ cyclooxygenase is not simply due to scavenging of peroxyl radicals generated at the active site of the enzymes. Robak *et al*[86] examined a series of flavonoids, isolated from plants, for their influence on soybean lipoxygenase activity, cyclooxygenase activity and inhibition of ascorbate- stimulated lipid peroxidation in rat liver microsomes. Most of the tested flavonoids stimulated cyclooxygenase when arachidonic acid was used as a substrate at 100 μM. A number of flavonoids were inhibitors of soybean lipoxygenase activity, and of lipid peroxidation. The most active inhibitors possessed vicinal hydroxyl groups in their B-ring.

An isoflavonoid glycoside containing OH groups at positions 3 and 4 of the B-ring, isolated from the roots of *Pueraria labata*, was found to inhibit enzymic (NADPH-induced) and non-enzymic (ascorbate or $H_2O_2+Fe^{2+}$- induced) lipid peroxidation in rat liver microsomes.[87] On the other hand, wogonin, a flavone with no OH substitution in the B-ring, inhibited only the enzymically- induced lipid peroxidation.[87] Formation of Fe^{2+} by NADPH-dependent cytochrome P450 reductase was inhibited by wogonin, but not by the isoflavonoid glycoside. The glycoside had no effect on terminating radical chain reaction during lipid peroxidation in the enzymatic system or in the linoleic acid hydroperoxide-induced peroxidation system, suggesting that its antioxidant activity was probably due to its ability to scavenge free radicals involved in the initiation of lipid peroxidation.

Laughton *et al*[88] found that both quercetin and myricetin were powerful inhibitors of iron-induced lipid peroxidation in rat liver microsomes. In these studies peroxidation was induced by adding Fe^{2+} (as ferrous ammonium sulfate), Fe^{3+} (as ferric chloride), Fe^{3+}-ascorbic acid, Fe^{3+}-EDTA or Fe^{3+}-ADP/NADPH. Myricetin possesses *o*-trihydroxy substitution (pyrogallol structure) in its B-ring. The inhibitory effect was particularly pronounced when lipid peroxidation was stimulated by adding Fe^{3+}/ascorbate. At low concentration, the phenols caused a "lag period" during the course of lipid peroxidation. This effect was attributed to their action as lipid-soluble chain-breaking inhibitors of the peroxidative process, scavenging intermediate peroxyl and alkoxyl radicals. At 100 μM, both quercetin and myricetin accelerated the generation of $^\bullet OH$ radicals from H_2O_2 in the presence of Fe^{3+}-EDTA. $^\bullet OH$ production was inhibited by catalase and SOD which prompted the authors to suggest a mechanism in which the phenols oxidize to produce $O_2^{\bullet -}$, which then induces $^\bullet OH$ generation from H_2O_2 in the presence of Fe^{3+}-EDTA. At concentrations up to 75 μM, quercetin and myricetin also accelerated bleomycin-dependent DNA damage in the presence of Fe^{3+}, which was suggested to be due to the reduction of the Fe^{3+}-bleomycin-DNA complex to the Fe^{2+} form. These phenols, however, caused no acceleration of microsomal lipid peroxidation in the presence of Fe^{3+} or other iron complexes. The authors contend that the chain-breaking antioxidant activity of the phenolics outweigh any iron-reducing activity. In view of their observed pro-oxidant effects, the author remark that these phenolics can not be classified simplistically as "antioxidants". At this juncture, it may be recalled that both α-tocopherol and ascorbate have simiiar pro-oxidant effects.[89-91]

Semisynthetic hydroxyethyl derivatives of flavonols have also been shown to display antioxidant action.[92] These derivatives are water-soluble. Several hydroxyethyl rutosides and 7,3',4'-trihydroxyethyl quercetin exhibited considerable inhibition of rat liver microsomal lipid peroxidation induced by $FeSO_4$ and ascorbate. They were less active than quercetin. They were also shown to be potent •OH scavengers and interacted with DPPH stable free radical. Increasing substitution on the phenolic groups resulted in a concomitant diminution in the observed inhibition of lipid peroxidation.

The antioxidant action of the flavonoids, silybin and (+)-cyanidanol-3 [(+)-catechin] was assessed in a peroxidation system consisting of linoleate and Fe^{2+}.[93] At a concentration of 200 mM, silybin (a water-soluble preparation of silybin as dihemisuccinate disodium salt) inhibited Fe^{2+}-induced linoleate peroxidation. The antioxidant effect exerted by (+)-cyanidanol-3 was far greater than that of silybin at concentrations ranging from 250 μM - 2.0 mM. At a concentration of 200 mM, the inhibitory action of silybin was comparable to that of butylated hydroxyanisole , while the antioxidant effect of (+)-cyanidanol-3 was similar to that obtained with butylated hydroxytoluene, one of the most powerful synthetic antioxidants. (+)-Cyanidanol-3 has been shown to have a powerful free radical scavenging activity and to inhibit lipid peroxidation in different experimental systems.[64,94-96] These include the inhibition of ethanol-induced enhancement of liver conjugated dienes[64] and of the chemiluminescence of rat liver *in situ*.[96]

Fraga *et al*[97] reported that (+)-catechin, eriodictoyl and myricetin, at low concentrations (IC_{50} 3-15 μM) inhibited the *tert*-butyl hydroperoxide initiated chemiluminescence of mouse liver homogenates that is associated with lipid peroxidation resulting from the formation of hemoprotein-catalyzed radicals following rupture of the hydroperoxide.[98] Administration of eriodictoyl and (+)-catechin to mice also depressed the enhancement of *in situ* liver chemiluminescence produced by CCl_4.[97] CCl_4 reacts with cytochrome P450 to initiate *in vivo* lipid peroxidation.[99] Both carbon- and oxygen- centered radicals[100] and excited species[101] are formed during this process. The observed inhibition of chemiluminescence was proposed to involve free radical- scavenging as well as excited species-quenching.

Sorata and coworkers[61] demonstrated that quercetin and rutin inhibited human erythrocyte lipid peroxidation accompanying photohemolysis. A number of flavonoids were observed to inhibit N-ethyl maleimide-induced lipid peroxidation in human platelets.[102] Very low IC_{50} values were observed and silymarin appeared to be particularly active. Utilizing phenazine methosulfate as an intracellular generator of oxygen free radicals, Maridonneau-Parini *et al*[62] reported a heterogenous effect of flavonoids on K^+ loss and lipid peroxidation induced by oxygen radicals in human erythrocytes. These authors suggested that the flavonoids could be classified according to the activity applied, for the most part, to both K^+ permeability and to membrane lipid peroxidation. This activity would include protective, toxic, biphasic, and inactive responses. Kappus *et al*[103] showed the inhibition of lipid peroxidation in isolated rat hepatocytes by (+)-cyanidanol-3.

Cholbi *et al*[104] described the activity of apigenin, luteolin, gardenin D, galangin, datiscetin and morin as well as catechin as inhibitors of CCl_4- induced rat liver NADPH-dependent microsomal lipid peroxidation. The polymethoxylated flavone, gardenin D possesses OH groups at 5- and 3'-positions, and OCH_3 groups at 6,7,8 and 4'-positions. Its potency was reported to be comparable to that of (+)-catechin, showing its strong inhibitory effect on cytochrome P450.

The flavonols, quercetin, rutin and morin and flavanones, naringin and hesperidin, were studied as chain-breaking antioxidants for the autoxidation of linoleic acid in cetyl trimethylammonium bromide micelles.[105] All the three flavonols exhibited antioxidant activities, while the two flavanones, naringin and hesperidin, did not suppress the oxidation appreciably. The 7-hydroxy group of the flavonoids is considered to be the first to dissociate and is thus the most likely site of attack by peroxyl radical.[106,107] The 7-hydroxy group is free for quercetin, rutin and morin, but in naringin and hesperidin, this phenolic site is blocked with a glycoside. Therefore, it was suggested that the former compounds exhibited active antioxidant activity, where as the latter were inactive.

Middleton, Drzewiecki and Kandaswami (unpublished results) examined the scavenging action of a wide range of flavonoids against DPPH radical. A number of flavonols, flavones and flavan-3-ols were active, although flavone, apigenin, naringin, naringenin, chrysin, etc. showed no activity. The 2,3-double bond and the 3-OH group appeared to intensify the radical scavenging potency at lower concentrations.

Bors and Saran[108] studied the radical scavenging efficiencies of different classes of flavonoids by employing the method of pulse radiolysis. Aroxyl radicals were generated by univalent oxidation of a member of flavonoids by azide (N_3) radicals at pH 11.5. Compounds with a saturated ring were predominantly attacked at the *o*-dihydroxy site at the B-ring and the semiquinones formed were quite stable. For a substance to act as an antioxidant, the stability of the radicals formed from it is of prime importance. Radicals derived from flavonoids with a 2,3-double bond and both 3- and 5- OH substituents (flavonols) apparently did not seem to possess a higher stability. The very high rate constant of formation and the relative stability of some of the aroxyl radicals led to the supposition that the biological function of flavonoids might be the scavenging of radicals. In a study dealing with the reaction of fatty acid peroxyl radicals, both kaempferol and quercetin turned out to be exceptionally good scavengers of linoleic acid peroxyl radicals.[109]

Bors *et al*[106] examined the radical scavenging and antioxidant potential of different classes of flavonoids, using the method of pulse radiolysis. Their studies demonstrated the effective radical scavenging capabilities of most flavonoids and indicated the existence of multiple mesomeric structures for aroxyl radical species of flavonoids. Three structural groups were important determinants for radical scavenging and for antioxidant potential: 1) the *o*-dihydroxy (catechol) structure in the B-ring, the obvious radical target site for all flavonoids with a saturated 2,3-double bond (flavan-3-ols, flavanones, cyanidin chloride) 2) the 2, 3-double bond in conjunction with a 4-oxo function and 3) the additional presence of both 3- and 5- OH groups for maximal radical-scavenging potential. The capacity of flavonoids to scavenge $O_2^{\bullet-}$, $^{\bullet}OH$ and lipid radicals has been frequently reported.[27,110-115] Flavonoids do react rapidly with $^{\bullet}OH$ because of the generally high reactivity of this radical with aromatic compounds.[106] In contrast, for $O_2^{\bullet-}$, even for the very efficient flavonol radical scavengers kaempferol and quercetin, only very low rate constants were found,[106] even though flavonoids have been considered as scavengers of this radical.[114,115] Bors *et al*[106] have questioned reports on the specific scavenging of different radicals by flavonoids. Sichel *et al*,[116] have recently reported the scavenger activity of some flavonoids against $O_2^{\bullet-}$ using ESR spectrometry. These authors suggest that the presence of hydroxyl groups in the B ring of flavonoids is essential for this scavenging activity. Cotelle *et al*[117] showed the formation of stable radicals from synthetic flavonoids by ESR spectroscopy.

Certain flavonoids have been shown to inhibit mitochondrial succinoxidase and NADH oxidase and other oxidase activities.[118-120] In a structure-activity investigation of 14 different flavonoids, four flavonoids, quercetogetin, quercetin, myricetin, and delphindin chloride, were shown to generate a cyanide-insensitive respiratory burst in the presence of isolated beef heart mitochondria and to autoxidize in buffer alone.[118] Subsequently, the same flavonoids were shown to autoxidize with the concomitant production of semiquinone radicals, $O_2^{\bullet-}$, $^{\bullet}OH$ and H_2O_2.[121] The inhibition of the above mitochondrial enzymes by flavonoid compounds was suggested to contribute to their antineoplastic activities. The inhibition of enzymes that catalyze oxidation-reduction reactions by flavonoids may involve flavonoid- generated ROS.[118-120]

The autoxidation of flavonoids such as quercetin and myricetin (having catechol and pyrogallol configuration in the B-ring, respectively) in aqueous media at pH 7.5 has been described.[122] This autoxidation resulted in the generation of $O_2^{\bullet}$, H_2O_2 and $^{\bullet}OH$. The autoxidation was, however, quite slow at pH 7.5 for quercetin. Such prooxidant effects are of interest in the context of tumor cell cytotoxicity, although not considered to have toxicological consequences.

ANTITOXIC EFFECTS OF FLAVONOIDS

Silymarin has been shown to have hepatoprotective effects *in vivo*. Both silymarin and silybin dihemisuccinate have been shown to be effective protective agents against the hepatotoxicity of CCl_4, phalloidin and α-amanitin.[123] With regard to its protective mechanism, it is considered that the flavonoid exerts a membrane-stabilizing action thus preventing or inhibiting lipid peroxidation.[124] Silymarin has been widely used in Europe in the treatment of alcoholic liver disease and diseases associated with increased vascular permeability and capillary fragility.[125]

It was reported that *in vivo* treatment with silymarin protected against lipid peroxidation and hemolysis induced in rat erythrocytes, when incubated with phenylhydrazine.[126] In addition, *in vivo* treatment with silybin dihemisuccinate was shown to inhibit the release of malondialdehyde induced by phenylhydrazine in the perfused rat liver.[127] Silymarin also prevented liver glutathione depletion and lipid peroxidation induced by an acute intoxication with ethanol in the rat.[128] These effects attest to the suggested action of the flavonoid as a cytoprotective agent. Intraperitoneal administration (50 mg/Kg) of silybin dihemisuccinate to rats inhibited lipid peroxidation, methemoglobin formation and osmotic fragility induced *in vitro* by phenylhydrazine in erythrocytes.[129] Effects on osmotic fragility was suggested to be to a consequence of the membrane-stabilizing properties of the flavonoid. These effects were also ascribed to the antioxidant properties of the flavonoid, since spontaneous or induced oxidative stress may labilize cell membranes. The observed novel pharmacological action of silybin dihemisuccinate, primarily used in the treatment of hepatic diseases, could have other therapeutic implications. Several drugs are metabolized to hydrazine derivatives producing not only liver damage, against which silybin has been shown to have a protective effect,[127] but also hematological disorders. Prophylactic or therapeutic treatment with the above flavonoids has been suggested to confer protection against these deleterious effects.[129]

Elucidation of the mechanism for the protective effect of silymarin against the hepatotoxicity of CCl_4 has received considerable interest. A short report showed decreased amounts of diene conjugates in rats pretreated by silymarin prior to the administration of CCl_4.[130] The possible mechanisms for the protective effect of silymarin against the hepatotoxicity of CCl_4 have recently been elucidated by Letteron *et al* [131] Intraperitoneal administration (800 mg/kg) of silymarin to mice protected the liver from CCl_4-induced lipid peroxidation and hepatotoxicity. Silymarin inhibited the metabolic activation of CCl_4 *in vivo*, as suggested by a decreased covalent binding of CCl_4 metabolites to hepatic lipids *in vivo*. Decreased metabolic activation of CCl_4 by cytochrome P450 will depress the initial formation of the trichloromethyl free radical and therefore diminish the initiation of lipid peroxidation. Silymarin (800 μg/mL) impaired the irreversible binding of CCl_4 metabolites to hepatic microsomal protein by only 21%, although it decreased by 72% the *in vivo* lipid peroxidation mediated by CCl_4 metabolites. Silymarin treatment *in vivo* diminished the irreversible binding of CCl_4 metabolites to hepatic lipids by only 39%, although it depressed by 60% the exhalation of ethane during the first hour after the administration of CCl_4. Silymarin (800 μg/mL) decreased by 70% *in vitro* lipid peroxidation mediated by CCl_4 metabolites, and decreased by 90% lipid peroxidation mediated by NADPH alone. In this system, lipid peroxidation is thought to be mediated by the reduction of iron to the ferrous state.[132] It has been earlier reported that silymarin can prevent lipid peroxidation mediated by the addition of Fe^{2+}-ascorbate, cumene hydroperoxide or *tert*-butylhydroperoxide, suggesting that they can act as chain-breaking antioxidants.[58,67,93,102] Letteron *et al*[131] concluded that silymarin prevented CCl_4- induced lipid peroxidation and hepatotoxicity in mice by a dual mechanism, by decreasing the metabolic activation of CCl_4 into free radicals as well as by scavenging free radicals. Ferer *et al* [133] showed that silymarin treatment corrected the decreased SOD activity of erythrocytes and lymphocytes in patients with alcoholic cirrhoses, thus exemplifying the therapeutic utility of the flavonoid.

The activity of intravenous S 5682, a purified fraction containing 90% diosmin (a flavone derivative) and 10% hesperidin (a flavanone derivative) was evaluated in the rat by measuring the degree of hyperglycemia provoked by an intravenous injection of alloxan, the metabolism of which produces ROS which are toxic to B cells of the pancreas. S 5682 produced a decrease in hyperglycemia in a dose-dependent manner (25 mg/kg and 50 mg/kg).[134] The authors suggest that the radical scavenging properties of S 5682 might explain its diverse pharmacological effects such as: 1) the reduction in the capillary permeability induced in the rat and rabbit by the injection of antigenic particles, the application of chloroform swabs or by irradiation, 2) the anti inflammatory and antiedematous effects seen in inflammatory granulomas in the rat.[134]

The flavonoids quercetin, kaempferol, catechin and taxifolin suppressed the cytotoxicity of $O_2^{\bullet -}$ and H_2O_2 on Chinese hamster V79 cells, as assessed with a colony formation assay.[135] Quercetin and kaempferol showed protective effects at 5-10 μM concentrations whereas much higher concentrations of catechin and taxifolin were necessary for the prevention of cytotoxicity. The protective activity was ascribed to the *o*-dihydroxy structure in the B-ring, or 3- and 5- OH groups with the 4-oxo function or the 2,3-double bond in conjunction with the 4-oxo function. The authors earlier suggested that the *o*-dihydroxy structure of polyphenols was essential for protection against H_2O_2-induced cytotoxicity in V79 cells, because antioxidants bearing only one phenolic OH, such as ferulic acid methyl ester and α-tocopherol, exhibited no protective effects.[136] The observation that kaempferol, lacking the above structure, showed protective effect seems to be an exception. The conversion of kaempferol to quercetin by *o*-hydroxylation under the experimental conditions utilized might explain this effect. The mutagenic effect of chrysotile asbestos fibers, zeolite and latex particles on human lymphocytes in whole blood was inhibited by antioxidant enzymes, SOD and catalase, and radical scavengers such as rutin, ascorbic acid and bemitil, indicating that the mutagenic effects of the particles was mediated by oxygen radicals.[137] Of the radical scavengers studied, rutin was the most effective inhibitor of the mutagenic effect of mineral fibers and dusts. The study of lucigenin- and luminol- amplified chemiluminescence of peritoneal macrophages stimulated by the above fibers and particles showed that their mutagenic action was probably mediated by different oxygen species. Rutin was more potent than ascorbate in inhibiting luminol-dependent chemiluminescence of peritoneal macrophages activated by chrysotile fibers or zeolite particles.[137]

Kantengwa and Polla[138] reported that erythrophagocytosis induced in human monocytes-macrophages the synthesis of stress proteins including the classical heat shock proteins and heme oxygenase. Quercetin and kaempferol inhibited this induction. The results suggest that 1) erythrophagocytosis-related oxygen radicals are involved in the induction of the stress response in phagocytic cells, 2) the induction of classical heat shock proteins appears, at least in part, to be dependent on protein kinase C; heme oxygenase, in contrast appears to be strictly dependent on reactive oxygen species, and 3) the effects of the flavonoids on heme oxygenase are linked to their scavenging activity rather than to protein kinase C modulation. Antioxidant catechins (flavans) isolated from Chinese green tea showed scavenging activity against H_2O_2 and $O_2^{\bullet -}$ (generated by the xanthine-xanthine oxidase system.[139] The flavans also prevented oxygen radical-induced cytotoxicity and inhibition of intercellular communication in cultured B6C3F1 mouse hepatocytes and keratinocytes (NHEK cells). Cytotoxicity and inhibition of intercellular communication represent two possible mechanisms by which tumor promoters produce their promoting effects.[140] The prevention of these two effects of prooxidants by tea flavans may suggest a mechanism by which these catechins inhibit tumor promotion *in vivo*.

The cytoprotective effect of three flavonoids, catechin, quercetin and diosmetin was recently investigated on iron-loaded rat hepatocyte cultures, considering two parameters, namely the prevention of iron-induced increase in lipid peroxidation and the inhibition of intracellular lactate dehydrogenase release.[141] The potency of these flavonoids for the above two aspects of cytoprotection was classified as follows: catechin > quercetin > diosmetin. The investigation of the capacity of the above flavonoids to remove iron from iron- loaded hepatocytes revealed that the iron-chelating capacity of the three compounds followed the same order as did the cytoprotective effect (catechin > quercetin > diosmetin). The authors suggest that this relationship has to be taken into consideration in further development of these protective flavonoids which could have important applications in human diseases, like the widely known silymarin used in liver dysfunction, and diosmin acting as vascular protectant. Some flavonoids have been reported to be able to mobilize iron from ferritin[142] and to be capable reducing Fe^{3+} to Fe^{2+} ions.[143] These considerations were thought to be of importance although some authors ruled out the possibility that the antiperoxidative action was related to an interaction of the flavonoids with iron ions.[58,144]

FLAVONOID ACTION IN RELATION TO CORONARY ARTERY DISEASE AND ISCHEMIA REPERFUSION

De Whalley *et al*[145] showed that certain flavonoids were potent inhibitors of the modification of low density lipoproteins (LDL) by mouse macrophages with IC_{50} values in the micromolar range (e.g. 1-2 μM for fisetin, morin and quercetin). Flavonoids also inhibited the cell-free oxidation of LDL mediated by $CuSO_4$. The flavonoids appeared to act by protecting LDL against oxidation caused by the macrophages, as they inhibit the generation of lipid hydroperoxides and protect α-tocopherol, a major lipophilic antioxidant carried in lipoproteins, from being consumed by oxidation in the LDL. The flavonoids protect the α-tocopherol (and possibly other endogenous antioxidants) in LDL from oxidation, maintain their levels for longer periods of time and delay the onset of lipid peroxidation. While the mechanisms by which flavonoids inhibit LDL oxidation are not certain, the following possibilities have been advanced. They may reduce the generation or release of free radicals in the macrophages or may protect the α-tocopherol in LDL from oxidation, by being oxidized by free radicals themselves in preference to it. Another possible mechanism is that the flavonoids could have regenerated active α-tocopherol by donating a hydrogen atom to the α-tocopheryl radical, which is formed when it transfers its own OH hydrogen atom to a lipid peroxyl radical to terminate the chain reaction of lipid peroxidation. Another possibility is the sequestration of metal ions like iron and copper by flavonoids thereby diminishing the engendered free radicals in the medium. Since some flavonoids at a concentration of only 10 μM completely inhibited the modification of LDL by 100 μM Cu^{2+}, a doubt was raised whether metal complexation by flavonoids could explain all their effects. Macrophage-associated free radicals could be derived from a number of sources, two of which are as byproducts of the action of 5-lipoxygenase and cyclooxygenase. No obvious correlations were apparent, however, between the known potencies of individual flavonoids in inhibiting these two enzymes and their potencies in inhibiting LDL modification. Flavone itself was not a good inhibitor of LDL modification by macrophages whereas the polyhydroxylated aglycone flavonoids were potent inhibitors. The importance of OH groups of the flavone nucleus could thus be discerned.

Negre-Salvayre *et al* [146] reported the protection of lymphoid cell lines against peroxidative stress induced by oxidized LDL utilizing a combination of α-tocopherol, ascorbic acid and the quercetin glycoside, rutin. These investigators have also shown that the cytotoxicity of oxidized LDL can be prevented by flavonoids in two ways, either by inhibiting the lipid peroxidation of LDL (induced by UV irradiation) or by blocking at the cellular level the cytotoxicity of previously oxidized LDL.[147] Their studies showed that: 1) probucol (25 μM) a synthetic antioxidant, was very effective in preventing UV-induced lipid peroxidation of LDL and their subsequent cytotoxicity on lymphoid cell lines (EBV-transformed cell lines), but it could not protect cells against the cytotoxicity of previously oxidized LDL, 2) vitamin E (100 μM) prevented poorly the lipid peroxidation of LDL, but it was able to abrogate the cellular oxidative stress and cytotoxicity induced by previously oxidized LDL, and 3) catechin (10 μM) exhibited two types of protective effects; it inhibited the peroxidation of LDL and their subsequent cytotoxicity and very effectively protected the cells against the toxicity of previously oxidized LDL. In subsequent studies, the above investigators showed that both quercetin and rutin exhibited similar effects as catechin did, viz. inhibiting the lipid peroxidation of LDL and blocking at the cellular level the cytotoxicity of previously oxidized LDL.[148] Flavone was completely inefficient in exerting any of these effects. The oxidation of LDL induced by UV-radiation, attacks mainly the lipid core of the LDL, in contrast to the cell- or copper- mediated oxidation which primarily attacks the LDL surface components.[149] The inhibition of LDL lipid peroxidation by the flavonoids was well correlated with the prevention of the cytotoxicity of oxidized LDL . In the protection of the cells by polyphenolic flavonoids, two lines of defense were inferred: 1) at high concentrations (IC_{50}, 10-20 μM for quercetin or rutin), an antioxidant effect inhibiting the lipoprotein oxidation and subsequent cytotoxicity; 2) at low concentrations (IC_{50}, 0.1 and 3 μM for quercetin or rutin) direct protection of cells against the cytotoxic effect of oxidized LDL. The intracellular mechanisms for this direct prevention of the cytotoxic effect of oxidized LDL are unknown. The authors suggest that quercetin acts at the cellular level by blocking the intracellular "transduction" of the cytotoxic "signal" contained in oxidized LDL. The authors suggest that the cellular mechanisms underlying the protective effect could be related to the various biological actions of quercetin such as: 1) antioxidant properties preventing oxidative attack of membrane lipids by sparing vitamin E or regenerating it as does ascorbic acid in the maintenance of α-tocopherol levels, 2) inhibition of lipoxygenases which are known to be stimulated by lipid peroxides and which can be involved in oxidative stress, as suggested by their role in LDL oxidation in cells, 3) inhibition of cellular enzymes involved in signal transduction. The above results suggest that dietary flavonoids or related compounds could be involved in the prevention of atherosclerosis not only by inhibiting LDL oxidation but also by increasing the cellular resistance to the deleterious effects of oxidized LDL. Recruitment of different flavonoids effective in directly protecting cells represents a novel approach in the prevention of atherosclerosis by nutritional intervention.

A recent study from the Netherlands showed an inverse correlation between dietary flavonoid intake and the incidence of coronary artery disease.[150] The individuals with the lowest dietary intake of flavonoids had the highest incidence of heart disease. Interestingly, the relative incidence of heart disease among men who had the highest intake of flavonoids was only a third of those who had the lowest intake of flavonoids. The result were the same even after adjustment for age, body fat, smoking, cholestrol, blood pressure, physical activity, coffee consumption and the intake of calories, vitamin C, vitamin E, β-carotene and dietary fibers. The main source of dietary flavonoids for the above individuals were apples, onions, and tea.

Consumption of diets high in saturated fat and cholestrol is associated with increased risk of coronary artery disease. However, epidemiological evidence indicates that heart disease is less in the French than expected, based on saturated fat intake and cholestrol levels. This unusual effect, known as the "French Paradox" has been attributed to the drinking of red wine. A recent study indicates that polyphenolic compounds (e.g. anthocyanin flavonoids and related compounds) present in wine could exert a protective effect.[151] Quercetin and phenolic compounds isolated from red wine effectively impaired copper ion-catalyzed oxidation of LDL. α-Tocopherol exhibited only 60% of the potency of wine phenolics or quercetin.[151]

Two flavonoids, quercetin and silybin, were reported to a exert protective effect by preventing the decrease in the xanthine dehydrogenase/oxidase ratio observed during ischemia-reperfusion in the rat.[152] The results indicated the conversion of xanthine dehydrogenase to xanthine oxidase during the early stages of kidney ischemia. The enzyme xanthine oxidase, implicated in tissue oxidative injury after ischemia-reperfusion, is a source of ROS and is formed from a dehydrogenase during ischemia.[153] The protective effect of quercetin and silybin on the xanthine dehydrogenase/oxidase ratio, observed in the above study, was postulated to be due to the inhibition of the dehydrogenase to oxidase transformation by the flavonoids. The inhibition of xanthine oxidase activity by flavonoids has been described earlier.[154] Ning *et al*[155] reported that flavone administration markedly improved functional recovery in the reperfused rabbit heart after a bout of global ischemia. The effects of the compound on postischemic recovery was proposed to be due to its stimulation of the cytochrome P450 system. Cytochrome P450 reductase, which transfers electrons from NADPH to cytochrome P450 during P450-dependent catalysis, is capable of reducing oxygen to yield $O_2^{\bullet -}$, and the oxygenated intermediates of P450 themselves decompose in a side reaction to release $O_2^{\bullet -}$.[156,157] It was advanced that flavone might be acting as an allosteric effector that improves catalytic efficiency, thereby diminishing detrimental ROS production.[155] Ning *et al*[155] have highlighted the potential utility of flavonoids as a means of enhancing myocardial ischemic tolerance or resistance to reperfusion injury, or both. They also point out the recent identification of an interesting flavonoid compound, puerarin (8-C-C-glycopyranosyl-1-4'-7-dihydroxyisoflavone), as an active ingredient of in *Radix puerariae*, a traditional Chinese medicinal herb that has been used for decades for the treatment of hypertension and angina pectoris in China.[158]

The vascular endothelium is extremely sensitive to oxidative damage mediated by ROS, released from inflammatory cells.[159,160] Of these metabolites H_2O_2 appears to be an important mediator of acute cellular injury in a variety of settings.[160] Such oxidative damage may play a role in the pathogenesis of atheroslerosis.[161] The flavan-3-ol compounds, gallocatechin-3-O-gallate and epicatechin-3-O-gallate, isolated from tea, were effective in preventing H_2O_2-induced injury to bovine endothelial cells in culture.[162] These observations suggest a possible role for these catechins in maintaining vascular homeostasis.

Gryglewski and coworkers studied the mechanism of the antithrombotic action of flavonoids.[163] Four flavonoids (quercetin, rutin, cianidanol and meciadonol) each inhibited platelet lipoxygenase activity and ascorbate-induced rat liver microsomal lipid peroxidation, whereas only quercetin and rutin stimulated cyclooxygenase and bound to platelet membranes. Quercetin and rutin were capable of dispersing platelet thrombi adhering to rabbit aortic endothelium *in vitro* and prevented platelets from aggregating over a blood-superfused collagen strip (adhesion-related phenomena). The *in vivo* counterpart of these experiments involved the infusion of quercetin and rutin into an extracorporeal stream of blood. Quercetin and rutin inhibited the deposition of platelet thrombi on the blood-superfused collagen strip at calculated plasma concentrations of 0.05 and 0.03 μM. Analogously, in the model for studying platelet-endothelium interactions, quercetin and rutin, when infused into the stream of blood which superfused a rabbit aortic endothelial surface, caused the disaggregation of preformed platelet thrombi, again at low concentrations. Clearly the expression and/or activity of platelet/ endothelium adhesion molecules are affected by the flavonoids. The authors concluded that flavonols are antithrombotic because they are bound selectively to mural platelet thrombi and, because of their free radical scavenging properties, modify damaged endothelial cells and permit normal prostacyclin and EDRF.[163]

FLAVONOIDS AND ASCORBIC ACID INTERACTIONS

Ascorbic acid is a universal component of plant cells. Ascorbic acid and flavonoids coexist in plants and thus the two are consumed together in the diet.[25,28] A large literature has accumulated concerning the interactions of flavonoids with ascorbic acid in biological systems.[19,25,29] Several flavonoids serve as antioxidants for ascorbic acid. One of the mechanisms for this effect is thought to be by chelating metals even though the free radical accepting action may be more important. The sparing effect of flavonoids on ascorbate oxidation may explain many of the interactions of flavonoids with ascorbic acid described in the voluminous literature on these compounds. Of historical importance is the observation that a mixture of two flavonoids called citrin (eriodictoyl and hesperidin) was considered to possess vitamin-like activity.[19,25] The term vitamin P was coined to indicate that this material had the property of decreasing capillary permeability (and fragility), prolonging the life of marginally scorbutic guinea pigs, and reducing the signs of hypovitaminosis C in experimental animals. The term vitamin P was subsequently abandoned, since citrin was not found to fulfill the classical function of a vitamin. Nevertheless, subsequent studies have shown that flavonoids had potent antioxidant-dependent vitamin C- sparing activity.[19,25] Human studies showed increased tissue concentration of ascorbic acid as well as increased urinary output of the vitamin.[25,30] Considerable evidence indicates that flavonoids may influence the metabolism of ascorbic acid although the basis of this impact is not understood.[19,25]

In vitro studies indicated that flavonoids had a considerable capacity for retarding the breakdown of ascorbate to dehydroascorbate.[164] One mechanism for this protection might involve the chelation of copper and other trace metals by flavonoid resulting in the retardation of metal-catalyzed oxidation of ascorbic acid. Another protective mechanism is based on the ability of flavonoids to act as free radical acceptors. Free radical formation is considered to be an all important phase of ascorbate oxidation . Clemtson and Anderson related ascorbate-protective capacity to the structure of the flavonoids.[19,29] They examined the effect of some 34 different flavonoids on the oxidation of ascorbic acid at physiological pH and concluded that significant antioxidant activity was confined to compounds possessing: 1) 3',4'-OH groups (couplet) of the B-ring and 2) the 3-hydroxy-4-carbonyl group (couplet) of the γ-pyrone ring. In confirmity with this, quercetin and rutin were found to have a greater ascorbic acid-protective capacity than the other flavonoids examined.[25] An apparent exception to the above generalization is hesperidin which did not conform to the prescribed pattern and yet had *in vitro* protective capacity and *in vivo* increased the tissue ascorbic acid concentrations.[165,166] It was reported that commercial samples of hesperidin contained other flavonoids as impurities.[29]

Thiol compounds such as glutathione are potential hydrogen donors for the reduction of dehydroascorbic acid to ascorbic acid.[25,167] Flavonoids such as quercetin, and hesperidin were shown to enhance the reduction of dehydroascorbic acid by glutathione. Parrot and Gazawe[167] reported that (+)-catechin potentiated the reduction of dehydroascorbic acid by glutathione. It has been suggested that certain flavonoids [flavan-3-ols (e.g. catechin)] designated as vitamin C_2 (or C_2 factor) and regarded as a second antiscorbutic factor, can potentiate the chemical reduction of dehydroascorbic acid.[25] While studying the reduction of dehydroascorbic acid to ascorbic acid by thiols in an aerobic environment, Regnault-Roger *et al*[168] reported that the flavan-3-ol configuration of C_2 factor compounds like epicatechin plays a specific role as a catalyst in this reaction, which might not be dependent on the redox potential alone. The possibility that flavonoids might stimulate the tissue reduction of dehydroascorbic acid was examined by Zloch.[169] Guinea pigs given a standard diet of dehydroascorbic acid with an without flavonoids (rutin, epicatechin). The tissue ascorbic acid content was 30-100% greater in the case of the flavonoid-treated group.

Several physiological interactions of ascorbic acid with plant flavonoids have been considered[25] such as 1) increase in ascorbic acid absorption, 2) stabilization of ascorbic acid, 3) reduction of dehydroascorbate to ascorbate and 4) metabolic sparing of ascorbic acid by flavonoids. Flavonoids have been considered to function as antioxidants and UV light filters in higher plants.[28] This antioxidant activity has been related to their protection against ascorbic acid oxidation . The oxidation of ascorbic acid by metal ions such as Cu^{2+} and copper-containing enzymes is well known.[170] The inhibitory effect of flavonoids on ascorbic acid oxidation is considered to be due to metal-flavonoid complexation and/or free-radical trapping by the flavonoids.[25] The protection of ascorbic acid by flavonoids could have important biological implications as emphasized by Hughes and Wilson.[25] Ascorbic acid metabolites can be mutagenic for mammalian cells[171]. An increased production of these metabolites could be a key factor in aging, according to the intrinsic mutagenesis theory of aging.[172] Flavonoids and other factors which suppress the breakdown of ascorbic acid could, therefore, function as antiaging factors. Ascorbate may also protect flavonoids from oxidation. Purified cyanidin 3-gentiobioside, cyanidin 3-rhamnoside and pelargonidin 3-glucoside were decolorized by low levels of H_2O_2 and horseradish peroxidase. Ascorbate added to this system inhibited the decolorization of the anthocyanins to one tenth the rate of the control, apparently by reducing an early oxidation product of anthocyanin breakdown.[28] This scheme has implications in flavonoid metabolism.

Sorata *et al*[173] studied the promoting effect of ascorbate on quercetin-induced suppression of photohemolysis in human erythrocytes. These authors suggested that the cooperation of quercetin with ascorbate in photohemolysis was due to the reduction of oxidized quercetin by ascorbate, resulting in the regeneration of the flavonol. Takahama's studies also suggested the reduction of oxidized quercetin to quercetin by ascorbate.[111] This author reported that the suppression of photooxidation of quercetin by ascorbate; during this suppression ascorbate was oxidized. Jan *et al*[174] reported that the antioxidative function of quercetin in inhibiting the photooxidation of α-tocopherol was enhanced by ascorbate. The enhancement was attributed to the reduction of oxidized quercetin by ascorbate. Takahama showed that the intermediates formed during the oxidation of flavonoids by the horseradish peroxidase-H_2O_2 system might be reduced by ascorbate; the oxidized product which could be reduced by ascorbate appeared to be an ortho-quinone derivative.[175]

An example with potential clinical relevance is the preservation of antiviral activity of quercetin in the presence of ascorbate which inhibits the oxidative degradation of the quercetin.[176] Maintenance of biological activity of other flavonoids by ascorbate is also suggested by the experiments of Kandaswami *et al*[177] who found that ascorbic acid augmented by about two-fold the antiproliferative effect of fisetin and quercetin on proliferation of HTB43 squamous cell carcinoma in tissue culture. Flavone had no effect indicating the requirement for hydroxylation; most likely the ascorbic acid had flavonoid protective activity. In other experiments (Middleton, Drzewiecki and Kandaswami, unpublished observations) it was demonstrated that low concentrations of ascorbic acid completely blocked the oxidation of quercetin in aqueous medium at pH 7.5 as determined spectrophotometrically over a 24 hour period. Considering the redox potentials for the reduction of ascorbic and metal ions,[170] ascorbic acid can itself reduce cupric and ferric ions. Metal ions like Cu^{2+} are known to oxidize flavonols such as quercetin in aqueous media.[178] Chelation of the vicinal hydroxyl groups of quercetin by Cu^{2+} would result in its conversion to quinone. The reduction by ascorbic acid of the quinone to the flavonol could have enhanced its biological activity in the above instances.

CONCLUSION

A large volume of studies have emphasized the potential health promoting and disease preventive effects of fruits and vegetables. The beneficial effects of fruits and vegetables have frequently been attributed to ascorbic acid and the carotenoids present in these foods. However, as stated elsewhere, fruits and vegetables contain a multitude of flavonoids and related phenolic compounds. The importance of flavonoids in disease prevention is being increasingly realized. The significance of these protective phytochemicals is emerging as an important issue. The natural antioxidants from plants are phenolic compounds. Flavonoids and related cinnamic acid derivatives, coumarins, tocopherols and phenolic acids constitute the most common antioxidants. Flavonoids can function as 1) metal chelators and reducing agents, 2) as scavengers of ROS, 3) as chain breaking antioxidants, and 4) as quenchers of the formation of singlet oxygen. Protection of ascorbic acid appears to be one of their important functions. They form complexes with metals. However, their primary antioxidant activity could be of greater value. In many of the studies reported, it is not certain whether flavonoids inhibit the formation of ROS or scavenge them. It is obvious that flavonoids react with $^{\bullet}OH$ and, therefore, can be very important chain breaking antioxidants. They could play an important role in conserving tocopherols in biological membranes. Flavonoids could also be important in protecting LDL from oxidation, thus reducing their atherogenicity. In general, flavonoids could potentially influence disease states in which lipid peroxidation products are intricately involved. Flavonoids can be important for protection against liver diseases, vascular disorders and heart diseases. Recent studies have identified isoflavonoids as constituents of plasma in healthy human subjects. No information is available as yet on the plasma concentrations of other flavonoids. Seeking such information is of crucial importance, since dietary exposure to flavonoids is substantial. Antioxidant flavonoids have been shown to have cancer preventive effects in experimental animals. They could have a marked influence on the metabolic activation of chemical carcinogens by virtue of their modulating effects on carcinogen metabolizing enzymes. Flavonoid inhibition of enzymes such as lipoxygenase and phagocyte NADPH oxidase involved in oxidative stress could play a role in the inhibition of tumor promotion. Plant flavonoids are a relatively unexplored group of potentially important nontoxic dietary antioxidants that warrant further intensive investigation.

ACKNOWLEDGMENTS

The financial assistance from the Cancer Research Foundation of America, the Department of Citrus, State of Florida, and Margaret Duffy and Robert Cameron Troup Memorial Fund is gratefully acknowledged.

REFERENCES

1. H. Sies. "Oxidative Stress," Academic Press, London (1985).
2. H. Sies, Oxidative stress: from basic research to clinical application. *Am. J. Med.* 91: 3C-31S (1991).
3. A. Blast, R.M.M. Guido, M.M. Haenen, J.A. Cees and J.A. Doelman, Oxidants and antioxidants: state of the art. *Am. J. Med.*, 91:3C-2S (1991).
4. B. Halliwell and J.M.C. Gutteridge, *in:* "Methods in Enzymology," L. Packer, and A.N. Glazer, ed., Academic Press, San Diego, Vol. 186, p. 1 (1990).
5. B. Halliwell, J.M.C. Gutteridge and C.E. Cross, Free radicals, antioxidants, and human disease: where are we now? *J. Lab Clin. Med.*, 119:598 (1992).
6. B. Halliwell, Reactive oxygen species in living systems: source, biochemistry, and role in human disease. *Am. J. Med.*, 91:3C-14S (1991).
7. N.I. Krinsky, Mechanism of action of biological antioxidants. *Proc. Soc. Exp. Biol. Med.*, 200:248 (1992).

8. J.B. Harborne, T.J. Mabry, M. Mabry. "The Flavonoids," Academic Press, New York (1975).
9. J.B. Harborne, T.J. Mabry. "The Flavonoids: Advances in Research," Chapman and Hall, London (1982).
10. J.B. Harborne. "The Flavonoids: Advances in Research Since 1980," Chapman and Hall, London (1988).
11. J.B. Harborne. "The Flavonoids: Advances in Research Since 1986," Chapman and Hall, London (1993).
12. J.B. Harborne, *in*: "Plant Flavonoids in Biology and Medicine: Biochemical Pharmacological and Structure-Activity Relationships," V. Cody, E. Middleton and J.B. Harborne, ed., Alan R. Liss, New York, p. 15 (1986).
13. J.B. Harborne, *in*: "Plant Flavonoids in Biology and Medicine: Biochemical, Cellular and Medicinal Properties," V. Cody, E. Middleton and J.B. Harborne, ed., Alan R. Liss, New York, p. 17 (1988).
14. J. Ebel, K. Hahlbrock, *in*: "The Flavonoids: Advances in Research," J.B. Harborne and T.J. Marby, ed., Chapman and Hall, London, p. 641 (1982).
15. R.R. Scheline. "Handbook of Mammalian Metabolism of Plant Compounds," CRC Press, Boca Raton (1991).
16. R. Brouillard and A. Cheminant, *in*: "Plant Flavonoids in Biology and Medicine: Biochemical, Cellular and Medicinal Properties," V. Cody, E. Middleton and J.B. Harborne, ed., Alan R. Liss, New York, p. 93 (1988).
17. K. Hermann, Flavonols and flavones in food plants: a review. *J. Food Technol.* 11:433 (1976).
18. J. Kuhnau, The flavonoids. A class of semi-essential food components: their role in human nutrition. *World Rev. Nutr. Diet* 24:117 (1976).
19. C.A.B. Clemetson. "Vitamin C," CRC Press, Boca Raton (1989).
20. W.S. Pierpoint, *in*: "Plant Flavonoids in Biology and Medicine: Biochemical, Pharmacological and Structure-Activity Relationships," V. Cody, E. Middleton and J.B. Harborne, ed., Alan R. Liss, New York, p. 125 (1986).
21. M.G.L. Hertog, P.C.H. Hollman, M.B.Katan and D. Kromhout, Intake of potentially anticarcinogenic flavonoids and their determinants in adults in The Netherlands. *Nutr. Cancer* 20: 21 (1993).
22. E. Middleton, Jr. and C. Kandaswami, *in*: "The Flavonoids: Advances in Research Since 1986," J.B. Harborne, ed., Chapman and Hall, London, p. 619 (1993).
23. E. Middleton, Jr. and C. Kandaswami, Effects of flavonoids on immune and inflammatory cell functions. *Biochem. Pharmacol.* 43:1167 (1992).
24. D.E. Pratt, *in*: "Phenolic Compounds in Food and Their Effects on Health II," M-T. Huang, C-H. Ho and C.Y. Lee, ed., American Chemical Society, Washington, p. 54 (1992).
25. R.E. Hughes and H.K. Wilson, *in*: "Progress in Medicinal Chemistry," G.P.Ellis and G.B. West, ed., Elsevier, Amsterdam, Vol. 14, p. 285 (1977).
26. C.H. Lea, P.A.T. Swoboda, On the antioxidant activity of the flavonols, gossypetin and quercetagetin. *Chem. Ind.* 1426 (1956).
27. J. Torel, J. Cillard and P. Cillard, Antioxidant activity of flavonoids and reactivity with peroxy radical. *Phytochemistry* 25:383 (1986).
28. J.W. McClure, Physiology and Functions of Flavonoids, *in*: "The Flavonoids," J.B. Harborne, T.J. Mabry and H. Mabry, ed., Academic Press, New York, p. 970 (1975).
29. C.A.B. Clemetson and L. Anderson, Plant polyphenols as antioxidants for ascorbic acid. *Ann. NY Acad. Sci.* 136:341 (1966).
30. E. Jones and R.E. Hughes, The influence of bioflavonoids on the absorption of vitamin C. *IRCS Med. Sci.* 12:320 (1984).
31. B.M. Babior and R.C. Woodman, Chronic granulomatous disease. *Semin. Hematol.* 27:247 (1990).
32. J.T. Curnutte and B.M. Babior, Chronic granulomatous disease. *Adv. Hum. Genet.* 16:229 (1987).
33. S.S. Weiss, Tissue destruction by neutrophils. *N. Engl. J. Med.* 320:365 (1989).
34. S.M. Michel and C. Bors, Reactions of NO with $0_2^{\bullet-}$ implications for the action of endothelium-derived relaxing factor (EDRF). *Free Radical Res. Commun.* 10:221 (1991).

35. J.S. Beckman, T.W. Beckman, J. Chen, *et al.*, Apparent hydroxyl radical production by peroxynitrite: implications for endothelial injury from nitric oxide and superoxide. *Proc. Natl. Acad. Sci. USA* 87:1620 (1990).
36. M.A. Marletta, Nitric oxide: Biosynthesis and biological significance. *Trends Biochem. Sci.* 14:488 (1989).
37. S. Moncada, R.M.J. Plamer and E.A. Higgs, Biosynthesis of nitric oxide from L-arginine: a pathway for the regulation of cell function and communication. *Biochem. Pharmacol.* 38:1709 (1989).
38. R.J. Gryglewski, R.M.J. Palmer and S. Moncada, Superoxide anion is involved in the breakdown of endothelium-derived vascular relaxing factor. *Nature (London)* 320:454 (1986).
39. F.R.M. Laurindo, P.L. daLuz, L. Uint, *et al.*, Evidence for superoxide radical-dependent coronary vasospasm after angioplasty in intact dogs. *Crculation* 83:1705 (1991).
40. J.C. Fantone and P.A. Ward, Role of oxygen-derived free radicals and metabolites in leukocyte-dependent inflammatory reactions. *Am. J. Path.* 107:397 (1982).
41. W.W. Busse, D.E. Kopp, and E. Middleton, Flavonoid modulation of human neutrophil function. *J. Allergy Clin. Immunol.* 73:801 (1984).
42. B.A. 'T Hart, T. Ram, I.P. Vai Ching, H. Van Di and R.P. Labadie, How flavonoids inhibit the generation of luminol-dependent chemiluminescence by activated human neutrophils. *Chem. Biol. Interactions* 73:323 (1990).
43. A.I. Tauber, J.R. Fay and M.A. Marletta, Flavonoid inhibition of the human neutrophil NADPH-oxidase. *Biochem. Pharmacol.* 33:1367 (1984).
44. P.C. Ferriola, V. Cody and E. Middleton. Protein kinase C inhibition by plant flavonoids. Kinetic mechanisms and structure-activity relationships. *Biochem. Pharmacol.* 38:1617 (1989).
45. J. Pincemail, C. Deby, A. Thirion, M. de Bruyn-Dister and R. Goutier, Human myeloperoxidase activity is inhibited *in vitro* by quercetin. Comparison with three related compounds. *Experientia* 44:450 (1988).
46. C.C. Winterbourn, Comparative reactivities of various biological compounds with myeloperoxidase-hydrogen peroxide-chloride, and similarity of the oxidant to hypochlorite. *Biochim. Biophys. Acta* 840:204 (1985).
47. V. Stolc, Characterization of iodoproteins secreted by phagocytosing human polymorphonuclear leukocytes. *J. Biol. Chem.* 254:1273 (1979).
48. W.D. Blackburn, L.W. Heck and R.W. Wallace, The bioflavonoid quercetin inhibits neutrophil degranulation, superoxide production, and the phosphorylation of specific neutrophil proteins. *Biochem. Biophys. Res. Commun.* 144:1229 (1987).
49. H. Ogasawara, T. Fujitani, G. Drzewiecki and E. Middleton, The role of hydrogen peroxide in basophil histamine release and the effect of selected flavonoids. *J. Allergy Clin. Immunol.* 78:321 (1986).
50. B. Halliwell and S. Chirico, Lipid peroxidation: its mechanism, measurement, and significance. *Am. J. Clin. Nutr.* 57:715S (1993).
51. H. Esterbauer, H. Zollner and R.J.Schaur, Hydroxyalkenals: cytotoxic products of lipid peroxidation. *ISI Atlas Sci. Biochem.* 1:311 (1988).
52. P.A. Cerutti, Prooxidant states and tumor promotion. *Science*, 227:375 (1985).
53. J.S. Bus and J.E. Gibson, Lipid peroxidation and its role in toxicology. *Rev. Biochem. Toxicol.* 1:125 (1979).
54. G.L. Plaa and H. Witschi, Chemicals, drugs, and lipid peroxidation. *Ann. Rev. Pharmacol.Toxicol.* 16:125 (1976).
55. R.G. Recknagel and E.A. Glende, Lipid peroxidation: a specific form of cellular injury. *Handbook of Physiol.* 9:591 (1979).
56. A.L. Tappel, Protection against free radical lipid peroxidation reactions. *Exp. Med. Biol.* 97:111 (1978).
57. A. Letan, The relation of structure to antioxidant activity of quercetin and some of its derivatives. II. Secondary (metal-complexing) activity. *J. Food Sci.* 31:518 (1966).
58. A. Bindoli, L. Cavallin and N. Siliprandi, Inhibitory action of silymarin of lipid peroxide formation in rat liver mitochondria and microsomes. *Biochem. Pharmacol.* 26:2405 (1977).
59. L. Cavallini, A. Bindoli and N. Siliprandi, Comparative evaluation of antiperoxidative action of silymarin and other flavonoids. *Pharmacol. Res. Comm.* 10:133 (1978).

60. U. Takahama, Suppression of lipid photoperoxidation by quercetin and its glycosides in spinach chloroplasts. *Photochem. Photobiol.* 38:363 (1983).
61. Y. Sorata, U. Takahama and M. Kimura, Protective effect of quercetin and rutin on photosensitized lysis of human erythrocytes in the presence of hematophorphyrin. *Biochim. Biophys. Acta* 799:313 (1984).
62. I. Maridonneau-Parini, P. Braquet and R.P.Garay, Heterogeneous effect of flavonoids on K^+ loss and lipid peroxidation induced by oxygen-free radicals in human red cells. *Pharmac. Res. Commun.* 18:61 (1986).
63. M. Young and C.P. Sieger, Inhibitory action of some flavonoids on enhanced spontaneous lipid peroxidation following glutathione depletion. *Planta Med.* 43:240 (1981).
64. L.A. Videla, V. Fernandez, A. Valenzuela and G. Ugarte, Effect of (+)-cyanidanol-3 on the changes in liver glutathione content and lipoperoxidation induced by acute ethanol administration in the rat. *Pharmacology* 22:343 (1981).
65. A. Muller and H. Sies, Role of alcohol dehydrogenase activity and of acetaldehyde in ethane and pentane production by isolated perfused rat liver. *Biochem. J.* 206:153 (1981).
66. L.A. Videla, A. Valenzuela, V. Fernandez and A. Kriz, Differential lipid peroxidative response of rat liver and lung tissues to glutathione depletion induced *in vivo* by diethyl maleate: effect of the antioxidant flavonoid (+)-cyanidanol-3. *Biochem. Int.* 10:425 (1985).
67. A. Valenzuela and R. Guerra, Differential effect of silybin on the Fe^{2+}-ADP and *t*-butyl hydroperoxide-induced microsomal lipid peroxidation. *Experientia* 42:139 (1986).
68. M. Naim, B. Gestetner, A. Bondi and Y. Birk, Antioxidative and antihemolytic activities of soybean isoflavones. *J. Agric. Food Chem.* 24:1174 (1975).
69. H.C. Jha, G. von Recklinghausen and F. Zilliken, Inhibition of *in vitro* microsomal lipid peroxidation by isoflavonoids. *Biochem. Pharmacol.* 34:1367 (1985).
70. B.J.F. Hudson and J.I. Lewis, Polyhydroxy flavonoid antioxidants for edible oils. Structural criteria for activity. *Food Chem.* 10:47-55 (1965).
71. T.H. Simpson and N. Uri, Hydroxyflavones as inhibitors of the aerobic oxidation of unsaturated fatty acids. *Chem. Ind.* 956 (1956).
72. A.C. Mehta and T.R. Seshadri, Flavonoids as antioxidants. *J. Sci. Ind. Research* 18B:24 (1959).
73. A. Heimann, W. Heimann, M. Gremminger and H.H. Holman, Quercetin as an antioxidant. *Fette u. Soifen* 55:394 (1953).
74. Y. Kimura, M. Kubo, T. Tani, S. Arichi and H. Okuda, Studies on *Scutellariae radix* IV. Effects on lipid peroxidation in rat liver. *Chem. Pharm. Bull. (Tokyo)* 29:2610 (1981).
75. Y. Kimura, H. Okuda, Z. Taira, N. Shoji, T. Takemoto and S. Arichi, Studies on *Scutellariae radix* IX. New component inhibiting lipid peroxidation in rat liver. *Planta Med.* 50:290 (1984).
76. J. Baumann, G.Wurm and F.V.Bruchhausen, Prostaglandin synthetase inhibition by flavonoids and phenolic compounds in relation to their $0_2^{\bullet-}$ - scavenging properties. *Arch. Pharm.* 313:330 (1980).
77. G.W. Burton and K.U. Ingold, Autoxidation of biological molecules. 1. The antioxidant activity of vitamin E and related chain-breaking phenolic antioxidants *in vitro*. *J. Am. Chem. Soc.* 103:6472 (1981).
78. A.K. Ratty and N.P. Das, Effects of flavonoids on nonenzymatic lipid peroxidation: structure-activity relationship. *Biochem. Med. Metab. Biol.* 39:69 (1988).
79. I.B. Afanas'ev, A.I. Dorozhko, A.V. Brodskii, V.A. Kostyuk and A.I. Potapovitch, Chelating and free radical scavenging mechanisms of inhibitory action of rutin and quercetin in lipid peroxidation. *Biochem. Pharmocol.* 38:1763 (1989).
80. B.A. Svingen, J.A. Buege, F.O. O'Neal and S.D Augst, The mechanism of NADPH-dependent lipid peroxidation. The propagation of lipid peroxidation. *J. Biol. Chem.* 254:5892 (1979).
81. E. Albano, K.A.K. Lott, T.F. Slater, A. Stier, M.S.R. Symons and P. Tomasi, Spin-trapping studies on the free-radical products formed by metabolic activation of carbon tetrachloride in rat liver microsomal fractions of isolated hepatocytes and *in vivo* in the rat. *Biochem. J.* 204:593 (1982).

82. H. Hara, T. Sukamoto, H. Ohtaka, K. Abe, Y. Tatumi, Y. Saito, A. Suzuki and G. Tsukamoto, Effects of baicalein and alpha-tocopherol on lipid peroxidation, free radical scavenging activity and 12-*0*-tetradecanoylphorbol acetate-induced ear edema. *Eur. J. Pharmacol.* 221:193 (1992).
83. A. Mora, M. Paya, J.L. Rios and M.J. Alcaraz, Structure-activity relationships of polymethoxyflavones and other flavonoids as inhibitors of non-enzymic lipid peroxidation. *Biochem. Pharmacol.* 40:793 (1990).
84. M.J. Alcaraz and M.L. Ferrandiz, Modification of arachidonic metabolism by flavonoids. *J. Ethnopharmacol.* 21:209 (1987).
85. M.J. Laughton, P.J. Evans, M.A. Moroney, J.R.S. Hoult and B. Halliwell, Inhibition of mammalian 5-lipoxygenase and cyclo-oxygenase by flavonoids and phenolic dietary additives. Relationship to antioxidant activity and to iron ion-reducing ability. *Biochem. Pharmacol.* 42:1673 (1991).
86. J. Robak, F. Shridi, M. Wolbis and M. Krolikowska, Screening of the influence of flavonoids on lipoxygenase and cyclooxygenase activity, as well as on nonenzymic lipid oxidation. *Pol. J. Pharmacol. Pharm.* 40:451 (1988).
87. T. Sato, A. Kawamoto, A. Tamuro, Y. Tatsumi and T. Fugii, Mechanism of antioxidant action of pueraria glycoside (PG)-1 (an isoflavonoid) and mangiferin (a xanthonoid). *Chem. Pharm. Bull.* 40:721 (1992).
88. M.J. Laughton, B. Halliwell, P.J. Evans and J.R.S. Hoult, Antioxidant and pro-oxidant actions of the plant phenolics quercetin, gossypol an myricetin. Effects on lipid peroxidation, hydroxyl radical generation and bleomycin-dependent damage to DNA. *Biochem. Pharmacol.* 38:2859 (1989).
89. S.R. Husain, J. Cillard and P. Cillard, α-Tocopherol prooxidant effect and malondialdehyde production. *J. Am. Oil Chem. Soc.* 64:109 (1987).
90. K. Yamamoto and E. Niki, Interaction of alpha-tocopherol with iron: antioxidant and prooxidant effects of alpha-tocopherol in the oxidation of lipids in aqueous dispersions in the presence of iron. *Biochim. Biophys. Acta* 958:19 (1988).
91. A.W. Girotti, J.P. Thomas and J.E. Jordan, Prooxidant and antioxidant effects of ascorbate on photosensitized peroxidation of lipids in erythrocyte membranes. *Photochem. Photobiol.* 41:267 (1985).
92. E. Rekka and P.N. Kourounakis, Effect of hydroxyethyl rutosides and related comounds on lipid peroxidation and free radical scavenging activity. Some structural aspects. *J. Pham. Pharmacol.* 43:486 (1991).
93. A. Valenzuela, R. Guerria and L.A. Videla, Antioxidant properties of the flavonoids silybin and (+)-cyanidanol-3: comparison with butylated hydroxyanisole and butylated hydroxytoluene. *Planta Med.* 52:438 (1986).
94. T.F. Slater, M.N. Eakins, *in*: "New Trends in the Therapy of Liver Disease," A. Bertelli, ed., Karger, Basel, p. 84 (1975).
95. L.A. Videla, Assessment of the scavenging action of reduced glutathione, (+)-cyanidanol-3 and ethanol by the chemiluminescent response of the xanthine oxidase reaction. *Experientia* 34:500 (1983).
96. L.A. Videla, C. Fraga, O. Koch and A. Boveris, Chemiluminescence of the *in situ* rat *Biochem. Pharmacol.* 32:2822 (1983).
97. C.G. Fraga, V.S. Martino, G.E. Ferraro, J.D. Coussio and A. Boveris, Flavonoids as antioxidants evaluated by *in vitro* and *in situ* liver chemiluminescence. *Biochem. Pharmacol.* 36:717 (1987).
98. A. Boveris, S.F. Llesuy and C.G. Fraga, Increased liver chemiluminescence in tumor-bearing mice. *Free Radical Biol. Med.* 1:131 (1985).
99. T.F. Slater, Free-radical mechanisms in tissue injury. *Biochem. J.* 221:1 (1984).100.
100. P.B. McCay, E.K. Lai, J.L. Poyer, C.M. DuBoise and E.G. Janzen, Oxygen- and carbon-centered free radical formation during carbon tetrachloride metabolism. Observation of lipid radicals in vivo and in vitro. *J. Biol. Chem.* 259:2135 (1984).
101. B. Chance, H. Sies and A. Boveris, Hydroperoxide metabolism in mammalian organs. *Physiol. Rev.* 59:527 (1979).
102. H.P. Koch and E. Loffler, Influence of silymarin and some flavonoids on lipid peroxidation in human platelets. *Methods Fund. Exp. Clin. Pharmacol.* 7:13 (1985).
103. H. Kappus, D. Koster-Albrecht and H. Remmer, 2-Hydroxyoestradiol and (+) -cyanidanol-3 prevent lipid peroxidation of isolated rat hepatocytes. *Arch. Toxicol.* 2:321 (1979).

104. M.R. Cholbi and M.J. Alcaraz, Inhibitory effects of phenolic compounds on CCl_4-induced microsomal lipid peroxidation. *Experientia* 47:195 (1991).
105. P-F. Wang and R-L. Zheng, Inhibition of the autoxidation of linoleic acid by flavonoids in micelles. *Chem. Phy. Lipids* 63:37 (1992).
106. W. Bors, W. Heller, C. Michel and M. Saran, *in*: "Methods in Enzymology," L. Packer and A.M. Glazer, ed., Academic Press, NY, p. 343 (1990).
107. T.J. Mabry, K.R. Markham and M.B.Thomas. "The Systematic Identification of Flavonoids Part 2," Spring-Verlag, Berlin (1970).
108. W. Bors and M. Saran, Radical scavenging by flavonoid antioxidants. *Free Rad. Res. Commun.* 2:289 (1987).
109. M. Erben-Russ, W. Bors, and M. Saran, Reactions of linoleic acid peroxyl radicals with phenolic antioxidants: a pulse radiolysis study. *Int. J. Radiat. Biol.* 52:393 (1987).
110. I. Ueno, M. Kohno, K. Haraikawa and I. Hirono, Interaction between quercetin and superoxide radicals. Reduction of the quercetin mutagenicity. *J.Pharmacobio-Dyn.* 7:798 (1984).
111. U. Takahama, $0_2^{\bullet -}$ dependent and -independent photooxidation of quercetin in the presence and absence of riboflavin and effects of ascorbate on the photooxidation. *Photochem. Photobiol.* 42:89 (1985).
112. S.R. Husain, J. Cillard and P. Cillard, Hydroxyl radical scavenging activity of flavonoids. *Phytochemistry* 26:2489 (1987).
113. U. Takahama, Oxidation products of kaempferol by superoxide anion radical. *Plant Cell Physiol.* 28:953 (1987).
114. J. Robak and R.J. Gryglewski, Flavonoids are scavengers of superoxide anions. *Biochem. Pharmacol.* 37:837 (1988).
115. A.I. Huguet, S. Manez and M.J. Alcaraz, Superoxide scavenging properties of flavonoids in a non-enzymic system. *Z. Naturforsch* 45C:19 (1990).
116. G. Sichel, C. Corsaro, M. Scalia, A.J. Bilio and R.P. Bonomo, *In vitro* scavenger activity of some flavonoids and melanins against $0_2^{\bullet -}$. *Free Rad. Biol. Med.* 11:1 (1991).
117. N. Cotelle, J.L. Bernier, J.P. Henichart, J.P.C. Catteau, E. Gaydou and J.C. Wallet, Scavenger and antioxidant properties of ten synthetic flavones. *Free Rad. Biol. and Med.* 13:211 (1992).
118. W.F. Hodnick, F.S. Kung, W.J. Roettger, C.W. Bohmont and R.S. Pardini, Inhibition of mitochondrial respiration and production of toxic oxygen radicals by flavonoids. A structure-activity study. *Biochem. Pharmacol.* 35:2345 (1986).
119. W.F. Hodnick, C.W. Bohmont, C. Capps and R.S. Pardini, Inhibition of the mitochondrial NADH-oxidase (NADH coenzyme Q oxido-reductase) enzyme system by flavonoids: a structure-activity study. *Biochem. Pharmacol.* 36:2873 (1986).
120. A.J. Elliott, S.A. Schreiber, C. Thomas and R.S. Pardini, Inhibition of glutathione reductase by flavonoids. A structure-activity study. *Biochem. Pharmacol.* 44:1603 (1992).
121. W.F. Hodnick, B. Kalyanaraman, C.A. Fritsos and R.S. Pardini, *in*: "Oxygen Radicals in Biology and Medicine," M.G. Simic, K.A. Taylor, J.F. Ward, C. VonSonntag, ed., Plenum Press, New York, p. 149 (1988).
122. A.T. Canada, E. Gianella, T.D. Nguyen and R.P. Mason, The production of reactive oxygen species by dietary flavonols. *Free Rad. Biol. Med.* 9:441 (1990).
123. G. Hahn, H.D. Lehmann, M. Kurten, H. Uebel, and G. Vogel, On the pharmacology and toxicology of silymarin, an antihepatotoxic active priciple from silybum marianum. *Arzneim.-Forsh. (Drug Res.)* 18:698 (1968).
124. A. Greimel and H. Koch, Silymarin - an inhibitor of horseradish peroxidase. *Experientia* 33:1417 (1977).
125. D. Perrissoud, *in*: "Plant Flavonoids in Biology and Medicine: Biochemical, Pharmacological and Structure-Activity Relationships," V. Cody, E. Middleton and J.B. Harborne, ed., Alan R. Liss, New York, p. 559 (1986).
126. A. Valenzuela, T. Barria, T. Guerra and A. Garrido, Inhibitory effect of the flavonoid silymarin on the erythrocyte hemolysis induced by phenylhydrazine. *Biochem. Biophys. Res. Commun.* 126:712 (1985).
127. A. Valenzuela and R. Guerra, Protective effect of the flavonoid silybin dihemisuccinate on the toxicity of phenylhydrazine on rat liver. *FEBS Lett.* 181:271 (1985).

128. A. Velenzuela, C. Lagos, K. Schmidt and L.A. Videla, Silymarin protection against hepatic lipid peroxidation induced by acute ethanol intoxication in the rat. *Biochem. Pharmacol.* 39:2209 (1985).
129. A. Valenzuela, R. Guerra and A. Garrido, Silybin dihemisuccinate protects rat erythrocytes against phenylhydrazine-induced lipid peroxidation and hemolysis. *Planta Medica* 53:402 (1987).
130. H.M. Rauen, H. Schriewer, U. Tegtbauer and J.E. Lasana, Silymarin prevents peroxidation of lipids in carbon tetrachloride-induced liver damage. *Experientia* 29:1372 (1973).
131. P. Letteron, G. Labre, C. Degott, A. Berson, B. Fromenty, M. Delaforge, D. Larrey and D. Pessayre, Mechanism for the protective effects of silymarin against carbon tetrachloride-induced lipid peroxidation and hepatotoxicity in mice. Evidence that silymarin acts both as an inhibitor of metabolic activation and as a chain-breaking antioxidant. *Biochem. Pharmacol.* 39:2027 (1990).
132. G. Labbe, V. Descatoire, P. Letteron, C. Degott, M. Tinel, D. Larrey, Y. Carrion-Pavlor, G. Amouyal and D. Pessayre, The drug methoxsalen, a suicide substrate for cytochrome P-450, decreases the metabolic activation, and prevents the hepatotoxicity of carbon tetrachloride in mice. *Biochem. Pharmacol.* 36:907 (1987).
133. J. Feher, A. Cornides, J. Pal, I. Lang and G. Csomos, Liver cell protection in toxic liver lesion. *Acta Physiol. Hungarica*, 73:285 (1989).
134. M. Lonchampt, B. Guardiola, N. Sicot, M. Bertrand, L. Perdix and J. Duhault, Protective effect of a purified flavonoid fraction against reactive oxygen radicals. *In vivo* and *in vitro* study. *Arzneim.-Forsch (Drug Res.)* 39:882 (1989).
135. T. Nakayama, M. Yamada, T. Osawa and S. Kawakishi, Suppression of active oxygen-induced cytotoxicity by flavonoids. *Biochem. Pharmacol.* 45:265 (1993).
136. T. Nakayama, T. Niimi, T. Osawa and S. Kawakishi, The protective role of polyphenols in cytotoxicity of hydrogen peroxide. *Mutat. Res.* 281:77 (1992).
137. L.G. Korkina, A.D. Durnev, T.B. Suslova, Z.P. Cheremisina, N.O. Daugel-Dauge and I.B. Afanas'ev, Oxygen radical-mediated mutagenic effect of asbestos on human lymphocytes: suppression by oxygen radical scavengers. *Mutat. Res.* 265:245 (1992).
138. S. Kantengwa and B.S. Polla, Flavonoids, but not protein kinase C inhibitors, prevents stress protein synthesis during erythrophagocytosis. *Biochem Biophys. Res. Comm.* 180:308 (1991).
139. R.J. Ruch, S-J. Cheng and J.E. Klaunig, Prevention of cytotoxicity and inhibition of intercellular communication by antioxidant catechins isolated from Chinese green tea. *Carcinogenesis* 10:1003 (1989).
140. J.E. Trosko and C.C. Chang, *in*: "Mechanisms of Tumor Promotion: Cellular Responses to Tumor Promoters," T.J. Slaga, ed., CRC Press, Boca Raton, Vol. 4, p. 119 (1984).
141. I. Morel, G. Lescoat, P. Cogrel, O. Sergent, N. Pasdeloup, P. Brissot, P. Cillard and J. Cillard, Antioxidant and iron-chelating activities of the flavonoids catechin, quercetin and diosmetin on iron-loaded rat hepatocyte cultures. *Biochem. Pharmcol.* 43:13 (1993).
142. R.F. Boyer, H.M. Clark and A.P. LaRoche, Reduction and release of ferritin iron by plant phenolics. *J. Inorg. Biochem.* 32:171 (1988).
143. O.I. Aruoma, *in*: "Free Radicals and Food Additives," O.I. Aruoma and B. Halliwell, ed., Taylor and Francis, London, p. 173 (1991).
144. A. Kapus and G.L. Lukacs, (+)-cyanidanol-3 prevents the functional deterioration of rat liver mitochondria induced by Fe_2^+ ions. *Biochem. Pharmacol.* 35:2119 (1986).
145. C.V. de Whalley, S.M. Rankin, R.S. Hoult, W. Jessup and D.S. Leake, Flavonoids inhibit the oxidative modification of low density lipoproteins by macrophages. *Biochem. Pharmacol.* 39:1743 (1990).
146. A. Negra-Salvayre, V. Reaud, C. Hariton and R. Salvayre, Protective effect of alpha-tocopherol, ascorbic acid and rutin against peroxidative stress induced by oxidized lipoproteins on lymphoid cell lines. *Biochem. Pharmacol.* 2:450 (1991).
147. A. Negra-Salvayre, Y. Alomar, M. Troly and R. Salvayre, Ultraviolet-treated lipoproteins as a model system for the study of the biological effects of lipid peroxides on cultured cells. III. The protective effect of antioxidants (probucol,

catechin, vitamin E) against the cytotoxicity of oxidized LDL occurs in two different ways. *Biochim. Biophys. Acta* 1096:291 (1991).

148. A. Negre-Salvayre and R. Salvayre, Quercetin prevents the cytoxicity of oxidized LDL on lymphoid cell lines. *Free Rad. Biol. Med.* 12:101 (1992).
149. A. Negra-Salvayre, M. Lopez, T. Levade, M.T. Pierraggi, N. Dousset, L. Douste-Blazy and R. Salvayre, Ultraviolet-treated lipoproteins as a model system for the study of the biological effects of lipid peroxides on cultured cells. II. Uptake and cytotoxicity of ultraviolet-treated LDL on lymphoid cell lines. *Biochim. Biophys. Acta* 1045:224 (1990).
150. M.G.L. Hertog, E.J.M. Feskens, P.C.H. Hollman, M.B. Katan and D. Kromhout, Dietary antioxi dant flavonoids and risk of coronary heart disease: the Zutphen Elderly Study. *Lancet* 342:1007 (1993).
151. E.N. Frankel, J. Kanner, J.B. German, E. Parks and J.E. Kinsella, Inhibition of oxidation of human low-density lipoprotein by phenolic substances in red wine. *Lancet* 341:454 (1993).
152. J. Sanhueza, J. Valdes, R. Campos, A. Garrido and A. Valenzuela, Changes in the xanthine dehydrogenase/xanthine oxidase ratio in the rat kidney subjected to ischemia-reperfusion stress: preventive effect of some flavonoids. *Res. Commun. Chem. Pathol. Pharmcol.* 78:211 (1992).
153. J.M. McCord, Oxygen-derived free radicals in postischemic tissue injury. *N. Engl. J. Med.* 312:159 (1986).
154. M. Iio, Y. Ono, S. Kai and M. Fukumoto, Effects of flavonoids on xanthine oxidation as well as on cytochrome C reduction by milk xanthine oxidase. *J. Nutr. Sci. Vitaminol.* 32:635 (1986).
155. X-H. Ning, X. Ding, K.F. Childs, S.F. Bolling and K.P. Gallagher, Flavone improves functional recovery after ischemia in isolated reperfused rabbit hearts. *J. Thorac. Cardiovasc. Surg.* 105:541 (1993).
156. R.E. White, M.J. Coon, Oxygen activation by cytochrome P-450. *Ann. Rev. Biochem.* 49:315 (1980).
157. B. Halliwell, J.M.C. Gutteridge. "Free Radicals in Biology and Medicine," Oxford, New York (1985).
158. L.L. Fan, D.D. O'Keefe, W.J. Powell, Pharmacologic studies on *Radix puerariae*: effect of puerarin on regional myocardial blood flow and cardiac hemodynamics in dogs with acute myocardial ischemia. *J. Chin. Med. (England)* 98:821 (1985).
159. T. Sacks, C.F. Moldow, P.R. Craddock, T.K. Bowers, H.S. Jacobs, Oxygen radicals mediate endothelial cell damage by complement-stimulated granulocytes. An *in vitro* model of immune vascular damage. *J. Clin. Invest.* 61:1161 (1978).
160. S.J. Weiss, J. Young, A.F. LoBuglio, A. Slivka and N.G. Nimch, Role of hydrogen peroxide in neutrophil-mediated destruction of cultured endothelial cells. *J.Clin. Invest.* 68:714 (1981).
161. T. Mazzone, M. Jensen and A. Chait, Human arterial wall cells secrete factors that are chemotactic for monocytes. *Proc. Natl. Acad. Sci. USA* 80:5094 (1983).
162. W-C. Chang and F-L. Hsu, Inhibition of platelet activation and endothelial cell injury by flavan-3-ol and saikosaponin compounds. *Prostg. Leukotr. Ess. Fatty Acids* 44:51 (1991).
163. R.J. Gryglewski, R. Korbut, J. Robak and J. Swies, On the mechanism of antithrombotic action of flavonoids. *Biochem. Pharmacol.* 36:317 (1987).
164. K.A. Harper, A.D. Morton and F.J. Rolfe, The phenolic compounds of black currant juice and their protective effect on ascorbic acid. *J. Food Technol.* 4:255 (1969).
165. K. Bhagvat, Effect of hesperedin and a factor in Bengal gram (*Cicer arietinum*) on growth of guinea-pigs. *Indian J. Med. Res.* 34:87 (1946).
166. H.K. Wilson and C. Price-Jones, The influence of an extract of orange peel on the growth and ascorbic acid metabolism of young guinea-pigs. *J. Sci. Food Agr.* 22:551 (1971).
167. J.L. Parrot and J.M. Gazave, Reduction de l'acide dehydroascorbique par le glutathion. Augmentation du rendement final de la reaction par la catechine. *C. R. Soc. Biol.* 145:821 (1951).
168. C. Regnault-Roger, J.M. Gazave and J. Devynck, Reduction of dehydroascorbic acid by thiols in presence of biocatalytic substances: polarographic study. *Int. J. Vitamin. Nutr. Res.* 52:158 (1982).

169. Z. Zloch, Effect of bioflavonoids on the utilization of the vitamin C activity of crystalline L-dehydroascorbic acid. *Int. J. Vitamin. Nutr. Res.* 43:378 (1973).
170. T.K. Basu and C.J. Schorah. "Vitamin C in Health and Disease," AVI Publishing, Westport (1982).
171. H.G. Stich, J. Karim, J. Moropatnick and L. Lo, Mutagenic action of ascorbic acid. *Nature (London)* 260:722 (1976).
172. M. Burnet. "Intrinsic Mutagenesis: A Genetic Approach to Aging, " MTP, Lancster, (1974).
173. Y. Sorata, U. Takahama and M. Kimura, Cooperation of quercetin with ascorbate in the protection of photosensitized lysis of human erythrocytes in the presence of hematoporphyrin, *Photochem. Photobiol.* 48:195 (1988).
174. C-Y. Jan, U. Takahama and M. Kimura, Inhibition of photooxidation of α-tocopherol by quercetin in human blood cell membranes in the presence of hematoporphyrin as a photosensitizer. *Biochim. Biophys. Acta* 1086:7 (1991).
175. U. Takahama, Spectrophotometric study on the oxidation of rutin by horseradish peroxidase and characteristics of the oxidized products. *Biochim. Biophys. Acta* 882:445 (1986).
176. R. Vrijsen, L. Everaert and A. Boeye, Antiviral activity of flavones and potentiation by ascorbate. *J. Gen. Virol.* 69:1749 (1988).
177. C. Kandaswmi, E. Perkins, D.S. Soloniuk, G. Drzewiecki and E. Middleton, Jr., Ascorbic acid-enhanced antiproliferative effect of flavonoids on squamous cell carcinoma *in vitro*. *Anti-Cancer Drugs* 4:91 (1993).
178. J.K. Kochi. "Organometallic Mechanisms and Catalysis," Academic Press, New York (1978).

PLASMA CLEARANCE AND IMMUNOLOGIC PROPERTIES OF LONG-ACTING SUPEROXIDE DISMUTASE PREPARED USING 35,000 TO 120,000 DALTON POLY-ETHYLENE GLYCOL

Mark G.P. Saifer[1], Ralph Somack[2] and L. David Williams[1]

[1]DDI Pharmaceuticals, Inc. 518 Logue Avenue, Mountain View, CA 94043 USA
[2]Applied BioSystems, Inc., 850 Lincoln Centre Drive, Foster City, CA 94404

ABSTRACT

Some biological properties of bovine and recombinant human Cu,Zn superoxide dismutase (bSOD and rhSOD)-poly-ethylene glycol (PEG) adducts prepared by coupling 1-9 strands of high molecular weight PEG (35,000-120,000 daltons) are compared to SOD adducts coupled with 7 or 15 strands of low molecular weight PEG (5,000 daltons).

Plasma clearance after i.v. injection was measured in mice and dogs. Conjugates of bSOD with 2 strands of PEG 40,000, 3 strands of PEG 72,000 or 1 strand of PEG 100,000 demonstrated half-lives of about 36 hours in mice, whereas the half-life of a conjugate with 7 strands of PEG 5,000 was about 24 hours. A PEG-bSOD with an average of 3.3 strands of PEG 41,000 was cleared from plasma with a terminal half-life of 36 to 48 hours after intraperitoneal injection in mice. PEG-SODs prepared from bSOD and rhSOD with 3 strands of PEG 50,000 each had plasma half-lives of approximately five days in dogs.

An enzyme immunoassay (ELISA) was employed to measure cross-reactivity with a rabbit antibody directed against bSOD. A series of bSOD adducts with 2 to 9 strands of PEG 35,000-120,000 were compared to PEG-bSODs with 7 or 15 strands of PEG 5,000. Attaching larger PEG strands was at least 3 times more effective in reducing antigenicity, compared to PEG 5,000.

Ability to induce sensitizing antibodies was measured using subcutaneous sensitization followed by i.v. or s.c. challenge in mice. Some bSOD conjugates with either 7 or 15 strands of PEG 5,000 induced sensitization reactions before the sixth challenge. Fewer than 1% of the animals tested with bSOD or rhSOD adducts with 3 or 4 strands of PEG 65,000, or 3 strands of PEG in the 30,000-50,000 molecular weight range, showed signs of anaphylaxis during six or seven challenges.

In a passive cutaneous anaphylaxis test (optimized to measure mouse IgE), neither bSOD coupled to 2 strands of PEG 120,000 nor bSOD coupled to 9 strands of PEG 35,000 induced detectable antibodies against either of these PEG-bSOD preparations or against bSOD; however, the adduct with 2 strands of PEG 120,000 reacted weakly with pre-formed antibodies to bSOD. The high molecular weight PEG-bSODs tested were not immunogenic, but were weakly antigenic, compared to bSOD.

KEY WORDS: Long-acting, superoxide dismutase, SOD, PEG, PEG-SOD.

INTRODUCTION

Bovine orgotein, a pharmaceutical preparation of the naturally-occurring enzyme Cu,Zn superoxide dismutase (SOD) has shown therapeutic potential in treating disorders where superoxide radicals are implicated, including inflammatory diseases (especially arthritis), oxygen toxicity and radiation injury; however, the short circulating half-life of SOD (25 minutes in man)[1] limits its effectiveness in conditions requiring sustained levels of SOD activity in the plasma. Although highly purified bSOD is well tolerated in man, hypersensitivity has been observed occasionally. The availability of rhSOD might reduce the incidence of immunization, but does not overcome the limitation of rapid clearance from the circulation.

Increasing the hydrodynamic radius of SOD by attaching water-soluble polymers extends the enzyme's half-life in animals.[2] Poly-ethylene glycol or poly-ethylene oxide (PEG) has most often been used[3]; however, the extent of PEG modification needed (usually 50% or more of the 20 lysine residues) often causes substantial reduction of SOD activity.[4-7]

Previously, we described the preparation of PEG-SODs containing high molecular weight PEG (41,000-72,000 daltons) and showed that by attaching fewer than 4 high molecular weight PEG strands, very large conjugates are produced which retain at least 90% of native SOD activity.[8] In the present report we compare the biological properties of these new conjugates to PEG-SODs containing 7 or 15 strands of low molecular weight PEG (5,000 daltons).

MATERIALS AND METHODS

Materials

Cu,Zn SOD derived from beef liver (bSOD) or from recombinant organisms (rhSOD) from DDI Pharmaceuticals, Inc. (Mountain View, CA) was coupled to high molecular weight PEG as described.[8]

Poly-ethylene oxide 21,000 (PEG 21K) and 45,000 (PEG 45K) were calibration standards purchased from Toyo Soda Manufacturing Company, Ltd. (Tokyo). Poly-ethylene glycol 35,000 (PEG 35K) was from Fluka Chemical Corp. (Ronkonkoma, NY) and poly-ethylene glycol (Polyox) 100,000 (PEG 100K) was from Union Carbide Corp., Specialty Chemicals Division (Danbury, CT). The apparent molecular weights of these compounds measured by steric exclusion HPLC based on commercial PEG calibration standards were 21,000, 46,000, 41,000 and a broad distribution with a mode at 50,000 daltons, respectively. All molecular weights reported are peak molecular weights estimated from steric exclusion HPLC,[8] using PEG (poly-ethylene oxide or poly-ethylene glycol) calibration standards. PEG 5K-bSOD compounds containing 7 and 15 polymer strands per molecule and demonstrating 52% and 43% of native activity were purchased from Enzon, Inc. (South Plainfield, NJ).

Sodium hydride and ethyl bromoacetate were from Aldrich Chemical Co. (Milwaukee, WI). Xanthine oxidase was from Calbiochem Corp. (La Jolla, CA). Cytochrome-c, horseradish peroxidase, *o*-phenylenediamine and nitroblue tetrazolium (NBT) were purchased from Sigma (St. Louis, MO). Mice were 17-32 gm Swiss Webster females (Simonsen Laboratories, Gilroy, CA). Dogs were 9-15 kg male and female, vaccinated, laboratory-bred beagles (CERM, Riom, France).

Methods

Analytical Methods for PEG-SOD Conjugates. PEG-SOD molecular weights and extents of conjugation (number of strands per 31,440 dalton SOD molecule) were determined by aqueous steric exclusion HPLC.[8] Protein was measured by the biuret method and SOD activity by the cytochrome-c assay.[9]

Preparation of PEG-SOD Conjugates Using High Molecular Weight PEG. PEG 21K, PEG 41K and PEG 46K were used without purification. PEG 100K was either used directly or fractionated by ultrafiltration[8] to obtain PEG starting materials with molecular weights ranging from 32,000 to 120,000. PEG-succinate esters were prepared by reacting PEG with succinic anhydride.[8] PEG-acetic acid ethers were synthesized by reaction with ethyl bromoacetate in dioxane in the presence of sodium hydride.[10] The PEG-acids were coupled to SOD after activation using N-hydroxysuccinimide (NHS) and dicyclohexylcarbodiimide.[8] Unless otherwise specified, the conjugates described in the present paper were derived from NHS-activated PEG-succinic acid esters.

Blood Clearance Studies. PEG-SOD elimination was measured by analysis of plasma or serum samples taken at intervals after injection of the test compound. SOD activity in the sample was measured by thin film agarose gel electrophoresis using NBT-riboflavin activity staining[11] and also by a soluble NBT-xanthine/xanthine oxidase spectrophotometric assay.[12] SOD activity in the gel assay was estimated by visual comparison with standards on the same gel. SOD activity in the spectrophotometric NBT assay was measured by comparison to a bSOD standard curve. Assays were performed in duplicate or triplicate and the results averaged. In studies with mice, solutions of the test materials were prepared in 0.9% saline and the dose introduced intravenously (5 mg/kg in 0.1-0.2 ml) through the tail vein, or into the peritoneal cavity (8 mg/kg in 0.2-0.3 ml). Each compound was tested in two mice. Blood samples were drawn from the tail into heparinized capillary tubes. In studies with dogs, solutions in 0.9% saline were injected via the cephalic vein and blood samples were drawn from either the cephalic or the jugular vein. After centrifugation, the plasma (heparinized tubes) or serum was withdrawn and stored frozen until assay.

Sensitization Testing. Sensitization in mice was assessed by two protocols: a) four subcutaneous injections (0.075 mg/mouse) followed by repeated challenges every 3 weeks; and b) repeated injections of 5 mg/kg every 2 weeks. Challenge injections were intravenous (Table 1 and some of Table 2) or subcutaneous (most of Table 2). Doses are listed in the tables. Test solutions were prepared in 0.9% saline. Concentrations of test materials were based on biuret protein assay. For Table 1, after each intravenous challenge, the mice were observed for 30 minutes and again 1 hour after injection. The occurrence of signs indicative of anaphylaxis was recorded. Test scores for each animal are the sum of such signs: 1 = excessive licking, rubbing or scratching, 2 = abdominal stretching, 4 = unusual quietness, 8 = sedation, 16 = convulsions and 32 = death. In later experiments (Table 2) mice were observed at intervals for up to two hours after injection, and death was taken as the indicator of anaphylaxis.

Antigenicity Studies. The cross-reactivity of PEG-bSODs with antibody directed against bSOD was studied using a solid-phase, competitive-binding, enzyme immunoassay (ELISA). Polystyrene tubes were coated with rabbit anti-bSOD antibodies. The test sample was added together with a constant level of horseradish peroxidase-labeled bSOD conjugate (bSOD-HRP). Non-bound SOD species were removed by aspiration and washing. *o*-Phenylenediamine (OPD) and H_2O_2 were then added to the tube. The enzyme reaction was stopped by the addition of 1N HCl. The intensity of the color at 492 nm is proportional to the amount of SOD-HRP bound, which is inversely related to the amount of bSOD antigen present. The amount of antigen present was calculated from a standard curve obtained using bSOD standards.

Passive Cutaneous Anaphylaxis (PCA) Test. Mice were sensitized to test compounds by protocol b) described under Sensitization testing: The bSOD sensitization and challenge doses were 5 mg/kg and a total of 3 subcutaneous injections were given, followed by intravenous challenge doses, all spaced 2 weeks apart. Serum from surviving animals was collected 10 days after the last challenge and pooled by group for the PCA test. Rats were intradermally injected in the shaved skin of the back with 0.1 ml of diluted, pooled serum and, after 48 hours, challenged intravenously with 0.5 mg of test antigen in 2 ml of 0.5% Evans Blue dye. The rats were sacrificed one hour after the challenge and vascular permeability was scored by the relative dye intensity at the intradermal serum injection sites.

RESULTS

SOD Clearance after Intravenous and Intraperitoneal Administration. After intravenous injection, the clearance of PEG-SODs with either 7 strands of PEG 5K or 3 strands of PEG 72K was tested in mice (Figure 1). The half-life measured for the PEG 72K-SOD conjugate is about 36 hours, compared to about 24 hours for the PEG 5K-SOD conjugate. Plasma half-lives of at least 36 hours were also observed for PEG-SODs containing either 2 strands of PEG 40K or 1 strand of PEG 100K (data not shown).

The clearance profile after intraperitoneal injection of a bSOD conjugate with an average of 3.3 strands of PEG 41K is shown in Figure 2. A major species was cleared with a half-life of 1.5-2 days and a second component was cleared with a half-life of 3-4 days. Activity was still detected in plasma by electrophoretic analysis[8] one week after injection. In dogs, clearance of PEG-bSOD and PEG-rhSOD adducts, each with approximately three strands of PEG 50K, was approximately first order, with half-lives of about five days (Figures 3, 4 and 5).

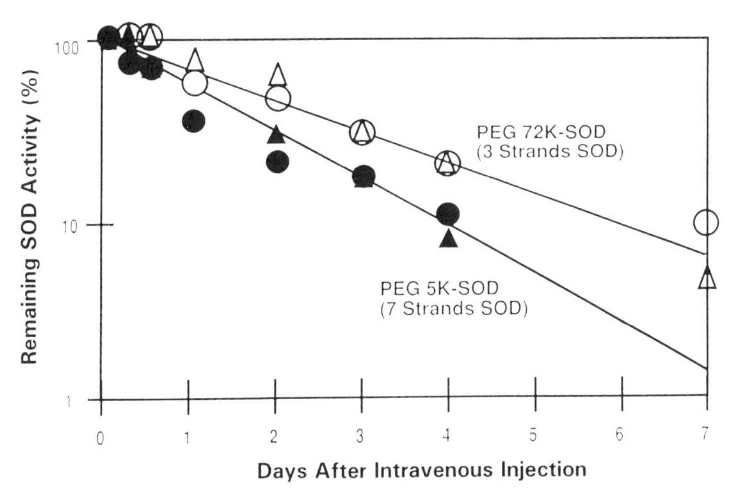

Figure 1. Clearance of PEG 5K-bSOD with 7 strands and PEG 72K-bSOD with 3 strands from mouse plasma after intravenous injection. Two animals in each group were injected with 5 mg SOD protein/kg via the tail vein. PEG 5K-SOD, 7 strands (●,▲); PEG 72K-bSOD, 3 strands (O,Δ). Blood was drawn at the indicated intervals and the plasma assayed for SOD activity by the NBT-agarose gel electrophoretic technique described in Methods. The half-lives observed are about 24 hours and 36 hours, respectively.

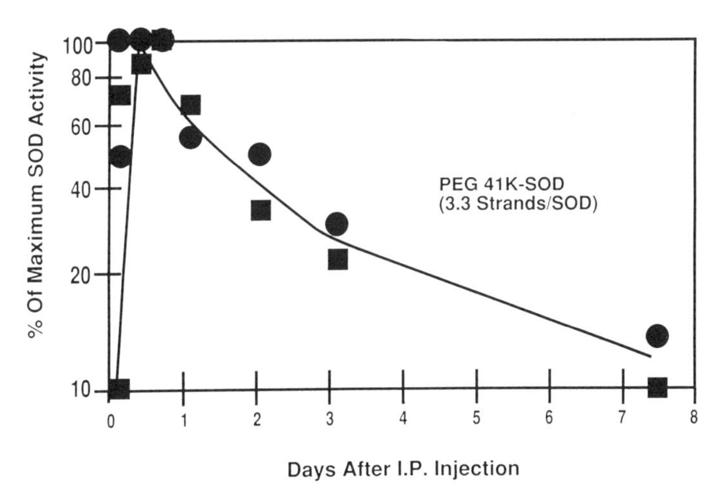

Figure 2. Clearance of PEG 41K-bSOD with 3.3 strands from mouse plasma after intraperitoneal injection. Two animals (■ and ●) were injected i.p. with 8 mg SOD protein/kg. Blood was drawn at the indicated intervals and the plasma assayed for SOD activity using the spectrophotometric NBT-xanthine/xanthine oxidase assay described in Methods.

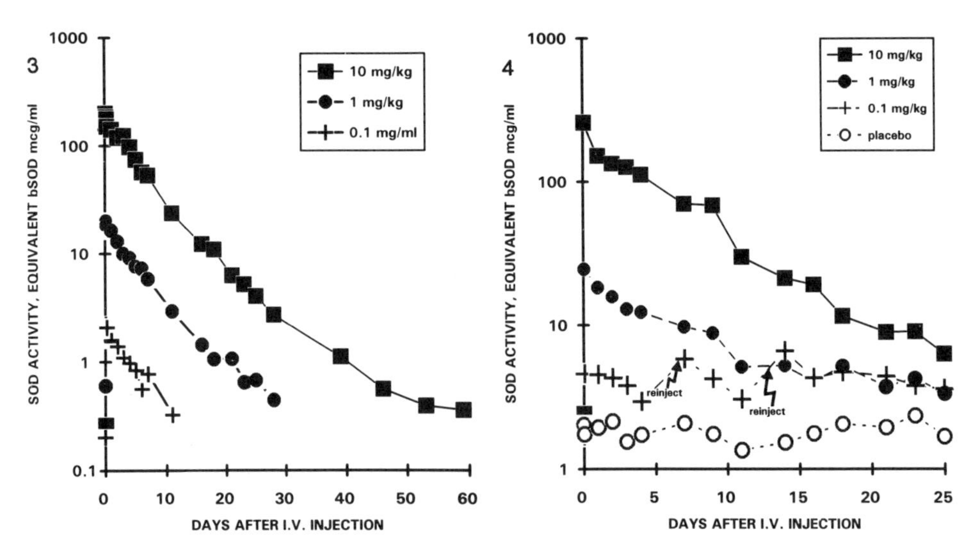

Figure 3. SOD activity in dog serum after intravenous injection of PEG-bSOD. Two male and two female beagle dogs at each of three dose levels, 10 (■), 1 (●) and 0.1 (+) mg/kg, were injected intravenously at 2 ml/kg with PEG 50K-bSOD with 3 strands PEG coupled per SOD. Blood was drawn at the indicated intervals and the serum analyzed by the spectrophotometric NBT-xanthine/xanthine oxidase assay described in Methods. The average from all four dogs per dose level is plotted at each point.

Figure 4. SOD activity in dog plasma after intravenous injection of PEG-rhSOD. Two male and two female beagle dogs at each of four dose levels, 10 (■), 1 (●), 0.1 (+) and 0 (O) mg/kg, were injected intravenously at 2 ml/kg with PEG 50K-rhSOD with 3 strands PEG coupled per SOD. Blood was drawn at the indicated intervals and the plasma analyzed by the spectrophotometric NBT-xanthine/xanthine oxidase assay described in Methods. The average from all four dogs per dose level is plotted at each point.

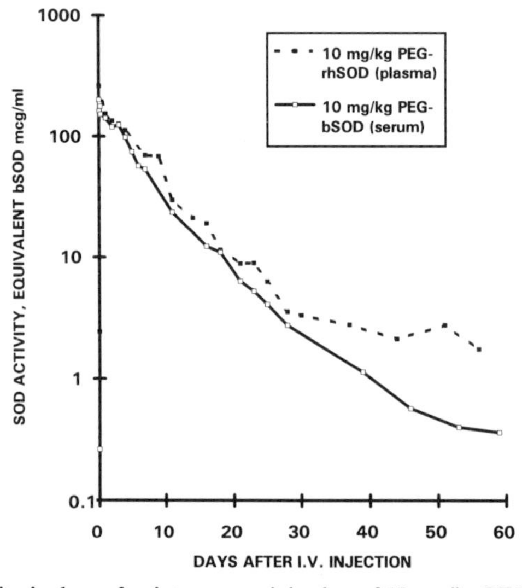

Figure 5. SOD activity in dogs after intravenous injection of 10 mg/kg PEG-bSOD or PEG-rhSOD. The 10 mg/kg PEG-bSOD serum data (□) from Figure 3 and the PEG-rhSOD plasma data (■) from Figure 4 are plotted together.

Sensitization Potential. Immunogenicity (ability to induce sensitizing antibodies) was measured using sensitization/challenge tests in mice. In Table 1, the summed test scores are shown for each group. Both the 7-stranded and 15-stranded PEG 5K-bSOD induced sensitization reactions before the sixth challenge. In contrast, none of the animals tested with either the 3-stranded or 4-stranded PEG 65K-bSOD showed signs of sensitization other than occasional mild signs (licking, rubbing or scratching). Similar results, summarized in Table 2, were obtained with many lots (20 of bSOD and 4 of rhSOD) of unmodified SOD and with 8 different adducts containing 3 PEG strands in the 35K-65K molecular weight range. Only one death occurred out of the 191 mice tested with high molecular weight PEG-SOD, whereas over 60% of bSOD recipients and over 40% of rhSOD recipients died by the sixth or seventh challenge dose.

Table 1. PEG-bSOD Immunogenicity Measured by a Mouse Sensitization/Challenge Test

Antigen	No. of Mice	Total Score at Challenge Number* 1	2	3	4	5	6	Number of Deaths
Native SOD	11	2	1	4	14	123++	11	2
PEG 5K-bSOD (15 strands)	10	0	4	10	143++	4	0	2
PEG 5K-bSOD (7 strands)	10	1	20	2	14	101+	49	1
PEG 65K-bSOD (4 strands)	10	0	2	1	1	2	1	0
PEG 65K-bSOD (3 strands)	10	1	2	1	1	2	2	0

* Sensitization reaction was scored for each animal as the sum of its symptom scores, as described in Methods. Total group scores at each intravenous challenge are shown.

\+ Indicates death of one mouse.

Antigenicity. A solid-phase enzyme immunoassay (ELISA) was employed to measure reactivity of PEG-SODs with rabbit antibody directed against bSOD. A series of bSOD adducts with 2 to 9 strands of high molecular weight PEG (21K-120K) were compared to PEG-bSODs with 7 or 15 strands of PEG 5K. Conjugation of 1/4 to 1/3 as many strands of high molecular weight PEG was sufficient to achieve the same or higher degree of reduction of antigenicity demonstrated by PEG 5K conjugates (Figure 6).

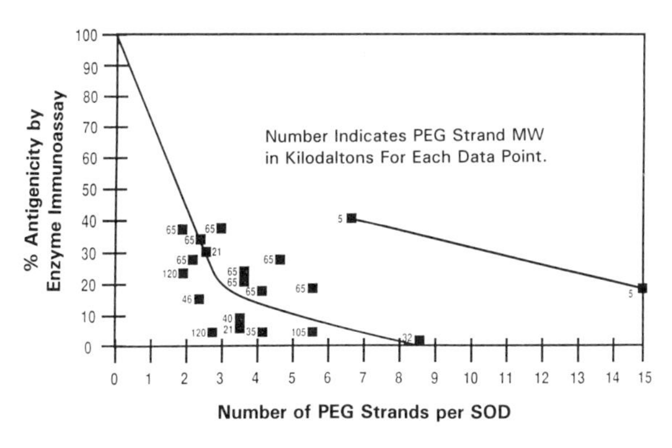

Figure 6. Cross reactivity of various PEG-bSODs, differing in PEG molecular weight and number of strands, with native Cu,Zn bSOD. PEG molecular weights are indicated by the number adjacent to each data point (■). All conjugates were synthesized via PEG succinate-active esters except the PEG 40K and PEG 120K conjugates which were prepared via PEG acetic acid esters. Antigenicity was measured by ELISA as described in Methods.

Table 2. Summary of Responses to SOD and PEG-SOD Injections Measured by Mouse Sensitization/Challenge Tests

A. Mice received four 0.075 mg/mouse sensitization injections in two weeks, then challenge injections every three weeks.

	Survival Statistics After Challenge Number					
	1	2	3	4	5	6
Usual Dose (mg/mouse) =	0.04	0.4	0.04	0.2	0.2	0.2
bSOD Initial n = 182 mice						
Cumulative Survival Rate	97%	68%	56%	32%	26%	24%
rhSOD Initial n = 40 mice						
Cumulative Survival Rate	98%	90%	90%	85%	78%	55%
PEG-SOD Three PEG-bSOD and two PEG-rhSOD lots. Initial n = 51 mice						
Cumulative Survival Rate	100%	100%	100%*	100%	100%	100%

*One mouse died 9 days after third challenge, not counted as dose-related.

B. Mice received repeated injections of 5 mg/kg every two weeks.

	Survival Statistics After Injection Number						
	1	2	3	4	5	6	7
bSOD Initial n = 272 mice							
Cumulative Survival Rate	100%	100%	98%	71%	44%	37%	34%
rhSOD Initial n = 51 mice							
Cumulative Survival Rate	100%	100%	100%	84%	80%	67%	57%
PEG-SOD Six PEG-bSOD and two PEG-rhSOD lots. Initial n = 140 mice							
Cumulative Survival Rate	100%	100%	99%	99%	99%	99%	99%

All PEG-SOD lots summarized in Table 2 have approximately three strands of PEG of 35,000 - 65,000 daltons coupled per SOD.

Table 3. Immunogenicity and Antigenicity of PEG-bSODs Measured by the Passive Cutaneous Anaphylaxis (PCA) Test in Rats[1]

Immunogen for Intradermal Mouse Serum	Intravenous Test Antigen			
	Native SOD	PEG-bSOD 2xPEG 120K[2]	PEG-bSOD 9xPEG 35K[3]	Saline
Native bSOD	1000	100	0	0
PEG-bSOD 2xPEG 120K[2]	0	0	0	0
PEG-bSOD 9xPEG 35K[3]	0	0	0	0
Saline	0	0	0	0

[1]Scoring is relative to intensity of dye deposited at native SOD antiserum sites challenged with native SOD = 1000.

[2]2xPEG 120K is a PEG-bSOD with an average of 2 PEG strands of molecular weight 120,000.

[3]9xPEG 35K is a PEG-bSOD with an average of 9 PEG strands of molecular weight 35,000.

Passive Cutaneous Anaphylaxis (PCA) Test. The PCA test optimized to measure mouse IgE is a measure of systemic anaphylaxis potential.[13] As shown in Table 3, neither PEG-bSOD containing two strands of PEG 120K nor PEG-bSOD containing nine strands of PEG 35K induced detectable antibodies against either of the PEG-bSOD preparations or against bSOD; however, the two-stranded 120K PEG-bSOD reacted weakly with antiserum from mice sensitized with bSOD.

DISCUSSION

PEG is well suited for modification of therapeutic proteins to reduce the potential for immune reactions and to enhance persistence in the blood; however, the degree of PEG modification of protein nucleophilic groups necessary to achieve these effects often markedly reduces enzyme activity.[4-7] In order to recover 90% or more of the native SOD activity, conjugation must be limited to fewer than 4 PEG strands per molecule of SOD whether low molecular weight PEG (5K) or high molecular weight PEG (40K-70K) is used.[8]

PEG-SODs containing a few strands of high molecular weight PEG demonstrated significantly longer plasma half-lives than PEG-SODs containing many more strands of low molecular weight PEG (5K). Activity of the 3-stranded PEG 72K adduct could still be detected in the plasma after 9 days, whereas the 7-stranded PEG 5K adduct could not be detected at 7 days, even though equal amounts of SOD protein had been injected (Figure 1). This difference in clearance rates may be explained by the greater specific enzyme activity and size of the adduct prepared with fewer strands of high molecular weight PEG (98% compared to 52% of native SOD activity and 220,000 compared to 30,000 daltons, based on PEG standardized HPLC, respectively). For comparison, Beauchamp *et al*[14] reported half-lives of 9-17 hours for PEG-bSODs with 18-19 PEG 4-5K strands. In contrast, the half-life of bSOD in rats and mice is about 6 minutes.[4,15] Pyatak *et al*[4] observed a biphasic clearance of an extensively modified PEG 5K-bSOD conjugate after i.v. injection with an early half-life of about 20 hours and 36% activity remaining after 72 hours. After i.p. injection, peak serum levels were reached in 4 hours.

After i.p. injection of 8 mg/kg of a PEG-bSOD with an average of 3.3 strands PEG 41K, peak blood levels were reached in less than 8 hours, elimination had a half-life of about 36 hours, and PEG-SOD activity could still be detected by gel electrophoresis even after a week (Figure 2). The large size of the adduct (about 145,000 daltons by PEG-standardized HPLC or 1,000,000 daltons by protein-standardized HPLC) apparently does not prevent its rapid absorption into the blood stream.

In the rat, Boccù *et al*[7] showed a relationship between i.v. serum clearance rates and degree of derivatization. While the native enzyme was cleared in less than 1 hour, elimination half-lives for adducts containing 3 and 18 PEG 5K strands were 4 and 25 hours, respectively. These adducts had about 70% and 50% native activity, respectively. An initial half-life of 2 hours in rats was reported for a SOD adduct made with an undisclosed PEG type and number of strands[16] and 35 hours for another adduct containing an undisclosed number of PEG 1.9K strands.[15] We have not investigated the persistence of adducts with a few strands of high molecular weight PEG in the rat. The observation that PEG-bSOD and PEG-rhSOD adducts with approximately three strands of PEG 50K coupled per SOD have nearly equal half lives in dogs (Figure 5) indicates that the known differences in stability of unmodified bSOD and rhSOD do not influence the clearance rates of these PEG adducts.

That fewer strands of high molecular weight PEG are more effective than more strands of low molecular weight PEG in reducing the potential of bSOD to induce sensitizing antibodies in the mouse (Table 1) can be explained by the ability of the longer PEG strands to more effectively shield the protein's antigenic sites from antigen-recognizing cells.

The decreased antigenicity of bSOD conjugated with PEG 1.9K has been reported[5]. That the incremental attachment of larger PEG strands is more effective in reducing antigenicity compared to adding smaller PEG strands (Figure 6) can be explained by the ability of the longer PEG strands to more effectively shield the protein's antigenic sites from antibody bound to a plastic surface.

The PCA test results (Table 3) indicated that the high molecular weight PEG-SODs were not immunogenic, as observed in the sensitization potential study, but were weak antigens, as seen in the ELISA test.

The advantage of attaching fewer strands of high molecular weight PEG is that larger adducts are produced per degree of protein modification. Consequently, there is less inactivation relative to size increase. These conjugates also demonstrate superior biological properties related to potential therapeutic applications, including longer circulating half-lives and reduced immunogenicity and antigenicity at each degree of substitution.

REFERENCES

1. W. Huber and K.B. Menander-Huber (1980) Orgotein. *Clinics in Rheumatic Diseases*, **6**:465-497.
2. F.F. Davis, A. Abuchowski, T. van Es, N.C. Palczuk, K. Savoca, R. H-L. Chen and P. Pyatak (1980) Soluble, nonantigenic polyethylene glycol-bound enzymes. In *Biomedical Polymers, Polymeric Materials and Pharmaceuticals for Biomedical Use* (E.P. Goldberg and A. Nakajima, eds.), Academic Press, London, pp. 441-442.
3. F.F. Davis, A. Abuchowski, T. van Es, N.C. Palczuk, H-L. Chen, K. Savoca and K. Wieder (1978) Enzyme-polyethylene glycol adducts: modified enzymes with unique properties. In *Enzyme Engineering*, Vol. **3**, (G.B. Braun, G. Manecke and L.B. Wingard, Jr., eds.), Plenum Press, New York, pp. 169-173.
4. P.S. Pyatak, A. Abuchowski and F.F. Davis (1980) Preparation of a polyethylene glycol-superoxide dismutase adduct and an examination of its blood circulating life and anti-inflammatory activity. *Research Communications in Chemical Pathology and Pharmacology*, **29**:113-127.
5. F.M. Veronese, R. Largajolli, E. Boccù, C.A. Benassi and O. Schiavon (1985) Surface modification of proteins. Activation of monomethoxy-polyethylene glycols by phenylchloroformates and modification of ribonuclease and superoxide dismutase. *Applied Biochemistry and Biotechnology*, **11**:141-152.
6. E. Boccù, R. Largajolli and F.M. Veronese (1982) Coupling of monomethoxy-poly-ethyleneglycols to proteins via active esters. *Zeitschrift für Naturforschung*, **38c**:94-99.
7. E. Boccù, G.P. Velo and F.M. Veronese (1982) Pharmacokinetic properties of polyethylene glycol derivatized superoxide dismutase. *Pharmacological Research Communications*, **14**:113-120.
8. R. Somack, M.G.P. Saifer and L.D. Williams (1991) Preparation of long-acting superoxide dismutase using high molecular weight polyethylene glycol (41,000-72,000 daltons). *Free Radical Research Communications*, **12-13**:553-562.

9. J.M. McCord and I. Fridovich (1969) Superoxide dismutase. An enzymatic function for erythrocuprein (hemocuprein). *Journal of Biological Chemistry*, **244**:6049-6055.
10. M. Leonard, J. Neel and E. Dellacherie (1984) Synthesis of monomethoxy-polyoxyethylene-bound haemoglobins. *Tetrahedron*, **40**:1581-1584.
11. C.O. Beauchamp and I. Fridovich (1971) Improved assays and an assay applicable to acrylamide gels. *Analytical Biochemistry*, **44**:276-287.
12. Y. Sun, L.W. Oberley and Y. Li (1988) A simple method for clinical assay of superoxide dismutase. *Clinical Chemistry*, **34**:497-500.
13. Z. Ovary (1964) Passive cutaneous anaphylaxis. *Immunological Methods* (J.F. Ackroyd, ed.), Blackwell, London, pp. 259-283.
14. C.O. Beauchamp, S.L. Gonias, D.P. Menapace and S.V. Pizzo (1983) A new procedure for the synthesis of polyethylene glycol-protein adducts: effects on function, receptor recognition and clearance of superoxide dismutase, lactoferrin and α_2-macroglobulin. *Analytical Biochemistry*, **13**:25-33.
15. J.M. McCord and K. Wong (1979) Phagocyte-produced free radicals: roles in cytotoxicity and inflammation. In *Ciba Foundation Symposium*, Vol **65,** Excerpta Medica, Amsterdam, pp. 343-360.
16. K. Grankvist, S. Marklund and I-B. Täljedal (1981) Superoxide dismutase is a prophylactic against alloxan diabetes. *Nature*, **294**:158-160.

CLINICAL TRIALS WITH DISMUTEC™ (PEGORGOTEIN; POLYETHYLENE GLYCOL-CONJUGATED SUPEROXIDE DISMUTASE; PEG-SOD) IN THE TREATMENT OF SEVERE CLOSED HEAD INJURY

J. Paul Muizelaar

Division of Neurosurgery
Medical College of Virginia/Virginia Commonwealth University
Richmond, VA 23298-0631

INTRODUCTION

Although the suspicion for a role of oxygen radicals in disease and degeneration was raised a long time ago, the first publications relating free radicals with traumatic brain injury appeared only in 1981.[1,2] Since then, however, a rapid development has taken place, culminating in a phase II human trial with a superoxide scavenger started in 1989[3] and several phase III clinical trials taking place at present. The initial laboratory experiments were concentrated on elucidating mechanisms for vascular damage after head injury. First, it was shown that in the early minutes after experimental fluid percussion injury in the cat, prostaglandin concentration in the brain increases, secondary to activation of the arachidonic acid pathway; this pathway leads to superoxide anion formation through the enzyme PGH synthase.[1] Pretreatment with radical scavengers or with indomethacin, which inhibits cyclo-oxygenase and thus deprives the prostaglandin hydroperoxidase of substrate, protects the endothelium of the cerebral microcirculation from the effects of experimental brain injury.[2] Subsequently, it was shown that immediately after the impact there is an increase in phospholipase C activity, which can release arachidonic acid in tissue, leading to oxygen radical formation.[4] Further work showed that the radicals are produced not only in cerebral blood vessel walls but also in leukocytes and macrophages that accumulate in the brain between 3 and 24 hours after experimental injury.[5] With the method used to identify the site of generation of the superoxide anion,[6] only the vessel walls could be implicated in this process, but this is possibly because of insufficient penetration of the reagents into brain tissue.

Apart from the clarification of superoxide anions role in mediating vascular damage after experimental head injury, later work at the Medical College of Virginia concentrated on possible clinical beneficial effects of scavenging the radicals, usually with i.v. SOD. In 1989 several beneficial effects were described. Zimmerman et al. noted a reduction of intracranial hypertension in a model of cryogenic injury topically treated with SOD and catalase;[8] Schettini et al. described an epidural balloon compression model with ischemia-reperfusion injury which showed considerable improvement in cerebral blood flow (CBF), brain edema, and histologic damage in dogs treated with i.v. SOD;[9] however, the first improved clinical outcome was described by Levasseur et al., who noted significantly better survival and less neurological deficits in rats subjected to fluid percussion injury treated with i.v. SOD.[10]

Most of the foregoing work in experimental head injury, both from the Medical College of Virginia and from other laboratories was summarized by Kontos.[7,11] Obviously, the time was ripe for a human clinical trial with a suitable superoxide scavenger, which was found in the form of pegorgotein (polyethyleneglycol-superoxide dismutase or PEG-SOD, Dismutec™).

PEGORGOTEIN

Product Description

Pegorgotein is bovine Cu^{++}/Zn^{++} superoxide dismutase (E.C. 1.15.1.1) modified by covalent addition of monomethoxy-PEG_{5000}. Superoxide dismutase is a homomeric dimer containing a total of 20 superficial lysines which are available for covalent addition of methoxy-PEG_{5000}. During the process of "pegating" superoxide dismutase, 9-13 monomethoxy-PEG_{5000} molecules are covalently added to the ϵ-amino group of an equal number of superficial lysines. The resulting "pegated" homodimer of superoxide dismutase retains superoxide dismutase activity and is significantly more stable to enzymatic hydrolysis *in vivo*. Pegation increases the plasma half-life of SOD activity from approximately 8 minutes to greater than 30 hours in rodents and approximately 100 hours in dogs (Sterling Winthrop, unpublished data). Thus, a single administration of pegorgotein has the potential to provide protection against superoxide-related toxicities for several days.

Toxicological Evaluation

Toxicological evaluation of pegorgotein in mouse and rat by single bolus injection demonstrated that the product was very well tolerated; the LD_{50} was greater than 1,000,000 U/kg.

Preliminary Human Investigations

Pegorgotein has been evaluated in healthy volunteers as well as in subjects with thermal injury (full thickness burns), subjects undergoing elective cardiac surgery or renal transplantation and subjects with multiple trauma. Single intravenous bolus doses of up to 20,000 U/kg are well tolerated in both healthy volunteers and subjects with acute disease states, such as renal transplantation or cardiac surgery. One healthy volunteer in a pharmacokinetic study who received pegorgotein developed an anaphylactoid reaction during the 60-minute infusion. The subject responded rapidly to standard therapeutic measures and had no clinical sequelae from the episode. High circulating levels of IgE in the patient suggested a predisposition to an allergic reaction. Western blot analysis failed to confirm any antibody specificity to pegorgotein. Up to three sequential infusions of pegorgotein are well tolerated in healthy volunteer studies. Pharmacokinetic studies in healthy volunteers demonstrated dose proportional increases in the maximal plasma superoxide dismutase activity, as assessed by the cytochrome oxidase assay, as well as proportionality of the area-under-the curve values. Dose independence of T_{max}, $t_{1/2}$ and the Cl was also noted. The plasma elimination half life in healthy volunteers is approximately 4.8 days.

A PHASE II CLINICAL TRIAL OF PEGORGOTEIN IN SEVERE HEAD INJURY

With support from the National Institutes of health (NIH grant 1 54600-1758-A1) and Sterling Winthrop a trial was designed to evaluate both the safety and efficacy of pegorgotein in patients with severe head injury. This was a randomized, third party blind study comparing placebo (normal saline) with pegorgotein administered intravenously in doses of 2000 U/kg, 5000 U/kg or 10,000 U/kg as a single bolus within 12 hours of injury. Primary endpoint was the Glasgow Outcome Scale (GOS)[12] at 3 months postinjury.

Clinical Material and Methods

Patients with severe head injuries admitted to the Medical College of Virginia (MCV), Richmond, Virginia, or to the Maryland Institute for Emergency Medical Services Systems (MIEMSS), Baltimore, Maryland, were evaluated for entry into the study. Eligibility criteria included a Glasgow Coma Scale (GCS)[13] score of 8 or less; the inability to follow commands after respiratory and circulatory resuscitation in the emergency room as necessary; an age of 15 years or older; and the administration of study medication within 12 hours of injury. Exclusion criteria included the known receipt of any investigational drug within 30 days of study enrollment; the inability to exclude pregnancy; the likelihood of brain death after resuscitation; a penetrating head injury; doubt as to the presence of head trauma; or refusal of a family member to provide informed consent.

Randomization and Administration. Subjects were randomly assigned in the order in which they were enrolled into the study to receive either placebo or pegorgotein according to a computer-generated randomization schedule as follows: 26 received 50 ml normal saline (placebo group), 26 received 2000 U/kg of pegorgotein, 25 received 5000 U/kg of pegorgotein, and 27 received 10,000 U/kg of pegorgotein. Separate randomization schedules were prepared for each of the two institutions. After the neurosurgical resident and/or research nurse had established that the patient met all entrance criteria, the investigational pharmacist was notified and a sealed envelope was opened by the pharmacist, who would then prepare the study drug in the required dose based on the estimated weight of the patient and adjust the volume by the addition of normal saline to provide a total volume of 50 ml. The 50-ml syringe containing the solution was then wrapped in a covering to hide the very slight bluish color of the pegorgotein and handed to the physician. The study drug was then administered as a bolus via peripheral venous access.

Patient Management and Data Collection. Patient management was similar in both institutions, as recently described.[14] In short, after intubation and stabilization of the patient in the emergency room and after administration of the study drug, a computerized tomography (CT) scan was obtained; patients with mass lesions causing 5 mm or more shift were treated surgically without delay. In the intensive care unit, all patients had monitoring of arterial blood pressure and ICP. Management of ICP below 20 mmHg was carried out in a stepwise fashion, starting with sedation, paralysis, cerebrospinal fluid (CSF) drainage if a ventricular catheter was employed, mannitol administration up to a serum osmolality of 320 mOsm, followed by hyperventilation (for the shortest interval possible), and barbiturate coma. Anticonvulsant drugs were used only in patients with a history of seizure or if a seizure occurred during the hospital course. Steroid drugs were not employed, except as a single dose before extubation.

Outcome at 3 and 6 months after injury was assessed using the Glasgow Outcome scale (GOS)[12] by a neurologist or rehabilitation specialist not associated with the acute treatment of the patient. In a few instances the study coordinator obtained the information by telephone.

Data were entered on data-collection forms, completed in an identical fashion by both institutions. These forms are a composite of the forms that have been used for our severely head-injured patients since 1976 as well as portions of the forms used by the Traumatic Coma Data Bank in which we have participated for 8 years.[15] These were then entered into our computerized database. The data collection was performed by the physicians and study nurses. The following data were collected: 1) Preadmission data concerning cause of injury, GCS score at the accident scene, and treatment prior to hospital admission; 2) admission data such as general demographic information, GCS score, results of neurological examination, associated injuries, CT findings, any therapy administered, surgical observations, time of injury, and time of study drug administration; 3) intensive care unit data such as serial neurological examination results, course of ICP, vascular reactivity therapy intensity level to maintain ICP at normal levels, and laboratory data including SOD levels in serum and CSF, electrolyte and arterial blood gas levels, coagulation studies, and chemistry profiles; and 4) outcome data.

At the Medical College of Virginia, studies of CBF, cerebral blood volume, arteriovenous differences of oxygen, brain water content mapping with magnetic resonance imaging, and pressure volume index data were undertaken intermittently, while ICP, mean arterial blood pressure, cerebral perfusion pressure, end-expiratory CO_2, and heart rate were entered continuously and processed by a VAX 3900 computer, located in the neuroscience intensive care unit.

Data Analysis and Statistical Techniques. It is well known that the outcome of severely head-injured patients is strongly correlated with age and GCS score, among other factors.[16,17] Thus, the use of these prognostic factors as covariables in testing the efficacy of the therapy could significantly increase the power. A multiple logistic regression model was applied,[17] incorporating age and GCS score as well as institution and institution by treatment effects, taking into account somewhat different treatment protocols and possible differences in the patient population between the two institutions. Each of the three dose levels of pegorgotein was compared with the placebo using the likelihood ratio procedure based on the logistic model. Dichotomized outcomes defined from the GOS score at 3 and 6 months postinjury were analyzed. The possible interaction effect on outcome of severity of injury and therapy was also analyzed.

RESULTS

Summary of Cases. A total of 104 patients were entered into the study, 59 at the MCV and 45 at the MIEMSS. Informed consent was refused in two cases, protocol violations occurred in seven, and follow-up data were not available for four patients at three months or for two patients at six months. Protocol violations included a GCS score of 9 or above in three cases, co-enrollment into another study in one, doubt about trauma being the cause of coma in one, a penetrating head injury in one, and an unknown time of injury in one. In Tables 1 to 5, the entire group of 104 patients is considered, except in the outcome data where 91 are evaluated at 3 months and 93 and 6 months.

Clinical features at entry into the study are presented in Table 1. Except for age, no differences were significant, either between the MCV and the MIEMSS patients or between the four treatment groups; mean age was significantly higher in the group receiving 10,000U/kg of pegorgotein than in the placebo group.

Complications are listed in Tables 2 and 3. Overall, the incidence of infectious complications (Table 2) and of complications of organ function (Table 3) do not appear to be influenced by the administration of pegorgotein. Although, considered on its own, the incidence of meningitis/ventriculitis was lower in the group receiving 10,000 U/kg, this was found not significant using a chi-squared test with Bonferroni adjustment for multiple testing.

Table 1. Data of 104 patients on admission to the study*

Feature	Treatment Group+	MCV Group	MIEMSS Group	Totals
Age (yrs)	Placebo	24.9 ± 8.3	26.1 ± 6.8	25.4 ± 7.6
	2,000 U/kg	26.6 ± 10.4	27.1 ± 8.3	27.0 ± 9.7
	5,000 U/kg	34.0 ± 14.7	30.8 ± 15.2	32.6 ± 14.7
	10,000 U/kg	35.6 ± 20.2	32.4 ± 15.0	34.2 ± 17.9
GCS score	Placebo	6.1 ± 1.4	5.0 ± 1.8	5.7 ± 1.6
	2,000 U/kg	5.3 ± 1.2	6.6 ± 1.9	5.7 ± 1.3
	5,000 U/kg	6.2 ± 1.4	6.5 ± 1.4	6.2 ± 1.4
	10,000 U/kg	5.8 ± 1.5	5.5 ± 1.5	5.6 ± 1.5
GCS motor score 1 & 2	Placebo	21.4%	45.5%	32.0%
	2,000 U/kg	35.7%	18.2%	28.0%
	5,000 U/kg	23.1%	18.2%	20.8%
	10,000 U/kg	20.0%	25.0%	22.2%
pupillary abnormalities	Placebo	35.7%	36.4%	36.0%
	2,000 U/kg	64.3%	36.4%	52.0%
	5,000 U/kg	25.0%	36.4%	30.4%
	10,000 U/kg	21.4%	41.7%	30.8%
time from injury to treatment (hrs)	Placebo	2.7 ± 0.9	4.5 ± 1.6	3.5 ± 1.5
	2,000 U/kg	2.5 ± 1.1	4.5 ± 1.4	3.3 ± 1.6
	5,000 U/kg	3.3 ± 1.9	4.4 ± 1.1	3.8 ± 1.7
	10,000 U/kg	4.2 ± 2.4	5.2 ± 1.8	4.6 ± 2.2

*There were 59 patients in the Medical College of Virginia (MCV) group and 45 in the Maryland Institute of Emergency Medical Services System (MIEMSS) group. GCS = Glasgow Coma Scale. Means are expressed ± standard deviation.

+Patients received placebo (50 ml normal saline) or pegorgotein in one of three dosages.

Table 2. Percentage of infectious complications in the four study groups

Infectious Complications	Placebo	Pegorgotein 2000 U/kg	5000 U/kg	10,000 U/kg
total cases	26	26	25	27
septicemia	46%	23%	20%	41%
pulmonary infection	73%	58%	44%	70%
ventriculitis/meningitis	15%	15%	12%	0%

Table 3. Percentage of organ dysfunction in the four study groups.

Infectious Complications	Placebo	Pegorgotein 2000 U/kg	5000 U/kg	10,000 U/kg
total cases	26	26	25	27
coagulopathy	27%	8%	12%	4%
electrolyte disturbance	65%	73%	80%	85%
kidney disorder	8%	4%	0%	4%
gastrointestinal disorder	8%	0%	8%	7%
pulmonary insufficiency	54%	42%	36%	44%
shock	42%	42%	28%	19%
cardiac complications*	15%	19%	20%	22%

*Includes myocardial infarction, congestive heart failure, and arrhythmia.

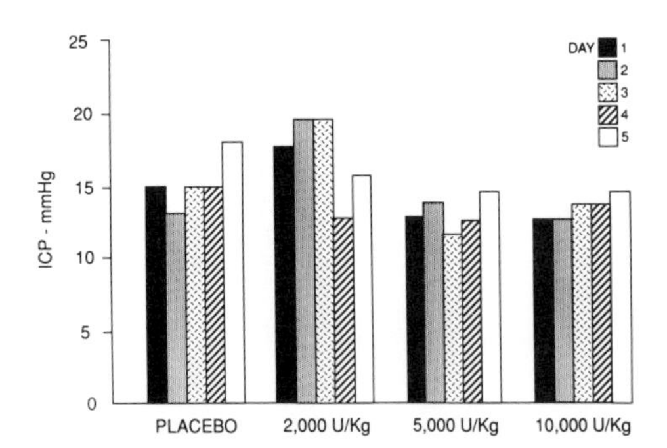

Fig. 1. Bar graph showing the average intracranial pressure (ICP) during the first 5 days following head injury (taken at 30-second intervals) in the placebo group and in three groups receiving pegorgotein. There were no statistically significant differences between results.

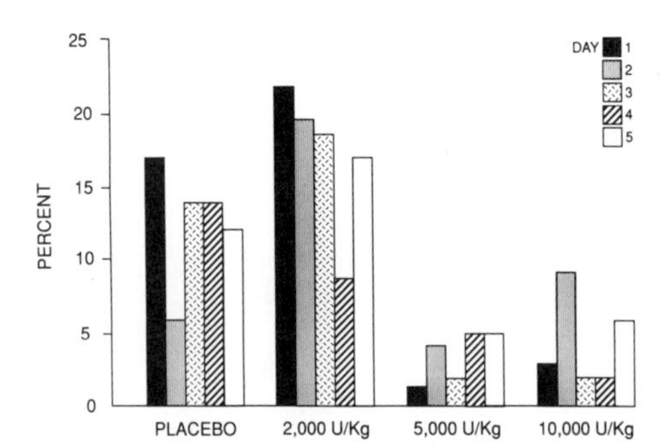

Fig. 2. Bar graph showing the mean percentage of time during which intracranial pressure was above 20 mmHg in the placebo group and in three groups receiving pegorgotein. Statistical significance: $p<0.05$ between the placebo group and the groups receiving 5000 U/kg and 10,000 U/kg on Days 1 and 4.

In clinical drug studies, death is generally considered an "adverse event," irrespective of its cause. Death is therefore considered here in all 104 patients (in contrast to the data concerning efficacy and outcome, where only patients fulfilling all entrance and evaluation criteria are taken into consideration). Within 3 months postinjury, death occurred in 7 of 26 patients in the placebo group, 8 of 26 in the group receiving 2000 U/kg, 7 of 25 in the group receiving 5000 U/kg, and 5 of 27 in the group receiving 10,000 U/kg; in no instance was death ascribed to the study drug by the blinded evaluators.

Figure 1 depicts the average ICP during the first five days following severe head injury in the MCV patients only; no significant differences were found between treatment groups. The percentage of time during which ICP is greater than 20 mmHg is considered a powerful prognostic factor.[18] This is indicated in Figure 2 for each group; between the placebo group and the group receiving 5000 U/kg of pegorgotein, the differences on Days 1 and 4 were significant ($p<0.01$ and $p<0.04$, respectively), and also between the placebo group and the group receiving 10,000 U/kg of pegorgotein, again at Days 1 and 4 ($p<0.08$ and $p<0.01$, respectively). Thus, even though mean or median values of ICP were not significantly different among the four groups, it appears that with the higher doses (5000 and 10,000 U/kg) of the study drug, it was easier to maintain ICP below the desired level of 20 mmHg. This was also evident when we examined the number of patients who received more than 1 gm/kg of mannitol on any given day (both MCV and MIEMSS patients). In the group receiving 10,000 U/kg, there were 33% and 36% fewer patients on Days 2 and 3, respectively, than in the placebo group receiving this amount of mannitol ($p<0.04$ for both days); all other differences among the pegorgotein dose and placebo groups on any of the first 5 days postinjury were not statistically significant.

Outcome at 3 Months and 6 Months. Outcome at 3 and 6 months after injury is shown in Table 4. Compared to the placebo group and the group receiving 2000 U/kg, there is a trend toward a better outcome in the group receiving 5000 U/kg, which is even more noticeable in the group receiving 10,000 U/kg; however, the difference is not statistically significant. If we combine the outcome categories "good," "moderate disability," and "severe disability," and compare them with "vegetative" plus "dead," we find that the difference between the placebo group and the group receiving 10,000 U/kg of pegorgotein was statistically significant, using a likelihood ratio test for treatment effect based on the logistic model controlling for age, GCS score, and institution ($p<0.03$ at 3 months and $p<0.04$ at 6 months, Table 5).

Table 4. Outcome of patients in the four study groups at 3 and 6 months postinjury.

Follow-Up Time & Group	Good	Moderate Disability	Severe Disability	Vegetative	Dead	Total
at 3 mos postinjury (91 patients)						
placebo	5	5	3	3	7	23
pegorgotein						
2,000 U/kg	6	3	3	1	8	21
5,000 U/kg	5	5	5	2	5	22
10,000 U/kg	6	9	5	0	5	25
at 6 mos postinjury (93 patients)						
placebo	10	4	2	1	8	25
pegorgotein						
2,000 U/kg	8	4	1	1	8	22
5,000 U/kg	8	3	5	1	5	22
10,000 U/kg	9	6	4	0	5	24

Table 5. Percentage of patients in a vegetative state or dead at 3 and 6 months postinjury

Study Group	3 Months	6 Months
placebo	43.5	36.0
pegorgotein		
2,000 U/kg	42.9	40.9
5,000 U/kg	31.8	27.2
10,000 U/kg	20.0+	20.8+

+Significant treatment effect compared to placebo ($p<0.04$, Wald test based on logistic model, controlling for age, Glasgow Coma Scale score, institution, and institution by treatment effects).

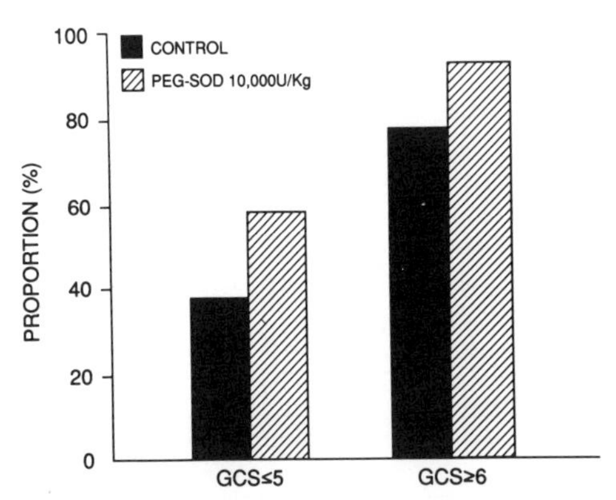

Fig. 3. Bar graph showing the mean percentage of patients in the combined outcome categories of good, moderate disability, and severe disability in the placebo group and in the group receiving 10,000 U/kg pegorgotein when considered separately by admission Glasgow Coma Scale (GCS) score ($\leq$ 5 and $\geq$ 6).

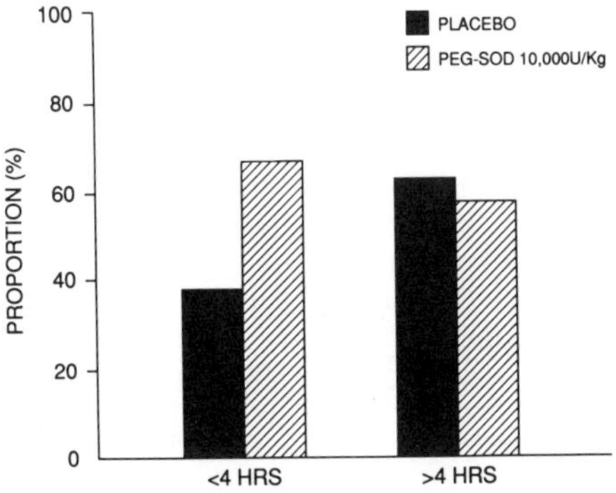

Fig. 4. Bar graph showing the percentages of patients in the combined outcome categories Good and Moderately Disabled at 3 months by time of administration of drug.

The interaction effect on outcome of severity of injury and therapy is depicted in Figure 3. The effect was not significant and the data suggest that the probable benefit of 10,000 U/kg of pegorgotein is independent of severity of injury defined by the GCS.

The interaction effect on outcome of the timing (within 4 hours or between 4-12 hours, postinjury) of administration of the therapy is shown in Figure 4.

PHASE III CLINICAL TRIALS OF PEGORGOTEIN IN SEVERE HEAD INJURY

After having established the unequivocal safety of i.v. administered pegorgotein in this patient group and in consideration of the promising results with the 10,000 U/kg group, it was decided to go ahead with full scale, phase III clinical trials. First, two such trails were started in the USA, in January 1993 and May 1993, respectively. These two trials, independent in their patient recruitment, are completely similar in design. Both are double-blind, placebo-controlled, randomized blocked designs, this time comparing placebo with 10,000 U/kg and 20,000 U/kg.

Patients eligible are those with a severe, closed head injury, with a GCS of 3-8 after resuscitation (systolic blood pressure $\geq$90 mmHg), 15-70 years. The drug is to be administered within 8 hours from injury. Pre-randomization stratification by age takes place with 15-45 years as the lower stratum and 46-70 as the upper one. Primary endpoint is GOS at 3 months, the five point scale being condensed into poor (dead and vegetative), fair (severe disability) and favorable (moderate disability and good). In the phase II trial the percentages in the placebo and 10,000 U/kg groups were 42 vs. 20 (poor), 13 vs 20 (fair), and 46 vs. 60 (favorable), respectively. With a Cochran-Mantel-Haenszel test with covariate testing for GCS, pupillary responses, age and timing of administration, 150 patients per group will have a 94% power with two-sided significance level of 0.05% to detect a similar difference in the phase III trials. Secondary endpoints will be mortality at 3 months, head injury specific mortality, GOS at 6 months and Disability Rating Scale (DRS)[19] at 3 and 6 months.

These two trials are presently ongoing and not unblinded as yet. However, we do know the entry characteristics and the percentages in each of the GCS categories being shown in Table 6. Also shown in Table 6 is the expected mortality derived from the Traumatic Coma Data Bank[20] for each of the post-resuscitation GCS categories. Combining the percentages of patients in each GCS category and the expected mortality, we can calculate an overall expected mortality of 37%. In actuality, the mortality amongst the 533 patients entered until January 1, 1994 is 23%. Thus, it seems likely from these preliminary data that pegorgotein reduces mortality from severe head injury.

Table 6. Percentage of patients in protocols SW 005 and 006 (January through December 1993, total number 533) in each of the admitting post-resuscitation GCS categories and the mortality in the Traumatic Coma Data Bank in each of these GCS categories.

	Admitting GCS					
	3	4	5	6	7	8
% of patients	16	17	14	22	22	8
Mortality (%) in TCDB*	78	56	40	21	18	11

*Mortality in TCDB from Marshall et al.

Another phase III trial is in the final planning stages. It will be conducted in Europe, and will be mostly similar to the American trials.

DISCUSSION

A distinction should be made between clinical trials that are explanatory and those that are pragmatic.[21] A combination is also possible; for instance in a clinical trail with aspirin to reduce stroke, in the pragmatic part one would look at strokes and mortality after one or five years of treatment, while in the explanatory part, one would look at the effect on platelet aggregation as a laboratory value or the effect on bleeding time in the study population. It should be emphasized that the past, phase II trial and the current phase III trials with pegorgotein are completely pragmatic. Thus, the specific cause for any beneficial effect the drug may have, remains mostly speculation.

The role that oxygen radicals play in the damage from ischemic injury at the neuronal level is the subject of intensive investigations (see the chapter by Pellegrini-Giampietro in this book). In a number of papers on oxygen radical mediated damage after traumatic injury, it is also assumed that this takes place at the neuronal level.[22-26] However, as noted in the introduction, the work done at the Medical College of Virginia has mostly focused in the vascular effects of oxygen radicals, and it is even possible that at least part of the radicals are generated from circulating blood elements such as leukocytes.[5,27] This distinction between vascular and neuronal damage is not inconsequential, since the penetration of pegorgotein into CSF with an intact blood brain barrier (BBB) is rather poor.[28] On the other hand, penetration is increased by approximately a factor of 6 with post-treatment after an experimental (fluid percussion) brain injury in rats.[29] No pegorgotein could be detected in the CSF taken from patients in the phase II trial, but CSF samples will be taken again in the phase III trials.

Even if the effect of pegorgotein would be mostly on cerebral blood vessels (which is not certain), then this would still provide a perfectly plausible explanation for its effect on outcome. In a model of focal ischemia, it was shown that pegorgotein may well exert its beneficial effects by raising blood pressure slightly and by promoting cerebral blood flow (CBF) during the ischemic phase.[30] One explanation for this finding is that SOD causes vasodilation by prolonging the half-life of endothelium derived relaxing factor (EDRF) and preventing the formation of toxic peroxynitrate.[31,32,33] It has also been shown that the superoxide anion is a direct vasoconstrictor, suggesting it may play a role in acute vasospasm.[34] The link between ischemia and traumatic brain injury is a very direct one: we have shown that CBF in the ischemic range, combined with a high arteriovenous difference of oxygen ($AVDO_2$) is a frequent finding very early after injury.[35,36] Moreover, we have shown that there is a strong correlation between CBF and outcome,[35,36,37] that the earlier ischemia is reversed, the better the outcome[37] and that spontaneous reversal of ischemia must take place at approximately four hours postinjury in practically all cases.[38] Thus, reversal of ischemia may explain at least part of the effects of pegorgotein, while the time course of ischemic (spontaneous resolution at 4 hours) may explain why pegorgotein works so much better when given within 4 hours after injury than when administered after this period (see Fig. 4).

Hopefully, the pragmatic phase II trials will soon be able to show that pegorgotein is indeed beneficial, whatever its mechanism. Nevertheless, knowledge of the mechanism of action is important, especially when one would combine pegorgotein with other drugs affecting parts of the pathophysiologic cascade different from that or those affected by pegorgotein.[39]

REFERENCES

1. E.F. Ellis, K.F. Wright, E.P. Wei, et al. Cyclo-oxygenase products of arachidonic acid metabolism in cat cerebral cortex after experimental brain injury. J. Neurochem. 37:892-896 (1991).
2. E.P. Wei, H.A. Kontos, W.D. Dietrick, et al. Inhibition by free radical scavengers and by cyclo-oxygenase inhibitor of pial arteriolar abnormalities from concussive brain injury in cats. Circ. Res. 48:95-103 (1981).
3. J.P. Muizelaar, A. Marmarou, H.F. Young, et al. Improving the outcome of severe head injury with the oxygen radical scavenger PEG-SOD: A phase II trial. J. Neurosurg. 78:375-382 (1993).
4. E.P. Wei, R.G. Lamb, and H.A. Kontos. Increased phospholipase C activity after experimental brain injury. J. Neurosurg. 56:695-698 (1982).
5. H.A. Kontos, and E.P. Wei. Superoxide production in experimental brain injury. J. Neurosurg. 64:803-807 (1986).
6. J.T. Povlishock, J.I. Williams, E.P. Wei, et al. Histochemical demonstration of superoxide in cerebral vessels. FASEB J. 2:A835 (1988).
7. H.A. Kontos. Oxygen radicals in experimental brain injury, in:Intracranial Pressure VII, J.T. Hoff, and A.L. Betz, eds., Springer-Verlag (1989), pp. 787-798.
8. R.S. Zimmerman, J.P. Muizelaar, and E.P. Wei. Reduction of intracranial hypertension with free radical scavengers, in J.T. Hoff, and A.L. Betz, eds., Springer-Verlag (1989), pp. 804-809.

9. A. Schettini, R. Lippman, and E.K. Walsh. Attenuation of decompressive hypoperfusion and cerebral edema by superoxide dismutase. J. Neurosurg. 71:578-587 (1989).
10. J.E. Levasseur, J.L. Patterson, N.R. Ghatak, et al. Combined effect of respirator-induced ventilation and superoxide dismutase in experimental brain injury. J. Neurosurg. 71:573-577 (1989).
11. H.A. Kontos, and J.T. Povlishock. Oxygen radicals in brain injury. Cent. Nerv. Syst. Trauma 3:257-263 (1986).
12. B. Jennett, and M. Bond. Assessment of outcome after severe brain damage. A practical scale. Lancet 1:480-484 (1976).
13. G. Teasdale, and B. Jennett. Assessment of coma and impaired consciousness. A practical scale. Lancet 1:81-84, (1974).
14. J. P. Muizelaar, A. Marmarou, J.D. Ward, et al. Adverse effects of prolonged hyperventilation in patients with severe head injury: a randomized clinical trial. J. Neurosurg. 75:731-739 (1991).
15. M.A. Foulkes, H.M. Eisenberg, J.A. Jane, et al. The Traumatic Coma Data Bank: design, methods, and baseline characteristics. J. Neurosurg. 75:S8-S13 (1991).
16. S.C. Choi, J.D. Ward, D.P. Becker, et al. Chart for outcome prediction in severe head injury. J. Neurosurg. 59:294-297 (1983).
17. S.C. Choi, J.P. Muizelaar, T.Y. Barnes, et al. Prediction tree for severely head-injured patients. J. Neurosurg 75:251-255 (1991).
18. A. Marmarou, R.L. Anderson, J.D. Ward, et al. Impact of ICP instability and hypotension on outcome in patients with severe head trauma. J. Neurosurg. 75:S59-S66 (1991).
19. M. Rappaport, K. Hall, K. Hopkins, T. Bellaza, and N. Cope. Disability rating scale for severe head trauma: Coma to communicating. Arch. Phys. Med. Rehabil. 63:118 (1982).
20. L.F. Marshall, T. Gautille, M.R. Klauber, H.M. Eisenberg, J.A. Jane, T.G. Luerssen, A. Marmarou, and M.A. Foulkes. The outcome from severe closed head injury. J. Neurosurg. 75(Suppl):528-536 (1991).
21. Schwartz, D., and Lellouch, J. Explanatory and pragmatic attitudes in therapeutical trials. J. Chronic Dis. 20:637-648 (1967).
22. E.D. Hall, J.M. McCall, P.A. Yonkers, R.L. Chase, and J.M. Braughler. A non-glucocorticoid steroid analog of methyl-prednisolone duplicates its high dose pharmacology in models of CNS trauma and neuronal membrane damage. J. Pharmacol. Exp. Ther. 242:137-142 (1987).
23. E.D. Hall, and M.A. Travis. Inhibition of arachidonic acid induced vasogenic brain edema by the non-glucocorticoid 21-aminosteroid V74006F. Brain Res. 451:350-352 (1988).
24. J. M. Braughler, and E.D. Hall. Central nervous system trauma and stroke. I. Biochemical considerations for oxygen radical formation and lipid peroxidation. Free Radical Biol. Med. 6:289-301 (1988).
25. R.V.W. Dimlich, P.A. Tornheim, R.M. Kindel, E.D. Hall, J.M. Braughler, and J.M. McCall. Effects of the 21-aminosteroid (U-74006F) on cerebral metabolites and edema after severe experimental head trauma. Adv. Neurol. 52:365-375 (1990).
26. E.D. Hall, P.K. Andrus, and P.A. Yonkers. Brain hydroxyl radical generation in acute experimental head injury. J. Neurochem. 60:588-594 (1993).
27. R. J. Schoettle, P.M. Kochanek, M.J. Magaru, M.W. Uhl, and E.M. Nemoto. Early polymorphonuclear leukocyte accumulation correlates with the development of post-traumatic cerebral edema in rats. J. Neurotrauma 7:207-217 (1990).
28. K. Yoshida, G.F. Burton, J.S. McKinney, et al. Brain and tissue distribution of polyethylene glycol-conjugated superoxide dismutase in rats. Stroke 23:865-869 (1992).

29. K. Yoshida, G.F. Burton, H.F. Young, et al. Brain levels of polyethylene glycol-conjugated superoxide dismutase following fluid percussion brain injury in rats. J. Neurotrauma 9:85-92 (1992).
30. N. Matsumiya, R.C. Koehler, J.R. Kirsch, et al. Conjugated superoxide dismutase reduces extent of caudate injury after transient focal ischemia in cats. Stroke 22:1193-1200 (1991).
31. G.M. Rubanyi, and P.M. Vanhoutte. Superoxide anions and hyperoxia inactivate endothelium-derived relaxing factor. Am. J. Physiol. 250:H822-H827 (1986).
32. J.J. Marshall, E.P. Wei, and H.A. Kontos. Independent blockade of cerebral vasodilation from acetylcholine and nitric oxide. Am. J. Physiol. 255:H847-H854 (1988).
33. J.S. Beckman, T.W. Beckman, J. Chen, et al. Apparent hydroxyl radical production by peroxynitrite: implications for endothelial injury from nitric oxide and superoxide. Proc. Natl. Acad. Sci. USA 87:1620-1624 (1990).
34. Z.S. Katusic, and P.M. Vanhoutte. Superoxide anion is an endothelium-derived contracting factor. Am. J. Physiol. 257:H33-H37 (1989).
35. G.J. Bouma, J.P. Muizelaar, S.C. Choi, et al. Cerebral circulation and metabolism after severe traumatic brain injury: the elusive role of ischemia. J. Neurosurg. 75:685-693 (1991).
36. G.J. Bouma, J.P. Muizelaar, W.A. Stringer, et al. Ultra-early evaluation of regional cerebral blood flow in severely head-injured patients using xenon-enhanced computerized tomography. J. Neurosurg. 77:360-368 (1992).
37. J.P. Muizelaar. Cerebral ischemia-reperfusion injury after severe head injury and its possible treatment with polyethyleneglycol-superoxide dismutase. Ann. Emerg. Med. 22:1014-1021 (1993).
38. M.L. Schröder, J.P. Muizelaar, and J. Kuta. Documented reversal of global ischemia immediately after removal of an acute subdural hematoma. Report of two cases. J. Neurosurg. 80:324-327 (1994).
39. S. Moon, and A.L. Bets. Interaction between free radicals and excitatory amino acids in the formation of ischemic brain edema in rats. Stroke 22:915-921 (1991).

A NEW METHOD FOR DETECTING LIPID PEROXIDATION BY USING DYE SENSITIZED CHEMILUMINESCENCE

Tadahisa Hiramitsu, Toyoko Arimoto, Takashi Ito, and Minoru Nakano

Photon Medical Research Center, Hamamatsu University School of Medicine
Hamamatsu, 431-31, Japan

Introduction

It has been suggested that iron induced-microsomal lipid peroxidation produces at least two excited species; one for $^1O_2$, and another for excited carbonyl(s), via splitting of a dimer of lipid peroxy radicals[1]. The carbonyls in excited triplet state are known to transfer their excitation energy to triplet sensitizers such as rose-bengal, eosin and 9:10-dibromoanthracene which have suitable molecular orbitals.

The present study was undertaken to verify the generation of excited carbonyls in microsomal phospholipid + methemoglobin reaction by the chemiluminescence method using rose-bengal as sensitizer and a single photon counter as chemiluminescence detector. In addition, usefulness of this chemiluminescence method for testing the ability of antioxidants to inhibit lipid peroxidation was investigated.

Principle

Unsaturated fatty acids in microsomal phospholipids undergo peroxidative cleavage, yielding carbonyl groups (L' = 0) in excited triplet state and/or singlet molecular oxygen ($^1O_2$), in the presence of a catalyst such as methemoglobin (MetHb).

Excited carbonyl transfers its energy to sensitizer in singlet ground state and the resulting dye in excited singlet state emits photons (*hv*) during its return to ground state.

Excited carbonyl(T1) + Sensitizer (S0)→ carbonyl (S0) + Sensitizer (S1)

Sensitizer (S1) → *hv* +Sensitizer (S0)

Results and Conclusion

The results are shown in Fig. 1 and 2.

Under our experimental conditions, O_2 uptake for lipid peroxidation (maximal rate) and chemiluminescence (maximal intensity) in the lipid peroxidation system increased with increasing the amount of phospholipid liposomes, but relative values were not coincident with each other. The chemiluminescence method is a very simple and reliable one to evaluate phospholipid peroxidation.

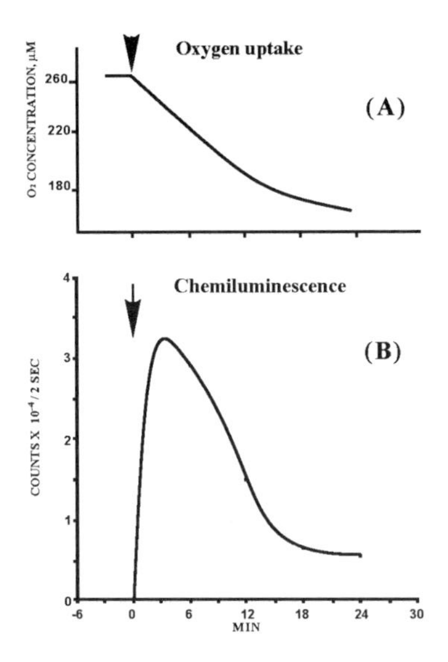

Fig.1. Oxygen uptake (A) and Chemiluminescence (B) during lipid peroxidation.
The system contained phospholipid (250 nmol of lipid phosphorus / ml), 1µM methemoglobin and 0.1M phosphate buffer at pH7.4 in a total volume of 2ml. The reaction was initiated by the addition of methemoglobin and the oxygen uptake was measured by a Clark electrode at 25°C. The initial rate of O_2 uptake, corresponding to maximal rate of O_2 uptake, was calculated from this figure. Rose-bengal (1µM) was added as a sensitizer. The reaction was initiated by the addition of methemoglobin and maintained at 25°C, without agitation. Maximal chemiluminescence intensity corresponds to maximal rate of photon emission.

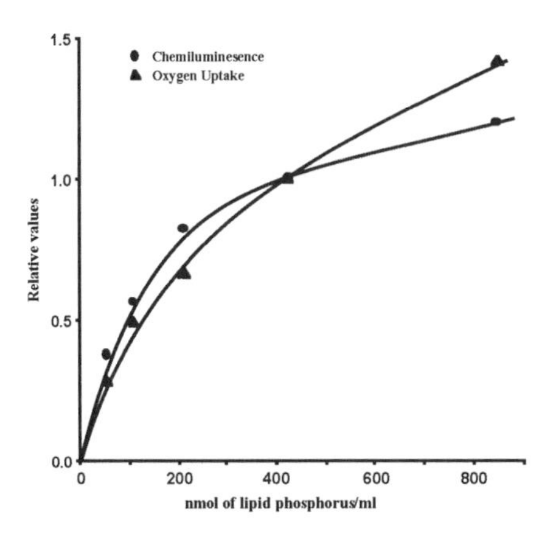

Fig. 2. Relationship between maximal chemiluminescence and initial rate of O2 uptake.
The reaction mixtures were essentially the same as described in Fig. 1 legend, except that phospholipid concentrations were varied. The values obtained with 420 nmol of lipid phosphorous / ml were taken as unity. Increasing amount of phospholipid increased initial rate of O2 uptake and also maximal chemiluminescence. However no complete parallelism was obtained between them.

References

1. K.Sugioka and M.Nakano, A possible mechanism of the generation of singlet molecular oxygen in NADPH-dependent microsomal lipid peroxidation. Biochim Biophys Acta 423:203 (1976).

CHEMILUMINESCENCE AND THE CHALLENGE TO ATTOMOLE DETECTION

John D. MacFarlane
JM Science Inc., Grand Island, NY 14072

Liquid chromatography (LC) has become a major tool for the determination of a wide variety of samples using refractive index, UV and fluorescence detectors. Unfortunately, the sensitivity and selectivity of these detection methods often do not meet the requirements of modern trace-level analysis. Fluorescence is frequently selected as the preferred method of detection if low concentration of analytes have to be determined. Since the range of compounds displaying strong native fluorescence is relatively small, derivatization and /or analyte conversion play an important role in LC with fluorescence detection. However, in many cases derivatization does not allow detection in the low or sub-nanogram/ml range which is increasingly demanded today.

We previously introduced a chemiluminescence detector for ultra-sensitive HPLC detection using a peroxyoxalate reagent system. Detection limits for this instrument were in the femtomole range for dansylated amino acids.

In this report, we describe improvements to this chemiluminescence detector that have contributed to even better levels of sensitivity. For instance, the incorporation of a unique photoelectron counting system results in extreme stability and detection limits to the attomole level for some compounds. The design of the spiral-type flow cell assures adequate mixing, reaction and detection time. The wide linear dynamic range between sample concentration and detector output along with the low cost per analysis and an easy-to-use rugged design make this detector an excellent choice for ultra-trace analysis.

These improvements have been applied to the detection of dansylated amino acids, drugs, carbohydrates, primary, secondary and polyamines, PAH's and fluorescamine-labelled catecholamines. Attomole detection limits for some of these compounds was easily demonstrated.

EVIDENCE AGAINST MALONDIALDEHYDE BOUND TO CELLULAR CONSTITUENTS IN PHOSPHOLIPID PEROXIDATION

Paolo Pedersini, Claudio Ceconi,[1] Anna Cargnoni, Palmira Bernocchi, Salvatore Curello,[1] and Roberto Ferrari[1]

Fondazione Clinica del Lavoro, Centro di Fisiopatologia Cardiovascolare "S. Maugeri", Gussago, Brescia, Italy
[1]Chair of Cardiology, University of Brescia, Spedali Civili, Brescia, Italy

The process of membrane phospholipids peroxidation is involved in the mechanism of injury in a variety of diseases in which oxygen-free radicals play a pathogenetic role. It can explain the loss of structural integrity and function that occurs during the early phases of myocardial post-ischemic reperfusion.[1,2]

Malondialdehyde (MDA) is considered one of the terminal breakdown products of spontaneous fragmentation of lipid peroxides. These breakdown reactions are extremely complex and yield a number of different products like, for example, 4-OH alkenals which have potential tissue toxicity in addition to MDA.[3]

The thiobarbituric acid (TBA) test, the first method used for MDA detection is based on the reactivity of colorless dialdehyde with thiobarbituric acid to produce a red adduct.[4] Although rapid and convenient, the TBA test lacks specificity because of the reactivity of thiobarbituric acid to substances other than MDA, including oxidised lipids, alkenals: moieties related to lipid peroxidation, but also some sugars, amino acids, urea: moieties unrelated to lipid peroxidation.[5,6]

To overcome these difficulties, HPLC assays suitable for application to specific quantitation of free MDA in tissue have been developed and applied in different pathophysiologic conditions.[7] The major criticism of free MDA quantitation by HPLC is that this technique might not detect MDA weakly bound to cellular constituents. It is known that MDA can react with amino and thiolic groups, e.g. of proteins. This MDA is not "free" under physiological conditions and therefore not detectable by direct HPLC method, while it is measurable after the acid heating steps of TBA test. Thus the HPLC method might underestimate the real extent of lipid peroxidation. This, in turn, can explain the failure of HPLC to detect any accumulation of MDA in myocardial tissue homogenates after post-ischemic reperfusion.[8,9]

The aim of this study was to assess whether specific chemical reactions designed to alter the amount of "free" MDA in tissue homogenates would be detectable by HPLC. Two different chemical reactions were employed.

Acidic catalysis. In physiological condition, MDA could react with amino groups; the compounds formed are able to undergo hydrolysis equilibrium and can be hydrolyzed by acidic catalysis.

$$R\text{-}NH_2 \sim\!\sim MDA \rightleftharpoons R\text{-}NH_2 + MDA$$

Exposure to reducing agent. MDA could react with sulphydril groups present in tissue homogenate giving a series of different products. Incubation with strong reducing agent ($NaBH_4$) restores reduced sulphydril groups.

$$R\text{-}S \sim\!\sim MDA \rightleftharpoons R\text{-}SH + MDA$$

We have previously shown that such conditions do not reduce MDA itself.

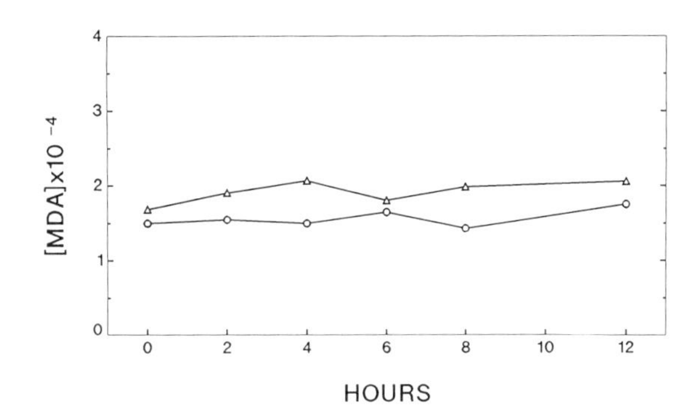

Figure 1. Time-course experiment of MDA concentration after incubation with acidic catalysis (O) and reducing agent (Δ).

Rabbit hearts were homogenized in phosphate buffer (pH=7.4) under nitrogen atmosphere after ischaemia and reperfusion and then supplied with butylated hydroxytoluene to avoid further oxidative processes. Samples were then divided into three sections: a) immediate de-proteinization with phenol (pH=7.4), chloroform-isoamilalcohol (24:1) and extraction of the excess of phenol in aqueous phase with diethylether; b) phenol buffered at pH=1.5 was added and vials were incubated at 40°C for differing periods (up to 12 hours). After incubation, samples were treated with chloroform-isoamilalcohol and then extracted with diethylether; c) phenol buffered at pH=7.4, and $NaBH_4$ were added. Samples were incubated and further treated as in b). HPLC separation conditions were performed according to our previous work.[7]

Table 1. Data representing the six experiments in which ischaemic and reperfused rabbit heart tissue were subjected to different incubation conditions.

	1	2	3	4	5	6
CONTROL	1.52	0.54	2.74	0.95	1.27	0.95
ACIDIC CATALYSIS	2.20	0.57	2.76	1.15	1.33	1.05
REDUCING AGENT	1.50	0.60	2.70	1.08	1.20	0.98

All data expressed in 10^{-10} mol MDA/mg of proteins.

From our experiments, we conclude that different chemical conditions are not able to generate significant additional amounts of MDA from heart tissue homogenate. These results suggest that free/bound MDA ratio is high, and they also confirm the accuracy of direct HPLC method for MDA quantitation that, in our experimental condition, is superior to TBA test. These findings strengthen previous studies in which no accumulation of MDA was found in tissue subjected to ischaemia and reperfusion damage.[7]

REFERENCES

1. P.B. Garlick, M.J. Davies, D.J. Hearse and T.F. Slater, Direct detection of free radicals in the reperfused rat heart using electron spin resonance spectroscopy, *Circ Res* 5:757 (1987).
2. R. Bolli, Oxygen-derived free radicals and post-ischemic myocardial infarction (stunned myocardium), *J Am Coll Cardiol* 12:239 (1988).
3. H. Esterbauer, R.J. Schaur and H. Zollner, Chemistry and biochemistry of 4-hydroxynonenal, malondialdehyde and related aldehydes, *Free Rad Biol Med* 11:81 (1991).
4. H.N. Ohkawa, N. Ohishi and K. Yagi, Assay for lipid peroxides in animal tissues by thiobarbituric acid reaction, *Anal Biochem* 95:351 (1979).
5. D.R. Janero and B. Burghardt, Analysis of cardiac membrane phospholipid peroxidation kinetics as malondialdehyde: nonspecificity of thiobarbituric acid-reactivity, *Lipids* 23:452 (1988).
6. D.R. Janero, Malondialdehyde and thiobarbituric acid-reactivity as diagnostic indices of lipid peroxidation and peroxidative tissue injury, *Free Rad Biol Med* 9:515 (1990).
7. C. Ceconi, A. Cargnoni, E. Pasini, E. Condorelli and R. Ferrari, Evaluation of phospholipid peroxidation as malondialdehyde during myocardial ischemia and reperfusion injury, *Am J Physiol* 260:H1057 (1991).
8. G.A. Fantini, T. Yoshioka and C. Ceconi, Use and limitationas of thiobarbituric acid reaction to detect lipid peroxidation, *Am J Physiol*, 263(*Heart Circ Physiol* 32-3):H981 (1992).
9. D. Lapenna, F. Cuccurullo and C. Ceconi, TBA test and "free" MDA assay in evaluation of lipid peroxidation and oxidative stress in tissue systems, *Am J Physiol*, 265(*Heart Circ Physiol* 34):H1030 (1993).

EVALUATION OF *IN VIVO* FREE RADICAL ACTIVITY DURING ENDOTOXIC SHOCK USING SCAVENGERS, ELECTRON MICROSCOPY, SPIN TRAPS, AND ELECTRON PARAMAGNETIC RESONANCE SPECTROSCOPY

Daniel J. Brackett,[1] Megan R. Lerner,[1] Michael F. Wilson,[3] and Paul B. McCay[2]

[1]Departments of Surgery and Anesthesiology, University of Oklahoma Health Sciences Center and the Veterans Affairs Medical Center
[2]Oklahoma Medical Research Foundation
Oklahoma City, Oklahoma 73190
[3]Department of Medicine, State University of New York
Millard Fillmore Hospitals
Buffalo, New York 14209

INTRODUCTION

Controversy exists concerning the involvement of free radical activity during sepsis and endotoxic shock as evidence has accumulated that both implicate and eliminate free radicals in the pathophysiologic events that occur during these disease processes. The majority of these data are from studies utilizing scavengers or inhibitors of lipid peroxidation to attenuate the activity resulting from excessive free radical generation and as such represent only *indirect* evidence of the presence of free radicals. This study employed methodology combining spin trapping techniques with electron paramagnetic resonance spectroscopy[1] to *directly* verify the presence of *in vivo* free radical generation in the conscious endotoxic rat.[2] Additionally, myocardial structure has been evaluated using transmission electron microscopy to assess the effect of a hydroxyl radical scavenger on endotoxin-induced injury to the myocardium.[3]

METHODOLOGY

Spin Trapping and Electron Paramagnetic Resonance Spectroscopy Ten male Sprague-Dawley rats (300 ± 12 g) were injected intraperitoneally with trimethoxy-a-phenyl-t-butyl-nitrone [$(MeO)_3$ PBN] at a dose of 75mg/kg or 5,5 dimethyl-l-pyrroline-N-oxide (DMPO) five min prior to intraperitoneal endotoxin (20 mg/kg) or saline (controls). Animals were guillotined at either 5 or 25 min after endotoxin and hearts and livers were removed and prepared for analysis. Samples were placed in an IBM ER-300 electron paramagnetic resonance spectrometer equipped with an X-band microwave bridge and evaluated for the presence of spin adducts. All spectra were obtained at 25 °C.

Transmission Electron Microscopy Male Sprague-Dawley rats were injected intravenously with endotoxin (20 mg/kg) and were perfused at one, two, or four hrs later with Karnovsky's fixative. At each of these times hearts were removed and processed for transmission electron microscopy. The effect of the hydroxyl radical scavenger dimethyl sulfoxide in this endotoxic model was evaluated by injecting the scavenger (6g/kg) intraperitoneally 20 min prior to the endotoxin challenge.

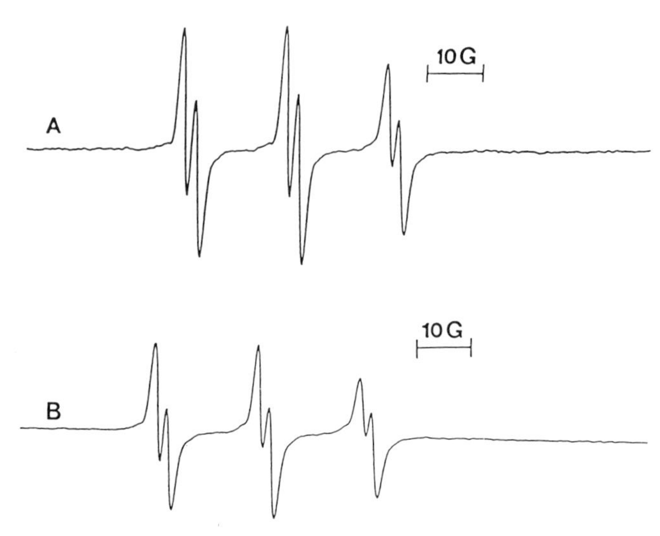

Figure 1. EPR signals of $(MeO)_3$PBN radical adducts generated *in vivo* in the heart (A) and liver (B) after endotoxin in conscious rats.

RESULTS AND DISCUSSION

Spin Trapping And Electron Paramagnetic Resonance Spectroscopy Myocardial and hepatic tissue collected 25 min after endotoxin from animals that received $(MeO)_3$ PBN generated spectra with nitrogen and β-hydrogen splitting constants of $a_N = 15.28$ G, $a^H_\beta = 2.00$ G which are consistent with the signals of $(MeO)_3$ PBN spin adducts of ortho demethylated carbon centered radicals (Figure 1; microwave power, 1.00 e+01 mW; modulation amplitude, 1.036 G; time constant, 1.28 ms; sweep width, 100 G; sweep time, 335.544 s; gain, 1.00 e+5; scale, 17 (A) and 18 (B). These adducts are similar to those detected in the hepatic metabolism of carbon tetrachloride which is known to induce lipid peroxidation and in heart and liver tissue from rats on ethanol diets. Tissue from matched control animals or removed 5 min after endotoxin did not produce detectable spectra. Lipid extracts from hearts and livers of DMPO treated rats generated spectra yielding hyperfine splitting constants of $a_N = 14.80$ G, $a^H_\beta = 20.27$ G and $a_N = 15.00$, $a^H = 23_\beta 04$ G which are consistent with DMPO carbon-centered radical adducts (spectra not shown).

Transmission Electron Microscopy Ultrastructure evaluation of myocardial tissue taken 4 hrs after endotoxin (Figure 2A & 2B) revealed high amplitude swelling of mitochondria (M), sarcomere relaxation with N-line formation (N), prominant I-bands (I), plasma membrane discontinuities, sarcotubular system dilation, both intra and intercellular edema, and depletion of glycogen stores. The pattern of injury observed in these hearts is indistinguishable from the pattern found in ischemic myocardium. Myocardial tissue removed at two hrs had significantly less damage with slight swelling and the presence of flocculent densities (FD) within the mitochrondria (Figure 2C). The tissue at one hr was indistinguishable from control myocardium. Importantly, the hydroxyl radical scavenger dimethyl sulfoxide significantly inhibited the damage at 4 hrs (Figure 2D).

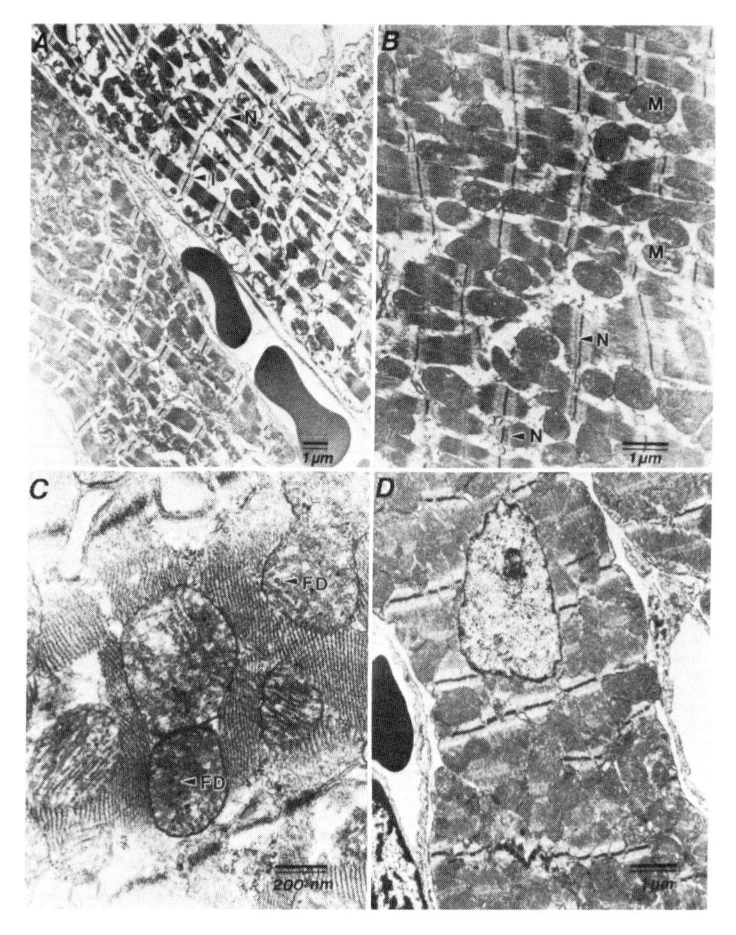

Figure 2. Micrographs of myocardium after endotoxin at 4 hrs (A & B), 2 hrs (C), and 4 hrs with DMSO (D).

Conclusions The detection of free radical generation *in vivo* in the heart and liver during endotoxemia supports studies based on data derived from indirect methodology that suggest free radical reactions do occur and are potential mediators of cellular injury during sepsis and endotoxic shock. Significant injury to the ultrastructure of the myocardium occurs during endotoxic shock and appears to be mediated in a large part by hydroxyl radicals. These findings provide direct evidence for *in vivo* endotoxin-induced free radical generating processes and implicates free radical generation as a significant mediator of organ failure during endotoxic shock.

REFERENCES

1. P.B.McCay and J.L. Poyer, General mechanisms of spin trapping *in vitro* and *in vivo*, *in*: "Handbook of Free Radicals and Antioxidants in Biomedicine" p187, CRC Press, Boca Raton.
2. D.J. Brackett, E.K. Lai, M.R. Lerner, M.F. Wilson, and P.B. McCay, Spin trapping of free radicals produced *in vivo* in heart and liver during endotoxemia, Free Rad Res Comms 7:315 (1989).
3. D.L. Stiers, D.J. Brackett, and M.F. Wilson, Dimethyl sulfoxide protection of endotoxin-induced myocardial injury, *in*: "Proceedings of the 46th Annual Meeting of the Electron Microscopy Society of America" p 320, G.W. Bailey, ed., San Francisco Press, Inc., San Francisco (1988).

INVOLVEMENT OF HYDROXYL RADICALS IN ENDOTOXIN-EVOKED SHOCK

Daniel J. Brackett,[1] Megan R. Lerner,[1] and Michael F. Wilson[2]

[1]Departments of Surgery and Anesthesiology, University of Oklahoma Health Sciences Center and the Veterans Affairs Medical Center Oklahoma City, Oklahoma 73190
[2]Department of Medicine, State University of New York Millard Fillmore Hospitals Buffalo, New York 14209

INTRODUCTION

There has been a myriad of publications recently evaluating the potential role of free radicals in the development and lethality of shock induced by endotoxin. Studies utilizing various types of compounds to interrupt the activity of free radicals or the injurious processes initiated by free radicals have not proven conclusive and the area remains controversial. This may be due to differences in the species of free radicals targeted by the inhibitory compounds selected and by the capacity of the compounds to effect intracellular free radical activity. Inhibitors of free radical activity that cross the cell membrane more effectively attenuate responses to endotoxin.[1] The hydroxyl radical is the most toxic of the free radical species and has been implicated as having a role in endotoxin-induced pulmonary damage and lethality in a study utilizing dimethyl sulfoxide.[2] Dimethyl sulfoxide (DMSO) is a hydroxyl radical scavenger that readily penetrates the cell membrane and therefore has the capacity to influence the actions of excessive intracellular hydroxyl radical generation. The studies described here were designed to exploit the properties of DMSO to evaluate the involvement of hydroxyl radical activity in hemodynamic, metabolic, and pathologic responses evoked by endotoxin.[3]

METHODOLOGY

Forty male, Sprague-Dawley rats weighing 307 ± 2 grams were initially anesthetized with 5% enflurane and then intubated and connected to a rodent respirator delivering 2.5 % enflurane. A combination thermistor/catheter was inserted into the right carotid artery for thermodilution cardiac output and arterial blood pressure measurements and blood sampling. A catheter was placed in the right jugular vein for injection of room temperature saline for generation of cardiac output curves and measurement of central venous pressure. Following catheter placement the animals were allowed to recover from anesthesia and were placed in chambers which permitted unrestrained movement and continuous monitoring. Animals received either endotoxin (40mg/kg, iv, LD 90_{24hrs}), DMSO

(6g/kg, ip) 15 min prior to endotoxin, or DMSO alone. Group numbers were 15, 14, and 11 respectively. Each instrumented rat was monitored for 4 hours after endotoxin. Macroscopic hemorrhage of the small intestine (a target organ in this model) was graded on a 5 point scale at the end of the study.

RESULTS AND DISCUSSION

Dimethyl Sulfoxide The administration of DMSO evoked an immediate, sustained *increase* in arterial blood pressure, heart rate, and systemic vascular resistance and *decrease* in cardiac output, cardiac stroke volume, and central venous pressure. These cardiovascular data suggest that DMSO alone produces a situation conducive to inadequate tissue perfusion. Blood lactate concentrations in this group were increased over control indicating a shift from aerobic to anaerobic metabolism. DMSO did not induce physiologically relevant changes in hematocrit, pH, PO_2, or PCO_2, but plasma glucose concentrations were significantly increased. These results are compatible with an increase in sympathetic outflow and/or circulating catecholamine concentrations.

Endotoxin Intravenous endotoxin produced an immediate, transient, hypotensive episode with blood pressure decreasing from 115 to 50 mmHg and returning to normal after 45 min. There were immediate, sustained *decreases* to 30% of control in cardiac output, 26% in cardiac stroke volume, and by 200% in central venous pressure and *increases* of 38% in heart rate and 200% in central venous pressure. These hemodynamic responses were not significantly different from those induced by DMSO. However, at 4 hrs plasma glucose concentrations were less than control and 30% of the values evoked by DMSO. Blood lactate values were 3 times greater than control and 2 times greater than the DMSO concentrations indicating a significantly greater compromise of tissue perfusion in the endotoxin challenged rats. Changes in blood gases and hematocrits were the same in these 2 groups. The small intestines of the endotoxin group were significantly hemorrhaged ranging from marked banding to extensive areas of lesion.

Dimethyl Sulfoxide And Endotoxin Pre-treatment with DMSO did not alter any endotoxin-induced hemodynamic response, except the transient, hypotensive episode which was blocked, strongly indicating that there was no improvement in blood flow to the tissue and that tissue perfusion was still severely compromised as a result of intense vasoconstriction and a profound reduction in cardiac output. However, the hypoglycemia was prevented and the hyerlacticemia was significantly attenuated inferring maintenance of aerobic metabolism. Most importantly, the hemorrhage of the small intestine was eliminated with occasional evidence of scattered petechiae.

Conclusions The ameliorating effects of DMSO with no evidence of improvement in tissue perfusion or peripheral blood flow can most logically be explained by the scavenging of hydroxyl radicals. Inhibition of hydroxyl radical activity would have decreased cellular injury resulting in the preservation of structure and metabolism as was demonstrated in the data from this study by indices of cellular function and prevention of tissue damage. Hydroxyl radicals may be important mediators of tissue injury in endotoxic shock.

REFERENCES

1. K. McKechnie, B.L. Furman, and J.R. Parratt, Modification by oxygen free radical scavengers of the metabolic and cardiovascular effects of endotoxin in conscious rats, Circ Shock 19:429 (1986).
2. R.E. Breen, R.S. Conneoo, and M.W. Harrison, The effect of dimethyl sulfoxide In endotoxin-induced pulmonary dysfunction: A biochemical and electron microscope study, Ann NY Acad Sci 411:324 1983).
3. D.J. Brackett, M.R. Lerner, and M.F. Wilson, Dimethyl sulfoxide antagonizes hypotensive, metabolic, and pathologic responses induced by endotoxin, Circ Shock 33:156 (1991).

THE ROLE OF NITRIC OXIDE IN ENDOTOXIN-ELICITED HYPODYNAMIC CIRCULATORY FAILURE

Daniel J. Brackett,[1] Megan R. Lerner,[1] and Michael F. Wilson[2]

[1]Departments of Surgery and Anesthesiology, University of Oklahoma
Health Sciences Center and the Veterans Affairs Medical Center
Oklahoma City, Oklahoma 73190
[2]Department of Medicine, State University of New York
Millard Fillmore Hospitals
Buffalo, New York 14209

INTRODUCTION

Endothelial derived relaxation factor has recently been identified as the free radical nitric oxide.[1] Nitric oxide is produced by endothelial cells, hepatocytes, Kupffer cells, neutrophils, and macrophages. All of these cells utilize L-arginine as a substrate for nitric oxide production[2] and each has been documented as having a role in the development of septic and/or endotoxic shock. Recent evidence has suggested that nitric oxide may be a significant mediator in the acute hypotension, cardiovascular dysfunction, and tissue injury associated with septic shock; therefore, inhibition of nitric oxide has been proposed as a potential therapeutic treatment in septic patients.[3] Our goal was to determine if the inhibition of nitric oxide synthesis attenuates the cardiovascular, metabolic, and pathological responses induced by endotoxin in the conscious rat. To accomplish this, N^G-monomethyl-l-arginine (L-NMMA), an inhibitor of both Ca^{2+}-dependent and Ca^{2+}-independent nitric oxide synthesis, was utilized.

METHODOLOGY

Sixteen male Sprague-Dawley rats (300 ± 15 g) were anesthetized with 5% isoflurane, an inhalation anesthesia that is rapidly metabolized, intubated, and maintained on 2% isoflurane delivered by a rodent respirator during surgery for catheterization. The tip of a cannula was advanced to the right atrium through the right external jugular vein for injection of saline to generate thermodilution cardiac output curves, for measurement of central venous pressure, and for infusion of endotoxin or equivolume saline. The tip of a combination thermistor/catheter was placed just distal to the aortic valve via the right common carotid artery for arterial blood sampling and measurement of aortic pressure and thermodilution curves. The instrumented rats were allowed to regain consciousness and were placed in monitoring chambers where their movement was completely unrestrained. Following a 60 min recovery period, 30 min of baseline measurements were acquired to assure stable, normal parameter values. At this time an intravenous infusion of L-NMMA at a concentration of 60 mg/kg (n=8) or saline (n=8) was delivered at 1 ml/min for 30 minutes. Ten minutes after the initiation of the L-NMMA infusion, endotoxin (20 mg/kg, LD-90_{24hrs}) was injected. Monitoring continued for 4 hours after endotoxin. At the end of the study the intensity of gross small intestinal hemorrhage characteristic of this model was graded on a 5 point scale.

RESULTS AND DISCUSSION

Saline Treatment When endotoxin was administered there was an immediate sustained, decrease in cardiac output to 30% of the control value that was associated with a sustained, 100% increase in systemic vascular resistance indicating intense vasoconstriction. Occurring concurrently was a precipitous fall in blood pressure which was transient, returning to near control after approximately 30 min as a result of the compensatory vasoconstriction. There was also a sustained increase in heart rate and decrease in central venous pressure and cardiac stroke volume. Plasma lactate concentrations increased 300% indicating a shift from aerobic to anaerobic metabolism and pH decreased (metabolic acidosis) consistent with compromised tissue perfusion suggested by the cardiovascular measurements. Endotoxin evoked a hyperglycemic response after 60 min followed by hypoglycemia. Total white blood cell counts decreased to 50% of control and neutrophil counts decreased to 30% of the control value. Additionally, there was intense hemorrhage of the small intestine at the end of the study.

L-NMMA Treatment Upon initiation of L-NMMA infusion, 10 min prior to the endotoxin challenge, there was a significant decrease in cardiac output to 70% and a 200% increase in systemic vascular resistance accompanied by a 40 mmHg rise in mean arterial pressure. The inhibition of nitric oxide caused a strikingly rapid attenuation of tissue perfusion. Following endotoxin the hypotensive episode was not prevented, but was only blunted, suggesting the involvement of another mediator(s) that is capable of reducing blood pressure independent of nitric oxide. Cardiac output was significantly lower and systemic vascular resistance was approximately doubled in this nitric oxide inhibited group compared to the untreated group. Hyperlacticemia, metabolic acidosis, and hypoglycemia were substantially more profound as a result of nitric oxide inhibition. These data strongly indicate that nitric oxide inhibition significantly exacerbates the compromise in tissue perfusion and potential for cellular damage that exists during hypodynamic endotoxic shock. Indeed, 5 of the 8 animals in this group died before the end of the study period whereas none of the animals in the untreated group died. There were no significant differences between groups in the other parameters discussed in the section above.

Conclusions Although there has been evidence that L-NMMA causes an increase in arterial blood pressure, our data demonstrates that the elevation in pressure is due to vasoconstriction associated with a pronounced decrease in cardiac output which can only compromise blood flow to peripheral tissue and enhance the potential for cellular injury. This study clearly establishes that when nitric oxide inhibition is applied to an already precarious hemodynamic situation such as hypodynamic endotoxic shock the results are disastrous. Contrary to other reports, nitric oxide inhibition with L-NMMA is not a beneficial intervention to ameliorate the development of endotoxic shock.

REFERENCES

1. R.M.J. Palmer, A.G. Ferrige, and S. Moncada, Nitric oxide release accounts for the biologic activity of endothelium-derived relaxing factor, Nature 327: 524 (1987).
2. S. Moncada, R.M.J. Palmer, and E.A. Higgs, Biosynthesis of nitric oxide from L-arginine: A pathway for the regulation of cell function and communication, Biochem. Pharmacol. 38: 1709 (1989).
3. A. Pitros, D. Bennett, and P. Vallance, Effect of nitric oxide synthase inhibitors on hypotension in patients with septic shock, Lancet 338: 1557 (1991).

IN VIVO EVALUATION OF THE ROLE OF NEUTROPHIL-DERIVED FREE RADICALS IN THE DEVELOPMENT OF ENDOTOXIC SHOCK

Daniel J. Brackett,[1,2,4] Megan R. Lerner,[1,4] Mike E. Gonce,[1] Paul B. McCay,[5] and Andre K. Balla[3]

Departments of [1]Surgery, [2]Anesthesiology, and [3]Pathology
University of Oklahoma Health Sciences Center
[4]Veterans Affairs Medical Center
[5]Oklahoma Medical Research Foundation
Oklahoma City, Oklahoma 73190

INTRODUCTION

Clinical and experimental data strongly suggests that a significant portion of the peripheral responses occurring during septic shock are initiated by endotoxin.[1] One of the peripheral responses initiated by endotoxin is the stimulation of neutrophils resulting from interaction with tumor necrosis factor derived from endotoxin activated macrophages. Stimulated neutrophils have the potential to marginate along endothelial cells and migrate to extravascular tissue. Once this has occurred activated neutrophils have the capacity to release enzymes, hypochlorous acid, and importantly hydrogen peroxide and superoxide ions resulting in cellular injury. Tumor necrosis factor is becoming increasingly recognized as a major mediator of sepsis and the effects of endotoxin. Evidence strongly suggests that a majority of tumor necrosis factor effects are a result of its action on neutrophils and oxidative metabolism.[2] Endotoxin has a dual action on neutrophil function since, in addition to the tumor necrosis factor mediated effect, this toxin has been shown to prime neutrophils for an enhanced oxidative burst and release of neutrophil-generated mediators of injury.[3] This study was designed to determine the significance of neutrophils in the development of endotoxic shock by challenging neutropenic rats with endotoxin.

METHODOLOGY

Ten male Sprague-Dawley rats were given vinblastine intravenously at a dose of 0.7 mg/kg and circulating neutrophils were counted each day for seven days. Neutrophil counts were significantly diminished at 3 days, were undetectable at days 4, 5, and 6, and were returning on the 7th day. An additional twenty rats were injected with either vinblastine (n = 10) or saline (n = 10). Five days after vinblastine or saline injection the responses of the animals to endotoxin were studied. Under isoflurane anesthesia a thermocouple-catheter combination was inserted into the right carotid artery for measurement of mean arterial pressure, heart rate, thermodilution cardiac output curves, and blood samples. The right jugular vein was catheterized for measurement of central venous pressure and for injection of 100 μl room temperature saline to generate cardiac output curves. The animals were then placed in chamber permitting continuous monitoring to regain consciousness. Following a recovery period and a 30 min period of control monitoring, conscious, unrestrained animals were challenged with endotoxin (20 mg/kg, LD $90_{24\ hrs}$) and then monitored for 4 hrs. At the end of the study the small intestine was graded for intensity of gross hemorrhage and lungs were processed for measurement of myeloperoxidase activity (an index of tissue neutrophil accumulation).

RESULTS AND DISCUSSION

Control Animals Administration of endotoxin to the rats treated with saline 5 days earlier resulted in the characteristic responses. There was an immediate, transient, hypotensive episode lasting 45 min before returning to near baseline levels. Endotoxin induced a precipitous decrease in cardiac output (67%) accompanied by an increase in systemic vascular resistance (250%) creating a situation of significantly compromised blood flow and intense vasoconstriction. These responses were sustained throughout the study. The suspected inadequate tissue perfusion was substantiated by a greater than 100% increase in blood lactate concentrations demonstrating a shift from aerobic to anaerobic metabolism. There were sustained increases in heart rate (29%) and cardiac stroke volume (72%) and there was an early, transient decrease in central venous pressure. ***Circulating neutrophil concentrations*** decreased to 24% for the entire monitoring period. Plasma glucose and tumor necrosis factor concentrations significantly increased early (60 and 90 min respectively) and then declined for the remainder of the study. Pulmonary myeloperoxidase activity was increased by 100% over control tissue and there was intense hemorrhage of the small intestine, a target organ in this model.

Neutropenic Animals The administration of endotoxin to vinblastine-treated neutropenic rats did not reveal significant, long-term modification in any of the cardiovascular or metabolic responses. Neutrophils were not detectable in vinblastine-treated animals at any time during the study. The early hypotensive episode was totally blocked due to an immediate, enhanced increase in systemic vascular resistance (vasoconstriction) and the early fall in central venous pressure was blunted. However, over the 4 hr monitoring period the cardiovascular responses were essentially the same. Neutropenia had no effect on lactate, glucose, or tumor necrosis factor responses induced by endotoxin and there was no alteration of intensity in the small intestinal hemorrhage. As would be predicted, there was no detectable myeloperoxidase activity in the lungs of the endotoxin-challenged neutropenic animals substantiating the validity of our counts of circulating neutrophil concentrations and the neutropenic model.

Conclusions The early effect of neutropenia on hemodynamics does not appear to have influenced the development of endotoxin-induced hypodynamic shock or the outcome in this model. Beginning at 30 min after endotoxin all parameters were essentially the same. There was no detectable indication that neutropenia improved compromised tissue perfusion or inhibited tissue injury. These data suggest that neutrophils are not essential mediators in this model of hypodynamic endotoxic shock.

REFERENCES

1. E.K. LeGrand, Endotoxin as an alarm signal of bacterial invasion: current evidence and implications, JAVMA 197: 454 (1990).
2. A. Kapp, G. Zeck-Kapp, and K. Blohm, Human tumor necrosis factor is a potent activator of the oxidative metabolism in human polymorphonuclear neutrophilic granulocytes, J. Invest. Dermatol. 92: 348 (1989).
3. L. A. Guthrie, L.C. McPhail, P.M. Henson, and J.R. Johnston, Priming of neutrophils for enhanced release of oxygen metabolites by bacterial lipopolysaccharide, J. ESP. Med. 160: 1656 (1984).

PHOTOSENSITIZED FORMATION OF 8-HYDROXY-2'-DEOXYGUANOSINE BY A CATIONIC meso-SUBSTITUTED PORPHYRIN

Thomas M. Nicotera and Robert J. Fiel

Departments of Biophysics and Experimental Biology
Roswell Park Cancer Institute
Elm and Carlton Streets
Buffalo, New York 14263-0001

INTRODUCTION

The meso-substituted N-methylpyridyl class of cationic porphyrins are known to have a high affinity for DNA (1). As such, they have been useful tools with which to probe nucleic acid structure and to study a broad range of interactions of cationic ligands with DNA. Activation of this class of porphyrins by visible light produces strand scission in the phosphodiester backbone of DNA at relatively low doses and some metallo-analogs are capable of producing strand scission in the dark. Other substituted porphyrins, hematoporphyrins and methylene blue were found to be less effective under similar conditions. The strand scission activity of cationic porphyrins is not clearly understood, although singlet oxygen has been implicated as a reactive species (1). In this report we have examined the mechanism of photo-induced base damage; specifically, the formation of 8-OHdG. Understanding their role in DNA damage may be useful in screening these compounds as potentially useful agents in photodynamic therapy.

MATERIALS AND METHODS

DNA Photolysis- Porphyrin was added to plasmid DNA (pGS81) derived from pBR322 (5 μg) in 0.05 M Tris, pH 8 in a 1.5 ml eppendorf microtube and placed 4 cm from a 15 watt fluorescent lamp at a light intensity of 0.90 Lux. The light exposure ranged from 4.0 x 10^{-5} to 4.8 10^{-4} joules/cm^2 (0.8 joules/cm^2/min).

HPLC-EC- DNA samples of 1.0 mg/ml in Tris buffer (50 mM, pH 8.0). The DNA was digested to nucleosides by incubation for 2 hours at 37°C with a mixture of DNAseI, snake venom phosphodiesterase I and alkaline phosphatase. The digested nucleosides were separated by HPLC and detected by UV light at 254 nm and 8-OHdG by electrochemical detection using a BAS amperometric detector with a glassy carbon electrode at a potential of 600 mV and a current of 0.2 nA. The mobile phase consisted of potassium phosphate buffer (0.05M, pH 5.5) containing acetonitrile (97.0:4.0, v/v) and pumped at a flow rate of 1.0 ml/min was used.

RESULTS AND DISCUSSION

The results demonstrate an increase in 8-OHdG formation with increasing cis-porphyrin concentration when exposed to light for 15 minutes while no substantial increase in 8-OHdG was observed with porphyrin alone. 8-OHdG also increased with increasing exposure to light at a constant dose of 2 μM cis-porphyrin. Hence, cis-dicationic porphyrin is a potent photo-inducer of 8-OHdG in DNA.

In order to examine the mechanism of 8-OHdG formation by cis-porphyrin and light, the effect of hydroxyl radical and singlet oxygen scavengers on its formation were evaluated. A linear decrease in 8-OHdG formation was observed as the fraction of D_2O increased. Use of hydroxyl radical scavengers mannitol and DMSO failed to inhibit 8-OHdG formation. Similar to its effect on the photo-induction of strand breaks, $MgCl_2$ (20mM), which is known to inhibit binding of this porphyrin to DNA, also inhibits the formation of 8-OHdG.

These results suggest that two mechanisms may be involved in the photo-induction of strand scission: one that requires oxygen and, based on the inhibitory effects of 1,3-diphenylisobenzofuran, involves singlet oxygen, and a second mechanism that is apparently independent of oxygen. It is apparent from the present work and that of others that the mechanism(s) by which photosensitizers damage DNA is complex and to some degree dependent on the particular photosensitizer, its ability to associate with DNA and the nature of that association. In the case of the photosensitizing action of cis-porphyrin, both strand breaks and base damage have been identified.

In a related investigation the authors suggested that a type I mechanism was the basis of inactivation of lambda phage by tetrapyridyl cationic porphyrin, ie., through the formation of a radical cation in guanine by electron transfer from the excited porphyrin (2). This is especially interesting in view of the GC specificity of the intercalating binding mode of these porphyrins. Studies in progress should provide more detailed insight into to the mechanism(s) of photoinduced lesions in DNA using cis-porphyrin.

REFERENCES

1. Fiel, R.J. *J. Biomolec. Struct. & Dynamics,* 6:1259 (1989).
2. Katsuri,C. and Platz,M.S *Photochem. Photobiol.* 56:427 (1992).

AGING ASSOCIATED DECLINES IN THE ANTIOXIDANT ENZYMES OF HUMAN TESTIS

Yong-Mei Yin, Abida K. Haque, Gerald A. Campbell
and Mary Treinen Moslen

Department of Pathology
University of Texas Medical Branch
Galveston, TX 77550

INTRODUCTION AND OBJECTIVES

One potential factor in aging is a decline in antioxidant protection that would facilitate progressive accumulation of damage by reactive oxygen species. Testicular atrophy is commonly found in aged males but little is known about effects of aging on the antioxidant protection systems of human testis. Our first objective was to determine if aging was associated with changes in testicular activities of aldehyde dehydrogenase, GSSG reductase, and GSH peroxidase. Aldehyde dehydrogenase detoxifies malondialdehyde, 4-hydroxynonenal and related products of lipid peroxidation. GSH peroxidase detoxifies H_2O_2 and fatty acid peroxide product of lipid peroxidation. GSSG reductase regenerates GSSG to GSH. Our second objective was to determine if observed changes in antioxidant protection were associated with morphological characteristics of testicular atrophy.

EXPERIMENTAL DESIGN

Testes were collected from adult males without infectious disease that were autopsied at UTMB within 24 hrs after death. Portions of testes were rapidly frozen for biochemistry or were fixed for routine histology. Detoxification enzymes were assayed by conventional kinetic methods. A color assisted computer analysis system was used to determine testicular morphometrics. At least 10 randomly chosen fields were evaluated for each case. Since alcoholic cirrhosis is a known factor in testicular atrophy, autopsy reports and liver histology were reviewed in order to exclude all cases of cirrhosis. Testes from 49 cases were evaluated for biochemical and histologic parameters. For purposes of analysis, samples were divided into 3 groups: young (<39 yr), middle-aged (40-64 yr), and old (>65 yr).

TABLE 1. Aging related declines in testicular enzyme activities

Enzymes	Young (N=11) $\leq$ 39 yrs	Middle-Aged (N=17) 40-64 yrs	Old (N=12) $\geq$ 64 yrs
Aldehyde Dehydrogenase	1.7 ± 0.5	1.3 ± 0.5 *	0.9 ± 0.3 *
GSSG Reductase	67.0 ± 11.6	53.2 ± 10.5 *	31.3 ± 18.2 *,‡
GSH-Peroxidase	54.5 ± 16.7	45.1 ± 11.2 *	36.2 ± 18.2 *
Protein	7.1 ± 1.2	7.1 ± 1.2	6.8 ± 1.1

Values are Mean ± SD, Statistical analysis by ANOVA and Scheffe's test
Enzyme activity units: μmoles/min/mg prot.
Protein unit: mg/ml

* Different from young group, $p<0.05$
‡ Different from middle-aged group, $p<0.05$

TABLE 2. Aging related changes in testicular morphology

	Young (N=18) $\geq$ 39 yrs	Middle-Aged (N=17) 40-64 yrs	Old (N=12) $\geq$ 65 yrs
Tubule Number (N/μ^2)	10.0 ± 1.2	10.8 ± 2.4	11.5 ± 1.9
Tubule Area (%)	81.4 ± 4.6	70.7 ± 7.6 *	58.2 ± 14.1 *,‡
Inner Diameters (μ)	220 ± 17	172 ± 17 *	143 ± 23 *,‡
Outer Diameters (μ)	236 ± 14	207 ± 21 *	205 ± 28 *
B.M. Thickness (μ)	6.8 ± 1.2	12.5 ± 4.2 *	25.8 ± 11.5 *,‡

Values are Mean ± SD, Statistics by ANOVA and Scheffe's test.

* Different from young group, $P<0.05$
‡ Different from middle-aged group, $P<0.05$

RESULTS AND CONCLUSIONS

As indicated in Table 1, aging was associated with declines in the activities of aldehyde dehydrogenase, GSSG reductase and GSH-peroxidase activities. Note that the 54% decline in GSSG reductase was progressive as indicated by the statistical differences of the old group from both the young and middle-aged groups. Table 2 shows that aging was associated with striking changes in testicular morphology, particularly a thickening of basement membranes. Note the progressive changes in tubular area, inner diameter, and basement membrane thickness. Further studies are needed to examine correlations between the biochemical and histological changes and to determine potential cause-effect relationships.

ACKNOWLEDGMENTS

Supported by the Seed Money Research Funding Program of the UTMB Center on Aging (YMY) and the John Sealy Foundation (MTM).

LIPID PEROXIDATION IN NORMAL PREGNANCY AND PREECLAMPSIA

Anna Cargnoni, Gina Gregorini,[1] Claudio Ceconi,[2] Rosario Maiorca[1], Roberto Ferrari[2]

Fondazione Clinica del Lavoro, Centro di Fisiopatologia Cardiovascolare "S. Maugeri", Gussago, Brescia, Italy
[1]Chair of Nephrology, University of Brescia, Spedali Civili, Brescia, Italy
[2]Chair of Cardiology, University of Brescia, Spedali Civili, Brescia, Italy

Increased levels of lipid peroxidation may contribute to the development of endothelial injury which is a key pathogenetic factor in preeclampsia[1]. Moreover, lipid peroxides per se inhibit prostacyclin formation without affecting thromboxane synthetic pathways[2]. In this way they could directly contribute to the generation of the thromboxane–prostacyclin imbalance reported in preeclampsia. Increased blood levels of lipid peroxidation products in preeclampsia have been reported by many authors[2–4]. Levels of lipid peroxidation were usually determined as malondialdehyde (MDA) content measured by the TBA–test.

This assay has recently been criticized because when applied to biological sample, it lacks specificity[5]. Conjugated dienes (CD) are thought to better reflect lipoperoxidation that occurs in vivo. CD are the first products of oxidative attack to the polyunsatured fatty acid molecules and reflect lipoperoxidation from early stage of the process; this is unlike MDA which is a measure of decomposition products.

We measured plasma levels of cholesterol triglycerides and CD in normal pregnancy and preeclampsia as lipids are potential oxidative substrates[6]. We hypothesized the existence of a correlation between levels of lipid and levels of CD in normal and pathological pregnancy.

Blood samples were collected from non–pregnant (group a), normally pregnant women at different stages of pregnancy (group b) and women with severe preeclampsia (group c). Lipids were extracted from plasma with a chloroform/methanol mixture (2:1), dried under nitrogen atmosphere and then redissolved in cyclohexane. CD were assayed in the extracted lipids by the second derivative absorbance spectrum[7]. Total cholesterol and triglycerides blood levels were measured by enzymatic assays.

As expected, in women with preeclampsia, weeks of pregnancy, fetal weight, systolic and diastolic blood pressure, serum uric acid, proteinuria and platelets count were all significantly different from normally pregnant women. Plasma levels of CD increased both in the normally pregnant (group b) and in women with preeclampsia (group c). In the normally pregnant group, plasma levels of CD were not related to the different stages of pregnancy.

Plasma levels of triglycerides and cholesterol increased to same extent in the normally pregnant or in patients with preeclampsia.

Table 1. Plasma levels of CD, triglycerides and cholesterol in the normally pregnant and in patients with preeclampsia.

	Group A n 17	Group B n 39	Group C n 11
CD (arbitrary units)	49.2±4.6	72.8±4.5[2]	76.8±12.5[1]
cholesterol (mg/dl)	191±6	270±9[3]	248±15[2]
triglycerides (mg/dl)	102±12	203±13[3]	285±24[3,4]

1 $p<0.05$ vs group A
2 $p<0.01$ vs group A
3 $p<0.001$ vs group A
3,4 $p<0.005$ vs group B

When all groups were considered in statistical analysis, there was a significant correlation between CD levels and triglycerides ($p<0.001$) and cholesterol ($p<0.001$) plasma concentrations.

These results suggest that: 1) CD levels increase during pregnancy, but no further changes occur in preeclampsia; 2) CD are related to the plasma content of cholesterol and triglycerides.

Thus, our study does not confirm the hypothesis that lipid peroxidation is a pathogenetic factor in preeclampsia. Further studies are needed, however, to clarify this important issue. It is possible that alteration of lipid metabolism in pregnant women could represent a potentially dangerous factor which becomes evident (perhaps as oxidizable substrate) in women with preeclampsia only.

REFERENCES

1. J.M. Roberts, R.N. Taylor, T.J. Musci, G.M. Rodgers, C.A. Hubel, and M.K. McLaughlin, Preeclampsia: an endothelial cell disorder, *Am J Obstet Gynecol* 161:1200 (1989).
2. A.M. Warso, and W.E.M. Lands, Lipid peroxidation in relation to prostacyclin and thromboxane physiology and pathophysiology. lipid peroxidation, prostacyclin and thromboxane, *Brithis Medical Bulletin* 39:277 (1983).
3. C.A. Hubel, J.M. Roberts, R.N. Taylor, and T.J. Musci, Lipid peroxidation in pregnancy: new perspectives on preeclampsia, *Am J Obstet Gynecol* 161:1025 (1989).
4. T.F. Slater. "Free Radical Mechanisms in Tissue Injury," Pion Limited, London (1972).
5. C. Ceconi, A. Cargnoni E. Pasini, E. Condorelli, S. Curello, and R. Ferrari, Evaluation of phospholipid peroxidation as malondialdehyde during myocardial ischaemia and reperfusion injury, *Am J Physiol* 260:1057 (1991).
6. J. Potter, and P. Nestel, The hyperlipidemia of pregnancy in normal and complicated pregnancies, *Am J Obstet Gynecol* 133:165 (1979).
7. F. Corongiu, and A. Milia, An improved and simple method for determining diene conjugated in auto-oxidized polyunsaturated fatty acids, *Chem Biol Interactions* 44:289 (1983).

ALKYLATION OF PURINE BASES BY CARBON-CENTERED RADICALS

Jae O. Kang

Department of Medical Laboratory Science
University of New Hampshire
Durham, NH 03824

Mono- and di-substituted hydrazines are found ubiquitously (1, 2). They are present in the environment since they are used as herbicides, chemical intermediates, and as high energy fuels for rockets. They are found in plants such as tobacco and mushrooms. They are also present in medicine. All substituted hydrazines tested for carcinogenicity in long-term animal studies have been found to be positive (1). 1,2-dimethylhydrazine (DMH) is a well known carcinogen that induces tumors specifically in the colon of rodents (3). Since DMH-induced lesions have characteristics that are remarkably similar to those of human colon cancer, this carcinogen is widely used as a model in the study of colon cancer. The ethyl derivative of DMH, 1,2-diethylhydrazine (DEH), is also a carcinogen. DEH produces cancer in the thymus, liver, and brain (4). The mechanism(s) by which these di-alkyl hydrazines induce tumors is not known. Kang et al. (5) have shown that in rats the metabolism of DMH and DEH produces methyl radicals ($\cdot CH_3$) and ethyl radicals ($\cdot CH_2CH_3$), respectively. Therefore, this investigation was conducted to explore the nature of the chemical interaction(s) between alkyl radicals and biomolecules using $\cdot CH_3$ and RNA as models.

RNA was incubated at 37°C in a system where $\cdot CH_3$ was produced during the oxidation of dimethylsulfoxide by hydroxyl radicals ($\cdot OH$); $\cdot OH$ was generated from a Fenton-type reaction between Fe^{2+}-EDTA and H_2O_2. After incubation, RNA was dialyzed against double-deionized water and hydrolyzed in either of two conditions: (1) in 0.1 M HCl at 70°C for 60 min; (2) 1.0 M HCl at 95°C for 60 min. The acid hydrolysates were analyzed by HPLC, using both a cation exchange column (Partisil 10 SCX, Whatman) and a RP analytical column (Specosil LC-18S, Spelco). Experimental details are described in Ref. 6.

Four new products were detected from the hydrolysates of $\cdot CH_3$-treated RNA: 8-methylguanine (8-MeG), 2-methyladenine (2-MeA), 8-methyladenine (8-MeA), and one highly unstable product of unknown structure. The yields of 2-MeA and 8-MeA were determined from the 0.1 M HCl-hydrolysates. The quantities of adenine, guanine, and 8-MeG were assayed from the 1.0 M HCl-hydrolysates because $\cdot CH_3$-treated RNA was not completely hydrolyzed in 0.1 M HCl. The peaks representing 2-MeA, 8-MeA, and the unstable product were not present in the chromatograms of the 0.1 M HCL-hydrolysates.

The effect of pH on the generation of the new products was significant. At pH 5.2, the yields of the new products (mean $\pm$ SD, n = 6) were: 8-MeG = 144 $\pm$ 18 mmol/mol guanine, 2-MeA = 6.5 $\pm$ 1.9 mmol/mol adenine, and 8-MeA = 0.5 $\pm$ 0.1 mmol/mol adenine. When pH was increased to 7.2 and to 9.2, the yield of 8-MeG was reduced to 39 mmol/mol guanine and 29 mmol/mol guanine, respectively. The yield of 2-MeA was 0.1 mmol/mol adenine at pH 7.2 and 0.05 mmol/mol adenine at pH 9.2. To the contrary, increasing pH from 5.2 to 7.2 augmented the production of 8-MeA by 80%. Between pH 7.2 and 9.2, 8-MeA was not affected.

Catalase, ethanol, and alpha-(4-pyridyl-1-oxide)-N-t-butylnitrone (POBN) significantly inhibited the production of methylated purine residues. Oxygen also effectively inhibited new product formation (6).

Acknowledgment: This work was supported by Grant CA 54443 from the National Cancer Institute.

REFERENCES

1. B. Toth. Synthetic and naturally occurring hydrazines as possible cancer causative agents. Cancer Res. 35:3693 (1975).
2. B. Kalyanaraman and B.K. Sinha. Free radical-mediated activation of hydrazine derivatives. Envir. Health Perspect. 64:179 (1985).
3. H. Druckrey. Organospecific carcinogenesis, in:"Topics in Chemical Carcinogenesis", eds. W. Nakahara, S. Takayama, T. Sugimura, and S. Odashima, University Park Press, Baltimore, pp. 87 (1971).
4. K.M. Pozharisski, Y.M. Kapustin, A.J. Likhachev, and J.D. Shaposhnikov. The mechanism of carcinogenic action of 1,2-dimethylhydrazine (SDMH) in rats. Int. J. Cancer 15:673 (1975).
5. J.O. Kang, G. Slater, A.H. Aufses Jr., and G. Cohen (1988). Production of ethane by rats treated with the colon carcinogen, 1,2-dimethylhydrazine. Biochem. Pharmacol. 37:2967 (1988).
6. J.O. Kang, K.S. Gallagher, and G. Cohen. Methylation of RNA purine bases by methyl radicals. Arch. Biochem. Biophys. 306:178 (1993).

CYTOCHROME BIOCHEMISTRY IN SHEEP RETINA FOLLOWING EXPOSURE TO OXYGEN

R. Stockton, J. Wilhelm and D. Armstrong

Departments of Ophthalmology and Medical Technology
SUNY, Buffalo

Introduction: The mitochondrial enzyme cytochrome c oxidase (COX) is abundant in the photoreceptor inner segments and the synaptic layers of the retina where it functions as essential enzyme in oxidative metabolism. Indeed, the neural retina has the highest known rates respiration and glucose oxidation (1,2). However, the greatest concentration of COX is found in the retinal pigment epithelium (RPE) where it presumably functions primarily as an antioxidant. Photoreceptor cells are highly sensitive to oxygen toxicity when oxygen tensions (PO2) are elevated by as little as two fold (3).The photoreceptor cell membrane is particularly susceptible to free radical attack because of its high ratio of docosahexaenoic acid (4). However O2 toxicity might be exacerbated by a direct effect of O2 upon the activity of COX which would diminish its efficiency in both its metabolic role and as an antioxidant.

Methods: Fresh sheep eyes were obtained from a local slaughter house immediately after death and placed in ice. within 2 hours, the neural retina and retinal pigment epithelium (RPE) removed separately to 0.2M phosphate buffer with 0.5% Triton X-100 an homogenized (5). Homogenates of neural retina and RPE were then centrifuged at 2400 or 800G respectively, the pellets re-suspended in the phosphate-Triton mixture. COX activity was determined through spectrophotometric analysis (6) following in vitro incubation of homogenates with hydroxyl radicals generated by the Fenton reaction (7). Kinetic analysis was by Linewever-Burk and Eadie-Hofstee plots (8).

Results: Homogenates of retina and RPE from adult sheep were incubated with OH radical and spectrophotometrically assayed to determine the COX activity. Table 1 shows the effect of accumulative oxidative damage upon COX activity and kinetics is as a function of incubation time. These data indicate that with increasing exposure to OH radical, COX in both retina and RPE rapidly decreases in activity (%COX). Fluorometric analysis determined that this effect was due to peroxidation (7). Concomitantly, COX affinity for oxygen decreases (increased Km) as does retinal ability to turn-over substrate (decreased Vmax). At 30 minutes,the COX activity in the RPE (61%) is more resistant to peroxidation than retinal COX (18%). But by 60 minutes, little to no activity remained in both tissues. A second difference between retinal and RPE COX is revealed through Eadie-Hofstee plot analysis. Retinal COX is shown to have a monophasic V/S to V relation, indicating a single catalytic site (Km= 14μM). In contrast, the RPE COX shows the commonly observed bi-phasic Eadie-Hofstee relation, indicating both a high affinity (Km= 0.65μM) and low affinity (Km= 28μM) catalytic sites (8).

Table 1. Change in Retinal and RPE COX activity (% COX) and kinetics during incubation with OH radical.

Incubation OH (min)	RETINA % COX	Vmax	Km	RPE % COX	V1max	Km1	V2max	Km2
0	100	242	14	100	28	0.65	180	28
30	18	43	20	61			110	19
60	11.5	30	21	0				
90	10	23	22					
120	6	13	21					

Vmax = nM oxidized cyt c /min /mg protien Km = μM

Conclusions: Our data show a distinct difference between the retinal and RPE forms of cytochrome c oxidase. The RPE species has high and low affinity binding sites similar to other tissues (9). In contrast, retinal COX is unusual in having a single catalytic site with a Km about midway between those of the RPE. Exposure of both retinal and RPE COX decreases activity and Vmax indicating that peroxidation of COX inhibits its ability to reduce oxygen. Initially, the retinal COX is much more susceptible to OH free radical attack than the RPE species. In essence, peroxidation decreases the capacity of COX to participate in normal respiration and to protect the retina from oxygen free radical damage. It is likely that retinal COX is primarily involved in oxidative metabolism, whereas in RPE, COX serves primarily as an antioxidant. Thus O2 toxicity may may initially involve an inhibition of aerobic metabolism in photoreceptors and other retinal neurons. With more prolonged exposure to elevated PO2, there may a direct peroxidation of cellular elements, especially photoreceptor outer segment membranes, exacerbated by a loss of antioxidant capacity in the RPE.

References

1. Cohen,L. and Noell, W. (1960). Glucose catabolism of rabbit retina before and after development of visual function. J. Neurochem. 5:253-276.
2. Winkler, B.S. (1981). Glycolytic and oxidative metabolism in relation to retinal function. J Gen. Physiol. 77:667-692.
3. Noell,W.K. (1958). Differentiatio, metabolic organization and viability in the visual cell. Arc. of Ophthal. 60:702-733.
4. Organisciak,D.T. et al. (1989). Intense light mediated chances in rod outer segment lipids and protiens. Prog. Clin. Biol. Res. 314:493-312.
5. Armstrong,D., Santangelo,G. and Connole,E. (1981). Distribution of peroxide regulating enzymes in the canine retina. Curr. Eye Res. 1:225-242.
6. Yonetani,T. and Ray,G.S. (1965). Studies on cytochrome oxidase. J. Biol. Chem. 240:3392-3398.
7. Bidlack,W.R. and Tappel,A.L. (1973). Fluorescent product of phospolipid peroxidation. Lipids 8:203-207.
8. Sinjorgo, K.M.C. et al. (1984). Biochim. et Biophys. acta. 767:48-56.
9. Cooper,C.E. (1990). The steady state kinetics of cytochrome c oxidation by cytochrome oxidase. Biochem. et Biophys. 1017:187-203.

CYTOCHROME OXIDASE ACTIVITY IN THE FETAL SHEEP RETINA

R. Stockton, J. Wilhelm, D. Armstrong, R. Klick, J. Cotter and J. Reynolds

Departments of Ophthalmology and Medical Technology
SUNY, Buffalo

Introduction: Oxidative metabolism in the retina is dependent upon mitochondrial cytochrome c oxidase (COX), which is the terminal enzyme in the respiratory chain. Because of its powerful reducing capacity, COX functions as critical antioxidant thereby protecting cell membranes from peroxidative free radical damage by superoxide, hydrogen peroxide, hydroxyl radical as well as other oxygen species. COX is most abundant in the retinal pigment epithelium (RPE) but there is also high COX activity ijn the photoreceptor inner segments and lesser activity in the synaptic and ganglion cell layers (1). COX is present in the fetal sheep retina by 120 days gestational age (2; term = 148d). However, the subunit composition, kinetics and activity of fetal COX, as compared to the adult, have not been studied previously (3). This pilot study was undertaken to elucidate some of the potential mechanisms of the retinal pathobiology seen in very low birth weight, premature infants (<1000gm) who are necessarily exposed to an increased PIO2 resulting from air breathing. In about 27% of these infants, the ensuing "Retinopathy of Prematurity" causes varying degrees of visual impairment totaling about 5000 cases per year in the U.S. (4).

Methods: In accord with approved ethical practices, 11 fetal sheep of gestational ages from 120 to 145 days were delivered from anesthetized ewes by cesarean section and then maintained under anesthesia while being mechanically ventilated at 100% FIO2 for up to 2.5 hours. Ewes and lambs were then euthanized, their eyes enucliated and the neural retina and retinal pigment epithelium (RPE) removed separately to 0.2M phosphate buffer with 0.5% Triton X-100 an homogenized (5). Homogenates of neural retina and RPE were then centrifuged at 2400 or 800G respectively, the pellets re-suspended in the phosphate-Triton mixture and spectrophotometrically scanned for cytochromes over the region of 530 to 640nm (6) to determine the molar concentrations of individual cytochromes (aa3,b,c,c1). Kinetic analysis was performed to yield the Km and turnover number (TN) for the COX species in the various tissues.

Results: Using spectrophotometric assay, we determined the molar concentrations of the individual cytochromes (aa3, b, c1, c) composing the COX complex in adult and fetal retina and retinal pigment epithelium (RPE). Table 1 shows that Adult and fetal tissues (Retina, RPE) differ significantly in both the molar concentrations and molar ratios of the individual cytochromes. Using kinetic analysis, Table 2 shows that fetal retina and RPE COX have a significantly lower Vmax and fetal retina has a lower turnover number (TN) compared to the adult. These data signify that fetal retinal and RPE COX is less able to detoxify oxygen and imply differing functional capacities in this COX as compared to the adult.

Table 1. COX concentrations - Adult vs Fetal.

Tissue	Molar Concentration**				Molar Ratio			
	aa3	b	c1	c	aa3:	b:	c1:	c
RETINA - Adult	0.23	0.30	0.17	0.21	1:	2:	1:	1
RETINA - Fetal	0.18	0.17	0.14	0.10	2:	2:	1:	1
RPE - Adult	1.94	4.74	3.02	2.47	1:	2:	2:	1
RPE - Fetal	0.71*	1.01*	0.42*	8.81*	2:	2:	1:	2

N= 14 Adult, 11 Fetal * *nM/mg protien * P= >0.01

Table 2. COX kinetics - Adult vs Fetal.

Tissue	V_{max}**	Turnover (M/s)
RETINA - Adult	181.0 ±87.7	11.9 ±3.7
RETINA - Fetal	66.5 ±21.8*	7.0 ±3.6*
RPE - Adult	168.2 ±86.3	1.8 ±0.3
RPE - Fetal	93.5 ±39.0*	2.3 ±0.4

** nM oxidized cyt c/ min/ mg protien * P= >0.01

Conclusions: The data from this pilot project show that there is a structural difference in the cytochrome oxidase complex between adult and fetal sheep in both retina and RPE COX species. Studies in adult sheep have shown that there is a functional difference between retinal and RPE COX species (7). Additionally, we have shown that peroxidation of COX decreases its Vmax as a function of time thereby inhibiting its ability to reduce oxygen (7). Presumably then, the exposure of the fetal lambs to about 2 hours of 100% FIO2 accounts for some of the decrease in Vmax. However,the lower molar concentrations of COX in both retina and RPE are real and suggest a lower functional capacity in the fetus. Given the structural differences, we also anticipate a real difference between fetus and adult in the O2 handling capacity of both retinal and RPE COX. Thus the combination of low COX concentrations, structural differences and high susceptibility to oxidative inhibition, suggests that fetal or very premature retinas may be very compromised in their abilities to both engage in oxidative metabolism and protect against peroxidation. These differences in fetal COX could contribute to some of the retinal pathology in human retinopathy of prematurity.

References

1. Kageyama,G.H. and Wong-Riley,M.T.T. (1984). The histochemical localization of cytochrome oxidase in the retina and lateral geniculate nucleus of the ferrett, cat and monkey... J. Neurosci. 4:2445-2459.
2. Dusse,J.L., Stockton,R.A., Cotter,J.R., Jockin,Y.M. and Reynolds,J.D. (1992). Cytochrome oxidase activity in the developing fetal retina. Invest. Ophthal. 33(4):p.1061.
3. Wilhelm,J., Armstrong,D., Dusse,J., Klick,R., Stockton,R. and Reynolds,J. (1992). Cytochrome biochemestry in the retina and RPE following exposure to Oxygen. Invest. Ophthal. 33(4):p.1185.
4. Palmer,E.A. et al. (1991). Incidence and early course of retinopathy of prematurity. Ophthalmology 98:1628-1640.
5. Armstrong,D., Santangelo,G. and Connole,E. (1981). Distribution of peroxide regulating enzymes in the canine retina. Curr. Eye Res. 1:225-242.
6. Yonetani,T. and Ray,G.S. (1965). Studies on cytochrome oxidase. J. Biol. Chem. 240:3392-3398.
7. Stockton,R.A., Wilhelm,J., and Armstrong,D. (1994). Cytochrome biochemistry in sheep retina following exposure to oxygen. In: D.Armstrong (ed). Free Radicals in Diagnostic Medicine:

AGE-RELATED PHOSPHOLIPID HYDROPEROXIDE LEVELS IN GERBIL BRAIN MEASURED BY HPLC-CHEMILUMINESCENCE ASSAY AND THEIR RELATION TO HYDROXYL RADICAL STRESS - CLINICAL IMPLICATIONS

Jue-Rong (John) Zhang, Paula K. Andrus and Edward D. Hall

CNS Diseases Research
The Upjohn Company
Kalamazoo, MI 49001
USA

INTRODUCTION

Oxidative stress and antioxidant efficiency are implicated in the aging process.[1] The formation of PCOOH (phosphatidylcholine hydroperoxide) through oxygen free radical induced lipid peroxidation in cell membrane causes oxidative damage in the brain. In the present study, PCOOH was directly quantified in the hippocampus, cortex and striatum from young (3 months), middle-aged (15 months) and old (20 to 24 months) gerbils by an HPLC-chemiluminescence assay[2] (Figure 1). The level of oxidative stress should be differentiated from oxidative damage. For example, an increase in hydroxyl radical or oxidized glutathione levels reflects an increase in oxidative stress, which may not induce oxidative damage. On the other hand, the oxidation of macromolecules such as DNA, protein and lipids (the formation of PCOOH) reflect the actual oxidative damage to the cell.

RESULTS AND CONCLUSION

PCOOH levels in hippocampus and cortex were found between 8.1 to 8.6 pmole/mg tissue, and no statistically significant difference was found across the age groups. In striatum, however, PCOOH levels were significantly higher in middle-aged and old gerbils compared to those in young ones. The regional comparison showed that PCOOH levels were significantly higher in striatum for all the age groups. Moreover, this regional difference increased with aging, from approximately 20% in young gerbil to 30% and 40% in middle-aged and old gerbil striatum, respectively. PCOOH to phospholipid ratio is approximately the same for all age groups at the level of 1.5/10,000.

The hydroxyl radical levels were also measured by the formation of its salicylate trapping product 2,3-DHBA[1] and used as a measure of oxidative stress. PCOOH as a function of the hydroxyl radical stress was calculated and expressed as PCOOH/2,3-DHBA, representing the oxidative damage as a function of oxidative stress. The higher this ratio,

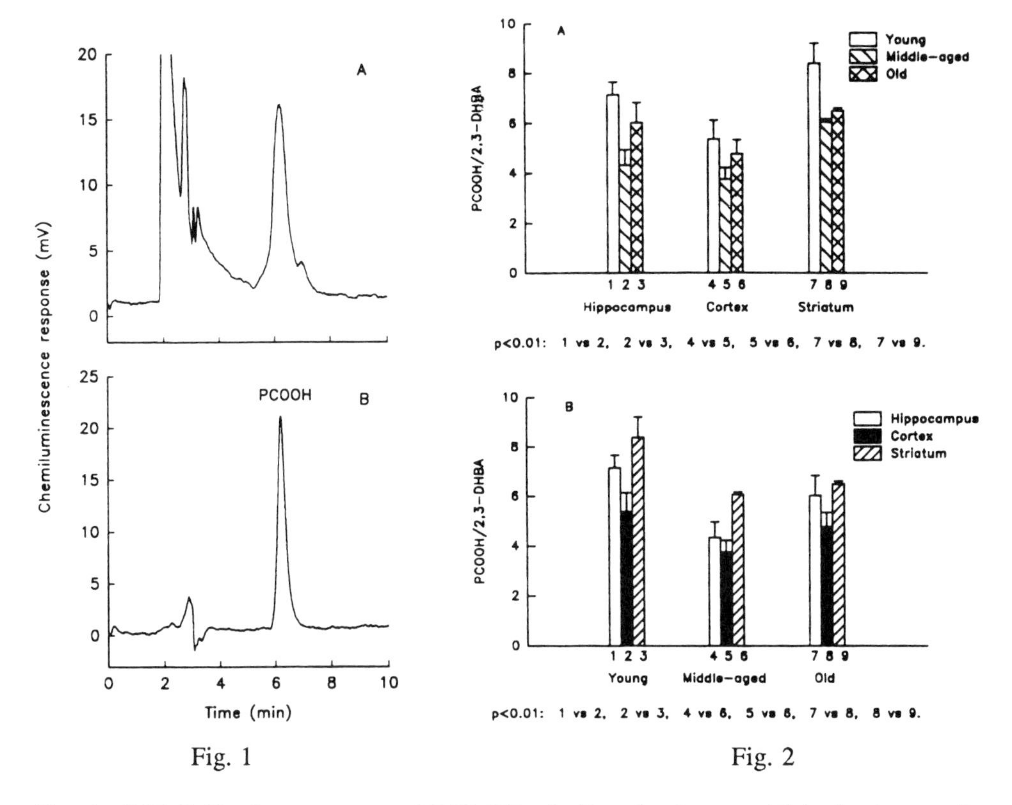

Fig. 1. HPLC-CL chromatogram of PCOOH. A. Samples from gerbil brain tissues. B. PCOOH standard.

Fig. 2. Susceptibility of tissue to oxidative stress as expressed by PCOOH/2,3-DHBA in gerbil brains. PCOOH/2,3-DHBA is abbreviated from (PCOOH/phospholipid $x10^{-4}$)/(2,3-DHBA/salicylate $x10^{-4}$) $x10^{-1}$, representing lipid peroxidative damage (the PCOOH level) under the same level of oxidative stress (the hydroxyl radical level). A. Comparison between age groups. B. Comparison between brain regions. Data are mean ± SD, n = 8.

higher in striatum than in hippocampus or cortex in all age groups, implying a greater susceptibility of striatum to oxidative stress.

PCOOH was also detected in the cat and dog blood plasma. The measurement and comparison of the oxidative damage to oxidative stress may apply to clinical antioxidant interventions.

References

1. J.-R. Zhang, P.K. Andrus, and E.D. Hall, Age-related regional changes in hydroxyl radical stress and antioxidants in gerbil brain, *J. Neurochem.* 61:1640 (1993).
2. J.-R. Zhang, P.K. Andrus, and E.D. Hall, Age-related phospholipid hydroperoxide levels in gerbil brain measured by HPLC-chemiluminescence and their relation to hydroxyl radical stress, *Brain Res.* (1994) in press.

LIPID PEROXIDATION AND DIABETIC COMPLICATIONS: EFFECT OF ANTIOXIDANT VITAMINS C AND E

J. Vinson, C. Hsu, C. Possanza, A. Drack, D. Pane,
R. Davis, C. Klock, K. Graser and X. Wang

Department of Chemistry
University of Scranton
Scranton, PA 18510

There are 3 possible mechanisms for the pathogenesis of diabetic complications; sorbitol pathway, protein glycation and lipid peroxidation.

SORBITOL PATHWAY

The sorbitol pathway, glucose → sorbitol, is increased in hyperglycemia leading to a buildup of sorbitol in tissues and organs. Aldose reductase inhibitor drugs are being developed to decrease sorbitol production and benefit diabetic complications, predominantly diabetic cataracts. Vitamin C decreased red blood cell and lens sorbitol concentrations *in vitro* when present with glucose in the medium. Also C given as a supplement to normal and diabetic subjects significantly decreased red blood cell sorbitol and normalized the sorbitol level in diabetic subjects[1].

GLYCATION

Glycation; glucose covalently bonded to glucose, followed by oxidative decomposition (AGE formation), is increased in diabetes and linked to oxidative stress. We have found that C and E inhibit glycation and AGE formation *in vitro* when albumin is incubated with glucose. *In vitro* lens glycation and AGE formation were also inhibited by C. Plasma glycation was significantly decreased in normal college-age and middle-age subjects supplemented with 1g of C. Other workers have shown that C[2] and E[3] supplementation diminished glycation in human diabetics.

LIPID PEROXIDATION

In diabetes hyperglycemia gives rise to oxidative stress; a decrease in the antioxidant defense (catalase, superoxide dismutase and glutathione peroxidase) and a subsequent increase in lipid peroxidation (LPO) compared to normals[4]. Human diabetes also produces an increase in plasma LPO[5]. Hyperglycemia causes LPO to be elevated in red blood cells[6], and LPO is higher in human diabetic lenses[7].

We have demonstrated that C in drinking water[8] or diet[9] inhibited the progression of galactose-induced cataracts in rats and also increased the regression on the cataracts when a normal diet was given after cataract formation[8]. We have found that C or E at physiological concentrations in the medium significantly decreases red blood cell LPO induced by the high glucose in the medium, $p < 0.05$ and 0.01 respectively[10]. Vitamin C given to streptozotocin-induced diabetic rats significantly lowered lens and plasma LPO and decreased the severity of the diabetic cataracts. Vitamin E given to rats decreased the severity of diabetic cataracts[11]. Vitamin C (500 mg/day) or E (800 IU/day) were given to normal subjects and plasma LPO was significantly reduced[9]. In a placebo-controlled double-blind crossover study, 1000 mg of C and 800 IU of E/day were given to diabetics for 2 months and plasma LPO was significantly diminished. The decrease was greater for E than for C, 66% vs 26%. Vitamin E also significantly lowered fasting plasma glucose (40%) in the diabetic subjects. Vitamin E at a much higher dose (2000 mg/day) has also been shown to decrease fasting plasma glucose in diabetics[12]. Vitamin E supplementation has also been found to improve glucose tolerance in diabetics by decreasing the oxidative stress[13].

CONCLUSION

Vitamins C and E have been shown to beneficially affect sorbitol, glycation and lipid peroxidation in both *in vitro* and supplementation studies with normal and diabetic humans. In animal models of diabetes, vitamins C and E have been shown to beneficially affect cataracts. These inexpensive, non-toxic antioxidant vitamins should prove to be useful supplements when given to diabetics in larger and long-term studies.

REFERENCES

1. J.A. Vinson, M.E. Staretz, P. Bose, H.M. Kassam, and B.S. Basalyga, *In vitro* and *in vivo* reduction of erythrocyte sorbitol by ascorbic acid, Diabetes 38, 1036 (1989).
2. P. Stolba, K. Hatle, A. Krnakova, M. Streda, and L. Starka, Effects of ascorbic acid on nonenzymatic glycation of serum proteins *in vitro* and *in vivo*, Diabetologia 30:585A (1987).
3. C. Ceriello, D. Giugliano, A. Quatraro, C. Donzella, G. Dipalo, and P.J. Lefebvre, Vitamin E reduction of protein glycosylation in diabetes, Diabetes Care 14:68 (1991).
4. S.A. Wohaieb and D.V. Godin, Alterations in free radical tissue-defense mechanisms in streptozocin-induced diabetes in rats. Diabetes 36:1014 (1987).
5. G. Gallou, A. Ruelland, B. Legras, D. Maugendre, H. Allannic and L. Cloarec, Clin. Chem. Acta 214:227 (1993).
6. S.K. Jain, Hyperglycemia causes membrane lipid peroxidation and osmotic fragility in human red blood cells, J. Biol. Chem. 35:21340 (1989).
7. F. Simonelli, Lipid peroxidation is higher in human diabetic lens and cataractous lens than normal lens, Exp. Eye Res. 49:181 (1989).
8. J.A. Vinson, C.J. Possanza, and A.V. Drack, The effect of ascorbic acid on galactose-induced cataracts, Nutr. Repts. Intl. 33:665 (1986).
9. J.A. Vinson, J.M. Courey, and N.P. Maro, Comparison of two forms of vitamin C on galactose cataracts, Nutr. Res. 12:915 (1992).

10. J.A. Vinson and C. Hsu, Effect of vitamins A, E and a citrus extract on *in vitro* and *in vivo* lipid peroxidation, Med. Sci. Res. 20:145 (1992).
11. W.R. Ross, M.O. Creighton, P.J. Stewart-DeHaan, M. Sanwal, M. Hirst and J.R. Trevithick, Can. J. Ophthalmol. 17:61 (1982).
12. M.L. Bierenbaum, F.J. Noonan, L.J. Machlin, S. Machlin, A. Stier, P.B. Watson, A.M. Naso, and A.I. Fleischmann, The effect of supplemental vitamin E on serum parameters in diabetics, post coronary and normal subjects, Nutr. Repts. Intl. 31:1171 (1985).
13. G. Paolisso, A. D'Amore, D. Giugliano, A. Ceriello, M. Varricchio and F. D'Onofrio, Pharmacologic doses of vitamin E improve insulin action in healthy subjects and non-insulin-dependent diabetics, Am. J. Clin. Nutr. 57:650 (1993).

VASCULAR COMPLICATIONS OF PATIENTS IN KUWAIT WITH TYPE 2 DIABETES MELLITUS (NIDDM) AND ELEVATED SERUM LIPID PEROXIDES

Richard J. Lanham,[1] Donald Armstrong,[2] and Nabila Abdella[3]

[1]Department of Medicine
School of Medicine
[2]Department of Medical Technology
School of Health Related Professions
Faculty of Health Sciences
State University of New York at Buffalo
Buffalo, NY 14225
[3]Department of Medicine
Faculty of Medicine
University of Kuwait
Safat, Kuwait 13110

INTRODUCTION

Vascular complications (VC) are important clinical sequalae of diabetes mellitus (DM). The affected vessels can be macrovascular (cerebral, coronary, peripheral) or microvascular (retinal, renal, neural) in caliber. In a pilot project, we have analyzed VC in 174 Kuwaitis with NIDDM. We were interested in a systems analysis of VC and report such an analysis, showing prevalence of combinations of VC and profiles of affected patients.

METHODS

The analysis was done in terms of clinical attributes and laboratory findings. The clinical attributes were: sex, age, duration of disease, body-mass index (BMI), and other VC. For the latter, we first analyzed all large vessels together (considered as a unit), all small vessels together (considered as a unit), each small vessel disease (considered alone), and then all possible combinations of these (e.g., large vessel disease alone, large vessel disease and nephropathy, large vessel disease and retinopathy, nephropathy and retinopathy together without any large vessel disease, etc.). We then analyzed which type of patient had 1, 2, 3 or 4 vessel disease without regard to the kind of vessels involved (large, small, or any particular small vessel).

The laboratory tests investigated were: fasting blood glucose, fructosamine, glycosolated hemoglobin, triglycerides (TG), cholesterol

(CHOL), and thiobarbituric acid reactive substance (TBARS), a measure of serum lipid peroxidation.[1,2]

RESULTS

Fifty-three percent of the patients were male. Their mean age was 52 years, with a range of 27-85. The duration of illness was about 9 years with a range of less than 1-27. The mean BMI was >30 with a range of 16-49.

Large disease alone was the VC most frequently seen with other combinations occurring less often or not at all. The specific frequencies are given in Table 1.

Table 1. Prevalence of specific combinations of vascular complications.

Combinations	Patients (%)	P
Large	28 (16.1)	
Large/neuro/retin	15 (8.6)	
Neuro/retin	13 (7.5)	
Large/neuro/retin/nephro	12 (6.9)	
Retin	10 (5.7)	
Neuro	09 (5.2)	
Large/nephro	04 (2.3)	
Large/retin	04 (2.3)	
Large/neuro/nephro	03 (1.7)	
Large/neuro	03 (1.7)	
Large/retin/nephro	02 (1.1)	
Nephro	01 (0.6)	
Retin/nephro	01 (0.6)	
Retin/nephro/neuro	01 (0.6)	
Neuro/nephro	0 (0)	
		<.001

Statistics: One-Sample Chi-Square Test.

Many clinicians believe that the occurrence of retinopathy and nephropathy in DM has a correlation that approaches 1. We did not find this the case, as the following table illustrates.

In further analyses we found that women had more VC than men. Youngest patients had the fewest complications while oldest patients had the most. Duration of illness was not directly associated with numbers of VC however with some types of VC (large and small combined) but not others (small VC considered alone); e.g.,nephropathy occurred early in disease; retinopathy and neuropathy presented late. BMI was associated with large, large and small, but not small VC considered as a unit or individually. A subset of these patients developed large VC, without accompanying small vessel disease very early in their course.

The findings of the laboratory analysis were that TG and TBARS were associated with the most VC, TG with large VC, CHOL with nephropathy, and fructosamine with neuropathy. The other tests did not correlate with VC.

Table 2. Probability of retinopathy and nephropathy occurring together. (Sensitivity and specificity.)

Combinations	Probability
Retinopathy given nephropathy	75%.
Non-retinopathy given non-nephropathy	71%.
Nephropathy given retinopathy	30%.
Non-nephropathy given non-retinopathy	95%.

Statistics: Kruskal and Goodman's Tau.

CONCLUSION

In this systems analysis of VC in Kuwaitis with NIDDM we found that there was a statistically significant difference in the prevalence of combinations of VC, that distinct clinical aspects of DM seemed to govern the appearance of particular VC's, that patients most affected with VC were females, especially when they were elderly and obese, and that TBARS and TG, not CHOL, appeared to be the laboratory tests associated with the greatest number of VC.

REFERENCES

1. D. Armstrong, N. Abdella, A. Salman, N. Miller, E. Abdel, E. Rahman, and M. Bojancyzk, Relationship of lipid peroxides to diabetic complications: comparison with conventional laboratory tests. *J. Diab. Comp.* 6:116 (1992).

2. N. Abdella and D. Armstrong, The TBA assay as a measure of lipid peroxidation in Arab patients with NIDDM, *Diabetic Res.* 15:173-177, 1991.

VASCULAR AND CELLULAR PROTEIN CHANGES PRECEDE HIPPOCAMPAL PYRAMIDAL CELL LOSS FOLLOWING GLOBAL ISCHEMIA IN THE RAT

T.M. Wengenack,[1] J.R. Slemmon,[2,1] J.M. Ordy,[3] W.P. Dunlap,[4] and P.D. Coleman[1]

Depts. of [1]Neurobiol. & Anat. and [2]Biochem., Univ. of Roch., Rochester, NY 14642, [3]Pittsford, NY 14534, [4]Psychology Dept., Tulane Univ., New Orleans, LA 70118

INTRODUCTION

Several recent studies have reported increased blood-brain barrier (BBB) permeability to large molecules such as immunoglobins and serum albumin following global ischemia.[1] These studies have been qualitative, however, using immunohistochemical methods with descriptive results. Furthermore, no studies have assessed the BBB permeability of hemoglobin following global ischemia. This is especially relevant since hemoglobin has been reported to generate free radicals and be neurotoxic in vitro and in vivo.[2] Free radicals, or reactive oxygen species, have been hypothesized to be a contributory factor in several neurodegenerative disorders, including ischemia.[3]

Previous studies in this laboratory, using quantitative methods of measuring endogenous brain proteins, have reported increased levels of hemoglobins in cerebellum of patients with Alzheimer's disease.[4] Therefore, the specific aim of this study was to compare quantitatively the levels of vascular-derived peptides and protein fragments, that may play a critical role in neuronal degeneration, with the postischemic histological progression of hippocampal neuron loss and levels of brain-derived peptides and protein fragments, as a measure of cellular events associated with neurodegeneration and gliosis following four-vessel occlusion (4-VO) global ischemia in the rat.

MATERIALS AND METHODS

Rats were sacrificed 1, 3, or 7 days after 30 min of 4-VO global ischemia or sham surgery. Crude peptide fractions of neutral, anionic, cationic, and high molecular weight peptide fragments and small proteins were isolated from supernatants of denatured hippocampal homogenates using gel and ion-exchange chromatography and then separated by reverse-phase (RP)-HPLC. Peaks that exhibited a 25% or greater change on any postischemic day relative to sham control values were sequenced and identified. For histological evaluations, CA1 neurons were counted in three regions of CA1, in two sections of the dorsal hippocampus, in two rats from each group.

RESULTS AND DISCUSSION

Histologically, the number of CA1 pyramidal neurons was not decreased significantly on postischemic day 1, but was on days 3 and 7 (data not shown). These results were consistent with those reported previously.[5] In the present study, there was also degeneration of CA3 pyramidal cells due to the severe, 30 min of ischemia used in this study. The RP-HPLC profiles of hippocampal peptides in the four peptide fractions is illustrated in Figure 1. The numbers indicate the peaks that exhibited significant postischemic changes. The most striking results were obtained for the protein fragments of vascular origin. Large increases were observed for numerous fragments of alpha and beta hemoglobin, as well as serum albumin (see Table 1). Increases in fragments of vascular proteins occurred on all three postischemic days, including day 1, before hippocampal pyramidal cell loss was histologically apparent. The largest increases were observed on day 3 when pyramidal cell loss became apparent. This is consistent with previous studies of serum proteins following global ischemia reporting increased parenchymal staining of various serum proteins beginning 24 hrs after ischemia.[1]

There were also significant postischemic changes in the levels of other hippocampal peptide fragments (see Table 1). These included fragments of cellular proteins such as: adenylate kinase, calmodulin, neurogranin, and ubiquitin. The levels of these protein

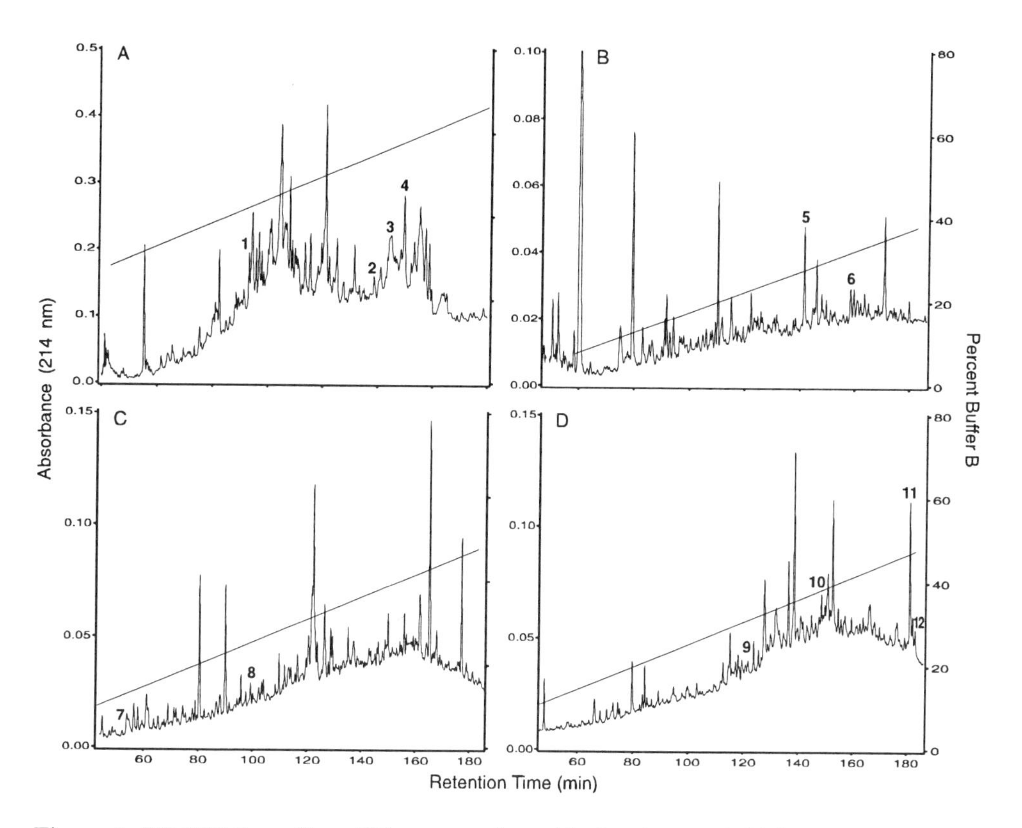

Figure 1. RP-HPLC profiles of hippocampal peptide fractions: A) high molecular weight peptides, B) neutral peptides, C) anionic peptides, D) cationic peptides. Numbers indicate peaks that showed a significant postischemic change.

Table 1. Summary of Significant Changes in Hippocampal Protein Fragments

Peak Identity	Peak Number*	Tryptic Sequence	% Change† Day 1	Day 3	Day 7
Alpha Hemoglobin	2	MFAAFPTTK	+26	+66‡	-10
Alpha Hemoglobin	4	TYFSHIDVSP	+39	-11	-55‡
Alpha Hemoglobin	5	LASVSTVLT	-15	+31‡	+29‡
Beta Hemoglobin	6	VVYPWTQRY	-6	+98‡	+58
Serum Albumin	3	TPVSEK	+20	+98‡	-11
Serum Albumin	10	DLGEQHFK	+1	+232§	+121‡
Adenylate Kinase	9	KVNAEGSVD	-60‡	-8	+52‡
Calmodulin	7	ADIDGDGQV	-43‡	-24	+52‡
Calmodulin	8	GERLTDEEV	-44§	-12	+36§
Neurogranin	1	IQASFRGHMAR	-13	-38‡	-24
Ubiquitin	11	TLSDYNIQK	-54‡	+6	+35‡
Ubiquitin	12	MQIFVK	-51‡	+41	+100‡

* Refer to peak numbers in Fig. 1.
† Relative postischemic change (%); individual 4-VO value (day 1, 3, or 7) v. control mean.
‡ Studentized outlier test, df=2, $p<0.05$; individual 4-VO value (day 1, 3, or 7) v. control mean.
§ Studentized outlier test, df=2, $p<0.01$; individual 4-VO value (day 1, 3, or 7) v. control mean.

fragments of cellular origin generally decreased on postischemic days 1 and 3, while they generally increased on day 7. The decreases in fragments of cellular origin on postischemic day 1, before neuron loss was histologically apparent, demonstrate a temporal dissociation between CA1 neuron loss and early peptide changes. These changes most likely reflect decreases in cellular metabolism and protein turnover due to decreased energy levels. Specific changes in fragments of calmodulin may reflect compensatory responses due to postischemic calcium influx. The significance of the present study is the evidence of increased fragmentation of hemoglobins, which would increase their permeability to the BBB. Since hemoglobin has been shown to generate free radicals, cause lipid peroxidation, and be neurotoxic, increased extravasation could significantly contribute to hippocampal neurodegeneration.

Acknowledgments

Supported by NIH AG 00107 and LEAD Award AG 09016 to P.D.C.

REFERENCES

1. R. Schmidt-Kastner, J. Szymas, and K.A. Hossmann, Immunohistochemical study of glial reaction and serum-protein extravasation in relation to neuronal damage in rat hippocampus after ischemia, *Neurosci.* 38:527 (1990).
2. S.M.H. Sadrzadeh, D.K. Anderson, S.S. Panter, P.E. Hallaway, and J.W. Eaton, Hemoglobin potentiates central nervous system damage, *J Clin Invest.* 79:662 (1987).
3. A. Schurr and B.M. Rigor, The mechanism of cerebral hypoxic-ischemic damage, *Hippocampus.* 2:221 (1992).
4. J.R. Slemmon, C.M. Hughes, G.A. Campbell, and D.G. Flood, Increased levels of hemoglobin-derived and other peptides in Alzheimer's disease cerebellum, *J Neurosci.* In press.
5. J.M. Ordy, T.M. Wengenack, P. Bialobok, P.D. Coleman, P. Rodier, R.B. Baggs, W.P. Dunlap, and B. Kates, Selective vulnerability and early progression of hippocampal CA1 pyramidal cell degeneration and GFAP-positive astrocyte reactivity in the rat four-vessel occlusion model of transient global ischemia, *Exp Neurol.* 119:128 (1993).

DIETARY ANTIOXIDANTS AND BREAST CANCER RISK: EFFECT MODIFICATION BY FAMILY HISTORY

C.B.Ambrosone[1], S.Graham[1], J.R.Marshall[1], R.Hellmann[1], T.Nemoto[2], J.L.Freudenheim[1]

[1] Department of Social and Preventive Medicine, State University of New York at Buffalo, Buffalo, New York 14214
[2] Department of Medicine, State University of New York at Buffalo, Buffalo, New York 14214

Introduction Since breast cancer in a first-degree relative has been shown with some consistency to be associated with breast cancer risk (1), it is possible that family history modifies the effects of other risk factors, with disease in women with a family history possibly resulting from a different etiologic pathway. Studies have indicated that risk associated with some reproductive factors varies significantly for women with and without a family history (2). Since antioxidants have been shown to have an inverse association with breast cancer risk (3), these factors may also vary in their effects for women with and without a family history of the disease.

Methods In this case-control study, we examined usual intake of dietary antioxidants and breast cancer risk, stratifying by family history of breast cancer. Interviews were conducted with 262 premenopausal women from Western New York with incident, primary, histologically confirmed breast cancer, and 273 frequency age and county matched, randomly selected controls. Detailed data on usual diet, health history and history of breast cancer among first degree relatives (FH) were obtained. Two indices were used for dietary intake of α-tocopherol. The first index was derived from USDA major food sources of vitamin E, mainly vegetable oils and their products, and some fruits (23 food sources). The second index was a more extensive measure, including all sources of foods with even trivial amounts of α-tocopherol (252 food sources). Two indices were used to separate the collinear carotenoids and vitamins C and E derived from fruits and vegetables in the model.

Results In women without a family history of breast cancer, intake of β-carotene was associated with a protective effect at all levels of intake. Neither α-tocopherol nor vitamin C was associated with decreased breast cancer among these women (Table 1). Among women with a family history of breast cancer, there was a significant protective effect associated with α-tocopherol intake, measured by index 1, but β-carotene and vitamin C did not significantly reduce breast cancer risk.

Table 1

Intake of selected nutrients and premenopausal breast cancer risk, adjusted for calories, quartiles of intake from lowest (1) to highest (4): Western New York 1986-1991

Quartiles	cases	controls	Odds Ratio (95% confidence interval) with family history	cases	controls	Odds Ratio (95% confidence interval) without family history
β-Carotene						
1	n=11	n= 6	1.00 *	n= 86	n= 61	1.00
2	n= 7	n= 5	1.60 (0.19-13.63)	n= 72	n= 80	0.61 (0.36-1.04)
3	n=13	n= 6	1.82 (0.32-10.27)	n= 58	n= 70	0.57 (0.33-0.99)
4	n= 7	n= 5	1.03 (0.13- 8.20)	n= 47	n= 83	0.43 (0.24-0.77)
	p(trend) = 0.37 **			p(trend) < 0.01		
Vitamin C						
1	n=11	n= 7	1.00	n= 81	n= 72	1.00
2	n=13	n= 5	3.61 (0.56-23.22)	n= 73	n= 74	0.99 (0.58-1.66)
3	n=10	n= 6	2.37 (0.49-15.65)	n= 65	n= 86	0.82 (0.48-1.40)
4	n= 4	n= 4	0.78 (0.24- 2.13)	n= 44	n= 62	0.71 (0.38-1.33)
	p(trend) = 0.66			p(trend) = 0.14		
α-Tocopherol (Index1)						
1	n= 7	n=28	1.00	n=100	n= 98	1.00
2	n= 2	n= 6	0.38 (0.04-4.11)	n= 66	n= 82	1.25 (0.77-1.60)
3	n= 8	n= 3	0.08 (0.02-0.56)	n= 69	n= 43	0.61 (0.35-1.05)
4	n= 5	n= 1	0.02 (0.00-0.57)	n= 59	n= 40	0.73 (0.40-1.32)
	p(trend) < 0.01			p(trend) = 0.18		
α-Tocopherol (Index2)						
1	n=16	n= 5	1.00	n= 66	n= 44	1.00
2	n= 6	n= 4	0.06 (0.005-0.79)	n= 67	n= 80	0.64 (0.36-1.16)
3	n= 7	n= 6	0.21 (0.02- 2.56)	n= 61	n= 86	0.49 (0.26-0.89)
4	n= 9	n= 7	0.07 (0.004-1.34)	n= 69	n= 84	0.48 (0.24-1.01)
	p(trend) = 0.09			p(trend) = 0.09		

* Odds ratios and 95% confidence intervals calculated by logistic regression, adjusted for age, education, age at menarche, age at first pregnancy, number of pregnancies and body mass index,
** p for trend derived from significance level of beta coefficient for adjusted independent variables

Conclusions These results suggest that biologic mechanisms of turmorigenesis may vary by family history of breast cancer. α-tocopherol may be a potential chemopreventive agent for premenopausal women with a family history of breast cancer, and β-carotene may be associated with decreased breast cancer risk for women without a family history. The observed difference in risk provided by different measures of α-tocopherol intake raises methodologic issues, and indicates that further research should address micronutrient quantification from dietary records.

References

1. J.R.Harris, M.E. Lippman, U. Veronesi, et al., Breast cancer, N Engl J Med.327:319-328 (1992).
2. C.Byrne, L.A.Brinton, R.W.Haile, et al., Heterogeneity of the effect of family history on breast cancer risk, Epidemiology.2:276-284 (1991).
3. D.J.Hunter, and W.C.Willett, Diet, body size, and breast cancer, Epidemiol Rev. 15:110-132 (1993).

EFFECT OF MAK-4 AND MAK-5 ON ENDOTHELIAL CELL AND SOYABEAN LIPOXYGENASE-INDUCED LDL OXIDATION

Hari M. Sharma, Atef N. Hanna, Lynda C. Titterington, and Ralph E. Stephens

Department of Pathology
College of Medicine
The Ohio State University
Columbus, OH 43210

INTRODUCTION

Excessive free radical formation is the biochemical basis for oxidant injury to cells. Free radicals have been implicated in the pathogenesis of a wide variety of diseases. MAK-4 and MAK-5 are herbal antioxidant mixtures[1,2]. In this study, we investigated the antioxidant effect of MAK-4 and MAK-5 on endothelial cell (EC)- and soyabean lipoxygenase (SLP)-induced LDL oxidation and SLP-induced linoleic acid oxidation.

METHODS

EC- and SLP-induced LDL Oxidation

Microvascular endothelial cells (EC) were isolated from the choriocapillaris of human eyes. The choroidal EC were isolated by mechanical and enzymatic methods in a cocktail of 0.5 mg/ml each of collagenase types I (Worthington) and B (Boehringer-Mannheim) and 1 mg/ml BSA in PBS. Washed cells were resuspended in Endothelial Cell Growth Medium (ECGM: M-199, 15mM HEPES, 90ug/ml Na-heparin, 150ug/ml ECGS, 10% FBS), plated onto fibronectin-coated (hFN, 25 ug/ml) tissue culture plates and incubated at 37°C, in 5% CO_2. Cultures were routinely subcultured at 80-90% confluence with a 1:1 solution of trypsin-EDTA (0.01%; 0.02%) and plated onto hFN-coated flasks in ECGM[3].

LDL (0.2 mg) was incubated with or without ECs or SLP in the presence or absence of various concentrations of alcoholic or aqueous extracts of MAK-4 or MAK-5 at 5% CO_2 and 95% air for 24 hrs. at 37°C. The degree of LDL oxidation was assessed by measuring thiobarbituric acid reactive substances (TBARS) and conjugated dienes. For the kinetic experiments, aliquots were taken at different times during the incubation period. MAK-4 and MAK-5 were supplied by MAPI, Inc. (Lancaster, MA).

SLP-induced Linoleic Acid Oxidation

A mixture containing linoleic acid with or without SLP in the presence or absence of various concentrations of alcoholic or aqueous extracts of MAK-4 or MAK-5 was time-scanned spectrophotometrically to measure the change in absorbance at 234 nm.

RESULTS

Both the aqueous and alcoholic extracts of MAK-4 and MAK-5 inhibited EC-induced LDL oxidation in a concentration-dependent manner. The agent concentrations (ug/2mL) which produce 50% inhibition (IC_{50}) of EC-induced LDL oxidation as assessed by measuring TBARS and conjugated dienes, respectively, were 100.4 $\pm$ 6.2 and 131.8 $\pm$ 44.5 for the aqueous extract of MAK-4, 57.5 $\pm$ 15.2 and 57.6 $\pm$ 17.0 for the alcoholic extract of MAK-4, 180.6 $\pm$ 17.9 and 188.9 $\pm$ 24.6 for the aqueous extract of MAK-5, and 7.1 $\pm$ 2.26 and 8.9 $\pm$ 2.9 for the alcoholic extract of MAK-5. The alcoholic extracts of MAK-4 and MAK-5 were significantly ($p<0.05$) more potent than the aqueous extracts. Moreover, the alcoholic extract of MAK-5 was significantly ($p< 0.05$) more potent than the alcoholic extract of MAK-4. Both the aqueous and alcoholic extracts of MAK-4 and MAK-5 caused prolongation in the lag phase and delay in the propagation phase of EC-induced LDL oxidation. Both the aqueous and alcoholic extracts of MAK-4 inhibited SLP-induced LDL oxidation in a concentration-dependent manner; the IC_{50} for the aqueous and alcoholic extracts of MAK-4 were 840.6 $\pm$ 196.6 and 246.4 $\pm$ 32.5, respectively. The aqueous extract of MAK-5 inhibited SLP-induced LDL oxidation with an IC_{50} of 503.0 $\pm$ 139.4; however, the alcoholic extract of MAK-5 did not inhibit SLP-induced LDL oxidation. The alcoholic and aqueous extracts of both MAK-4 and MAK-5 had no effect on SLP-induced linoleic acid oxidation.

DISCUSSION AND CONCLUSIONS

In this work we demonstrate antioxidant properties of MAK-4 and MAK-5 in inhibiting SLP- and EC-induced LDL oxidation. The mechanism for these properties is most likely due to an additive or synergistic effect of various antioxidants. The alcoholic extracts of MAK-4 and MAK-5 are more potent antioxidants than the aqueous extracts. The alcoholic extract of MAK-5 inhibited EC-induced LDL oxidation but did not inhibit SLP-induced LDL oxidation, indicating that different antioxidants have different potencies depending on the oxidizing system. MAK-4 and MAK-5 have no effect on the ability of SLP to oxidize linoleic acid. These results suggest MAK-4 and MAK-5 inhibit EC- and SLP-induced LDL oxidation by scavenging free radicals rather than inhibiting the lipoxygenase enzyme activity, and may be useful in preventing vascular damage that leads to atherosclerosis.

ACKNOWLEDGMENTS

This work was supported by Maharishi Ayur-Ved Foundation of America and the Department of Pathology, The Ohio State University.

REFERENCES

1. H.M. Sharma, A.N. Hanna, E.M. Kauffman, and H.A.I. Newman, Inhibition of human low density lipoprotein oxidation in vitro by Maharishi Ayur Veda herbal mixtures, *Pharmacol. Biochem. Behav.* 43:1175 (1992).
2. C. Dwivedi, H.M. Sharma, S. Dobrowski, and F.N. Engineer, Inhibitory effects of Maharishi-4 and Maharishi-5 on microsomal lipid peroxidation, *Pharmacol. Biochem. Behav.* 39:648 (1991).
3. L.E. Lantry, A.W. Fryczkowski, and R.E. Stephens, A human microvascular endothelial cell model of angiogenesis in diabetes, *In Vitro Cell. Dev. Biol.* 27A:164A (1991).

PREVENTION OF OXIDANT STRESS BY STUDENT RASAYANA (SR)

Atef N. Hanna, Ellen M. Kauffman, Howard A.I. Newman, and Hari M. Sharma

Department of Pathology
College of Medicine
The Ohio State University
Columbus, OH 43210

INTRODUCTION

Free radicals are linked to many pathological events. The central nervous system contains a high concentration of polyunsaturated fatty acids which are prone to oxidative damage. Brain functions are inversely related to the level of free radicals[1]. Arachidonic acid and its lipoxygenase metabolites may act as second messengers in long-term potentiation, a process associated with learning[2]. Children taking SR, a natural herbal mixture which contains various antioxidants, showed higher intelligence levels compared to the control group[3]. We investigated whether the mechanism by which SR increases intelligence is through scavenging free radicals and/or enhancing the activity of lipoxygenase.

METHODS

We evaluated the antioxidant activity of SR (using TBARS measurement to assess the degree of lipid peroxidation) in different systems such as Cu^{+2}-catalyzed LDL oxidation[4], endothelial cell (EC)-induced LDL oxidation[4], soyabean lipoxygenase (SLP)-induced LDL oxidation[4], enzymatic (NADPH)- and nonenzymatic (ascorbate-Fe^{+3})-induced microsomal lipid peroxidation and toluene-induced microsomal lipid peroxidation in vivo. Aqueous and/or alcoholic extract of SR effects on SLP were tested using SLP-induced oxidation of linoleic acid as assessed by measuring the increase in absorbance at 234 nm. MAPI, Inc. (Lancaster, MA) supplied SR.

RESULTS

Both the aqueous and alcoholic extracts of SR inhibited Cu^{+2}-catalyzed and EC-induced LDL oxidation in a concentration-dependent manner. The agent concentrations (ug/2mL) which inhibited 50% (IC_{50}) of Cu^{+2}-catalyzed LDL oxidation and EC-induced

LDL oxidation, respectively, were 46.0 ± 10.1 and 236 ± 30.0 for the aqueous extracts, and 20.4 ± 6.45 and 25.3 ± 7.71 for the alcoholic extracts. Both the aqueous and alcoholic extracts of SR (as well as vitamin E) inhibited the enzymatic- and nonenzymatic-induced microsomal lipid peroxidation in a concentration-dependent manner. The IC_{50}s for the enzymatic-induced microsomal lipid peroxidation were 187 ± 52.7 (aqueous SR extract), 32.3 ± 17.8 (alcoholic SR extract) and 12.0 ± 5.65 (vitamin E). The IC_{50}s for the nonenzymatic-induced microsomal lipid peroxidation were 1992 ± 174 (aqueous SR extract) and 54.8 ± 8.18 (vitamin E). The SR aqueous extract inhibited SLP-induced LDL oxidation in a concentration-dependent manner with an IC_{50} of 311 ± 59.4. However, the SR alcoholic extract enhanced the SLP-induced LDL oxidation in a concentration-dependent manner. Simultaneous addition of the aqueous and alcoholic extracts resulted in the inhibition of SLP-induced LDL oxidation. The alcoholic extract of SR increased the ability of SLP to oxidize linoleic acid. Simultaneous addition of both the aqueous and alcoholic extracts of SR did not change the enhancing effect of the alcoholic extract on SLP-induced linoleic acid oxidation.

Experiments, in vivo, showed that feeding rats regular chow supplemented with 2% SR (w:w) significantly ($p < 0.05$) inhibited toluene-induced microsomal lipid peroxidation in the rat brain.

DISCUSSION AND CONCLUSIONS

It has been shown that brain functions vary inversely to the level of free radicals[1], and arachidonic acid plus its lipoxygenase metabolites might act as second messengers in long-term potentiation[2], and SR increases intelligence in children[3]. We found that both the aqueous and alcoholic extracts of SR have antioxidant activity as demonstrated by the inhibition of Cu^{+2}-catalyzed and EC-induced LDL oxidation, but in both these systems the alcoholic extract was more potent. Although the aqueous extract of SR inhibited SLP-induced LDL oxidation, the alcoholic extract enhanced it. This enhancement could be due to induction of the activity of SLP, with a subsequent increase in free radical production and/or the alcoholic extract itself increasing free radicals. We found the alcoholic extract of SR enhanced SLP-induced linoleic acid oxidation, and the aqueous extract abolished the alcoholic extract-induced SLP-catalyzed LDL oxidation but had no effect on alcoholic extract-induced SLP-catalyzed linoleic acid oxidation. Furthermore, inhibition, in vivo, of toluene-induced microsomal lipid peroxidation verified SR antioxidant activity. We conclude that SR enhances brain functions through scavenging free radicals and increasing the activity of lipoxygenase, which is related to long-term potentiation.

REFERENCES

1. J.M. Carney, P.E. Starke-Reed, C.N. Oliver, R.W. Landum, M.S. Cheng, J.F. Wu, and R.A. Floyd, Reversal of age-related increase in brain protein oxidation, decrease in enzyme activity, and loss in temporal and spatial memory by chronic administration of the spin-trapping compound N-tert-butyl-alpha-phenylnitrone, *Proc. Natl. Acad. Sci. USA* 88:3633 (1991).
2. J.H. Williams, M.I. Errington, M.A. Lynch, and T.V.P. Bliss, Arachidonic acid induces a long-term activity-dependent enhancement of synaptic transmission in hippocampus, *Nature* 341:739 (1989).
3. S.I. Nidich, P. Morehead, R.J. Nidich, D. Sands, and H. Sharma, The effect of the Maharishi Student Rasayana food supplement on non-verbal intelligence, *Pers. Indiv. Diff.* (in press).
4. D. Steinberg, S. Parthasarathy, T.E. Carew, J.C. Khoo, and J.L. Witztum, Beyond cholesterol:modifications of low-density lipoprotein that increase its atherogenicity, *N. Engl. J. Med.* 320:915 (1989).

BIOCHEMICAL CHANGES INDUCED BY MAHARISHI AMRIT KALASH (MAK-4) AND MA-208 IN DIET-INDUCED HYPERCHOLESTEROLEMIC RABBITS

Jae Y. Lee, John A. Lott, and Hari M. Sharma

Department of Pathology
College of Medicine
The Ohio State University
Columbus, OH 43210

INTRODUCTION

Free radicals and reactive oxygen species (ROS) are involved in the pathogenesis of various disorders, including atherosclerosis, which could possibly be prevented by the use of antioxidant substances.[1] MAK-4 and MA-208 are antioxidant herbal mixtures. The purpose of this study was to evaluate the in vivo antioxidant properties of MAK-4 and MA-208 in diet-induced hypercholesterolemic rabbits.

MATERIALS AND METHODS

Twenty-four New Zealand White rabbits were divided into two groups, a control group of 12, fed normal rabbit chow, and a test group of 12, fed normal rabbit chow supplemented with 4% MAK-4 and 0.4% MA-208. These diets were fed during a "conditioning" period of 4 weeks, then both groups were switched to normal rabbit chow supplemented with 0.3% cholesterol for a "cholesterol feeding" period of 12 weeks, followed by a "regression" period of 12 weeks during which the 0.3% cholesterol was withdrawn from the diet. Blood was drawn at six-week intervals, and serum was assayed for superoxide dismutase (SOD), cholesterol, triglycerides, HDL, CK, LD, CK isoenzymes and LD isoenzymes. Serum lipid peroxide was assessed by measuring thiobarbituric acid-reactive substances (TBARS). Also, TBARS and electrophoretic mobilities were assayed in low density lipoprotein (LDL) and very low density lipoprotein (VLDL) fractions isolated by ultracentrifugation. MAK-4 and MA-208 were supplied by MAPI, Inc. (Lancaster, MA).

RESULTS AND DISCUSSION

Plasma cholesterol was markedly elevated in both groups of rabbits during the cholesterol feeding period, then decreased during the regression period, without any differences between the two groups. HDL and SOD followed the same pattern as cholesterol.

Lipid peroxide is an indicator of peroxidation of lipid membranes and a reflection of the production of malondialdehyde. Lipid peroxide was significantly higher in the control group during the experiment. Total CKs in both groups were the same and did not change during the conditioning and regression periods, but the total CK of the control group was 3.3-fold higher ($p<0.05$) after 12 weeks of feeding the cholesterol-supplemented diet. CK-BB and CK-MM were increased 4.8-fold ($p<0.05$) and 1.4-fold ($p<0.17$), respectively, in the control group as compared to the test group during the cholesterol feeding period. CK-MB was significantly higher in the control group, however the total activity was too low to make any meaningful interpretations. LD activity was elevated (5.4-fold, $p<0.08$) as well as LD-1 (5.9-fold, $p<0.05$) in the control group after the cholesterol feeding period, however the percentage of each isoenzyme of the total LD was not significantly changed. The concentration of TBARS in LDL and VLDL in the test group was 10-fold ($p<0.001$) and 8-fold ($p<0.001$) lower, respectively, than that of the control group during the cholesterol feeding period. The electrophoretic mobilities of LDL and VLDL in the control group were greater than that of the test group by 73% ($p<0.001$) and 68% ($p<0.001$), respectively.

In conclusion, our experiments show that a MAK-4- and MA-208-supplemented diet prevents lipid peroxidation in vivo, and protects cell membranes from free radical-induced damage. These studies suggest further experimentation to evaluate the anti-atherosclerotic effect of MAK-4 and MA-208.

ACKNOWLEDGMENTS

This work was supported by Maharishi Ayur-Ved Foundation of America and the Department of Pathology, The Ohio State University.

REFERENCE

1. H.M. Sharma. "Freedom from Disease", Veda Publishing, Toronto (1993).

IN VITRO ERGOTHIONEINE ADMINISTRATION FAILED TO PROTECT ISOLATED ISCHAEMIC AND REPERFUSED RABBIT HEART

Anna Cargnoni, Palmira Bernocchi, Claudio Ceconi[1],
Salvatore Curello[1], and Roberto Ferrari[1]

Fondazione Clinica del Lavoro, Centro di Fisiopatologia
Cardiovascolare "S. Maugeri", Gussago, Brescia, Italy
[1]Chair of Cardiology, University of Brescia, Spedali Civili, Brescia, Italy

Ergothioneine is a thiol histidine.

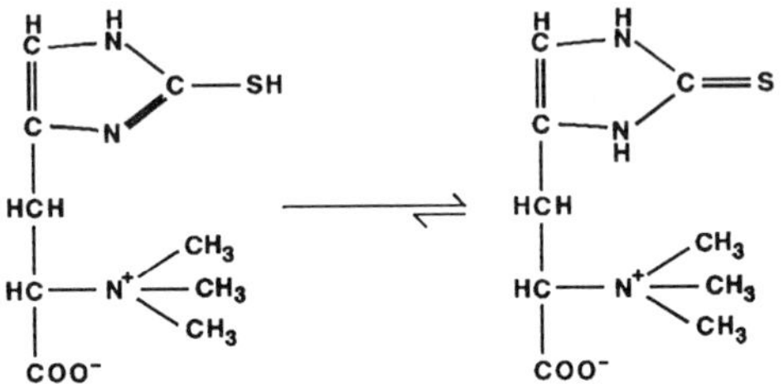

Figure 1.

In aqueous solutions, it is predominantly present as a thione rather than as the thiol, but it shows some properties of both these chemical groups. Ergothioneine represents the main source of thiols in some species of fungus but is also present, although not synthesized, in animal and human tissues[1,2]. Whereas it could represent an antioxidant molecule of biological origin, ergothioneine has stirred little interest in scientific circles. Recently, Arduini, Eddy and Hochstein[3] proposed ergothioneine as a potential protective agent against damage associated with global ischaemia followed by reperfusion.

The aim of this study was to evaluate the effects of ergothioneine on functional, metabolic and oxidative-myocardial parameters during post-ischaemic reperfusion. We measured reduced/oxidized glutathione (GSH and GSSG) and reduced/oxidized pyridine coenzymes (NADH and NAD) as an index of cellular redox status.

Isolated and Langendorff perfused rabbit hearts were used[4]. After an equilibration period (30 min, 22 ml/min), they were aerobically perfused for 60 min and subjected to global ischaemia (45 min, 0 ml/min) and then reperfused for 30 min (22 ml/min). All hearts were paced (180 b/min) and maintained at 37°C. The hearts were then randomly divided into two groups: aerobic control and ischaemic–reperfused. The ischaemic hearts received: a) saline vehicle (n=8), as ischaemic control group; b) ergothioneine 10^{-5}M (n=4); and c) ergothioneine 10^{-4}M (n=6). Ergothioneine was introduced into the perfusion solution before (60 min) and after ischaemia.

During each experiment, myocardial damage was evaluated for: mechanical function (developed isovolumetric pressure); creatinekinase (CK) and lactate release; and for myocardial content of: GSH and GSSG; NADH and NAD; and high energy phosphates (ATP and CP).

Ergothioneine (10^{-5} and 10^{-4}M) did not revert the following: (1) the depressed recovery of developed pressure on reperfusion (14.4±2.3 mmHg in group a; 10.3±2.9 and 12.5±2.3 mmHg in groups b and c respectively); (2) the ischaemic and reperfusion induced rise in diastolic pressure: (44.3±4.4 mmHg in group a, 49.8±5.8 and 48.0±7.7 mmHg in groups b and c respectively); (3) the increased CK and lactate release; (4) ATP and CP reduced restoration after post–ischaemic reperfusion; (5) the accumulation of myocardial GSSG (considered as index of oxidative stress); and (6) the decreased ratio oxidized vs reduced pyridine coenzymes (NAD/NADH as index of aerobic metabolism).

In our experimental model, ergothioneine failed to protect myocardium against ischaemia and reperfusion damage, even though this model revealed to be adequate for testing the protective effect of other thiol compounds such as: N–acetylcysteine[5] and dimercaptopropanol[6]. Most likely, the inefficacy of ergothioneine is due to the poor biological availability of the thiol–group so that its possibility of working as reducing agent or as glutathione cofactor is blocked.

REFERENCES

1. D.B. Melville, Ergothioneine, *Vitam Horm* 17:155 (1958).
2. P.E. Hartman, Ergothioneine as antioxidant, *in:* Methods in Enzymol, L. Packer, A.N. Glazer, eds., Academic Press Inc, San Diego et al. (1990).
3. A. Arduini, L. Eddy and P. Hochstein, The reduction of ferryl myoglobin by ergothioneine: a novel function for ergothioneine. *Arch Bioch Bioph* 281:42 (1990).
4. R. Ferrari, C. Ceconi, S. Curello, C. Guarnieri, C.M. Caldarera, A. Albertini and O. Visioli, Oxygen mediated myocardial damage during ischaemia and reperfusion: role of the cellular defences against oxygen toxicity, *J Mol Cell Cardiol* 17:937 (1985).
5. C. Ceconi, S. Curello, A. Cargnoni, R. Ferrari, A. Albertini and O. Visioli, The role of glutathione states in the protection against ischemic and reperfusion damage: effect of N–acetylcysteine, *J Mol Cell Cardiol* 20:5 (1988).
6. C. Ceconi, S. Curello, A. Cargnoni, G.M. Boffa and R. Ferrari, Antioxidant protection against ischaemia and reperfusion heart damage during cardiac ischaemia and reperfusion: effect of dimercapto–propanol, *Cardioscience* 1:191 (1990).

PENTOXIFYLLINE INTERFERES WITH POTENTAL SOURCES OF FREE RADICAL GENERATION DURING ENDOTOXEMIA

Megan R. Lerner,[1,3] Andre K. Balla,[2] Michael F. Wilson,[4] and Daniel J. Brackett[1,3]

[1]Departments of Surgery and Anesthesiology
[2]Department of Pathology
University of Oklahoma Health Sciences Center
[3]Veterans Affairs Medical Center
Oklahoma City, Oklahoma 73190
[4]Department of Medicine, State University of New York
Millard Fillmore Hospitals
Buffalo, New York 14209

INTRODUCTION

Increased concentrations of tumor necrosis factor concentrations, activated neutrophils, and inadequate oxygen delivery to tissue are associated with free radical generation and also with endotoxic shock. Pentoxifylline suppresses the synthesis of tumor necrosis factor in endotoxin stimulated macrophages[1] and when administered *in vivo* significantly attenuates circulating tumor necrosis factor and survival in mice challenged with endotoxin.[2] Pentoxifylline also inhibits neutrophil activation induced by tumor necrosis factor.[3] Additionally, pentoxifylline possesses antithrombotic activity and induces synthesis of endothelial cell prostacyclin, both of which should promote tissue perfusion during endotoxin-induced low flow conditions. This study was designed to determine the effect of pentoxifylline on the development of cardiovascular, metabolic, and pathologic responses to endotoxin.

METHODOLOGY

Twenty conscious, unrestrained, instrumented male Sprague-Dawley rats weighing 302 ± 12 g were used in this study. For instrumentation the animals were anesthetized with isoflurane which was maintained at 2% and delivered with a rodent respirator. A thermocouple-catheter combination was placed in the right carotid artery with the tip immediately distal to the aortic valve for measurement of arterial pressure and thermodilution curves and for blood sampling. The right jugular vein was catheterized for measurement of central venous pressure and injection of room temperature saline for generation of cardiac output curves. Catheters were tunneled under the skin and exteriorized through an incision at the back of the neck. The unrestrained animals were allowed to regain consciousness in a monitoring chamber where parameters could be continuously measured. Following a 60 min recovery period 30 minutes of control measurements were obtained to assure stable, normal values.

At this time pentoxifylline at a dose of 100 mg/kg or an equivalent volume of saline was administered intravenously. Sixty minutes later endotoxin (20 mg/kg, LD-$90_{24\,hrs}$) was given intravenously and all parameters were monitored for 4 hours. When the study concluded, the small intestine was graded macroscopically on a 5 point scale for intensity of hemorrhage which is characteristic of this model of endotoxic shock.

RESULTS AND CONCLUSIONS

Saline Treatment Endotoxin administered intravenously to this group of conscious rats, pretreated with saline rather than pentoxifylline, produced an immediate decrease in cardiac output to 34% of the control value which was sustained for the 4 hour monitoring period. Occurring concurrently with this cardiac output response was an immediate increase in systemic vascular resistance, but it did not reach its maximum until 30 min after endotoxin. This slower, extended response resulted in a 30 min hypotensive episode which reached an immediate nadir of 65 mmHg and then returned to near normal levels as vasoconstriction (resistance) reached its maximum. There was a 43%, sustained increase in heart rate and a 74% decrease in cardiac stroke volume; both occurred immediately after endotoxin. Measurements of plasma lactate concentrations revealed a 350%, sustained increase indicative of a shift from aerobic to anaerobic metabolism which correlated with evidence of metabolic acidosis (decreased pH) and with the compromise of tissue perfusion suggested by the cardiovascular data. Hypoglycemia occurred at 4 hrs and there was an increase in hematocrit. A significant increase in tumor necrosis factor over the normal concentration of 0.46 ng/ml occurred at 90 min and circulating neutrophil concentrations were reduced to 20% of control throughout the study period. Gross examination of the small intestine at the end of study revealed intense hemorrhage characteristic of this model.

Pentoxifylline Treatment Pentoxifylline treatment did not have a significant effect on the early hypotensive episode or the tachycardia induced by endotoxin. However, the severe increase in systemic vascular resistance was totally prevented and the decreases in cardiac output and cardiac stroke volume were significantly blunted. These hemodynamic data implying improved tissue perfusion were substantiated by inhibition of the increase in lactate concentrations and prevention of the fall in pH. The fall in glucose and the rise in hematocrit and tumor necrosis factor were also completely blocked by pentoxifylline. The decrease in circulating neutrophils 4 hrs after endotoxin was prevented implying a reduction of neutrophil adhesion to the endothelium and a diminishing of the potential for neutrophil migration into the tissue. Importantly, the hemorrhage of the small intestine was essentially eliminated.

Conclusions The profound, beneficial effect of pentoxifylline to avert tissue injury during endotoxic shock appears to be due to the maintenance of adequate tissue perfusion and systemic blood flow and the reduction of neutrophil adhesion to the endothelium. The improvement in these critical parameters may have resulted at least in part from the capacity of pentoxifylline to essentially block endotoxin evoked tumor necrosis factor synthesis. These data suggest that pentoxifylline is worthy of consideration as part of the therapeutic regimen for the treatment of certain types of septic shock.

REFERENCES

1. R.M. Stieter, D.G. Remick, P.A. Ward, R.N. Spengler, J.P. Lynch , J. Larrick, and S.L. Kuhkel, Cellular and molecular regulation of TNF-alpha production by pentoxifylline, Biochem. Biophys. Res. Comm. 155: 1230 (1988).
2. U.F. Schade, Pentoxifylline increases survival in murine endotoxin shock and decreases formation of tumor necrosis factor, Circ. Shock 31: 171 (1990).
3. H. Zheng, J.J. Crowley, J.C. Chan, H. Hoffmann, J.R. Hatherill, A. Ishizaka, and T.A. Raffin, Attenuation of TNF-induced endothelial cell cytotoxicity and neutrophil chemiluminescence, Am. Rev.Respir. Dis.42:1073 (1990).

CONTRIBUTORS

Patricia Abello, M.D. *
Research Fellow
Department of Surgery
The Johns Hopkins University, School of Medicine
601 N. Broadway/Blalock 685
Baltimore, MD 21205
(410)550-6979▪FAX 955-0834

Christine B. Ambrosone, M.S.
Department of Social and Preventive Medicine
School of Medicine and Biomedical Sciences
University at Buffalo
270 Farber Hall
Buffalo, NY 14214

Robert Anderson, Ph.D.,M.D. *
Professor of Ophthalmology
Cullen Eye Institute
Baylor College of Medicine
Houston, TX 77030
(713)798-5958▪FAX 798-4364

Donald Armstrong, Ph.D.,D.Sc. *
Professor and Chairman
Department of Clinical Laboratory Science
University at Buffalo and
Research Professor of Experimental Pathology
Roswell Park Cancer Institute
462 Grider Street
Buffalo, NY 14215
(716)898-5124▪FAX 898-5114

William J. Bettger, Ph.D.
Department of Nutritional Sciences
University of Guelph
Guelph, Ontario N1G 2W1
Canada

Jeffrey Blumberg, Ph.D. FACN *
Associate Director and Professor
Chief, Antioxidants Research Laboratory
USDA Human Nutrition Research Center on Aging
at Tufts University
Boston, MA 02111
(617)556-3334▪FAX 556-3295

Daniel Brackett, Ph.D.
Associate Professor and Director of Research
Department of Surgery, College of Medicine
University of Oklahoma Health Sciences Center
P.O. Box 26901
South Pavilion, Room 4SP-310
Oklahoma City, OK 73190
(405)271-5781▪FAX 271-3919

Timothy G. Buchman, M.D., Ph.D.
Department of Surgery
The Johns Hopkins University
School of Medicine
601 N. Broadway/Blalock 685
Baltimore, MD 21205

Gregory B. Bulkley, M.D.
Department of Surgery
The Johns Hopkins University
School of Medicine
601 N. Broadway/Blalock 685
Baltimore, MD 21205

Anna Cargoni, Ph.D.
Fondazione Clinical del Lovora
Centro di Fiscopatologia Cardiovascolare "S. Mangeri"
Gussago, Brescia
Italy

Jean Chaudiere, Ph.D.
Chief Executive Officer
Center for Research
Bioxytech S. A.
Z.A. des Petits Carreaux
2 av.des Coquelicots
94385 Bonneuil-sur-Marne
Cedex
France
33 1 49 80 45 64▪FAX 33 1 49 80 01 66

H.H. Draper, Ph.D. *
Professor Emeritus
Department of Nutritional Sciences
University of Guelph
Guelph, Ontario N1G 2W1
Canada
(519)824-4120 x-3740, 6688▪FAX (519)763-5902

Leonard Feld, M.D., Ph.D. *
Professor and Director
Division of Pediatric Nephrology
Childrens Kidney Center
The Childrens Hospital of Buffalo
219 Bryant Street
Buffalo, NY 14222
(716) 878-7275▪FAX(716)878-7914

Roberto Ferrari, M.D., Ph.D. *
Professor and Chair
Divisione di Cardiologia
Facolta di Medicine e Chirgia
Universita Degli Studi Brescia
Spedali Civili
25100 Brescia
ITALY
011-39-30-399-5576▪FAX 252-2362

Claude Gagnon, M.Sc., Ph.D. *
Director, Urology Research Laboratory
Royal Victoria Hospital and
Professor, Department of Surgery
Division of Urology
Royal Victoria Hospital
McGill University
687 Pine Avenue West, Room H6.47
Montreal, Quebec H3A 1A1
Canada
514) 842-1231 x-5429▪FAX 843-5429

Consuelo Guerri, Ph.D. *
Investigator
Instituto de Investigaciones Citologicas
Fundacion Valencia de Investigaciones Biomedicas
Amedeo de Saboya, 4
46010 Valencia
Spain
(011-34-6)369-8500■FAX 360-1453

Stephan M. Hahn, M.D.
Radiation Oncology Branch
National Cancer Institute
National Institutes of Health
Bldg. 10, Room B3-B69
9000 Rockville Pike
Bethesda, MD 20892-4200

Atif Hanna, Ph.D.
Department of Pathology
Room M376 Starling Loving Hall
The Ohio State University
320 West Tenth Avenue
Columbus, OH 43210

Tadahisa Hiramitsu, M.D., Ph.D.
Professor
Photon Medical Research Center
Hamamatsu University School of Medicine
Hamamatsu 431-31
Japan
011 815 3435-2390■FAX 435-2390

Paul Jennings, DM, M.R.C.P. *
York Diabetes Center
The York District Hospital
Wigginton Road
York YO3 7HE
United Kingdom
011-44-904-631-313■FAX 011-44-904-631-121

Chithan Kandaswami, Ph.D., C.N.S.
Principal Scientist
Division of Allergy and Clinical Immunology
Department of Medicine and Biomedical Sciences
University at Buffalo
Buffalo General Hospital
100 High Street
Buffalo, NY 14203

Jae Kang, Ph.D.
Associate Professor
Department of Medical Laboratory Science
University of New Hampshire
Hewitt Hall, Room 207
Durham, NH 03284-3653

Frank L. Kretzer, Ph.D.
Cullen Eye Institute
Baylor College of Medicine
Houston, TX 77030

C. Murali Krishna, Ph.D.
Radiation Oncology Branch
National Cancer Institute
National Institutes of Health
Bldg. 10, Room B3-B69
9000 Rockville Pike
Bethesda, MD 20892-4200

Eve de Lamirande, Ph.D.
Division Urologie
Departement de Chirugrgie
Hospital Royal Victoria, Local H6-47
687 Avenue des Pins Quest
Montreal, Quebec H3A 1A1
Canada

Richard Lanham, M.D.
Department of Medicine
University at Buffalo
462 Grider Street
Buffalo, NY 14215

Jae Lee, Ph.D.
Department of Pathology
Room M376 Starling Loving Hall
The Ohio State University
320 West Tenth Avenue
Columbus, OH 43210

Jonathan Leff, M.D. *
Associate Director
Pulmonary Immunology
Merck and Company, Inc.
P.O. Box 2000 WBD-330
Rahway, NJ 07065-0914
(908)750-8084▪FAX 750-8232

Edward Lesnefsky, M.D. *
Associate Professor
Department of Medicine
Division of Cardiology
Cleveland VA Medical Center
Case Western Reserve University
2074 Abington Road
Cleveland, OH 44106
(216) 844-8904▪FAX 844-8954

John MacFarlane, Ph.D.
JM Science, Inc.
5820 Main Street, Suite 300
Buffalo, NY 14221-5734

Paul B. McCay, Ph.D.
Free Radical Biology and Aging Program
Oklahoma Medical Research Foundation
Oklahoma City, OK 73190

Eugene Means, M.D. *
Director, Clinical Development
7217-2583
Upjohn Pharmaceutical Co.
Kalamazoo, MI 49001
(616)385-8425▪FAX 385-5473

James Mitchell, Ph.D. *
Deputy Chief
Radiation Oncology Branch
National Cancer Institute
National Institutes of Health
Bldg. 10, Room B3-B69
9000 Rockville Pike
Bethesda, MD 20892-4200
(301)496-7511 or 496-5457▪FAX 480-5439

Carmina Montoliu, B.Sc.
Instituto de Investigaciones Citologicas
Fundacion Valencia de Investigaciones Biomedicas
Amadeo de Saboya, 4
46010-Valencia
Spain

Mary Treinen Moslen, Ph.D. *
Professor
Department of Pathology
University of Texas Medical Branch
Galveston, TX 77550
(409)772-3650▪FAX 772-3606

J. Paul Muizelaar, M.D.
Lind Lawrence Professor
Division of Neurological Surgery
Medical College of Virginia
Virginia Commonwealth University
P.O. Box 980631
Richmond, VA 23298-0631
(804)786-9165▪FAX (804)371-0374

Thomas Nicotera, Ph.D. *
Cancer Research Specialist
Biophysics Department
Roswell Park Cancer Institute
Elm and Carlton Street
Buffalo, NY 14263-0001
(716)845-8294▪FAX 845-8899

Peter O'Brien, Ph.D. *
Professor
Faculty of Pharmacy
University of Toronto
Toronto, Ontario M55 2S2
Canada
(416) 978-2716▪FAX 978-8511

Paolo Pedersini, Ph.D.
Fondazione Clinical del Lovora
Centro di Fiscopatologia Cardiovascolare "S. Mangeri"
Gussago, Brescia
Italy

Domenico Pellegrini-Giampietro, M.D., Ph.D. *
Professor
Department of Preclinical and Clinical Pharmacology
University of Florence
Viale G.B. Morgani, 65
50134 Florence
Italy
011-39-55-423-7433▪FAX 55-436-1613

Alice Pentland, M.D. *
Associate Professor
Division of Dermatology
Department of Medicine
Washington University School of Medicine
Campus Box 8123
660 S. Euclid Avenue
St. Louis, MO 63110
(314)362-8180▪FAX 362-8159

Laurence M. Rapp, Ph.D.
Cullen Eye Institute
Baylor College of Medicine
Houston, TX 77030

Peter D. Reaven, M.D. *
Assistant Professor
Division of Endocrinology and Metabolism
Department of Medicine, 0682
University of California San Diego
LaJolla, CA 92093-0682
(619)534-0569▪FAX 546-9828

Jaime Renau-Piqueras, Ph.D.
Head
Electron Microscopy Section
Centro Investigacion La Fe
Avd. Campanar, 21
46009-Valencia
Spain

Mark G.P. Saifer, Ph.D. *
Director of Research
DDI Pharmaceuticals, Inc.
518 Logue Avenue
Mt. Pier, CA 94043
(415)964-7676▪FAX(415)967-5243

Wolfgang Schalch, Ph.D. *
Director
Human Nutrition Research
Roche Vitamins and Fine Chemicals
Hoffman-LaRoche Inc.
340 Kingland St.
Nutley, NJ 07110-1199
(201)909-8351▪FAX (201)909-8414

Hari Sharma, M.D., FRCPC
Professor
Department of Pathology
College of Medicine
The Ohio State University
320 West Tenth Avenue
Columbus, OH 43210

Ralph Somack, Ph.D.
DDI Pharmaceuticals, Inc.
518 Logue Avenue
Mt. Pier, CA 94043

Richard Stockton, Ph.D.
Department of Ophthalmology
School of Medicine and Biomedical Sciences
University at Buffalo
143 Cary Hall
Main Street
Buffalo, NY 14214

Joe Vinson, Ph.D.
Department of Chemistry
University of Scranton
Scranton, PA 18510-4626

Wayne R. Waz, M.D.
Division of Pediatric Nephrology
Childrens Kidney Center
The Childrens Hospital of Buffalo
219 Bryant Street
Buffalo, NY 14222

Peter Weber, M.D., Ph.D.
Human Nutrition Research
Roche Vitamins and Fine Chemicals
Hoffman-LaRoche Inc.
340 Kingland St.
Nutley, NJ 07110-1199

Thomas Wengenack, Ph.D.
Department of Neurobiology and Biochemistry
Molecular Biology and Molecular Neurobiology Laboratory
May Clinic and Mayo Foundation
Rochester, MN 55905

L. David Williams, Ph.D.
DDI Pharmaceuticals, Inc.
518 Logue Avenue
Mt. Pier, CA 94043

Michael Wilson, M.D., FA
Professor and Director of Cardiovascular Medicine
Department of Medicine
The Millard Fillmore Hospitals/University at Buffalo
3 Gates Circle
Buffalo, NY 14209

Kunio Yagi, M.D., Ph.D. *
Professor and Director
Institute for Applied Biochemistry
Yagi Memorial Park
Mitake, Gifu 505-01
Japan
(011-57)467-5500▪FAX 467-5310

Jue-Rong Zhang, M.D., Ph.D.
CNS Diseases Research
The UpJohn Co.
301 Henrietta Street
Kalamazoo, MI 49001

* Symposium Presenter

INDEX